Electronic Packaging: Design, Materials, Process, and Reliability

Electronic Packaging and Interconnection Series
Charles M. Harper, Series Advisor

ALVINO *Plastics for Electronics*
GARROU & TURLIK *Multichip Module Technology Handbook*
G. DI GIACOMO *Reliability of Electronic Packages and Semiconductor Devices*
GINSBERG AND SCHNOOR *Multichip Module and Related Techniques*
HARMAN *Wire Bonding in Microelectronics, 2/e*
HARPER *Electronic Packaging and Interconnection Handbook, 2/e*
HARPER AND SAMPSON *Electronic Materials and Processes Handbook, 2/e*
HARPER AND SERGENT *Hybrid Microelectronics Handbook, 2/e*
HWANG *Modern Solder Technology for Competitive Electronics Manufacturing*
JAVITZ *Printed Circuit Board Materials Handbook*
KONSOWSKI AND HOLLAND *Electronic Packaging of High-Speed Circuitry*
LAU *Ball Grid Array Technology*
LAU *Flip Chip Technologies*
LAU *Solder Joint Reliability of BGA, CSP, Flip Chip, and Fine Pitch SMT Assemblies*
LICARI *Multichip Module Design, Fabrication, and Testing*
LUDWIG-BECKER *Electronics Quality Management Handbook*
MROCZKOVSKI *Electronic Connector Handbook*

Related Books of Interest

BOSWELL *Subcontracting Electronics*
BOSWELL AND WICKAM *Surface Mount Guidelines for Process Control, Quality, and Reliability*
BYERS *Printed Circuit Board Design with Microcomputers*
CHEN *Computer Engineering Handbook*
CHRISTIANSEN *Electronics Engineers' Handbook, 4/e*
COOMBS *Electronics Instrument Handbook, 2/e*
COOMBS *Printed Circuits Handbook, 4/e*
GINSBERG *Printed Circuits Design*
JURAN AND GRYNA *Juran's Quality Control Handbook*
JURGEN *Automotive Electronics Handbook*
MANKO *Solders and Soldering, 3/e*
RAO *Multilevel Interconnect Technology*
SZE *VLSI Technology*
VAN ZANT *Microchip Fabrication, 3/e*

To order or receive additional information on these or any other McGraw-Hill titles, please call 1-800-822-8158 in the United States. In other countries, contact your local McGraw-Hill representative.

Electronic Packaging: Design, Materials, Process, and Reliability

John H. Lau
Express Packaging Systems, Inc.

C. P. Wong
Georgia Institute of Technology

John L. Prince
University of Arizona

Wataru Nakayama
University of Maryland

McGraw-Hill

New York San Francisco Washington, D.C. Auckland Bogotá
Caracas Lisbon London Madrid Mexico City Milan
Montreal New Delhi San Juan Singapore
Sydney Tokyo Toronto

Library of Congress Catalog Card Number: 98-065030

McGraw-Hill
A Division of The **McGraw·Hill** Companies

Copyright © 1998 by The McGraw-Hill Companies, Inc. All rights reserved. Printed in the United States of America. Except as permitted under the United States Copyright Act of 1976, no part of this publication may be reproduced or distributed in any form or by any means, or stored in a data base or retrieval system, without the prior written permission of the publisher.

1 2 3 4 5 6 7 8 9 0   DOC/DOC   9 0 2 1 0 9 8

ISBN 0-07-037135-0

The sponsoring editor for this book was Stephen Chapman and the production supervisor was Claire Stanley. It was set in Century Schoolbook by North Market Street Graphics.

Printed and bound by R. R. Donnelley & Sons Company.

McGraw-Hill books are available at special quantity discounts to use as premiums and sales promotions, or for use in corporate training programs. For more information, please write to Director of Special Sales, McGraw-Hill, 11 West 19th Street, New York, NY 10011. Or contact your local bookstore.

Information contained in this work has been obtained by The McGraw-Hill Companies Inc. ("McGraw-Hill") from sources believed to be reliable. However, neither McGraw-Hill nor its authors guarantees the accuracy or completeness of any information published herein and neither McGraw-Hill nor its authors shall be responsible for any errors, omissions, or damages arising out of use of this information. This work is published with the understanding that McGraw-Hill and its authors are supplying information but are not attempting to render engineering or other professional services. If such services are required, the assistance of an appropriate professional should be sought.

 This book is printed on recycled, acid-free paper containing a minimum of 50% recycled, de-inked fiber.

To our parents for their encouragement
To our wives for their support
To our children for their understanding
*To the ones who published peer-reviewed papers
 for sharing their knowledge*

Contents

Preface xiii
Acknowledgments xvii
About the Authors xxi

Chapter 1. Introduction   1

  1.1   IC Trends   2
  1.2   Packaging Technology Update   6
         1.2.1   Area-Array Flip-Chip Technology   11
         1.2.2   BGA Technology   11
         1.2.3   TCP   11
         1.2.4   TSOP and PQFP   11
         1.2.5   CSP and DCA   12
  1.3   Summary   13
  1.4   References   14

Chapter 2. Basic Electrical Effects in Electronic Packaging and Interconnects   17

  2.1   Basic Electrical Parameters and Phenomena in Electronic Packaging   17
         2.1.1   Introduction   22
         2.1.2   Intrinsic Delay   28
         2.1.3   Characteristic Impedance   30
         2.1.4   Maxwell's Equations, Capacitance, Inductance, and Resistance   31
  2.2   Basic Transmission Line Characteristics   43
         2.2.1   Introduction   43
         2.2.2   Distributed versus Lumped-Element Simulation for Transmission Lines   45
         2.2.3   Lossless Transmission Line Equations   52
         2.2.4   Lossy Transmission Lines   75
         2.2.5   Coupled Noise (Cross Talk)   77
  2.3   Power and Ground Bus Disturbances   83
         2.3.1   Introduction   86
         2.3.2   Present and Future Concerns   88
         2.3.3   SSO Noise Modeling   90
         2.3.4   Effects of Decoupling Capacitors   104

Chapter 3. Thermal Design, Analysis, and Measurement of Electronic Packaging — 111

- 3.1 Introduction — 111
- 3.2 Heat Diffusion from Junctions to Systems — 113
  - 3.2.1 Definition of Thermal Resistance — 113
  - 3.2.2 Heat Diffusion from Transistor Junctions — 116
  - 3.2.3 Heat Conduction from Integrated Circuits — 119
  - 3.2.4 Heat Flow Across Interfaces — 121
  - 3.2.5 Heat Conduction in Wiring Substrates — 125
  - 3.2.6 Forced Convective Heat Transfer from Flat Surfaces — 126
  - 3.2.7 Forced Convective Heat Transfer from Modules on Board — 130
  - 3.2.8 Conjugate Heat Transfer on Heat Spreader — 134
  - 3.2.9 Flow of Coolant in System Box — 136
  - 3.2.10 Natural Convection from Flat Surfaces — 139
  - 3.2.11 Natural Convection Heat Transfer in Constrained Space — 142
  - 3.2.12 Radiation Heat Transfer — 145
  - 3.2.13 Liquid Cooling — 146
- 3.3 Thermal Analysis and Design — 154
  - 3.3.1 Objectives of Thermal Design — 154
  - 3.3.2 Breadth and Depth of Thermal Analysis — 155
  - 3.3.3 Possible Cooling Modes — 156
  - 3.3.4 Scaling — 157
  - 3.3.5 Modeling — 161
  - 3.3.6 Formulation — 165
  - 3.3.7 Solution — 166
- 3.4 Hardware Devices for Cooling — 168
  - 3.4.1 Air-Cooled Heat Sinks — 168
  - 3.4.2 Fans and Blowers — 171
  - 3.4.3 Liquid-Cooling Hardware — 173
  - 3.4.4 Heat Pipes — 175
  - 3.4.5 Thermoelectric Coolers — 176
- 3.5 Measurements — 177
  - 3.5.1 Temperature Measurements — 178
  - 3.5.2 Air-Flow Measurements — 180
  - 3.5.3 Thermal Resistance Measurements — 183
- 3.6 Summary — 184
- 3.7 Nomenclature — 187
- 3.8 References — 190

Chapter 4. Mechanical Design, Analysis, Measurement, and Reliability of Electronic Packaging — 193

- 4.1 Introduction — 193
- 4.2 Thermal Mechanical Boundary Value Problems — 195
  - 4.2.1 Governing Equations for Electronic Packaging — 195
  - 4.2.2 Example I—Deflection of a Printed Circuit Board — 197
  - 4.2.3 Example II—Thermal Stress in a Package Subjected to an Instantaneous Heat Source — 201
  - 4.2.4 Example III—Thermal Stress in a Package Subjected to a Harmonic Heat Source — 208
- 4.3 Analysis of Stress — 212
  - 4.3.1 Three-Dimensional Stress State — 212
  - 4.3.2 Transformation of Coordinates — 213

4.3.3	Principal Stress and Direction in Three Dimensions	213
4.3.4	Example I—Transformation of Stress in Three Dimensions	215
4.3.5	Example II—Principal Stresses in Three Dimensions	215
4.3.6	Two-Dimensional Stress State	216
4.3.7	Example III—Principal Stress in Two Dimensions	217

4.4 Analysis of Strains 217
    4.4.1 Infinitesimal Strains in Cartesian Coordinates 217
    4.4.2 Infinitesimal Strains in Cylindrical Coordinates 222

4.5 Mechanical Behavior of Packaging Materials 222
    4.5.1 Stress-Strain Relation—Elastic 222
    4.5.2 Stress-Strain Relation—Plastic 225
    4.5.3 Stress-Strain Relation—Creep 233
    4.5.4 Creep of Solders Under Combined Loads 239
    4.5.5 TMA, DMA, DSC, and TGA of 63Sn37Pb Solder 251
    4.5.6 Effect of Underfill Materials on the Mechanical and Electrical Performance of a Functional Solder-Bumped Flip-Chip Device 260

4.6 Finite Element Analysis 271
    4.6.1 Nonlinear Finite Element Formulation 271
    4.6.2 Example I—Stress and Deflection Analysis of Partially Routed Panels for Depanelization 275
    4.6.3 Example II—Thermal Stress Analysis of a Plastic Leaded Package 282
    4.6.4 Example III—Mechanical Behavior of Microstrip Structures Made from $YBa_2Cu_3O_{7-x}$ Superconducting Ceramics 288
    4.6.5 Example IV—Design for Plastic Ball Grid Array Solder Joint Reliability 290
    4.6.6 Example V—Solder Joint Reliability of Plastic Ball Grid Array Packages with Solder-Bumped Flip Chip 294

4.7 The Roles of DNP (Distance to Neutral Point) on Solder-Joint Reliability of Area-Array Assemblies 301
    4.7.1 Example I—Chip Scale Package on PCB 306
    4.7.2 Example II—Solder-Bumped Flip Chip on PCB (DCA) 310
    4.7.3 Example III—Ceramic Ball Grid Array (CBGA) on PCB 312
    4.7.4 Example IV—Plastic Ball Grid Array (PBGA) on PCB 313

4.8 Mechanical Vibration of Electronic Packaging 314
    4.8.1 Horizontal Vibration of Packages with Heat Sinks 315
    4.8.2 Vertical Vibration of PCBs 322

4.9 Design, Analysis, and Measurement of Hermetic Microwave Packages 326
    4.9.1 Design and Fabrication of Packages 327
    4.9.2 Hermeticity Performance of Packages 329
    4.9.3 Thermal Stress Performance of Packages 334
    4.9.4 Qualification Tests of Packages 337
    4.9.5 Summary 338

4.10 Design, Analysis, and Measurement of a Low-Cost Chip Size Package—NuCSP 339
    4.10.1 NuCSP Design Concept 339
    4.10.2 NuCSP Design Examples 340
    4.10.3 Electrical Performance of NuCSP 344
    4.10.4 Thermal Performance of NuCSP 344
    4.10.5 Solder-Joint Reliability of NuCSP on PCB 345
    4.10.6 Summary 350

4.11 Design of Electrical and Thermal Enhanced Plastic Ball Grid Array Packages—NuBGA 352
    4.11.1 NuBGA Design Concepts 353
    4.11.2 NuBGA Design Examples 354
    4.11.3 NuBGA Package Family 357

Contents

4.12 Analysis and Measurement of NuBGA—Electrical Performance ... 357
    4.12.1 NuBGA Package Parasitic Parameters (R, L, C) ... 358
    4.12.2 NuBGA Electrical Analysis and Comparison with Other Packages ... 361
4.13 Analysis and Measurement of NuBGA—Thermal Performance ... 364
    4.13.1 Finite-Element Analysis of NuBGA with a Partial Heat Spreader ... 366
    4.13.2 Experimental Measurement of NuBGA with a Partial Heat Spreader ... 368
    4.13.3 Experimental Measurement of NuBGA with a Full-Size Heat Spreader ... 373
    4.13.4 Finite-Element Analysis of NuBGA with Various Sizes of Heat Spreaders ... 373
    4.13.5 NuBGA with Different Sizes of Heat Sinks ... 377
    4.13.6 Thermal Impedance Measurement of NuBGA ... 379
    4.13.7 Solder-Joint Reliability of NuBGA on PCB ... 382
    4.13.8 Summary ... 382
4.14 Summary ... 383
4.15 Acknowledgments ... 384
4.16 References ... 385

Chapter 5. Polymers for Electronic Packaging: Materials, Processes, and Reliability ... 393

5.1 Introduction ... 393
5.2 Purpose of Electronic Packaging ... 398
    5.2.1 Protection and Encapsulation ... 398
    5.2.2 Moisture Permeability ... 399
    5.2.3 Mobile Ion Contaminants ... 399
    5.2.4 Light and Alpha-Particle Radiations ... 400
    5.2.5 Hostile Environments ... 400
5.3 Prepackaging Cleaning ... 401
    5.3.1 Conventional Solvent and Aqueous Cleaning ... 401
    5.3.2 Reactive Oxygen Cleaning ... 401
    5.3.3 Reactive DC-Hydrogen Plasma Cleaning ... 402
5.4 Polymeric Materials for Microelectronic Packaging ... 402
    5.4.1 Interlayer Dielectrics ... 403
    5.4.2 Passivation Materials for IC Device Packaging ... 403
    5.4.3 Organic Encapsulants ... 404
    5.4.4 High-Performance No-Flow and Fast-Flow Underfills for Low-Cost Flip-Chip Applications ... 427
    5.4.5 Conclusions on No-Flow Underfills ... 442
5.5 Material Process Techniques ... 444
    5.5.1 Cavity-Filling Processes ... 444
    5.5.2 Saturation and Coating Processes ... 447
5.6 Anisotropically Conductive Adhesives for Microelectronic Assembly ... 448
    5.6.1 Materials and Components ... 449
    5.6.2 Characterization of ACAFs ... 452
    5.6.3 ACAF Assembly ... 459
    5.6.4 Accelerated Life Testing ... 461
    5.6.5 Conclusions on ACAF ... 461
5.7 High-Performance Plastic Packages with Reliability Without Hermeticity: A Case Study in Hermetic versus Nonhermetic Packaging ... 463
    5.7.1 Hermetic Packages ... 464
    5.7.2 Nonhermetic Packages—A Case Study on Flip-Chip Processes and Reliability ... 464

		5.7.3 Results and Discussions	469
		5.7.4 Conclusions	474
	5.8	Reliability Testing	474
		5.8.1 Correlation of 85°C/85%RH to HAST	475
		5.8.2 Recent Advances in Materials Packaging	476
	5.9	Conclusions and Future Developments	481
5.10		Acknowledgments	482
5.11		References	482

Index 487

Preface

The invention of bipolar transistors by Bardeen, Brattain, and Shockley at Bell Laboratories in 1947 foreshadowed the development of generations of computers yet to come. The invention of the silicon integrated circuit (IC) by Jack Kilby of Texas Instruments in 1958 and by Robert Noyce and Gordon Moore of Fairchild Semiconductor in 1959 excited the development of generations of scale integrations: small (SSI); medium (MSI); large (LSI); very large (VLSI); ultra large (ULSI); giga (GSI); and others yet to come (. . . SI).

The IC chip is not an isolated island. It must communicate with other chips in a circuit through an input/output (I/O) system of interconnects. Furthermore, the chip needs to be powered up and its embedded circuitry is delicate, requiring the package to both carry and protect it. Consequently, the major functions of electronic packaging are: (1) to provide a path for the electrical current that powers the circuits on the chip; (2) to distribute the signals onto and off of the silicon chip; (3) to remove the heat generated by the circuits; and (4) to support and protect the chip from hostile environments.

In most electronics products, there are four major IC devices: the microprocessor, the ASIC (application-specific IC), the cache memory, and the main memory. Today, microprocessors are expected to require over 1000 package-pin counts and to perform at over 400 MHz on-chip clock frequency. To prevent system data bottlenecks, the cache memories are expected to perform at high speeds similar to those of microprocessors. Even the ASICs are expected to run at up to 500 MHz on-chip clock frequency and have up to 900 package-pin counts. (Some telecommunication products' ASICs need more than 1000 pin counts.) Within a decade, multi-GHz on-chip clock frequencies and several thousand package-pin counts are predictable. Packaging and interconnection technology will be hard-pressed to meet all these future requirements.

Consequently, there was an urgent need, both in industry and in universities, to create a comprehensive book on all aspects and the current state of knowledge in electronic packaging and interconnections. This book should be written for everyone who can quickly learn about the basics and problem-solving methods, understand the trade-offs, and make system-level decisions. Also, because of the multidisciplinary nature of packaging and interconnection, this book should include the solutions of electrical, thermal, mechanical, chemical, material, process, and reliability problems in packaging. To meet this need, we—John Lau (a civil and applied mechanist), Wataru Nakayama (a thermal and mechanical scientist), John Prince (an electrical and electronics scientist), and C. P. Wong (a chemical and material scientist)—have decided to write this book.

This book is organized into five basic parts. Chapter 1 briefly discusses the IC trends and updates the electronic packaging and interconnection technologies. The second part (Chap. 2) presents the basic electrical effects in electronic packaging and interconnections. The third part (Chap. 3) discusses the thermal design, analysis, and measurement of electronic packaging and interconnections. The fourth part (Chap. 4) provides the mechanical design, analysis, measurement, and reliability of electronic packaging. The last part (Chap. 5) gives the materials, process, and reliability of polymers for electronic packaging.

For whom is this book intended? Undoubtedly it will be of interest to three groups of specialists: (1) those who are active or intend to become active in research and development of electronic packaging and interconnections; (2) those who have encountered practical electronic packaging and interconnection problems and wish to understand and learn more methods of solving such problems; and (3) those who have to choose a reliable, creative, high-performance, robust, and cost-effective packaging technique for their interconnect system. This book also can be used as a text for college and graduate students who could become our future leaders, scientists, and engineers in the electronics industry.

We hope this book will serve as a valuable source of reference to all those faced with the challenging problems created by the ever increasing IC speed and density and the reduction of product size and weight. We also hope that it will aid in stimulating further research and development of electrical and thermal designs, materials, process, manufacturing, electrical and thermal management, testing, and reliability, and more sound use of electronic packaging and interconnection in electronic products.

The organizations that learn how to design and manufacture electronic packages in their interconnect systems have the potential to make major advances in the electronics industry and to gain great benefits in cost, performance, quality, size, and weight. It is our hope that the information presented in this book may assist in removing road blocks, avoiding unnecessary false starts, and accelerate design, material, and process development of electronic packaging and interconnection. We refuse to hear that electronic packaging and interconnection will be the bottleneck of high-speed electronic products. Rather, we want to consider this as the golden opportunity to make a major contribution to the electronics industry by developing innovative, cost-effective, and reliable electronic packages. It is an exciting time for electronic packaging and interconnection!

John H. Lau, Ph.D., IEEE Fellow
C. P. Wong, Ph.D., IEEE Fellow
John L. Prince, Ph.D., IEEE Fellow
Wataru Nakayama, Ph.D., ASME Fellow

Acknowledgments

Development and preparation of *Electronic Packaging: Design, Materials, Process, and Reliability* was facilitated by the efforts of a number of dedicated people at McGraw-Hill and North Market Street Graphics. We would like to thank them all, with special mention to Wayne Coleson of North Market Street Graphics and Julie Lacus, Claire Stanley, and Fred Bernardi of McGraw-Hill for their unswerving support and advocacy. Our special thanks to Steve Chapman (Executive Editor of Electronics and Optical Engineering) of McGraw-Hill who made our dream of this book come true by effectively sponsoring the project and solving many problems that arose during the book's preparation. It has been a great pleasure and fruitful experience to work with them in transferring our messy manuscripts into a very attractive printed book.

The material in this book has clearly been derived from many sources including individuals, companies, and organizations, and we have attempted to acknowledge, in the appropriate parts of the book, the assistance that we have been given. It would be quite impossible for us to express our thanks to everyone concerned for their cooperation in producing this book, but we would like to extend due gratitude. Especially, we want to thank several professional societies and publishers for permitting us to reproduce some of their illustrations and information in this book. For example: the American Society of Mechanical Engineers (ASME) Conference Proceedings and Transactions (e.g., *Journal of Electronic Packaging*); the Institute of Electrical and Electronic Engineers (IEEE) Conference Proceedings and Transactions (e.g., *Components, Packaging, and Manufacturing Technology*); the International Microelectronics and Packaging Society (IMAPS) Conference Proceedings and Transactions (e.g., *Microcircuits & Electronic Packaging*); American Society of Metals (ASM) Conference Proceedings and books (e.g., *Electronic Materials Handbook,* vol. 1, *Packaging*); the Surface Mount Technology Association (SMTA) Conference Proceedings and journals (e.g., *Journal of Surface Mount Technology*); the National Electronic Packaging Conferences (NEPCON) and Proceedings; the *IBM Journal of*

Research and Development; Electronic Packaging & Production; Circuits Assembly; Surface Mount Technology; Connection Technology; Solid State Technology; Circuit World; Microelectronics International; and *Soldering and Surface Mount Technology.*

John Lau wants to thank his former employer, Hewlett-Packard Company, for providing him an excellent working environment that has nurtured him as a human being, fulfilled his job satisfaction, and enhanced his professional reputation. He also wants to thank his eminent colleagues (the enumeration of whom would not be practical here) at Hewlett-Packard Company, Express Packaging Systems, Inc., and throughout the electronics industry for their useful help, strong support, and stimulating discussions. Working and socializing with them have been a privilege and an adventure. He learned a lot about life and electronic packaging from them.

Wataru Nakayama wants to thank CALCE Electronic Packaging Research Center (EPRC) and the Department of Mechanical Engineering, both in the University of Maryland, for providing him with a precious opportunity to work in the United States, thereby expanding his professional scope on microelectronic packaging. He also extends his sincere thanks to his colleagues at CALCE EPRC, University of Maryland, his former colleagues at Hitachi, Ltd., his former colleagues and students at the Tokyo Institute of Technology, his professional colleagues at ASME, JSME, and IEEE, to all of whom he is indebted for their guidance, assistance, and thoughtful support.

John Prince wishes to acknowledge the funding and other support of the Semiconductor Research Corporation over more than a decade (1984–1998). The Semiconductor Research Corporation "caught the vision" of packaging early in the 1980s and was brave enough to fund our work at the University of Arizona at a time when many in the industry thought packaging electrical performance was not and would not be important. He would also like to acknowledge the benefit of working directly with his colleague and good friend, Dr. Andreas Cangellaris, at Arizona for a decade (1987–1997). Finally, he wishes to acknowledge each of the many students he worked with over the years. As is common in these situations, education never proceeds as a one-way street when one works with students.

C. P. Wong gratefully acknowledges his former colleagues at AT&T Bell Labs (currently Lucent Technology Bell Labs) for all their collaborations. Lucent Technology's equipment donated to his research group at Georgia Institute of Technology is also greatly appreciated. The author would also like to thank his colleagues at PMSE and the NSF-funded Packaging Research Center at Georgia Institute of Technology for their collaborations.

Lastly, John Lau wants to thank his daughter (Judy) and his wife (Teresa) for their love, consideration, and patience by allowing him to work on many weekends for this book. Their simple belief that he is making a small contribution to the electronics industry was a strong motivation for him. Wataru Nakayama wants to thank his wife (Michiko) for her continual and warm encouragement rendered to him to work on this book. John Prince wishes to thank his wife (Martha) for bearing with him during the writing of this book. He also wishes to thank his assistant Betsey Lyons and his students Abdul-Rahman Yaghmour, Samil Hasan, and Cheng Jiao, for the help they gave in putting his portion of the book together. C. P. Wong gratefully appreciates his wife (Lorraine) and children (Michelle and Dave) for their understanding and support as he worked on this book.

John H. Lau, Ph.D., IEEE Fellow
C. P. Wong, Ph.D., IEEE Fellow
John L. Prince, Ph.D., IEEE Fellow
Wataru Nakayama, Ph.D., ASME Fellow

About the Authors

JOHN H. LAU is the president of Express Packaging Systems (EPS), Inc., in Palo Alto, California. His current interests cover a broad range of electronic packaging and manufacturing technology.

Prior to founding EPS in November 1995, he worked for Hewlett-Packard Company, Sandia National Laboratory, Bechtel Power Corporation, and Exxon Production and Research Company. With more than 27 years of R&D and manufacturing experience in the electronics, petroleum, nuclear, and defense industries, he has authored and coauthored over 100 peer-reviewed technical publications, and is the author and editor of 11 books: *Solder Joint Reliability; Handbook of Tape Automated Bonding; Thermal Stress and Strain in Microelectronics Packaging; The Mechanics of Solder Alloy Interconnects; Handbook of Fine Pitch Surface Mount Technology; Chip On Board Technologies for Multichip Modules; Ball Grid Array Technology; Flip Chip Technologies; Solder Joint Reliability of BGA, CSP, Flip Chip, and Fine Pitch SMT Assemblies; Electronic Packaging: Design, Materials, Process, and Reliability;* and the forthcoming *Chip Scale Package (CSP): Design, Materials, Process, Reliability, and Applications.*

John served as one of the technical editors of the *IEEE Transactions on Components, Packaging, and Manufacturing Technology* and *ASME Transactions, Journal of Electronic Packaging*. He has also served as general chairman, program chairman, and session chairman, and invited speaker of several IEEE, ASME, ASM, MRS, IMAPS, SEMI, NEPCON, and SMI International conferences. He received a few awards from ASME and IEEE for best papers and technical achievements and is an IEEE Fellow. He is listed in *American Men and Women of Science* and *Who's Who in America*.

John received his Ph.D. in Theoretical and Applied Mechanics from the University of Illinois, an M.A.Sc. in Structural Engineering from the University of British Columbia, a second M.S. in Engineering Mechanics from the University of Wisconsin, and a third M.S. in Management Science from Fairleigh Dickinson University. He also has a B.E. in Civil Engineering from National Taiwan University.

DR. C. P. WONG received his B.S. in Chemistry from Purdue University and his Ph.D. in Organic/Inorganic Chemistry from the Pennsylvania State University. After his doctoral study, he was awarded two years as a postdoctoral scholar with Nobel Laureate Professor Henry Taube at Stanford University, where he conducted studies on electron transfer and reaction mechanism of metallocomplexes. He was the first person to synthesize the first known lanthanide and actinide porphyrin complexes, which represent a breakthrough in metalloporphyrin chemistry.

He joined AT&T Bell Laboratories in January 1977 as MTS. He has been involved with the R&D of polymeric materials (inorganic and organic) for electronic applications. He became senior MTS in 1982, a distinguished MTS in 1987 and an AT&T Bell Laboratories Fellow in 1992. In January 1996 he joined the Georgia Institute of Technology as a professor of Materials Science and Engineering and a research director of GIT's Packaging Research Center. His research interests lie in the fields of polymeric materials, high Tc ceramics, materials reaction mechanism, IC encapsulation, in particular, hermetic equivalent plastic packaging, electronic manufacturing packaging processes, PWB, and components reliability. He is one of the pioneers who demonstrated the use of silicone gel as a device encapsulant to achieve reliability without hermeticity in plastic IC packaging. He received the Best Paper Award in 1981 at ISHM's Annual Meeting; the AT&T Bell Laboratories Distinguished Technical Staff Award in 1987; a 1992 AT&T Bell Laboratories Fellow Award; the IEEE Components, Packaging and Manufacturing Technology (CPMT) Society Outstanding Paper Awards in 1990, 1991, and 1994; the CPMT Society Board of Governors Distinguished Service Award in 1991; the IEEE Technical Activities Board (TAB) Distinguished Service Award in 1994; and the 1995 CPMT Society's Outstanding Sustained Technical Contributions Award. He holds over 38 U.S. patents, numerous international patents, has published over 120 technical papers, and has delivered 100 keynotes and presentations in the related area. He is the editor and an author of the Academic Press textbook on *Polymers for Electronic and Photonic Applications* in 1993 and is an associate editor of the *IEEE Transactions on CPMT* (1995–).

Dr. Wong is a Fellow of the IEEE and AT&T Bell Labs, and a member of the Sigma Xi, Phi Lambda Upsilon, National Honorary Chemical Society, and Materials Research Society. He was the program chairman of the IEEE 39th ECTC. He was elected to the Board of Governors of the CPMT Society from 1987 to 1989, served as its technical vice president (1990 and 1991) and president (1992 and 1993). He currently chairs the IEEE Technical Activities Board, Steering Committee on Design and Manufacturing Engineering (1995–).

JOHN L. PRINCE received his BSEE from Southern Methodist University (1965), and as an NSF Graduate Fellow received his MEE (1968) and Ph.D. (1969) in Electrical Engineering from North Carolina State University.

He is currently professor of Electrical and Computer Engineering and director of the Center for Electronic Packaging Research at the University of Arizona. He came to the University of Arizona in 1983. He has been principal investigator of the Semiconductor Research Corporation (SRC) Program in VLSI Packaging and Interconnection Research at the university since 1984. In 1991–1992, he was acting director, Packaging Sciences at SRC. He has extensive industrial experience, including silicon device, circuit, and reliability work at Texas Instruments from 1970 to 1975, silicon research at the Research Triangle Institute from 1968 to 1970, and medical device design and reliability at Intermedics, Inc., from 1980 to 1983. He was associate professor and professor at Clemson University from 1975 to 1980. His research there concentrated on performance and characterization of silicon CMOS devices at high temperatures, up to 300+°C, and on solar cell reliability.

His current research interests center on developing modeling and simulation techniques for switching noise in packages and MCMs, on modeling and simulation techniques for mixed-signal system packaging, and on developing high-frequency measurements on packaging structures. He is active in consulting work in both the reliability and packaging areas. He is active as a member of the Technical Program Committee and session chair for the Electronic Component and Technology Conference, the Topical Meeting on Electrical Performance of Electronic Packaging, and the VLSI Packaging Workshop of Japan. He was a consultant member, Advisory Group on Electron Devices, Working Group B, Office of the Undersecretary of Defense for Research and Engineering, from 1979 to 1983. He is a member (1993–) of the National Technology Roadmap for Semiconductors Technical Working Group on Assembly and Packaging.

He has been recognized for his work in electronic packaging by the 1988 Award for the Advancement of Semiconductor Technology "for outstanding achievement in assembly and packaging for semiconductor packaging," by *Semiconductor International* magazine, by being named Fellow of the IEEE (1990) "for contributions to the development of computer-aided design tools for electronic packaging," and by being named Arizona Innovator of the Year (1991).

He is the author or coauthor of over 150 papers in the field of electronic packaging, and 30 papers in the fields of device physics, process development, and reliability.

DR. WATARU NAKAYAMA received his Dr. of Engineering from Tokyo Institute of Technology in 1966. He worked for Hitachi, Ltd. as a heat transfer specialist from 1970 to 1993. From 1989 to 1996, he taught at Tokyo Institute of Technology as a professor. He is currently associated with CALCE Electronic Packaging Research Center, University of Maryland.

At Hitachi, he conducted and supervised heat transfer research for various products. The list of projects in which he participated includes the development of enhanced heat transfer tubes, high-performance heat exchangers for refrigeration machines and air-conditioners, cooling of generators and transformers, crystal growth, and cooling computers. Due to his significant contributions to the progress of engineering for various products, he was promoted to the rank of Honorary Engineer at the end of his association with the company. At Tokyo Institute of Technology he taught and conducted research on power and thermal management of computers. At CALCE Electronic Packaging Research Center, University of Maryland, he is working on the development of international engineering education programs and research projects for integrated microelectronic packaging design.

Dr. Nakayama authored and coauthored 150 papers. He received prominent awards, including ASME Heat Transfer Best Paper Award in 1981, ASME Heat Transfer Memorial Award in 1992, ASME Electrical & Electronic Packaging Division Award in 1996, and ICHMT Fellowship Award in 1996.

He has been active in international professional societies. He served as chairman of the Heat Transfer Society of Japan in 1994, chairman of ASME Japan in 1990/92, and chairman of JSME Thermal Engineering Division in 1990, and is currently serving as a member of the Executive Committee of the International Center for Heat and Mass Transfer. He co-organized a number of international conferences—among them, Japan–United States Joint Seminar on Computers in Heat Transfer Science in 1991 and the International Intersociety Electronic Packaging Conference (InterPack) in 1995. He delivered keynote speeches and seminars at conferences and universities, including the keynote lecture at the 7th International Heat Transfer Conference in 1982 and the Hawkins Memorial Lecture at Purdue University in 1988. He is an ASME Fellow and IEEE Senior Member.

Chapter 1

Introduction

The invention of the bipolar transistors by Bardeen, Brattain, and Shockley at Bell Laboratories in 1947 foreshadowed the development of generations of computers yet to come. The invention of the silicon integrated circuit (IC) by Jack Kilby of Texas Instruments in 1958 and by Robert Noyce and Gordon Moore of Fairchild Semiconductor in 1959 excited the development of generations of scale integrations—small (SSI), medium (MSI), large (LSI), very large (VLSI), ultra large (ULSI), giga (GSI), and others yet to come.

Fifty years ago, the average family probably owned five active electronic devices. Today, the average family owns several million transistors, and it is probable that by the year 2000, each family will own more than one billion transistors.

There was no semiconductor industry until the early 1950s; in 1996, the total worldwide merchant semiconductor usage was $144 billion! ICs were first marketed in 1961; in 1996, about $110 billion worth of IC chips were sold. These results have produced a global revolution of major proportions that affects the lives of people everywhere.

The IC chip is not an isolated island. It must communicate with other IC chips in a circuit through an input/output (I/O) system of interconnects. Furthermore, the IC chip and its embedded circuitry are delicate, requiring the package to both carry and protect it. Consequently, the major functions of the electronics package are: (1) to provide a path for the electrical current that powers the circuits on the IC chip; (2) to distribute the signals onto and off the IC chip; (3) to remove the heat generated by the circuits on the IC chip; and (4) to support and protect the IC chip from hostile environments.

Figure 1.1 shows a schematic representation of the electronic package hierarchy. Although the wafer is not typically included in the packaging hierarchy, it is included in Fig. 1.1 to show where the IC chip

2 Chapter One

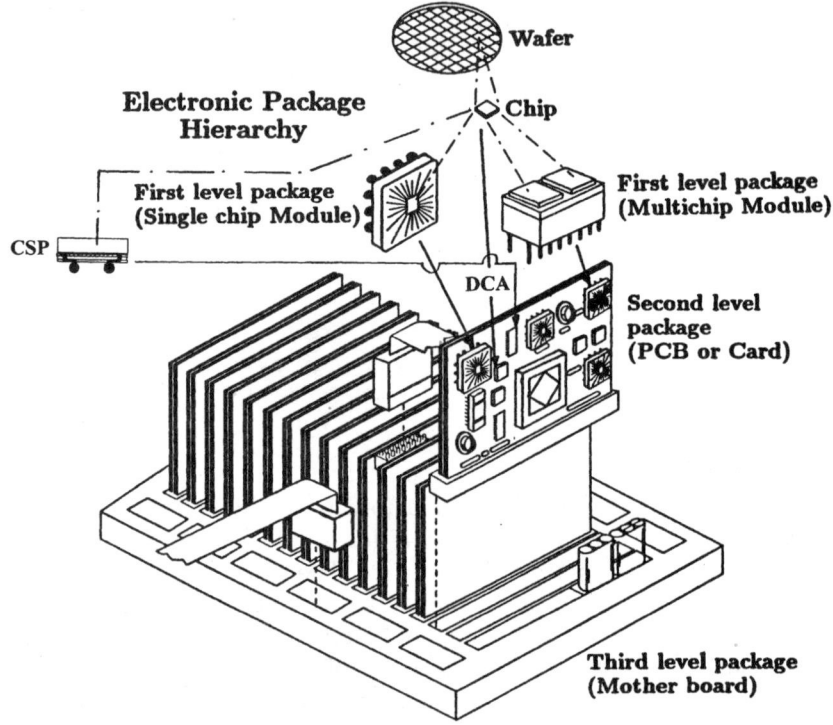

Figure 1.1 First three levels of electronics packaging. Although not part of the packaging hierarchy, the wafer and chip show the origin of the IC.

originates. Packaging focuses primarily on how chips may be packaged cost-efficiently and reliably [1–48].

Packaging is an art based on the science of establishing interconnections ranging from zero-level packages (i.e., chip-level connections, e.g., gold and solder bumps), to first-level packages (either single- or multichip modules), second-level packages [e.g., printed circuit board (PCB)], and third-level packages (e.g., motherboard). (See Fig. 1.1.) In this chapter, the IC trends are briefly discussed first, followed by a brief update on packaging technologies for ICs.

1.1 IC Trends

The past decade has witnessed an explosive growth in IC density, that is, number of transistors per chip and chip size. One of the key reasons is the advance of the CMOS (complementary metal-oxide-semiconductor) process (Fig. 1.2), which has very fine feature sizes and high yields (Fig. 1.3). Today, 0.3 µm is in volume production. According to Toshiba's CMOS road map, 0.25-µm CMOS technology is forecast for volume production by the year 2000 (see Table 1.1). For more information on IC trends, please read the beginning portion of Chapter 2.

Introduction 3

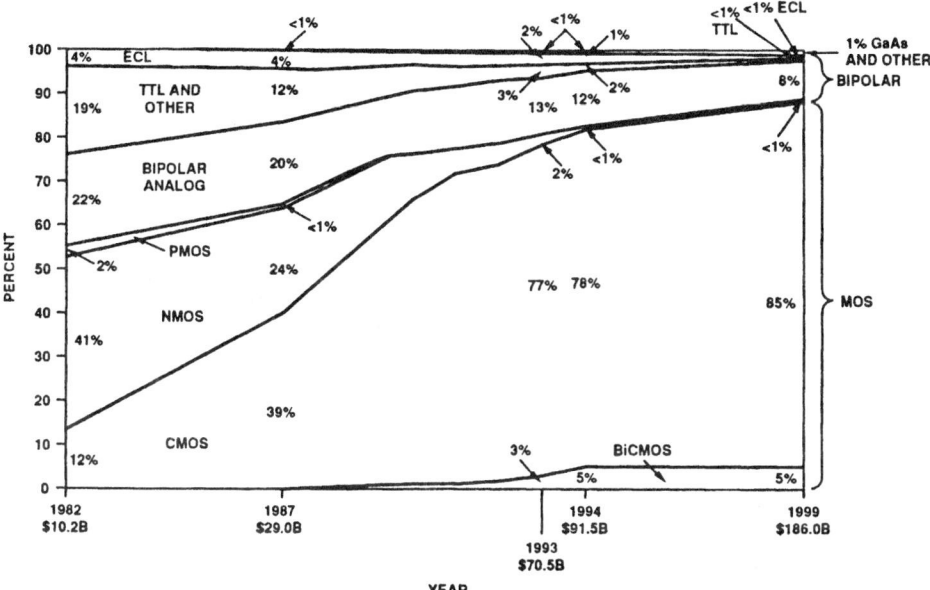

Figure 1.2 IC process technology trends.

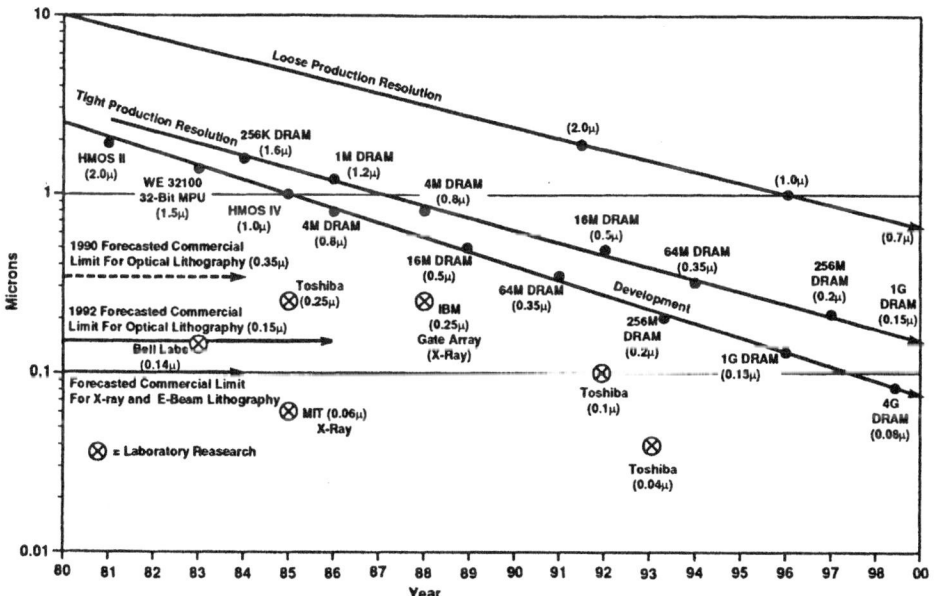

Figure 1.3 IC feature size trends.

TABLE 1.1 Toshiba's Complementary Metal-Oxide-Semiconductor (CMOS) Scaling Trend

	1993–1994	1996–1997	1999–2000
Line widths	0.5 µm	0.35 µm	0.25 µm
Power supply	3.3 V	3.3 V	2.0–2.5 V
Gate oxide thickness	11 nm	9 nm	6–7 nm
Gate length (n-channel/p-channel)	0.5/0.6 µm	0.35/0.35 µm	0.25/0.25 µm
Switching speed (fan-out = 1)	95 ps	68 ps	45 ps
Drain structure			
NMOS	FOLD*	Graded diffused drain	Conventional
PMOS	Conventional	Lightly doped drain	Conventional
Gate structure	n+ Polysilicon	n+ Polysilicon NMOS p+ Polysilicon PMOS	n+ Polysilicon NMOS p+ Polysilicon PMOS

* FOLD: Fully Overlapped Lightly Doped Drain.

Because of the explosive growth in portable electronics products, the operating voltage (power supply) on the IC chip has been reducing from 5 V to either 3 or 3.3 V now, and eventually to 2.5, then 1.5 V (Fig. 1.4 and Table 1.1). One of the key reasons is that the power consumption is proportional to the square of the operating voltage—that is, the

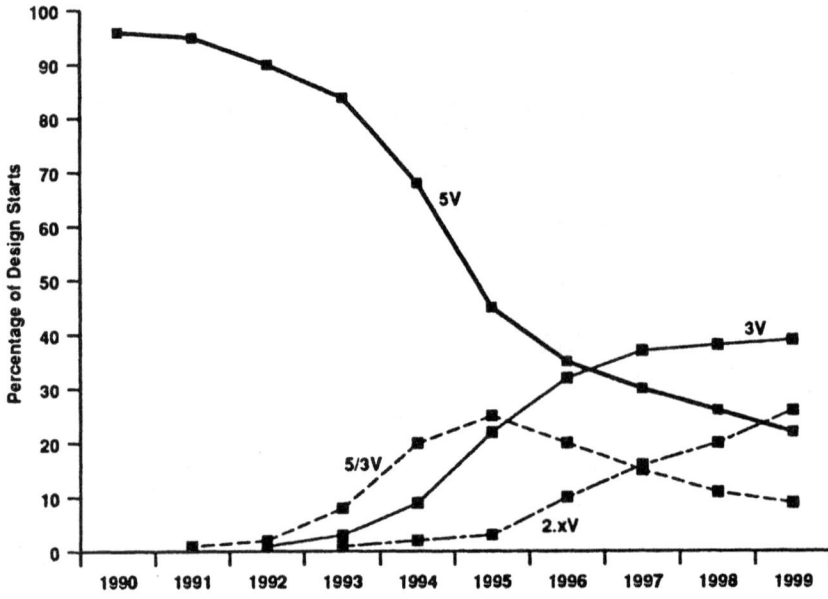

Figure 1.4 Transition from 5-V to 3-V operating voltage.

lower the operating voltage, the more the battery life for your portable electronics products.

In most electronic products, there are four major IC devices: the microprocessor, the ASIC (application-specific IC), the cache memory, and the main memory. For example, a personal computer usually has one microprocessor; a few cache memories [e.g., fast SRAM (static random access memory)]; a few ASICs (e.g., video, sound, data path, high-speed memory controller, NuBus controller, and I/O controller); and many system memories, such as ROM (read only memory), which contains permanent code used by software applications, and DRAM (dynamic random access memory) to store the information while the power is turned on.

The microprocessor is the brain of a computing system. Some well-known ones are: Intel's CISC (complex instruction-set computing)-based microprocessors family (e.g., the Pentium and Pentium-Pro); IBM, Motorola, and Apple's RISC (reduced instruction-set computing)-based PowerPC microprocessors; Hewlett-Packard's RISC-based PA8000; Digital Equipment Corporation's RISC-based Alpha chipset; Silicon Graphics' RISC-based MIPS; AMD's CISC-based K5 and K6; Cyrix's CISC-based 6X86; and Sun Microsystems' RISC-based UltraSparc. Both RISC-based and CISC-based microprocessors are expected to require over 1000 package pin counts and to perform over 400-MHz on-chip clock frequency (Fig. 1.5).

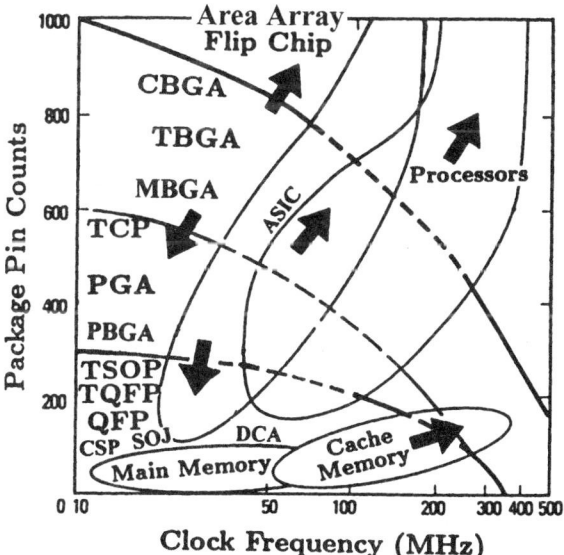

Figure 1.5 IC devices and packaging trends.

The SRAMs for cache memories are expected to perform at a high speed similar to that of microprocessors to prevent system data bottlenecks. Even the ASICs are expected to run faster than 200-MHz on-chip clock frequencies and have up to 900 package pin counts. (The ASICs of some telecommunication products need more than 1000 pin counts.) Packaging technology will be hard-pressed to meet all these future requirements. By the year 2000, packaging will become the bottleneck of high-speed computing if the advance of packaging technologies cannot keep pace with semiconductor IC technologies. Thus, there is a golden opportunity to make a major contribution to the electronics industry by developing innovative and useful electronics packages.

1.2 Packaging Technology Update

Different packaging technologies are required for different semiconductor IC devices and applications [1–8]. Figure 1.5 shows some well-known packaging technologies for four major IC devices—ASIC, microprocessor, cache, and system memory.

Figure 1.5 is divided into four regions (with three lines) of applications. The first region is for low-pin-count IC devices, and the packaging technologies are, for example, PQFP (plastic quad flat pack), SOP (Swiss outline package), SOJ (Swiss outline J-leaded), SOIC (small outline IC), PLCC (plastic leaded chip carrier), TQFP (thin quad flat pack), TSOP (thin small outline package), DCA (direct chip attach on PCB), and CSP (chip scale package). Some examples of DCA are shown in Fig. 1.6 for a solder-bumped flip chip on PCB; in Fig. 1.7 for a stud gold bumped flip chip on PCB with isotropic conductive adhesive; in Fig. 1.8 for a stud gold bumped flip chip on PCB with anisotropic conductive adhesive (ACA); and in Fig. 1.9 for a gold bumped flip chip on PCB with anisotropic conductive film (ACF). ACA consists of thermosetting adhesive and conductive particles and it looks like paste. ACF consists of thermosetting adhesive, conductive particles, and release film and it looks like paper (see Fig. 1.10). Figure 1.11 shows an example of CSP, a 3-D CSP.

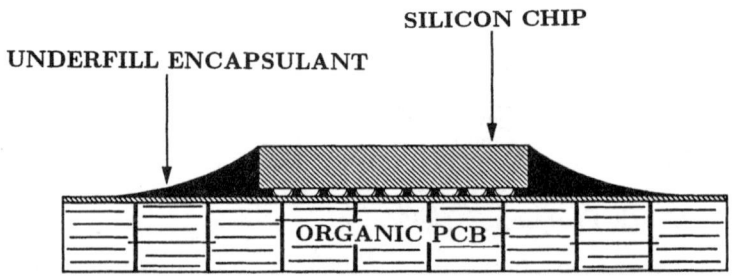

Figure 1.6 Solder-bumped flip chip on PCB (DCA).

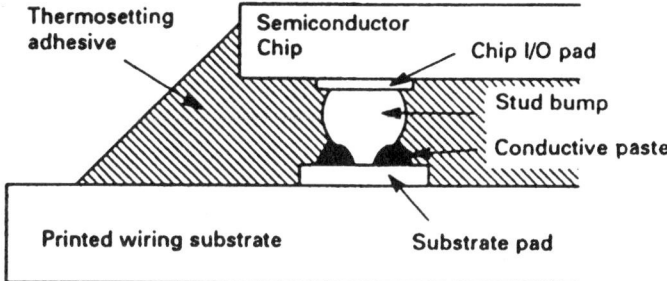

Figure 1.7 Solderless flip chip with stud bump and conductive paste (DCA).

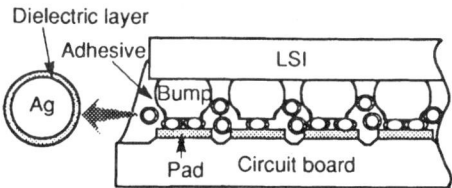

Figure 1.8 Solderless flip chip with anisotropic conductive adhesive (ACA) on PCB (DCA).

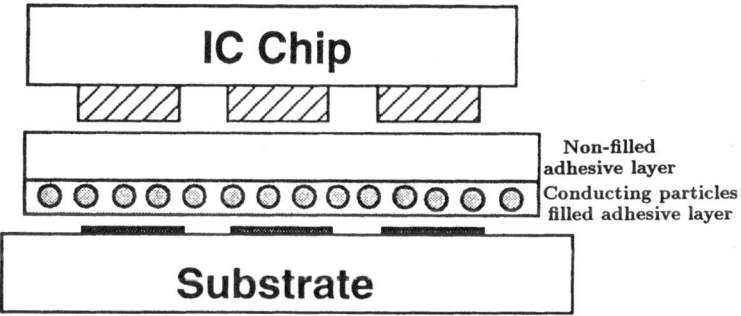

Figure 1.9 Solderless flip chip with anisotropic conductive film (ACF) on PCB (DCA).

The packaging technologies such as the TCP (tape carrier package), PPGA (plastic pin grid array), CPGA (ceramic pin grid array), and PBGA (plastic ball grid array) will meet the needs for the second region of applications. Some examples of PBGA are shown in Fig. 1.12 for a wire-bonded PBGA and in Fig. 1.13 for a set of solder-bumped PBGAs; for higher-pin-count and performance IC devices (the third region of applications), CBGA (ceramic ball grid array) in Fig. 1.14, TBGA (tape ball grid array) in Figs. 1.15 and 1.16, and MBGA (metal ball grid array) in Figs. 1.17 and 1.18—all are cost-effective packaging technologies. For the fourth region of applications (very high pin-count and performance IC devices), area-array solder-bumped flip-chip technology is the solution.

8   Chapter One

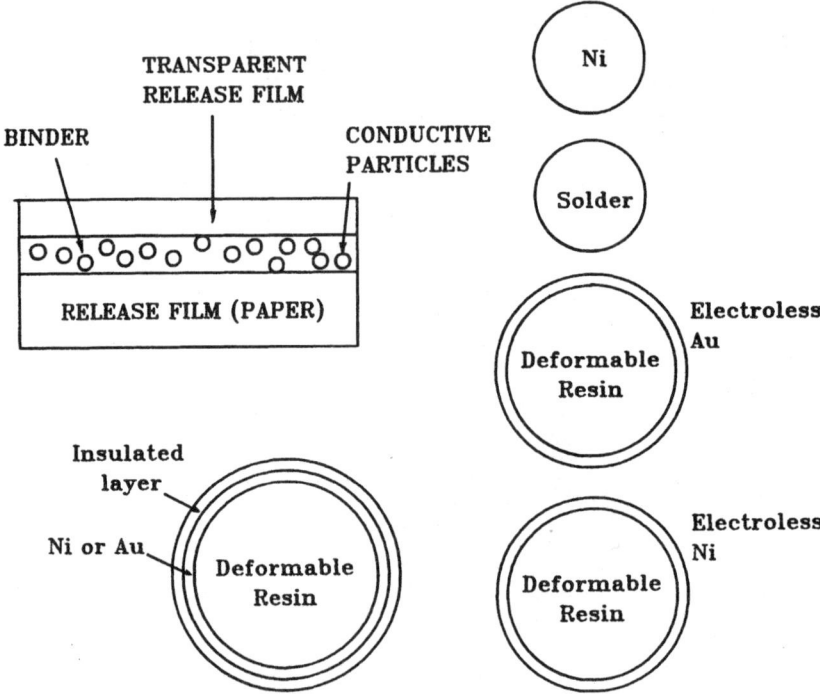

Figure 1.10   Some anisotropic conductive films (ACF).

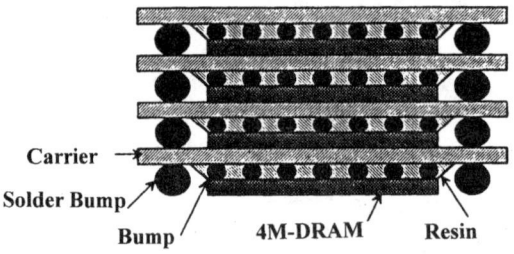

Figure 1.11   NEC's 3-D stack-up chip scale package (CSP) for memory devices.

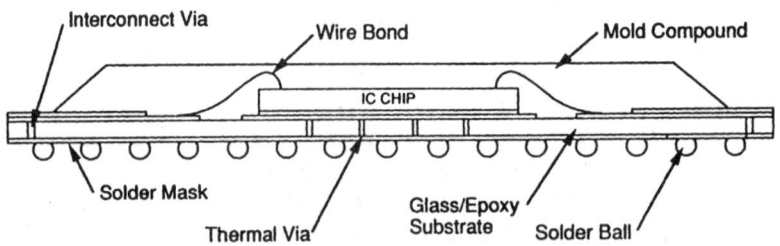

Figure 1.12   PBGA with wire-bonding chip on organic substrate.

Introduction 9

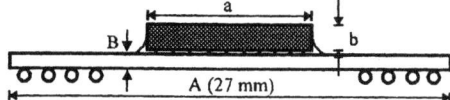

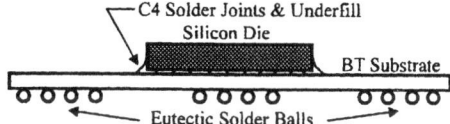

Figure 1.13 PBGA with solder-bumped flip chip on organic substrate.

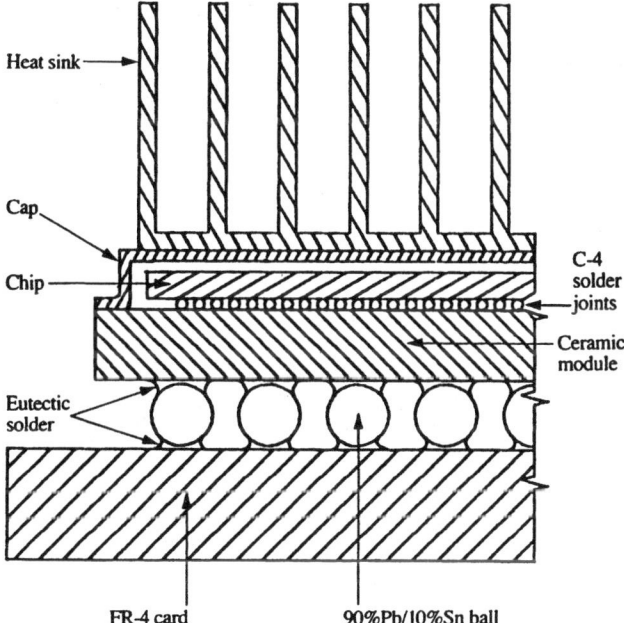

Figure 1.14 CBGA with heat sink (C-4: controlled-collapse chip connection).

10   Chapter One

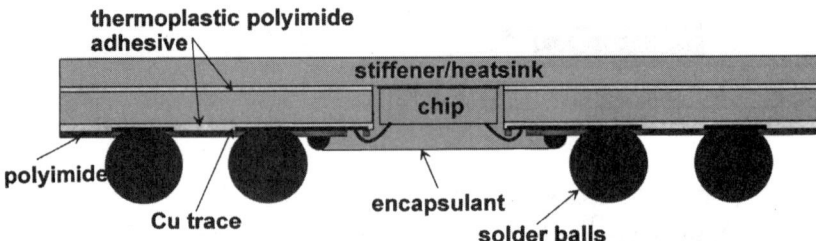

Figure 1.15 3M's single metal layer cavity down wire-bonding TBGA.

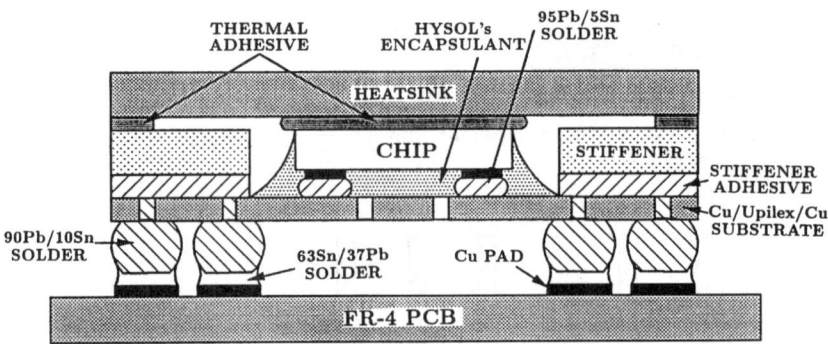

Figure 1.16 IBM's area array solder-bumped flip-chip TBGA.

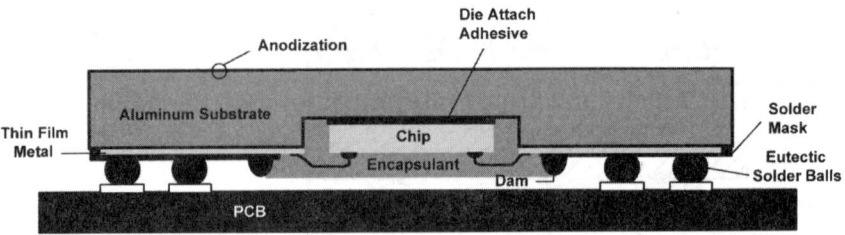

Figure 1.17 Olin's metal BGA (MBGA) with wire bonding.

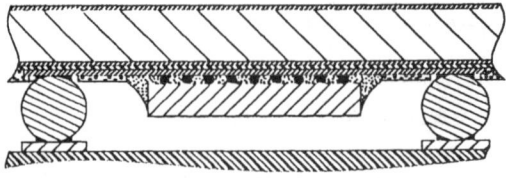

Figure 1.18 IBM's MBGA with solder-bumped flip chip.

1.2.1 Area-array flip-chip technology

For high-price and high-speed microprocessors and ASICs, complex IC designs require very high I/O and performance packages. For these types of ICs and subsystems, area-array solder-bumped flip-chip technology provides a viable answer to performance needs (Fig. 1.5). Usually, they are packaged in a CPGA, a PPGA, a CBGA (Fig. 1.14), a TBGA (Fig. 1.16), or an MBGA (Fig. 1.18) in a single-chip format. Recently, PBGAs with solder-bumped flip chip (Fig. 1.13) also have been considered for housing the high-price, high-speed microprocessors and ASICs.

1.2.2 BGA technology

There are many different kinds of BGAs [3]. Depending on their substrates, there are CBGA (Fig. 1.14), MBGA (Figs. 1.17 and 1.18), TBGA (Figs. 1.15 and 1.16), and PBGA (Fig. 1.13). Figure 1.5 shows that for high I/O and performance ASICs and microprocessers, CBGA, MBGA, and TBGA could meet the high pin count (>500), power, and clock-frequency requirements but with higher costs.

Due to the fine-pitch and pin-count limitations of PQFP and the high cost of CBGAs, MBGAs, and TBGAs, the PBGA is cost-effective between 250 and 500 package pin counts. The major difference between PQFP and PBGA is that PQFP has a leadframe and PBGA has an organic substrate. The leadframe has been standardized for fanning the circuitry for more than 10 years, while the PBGA's substrate is still custom designed. Today, the leadframe is cheaper than the organic substrate. Thus, in order for the PBGA packages to be popular, its substrate has to be standardized and at a lower cost.

1.2.3 TCP

TCP offers smaller pitches, thinner package profiles, and smaller footprints on the PCB. TCP can provide moderate-performance solutions for applications (e.g., ASICs and microprocessors) with up to about 600 package pin counts. (Intel uses TCP to house its Pentium microprocessors for all portable computers.) It is noted that unless volume production is very high, TCP may not be cost-effective due to the very high development cost of these custom-designed packages. Also, TCPs suffer the same drawbacks (e.g., peripheral chip carrier, long lead length, handling, board level manufacturing yield loss) as the PQFPs.

1.2.4 TSOP and PQFP

Up to 208-pin (0.5-mm pitch and 28-mm body size), 240-pin (0.5-mm pitch, 32-mm body size), and 304-pin (0.5-mm pitch and 40-mm body size), PQFPs [1, 8] are the most cost-effective packages for SMT (surface mount technology). They have been used extensively for ASICs, and low-performance and low-pin-count microprocessors. Sometimes, they are used to house one or more cache memories. Currently, the price for the PQFP packages is less than one penny a pin. For example, the listing price for the 208-pin PQFP is less than $1.50.

TSOP [1, 8] is a very low profile plastic package, which is specifically designed for SRAM, DRAM, and flash-memory (which retains information even when the power is turned off) devices for space-limit applications. Right now, the price for TSOP is less than one penny a pin.

1.2.5 CSP and DCA

One of the most cost-effective packaging technologies is DCA [1, 2, 4] (see examples in Figs. 1.6 through 1.9). However, because of the infrastructure and cost in supplying the known good die (KGD) and the cost and availability of the corresponding fine line and spacing PCB (Fig. 1.19), most in the industry are still working on these issues. However, it should be noted that, with the sequential build-up of PCB technologies such as the DYCOstrate, plasma etched redistribution layers (PERL), surface laminar circuits (SLC), film redistribution layer (FRL), interpenetrating polymer build-up structure system (IBSS),

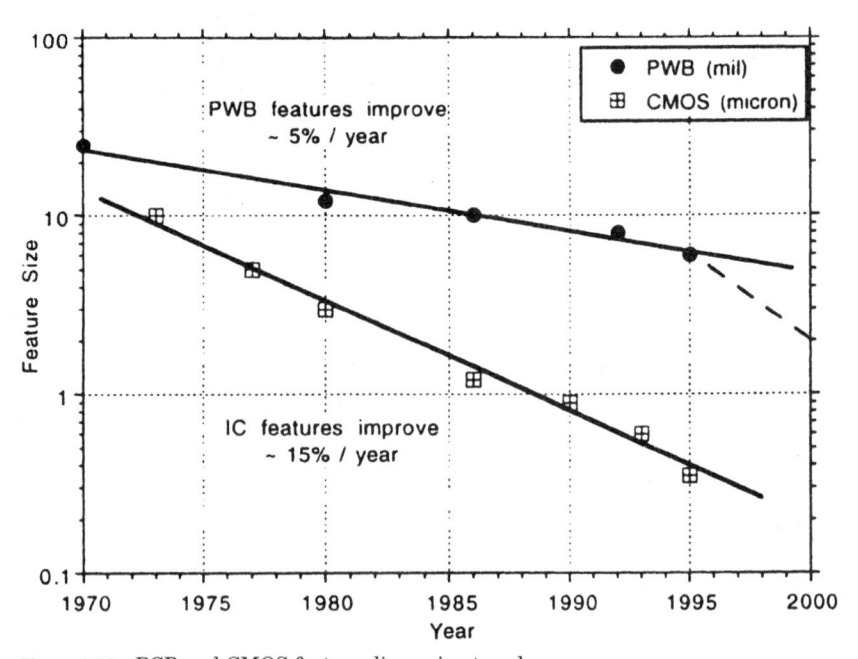

Figure 1.19 PCB and CMOS feature dimension trends.

high-density interconnect (HDI), conductive adhesive bonded flex, sequential bonded films, sequential bonded sheets, and microfilled via technology, 0.5-mm (2-mil) line width and space PCBs should be available by the turn of this century.

In the meantime, a new class of technology called chip scale package (CSP) has surfaced. There are more than 40 different CSPs reported today [10] and most of them are used for SRAMs, DRAMs, flash memories and not-so-high-pin-count ASICs and microprocessors [49–69]. The unique feature of most CSPs is the use of a substrate (carrier or interposer) to redistribute the very fine-pitch (as small as 0.075 mm) peripheral pads on the chip to much larger-pitch (1 mm, 0.75 mm, and 0.5 mm) area-array pads on the PCB.

The advantages of CSP versus DCA are: with the carrier (interposer), the CSP is easier to test-at-speed and burn-in for KGD, to handle, to assemble, to rework, to standardize, to protect the die, to deal with die shrink, and it is subject to less infrastructure constraints. On the other hand, the advantages of DCA are that it has better electrical and thermal performance, and less weight, size, and cost. It should be pointed out that die shrink is the biggest enemy of DCA!

1.3 Summary

A brief overview of IC devices (e.g., microprocessor, ASIC, cache memory, and system memory) trends has been presented. Also, various packaging technologies such as the flip chip, BGA, TCP, CSP, and DCA for IC devices have been briefly discussed.

The area-array solder-bumped flip-chip technology is suitable for very high I/O and performance single-chip modules for the microprocessors and ASICs. CBGAs, MBGAs, and TBGAs are cost-effective packages for microprocessors and ASICs with more than 500-pin counts and high-power dissipation. Usually, the area-array solder-bumped flip chip is packaged in a CBGA, a MBGA, or a TBGA.

In general, PBGAs are cost-effective for packaging the ASICs (and sometimes the microprocessors) with 250 to 500 pin counts and not-so-high power dissipation (less than 10 W). In some cases, PBGAs with lower pin counts (<200) are also cost-effective (for performance purpose) for housing a cache or a few fast SRAMs. For thermal-enhanced PBGAs, the back of the chip is attached to a heat spreader, which could be mounted with a heat sink. In that case, its power dissipation could be as high as 30 W and could be used to house microprocessors and ASICs.

CSPs are usually used to package memories and low-power (<1 W) and pin-count (<200) ASICs, even though higher-pin-count and power ICs have been tried in laboratories. Except in a handful of vertically integrated companies, DCAs are still waiting for their time when the

KGD and high density micro-via build-up PCBs are commonly available at reasonable costs.

1.4 References

1. Lau, J. H., and Y.-H. Pao. *Solder Joint Reliability of BGA, CSP, Flip Chip, and Fine Pitch SMT Assemblies.* New York: McGraw-Hill, 1997.
2. Lau, J. H. *Flip Chip Technologies.* New York: McGraw-Hill, 1996.
3. Lau, J. H. *Ball Grid Technology.* New York: McGraw-Hill, 1995.
4. Lau, J. H. *Chip On Board Technologies for Multichip Modules.* New York: Van Nostrand Reinhold, 1994.
5. Lau, J. H. *Solder Joint Reliability, Theory and Applications.* New York: Van Nostrand Reinhold, 1991.
6. Lau, J. H. *Handbook of Tape Automated Bonding.* New York: Van Nostrand Reinhold, 1992.
7. Lau, J. H. *Thermal Stress and Strain in Microelectronics Packaging.* New York: Van Nostrand Reinhold, 1993.
8. Lau, J. H. *Handbook of Fine Pitch Surface Mount Technology.* New York: Van Nostrand Reinhold, 1994.
9. Frear, D., H. Morgan, S. Burchett, and J. Lau. *The Mechanics of Solder Alloy.* New York: Van Nostrand Reinhold, 1994.
10. Lau, J. H., and R. Lee. *Chip Scale Packages (CSP): Design, Materials, Process, Reliability, and Applications.* New York: McGraw-Hill, to be published in 1998.
11. Wong, C. P. *Polymers for Electronic and Photonic Applications.* San Diego, Calif.: Academic Press, 1993.
12. Senthinathan, R., and J. L. Prince. *Simultaneous Switching Noise of CMOS Devices and Systems.* New York: Kluwer Academic Publishers, 1994.
13. Tummala, R. R., and E. Rymaszewski. *Microelectronics Packaging Handbook.* New York: Van Nostrand Reinhold, 1989.
14. Seraphim, D. P., R. Lasky, and C. Y. Li. *Principles of Electronic Packaging.* New York: McGraw-Hill Book Company, 1989.
15. Vardaman, J. *Surface Mount Technology, Recent Japanese Developments.* New York: IEEE Press, 1992.
16. Hwang, J. S. *Solder Paste in Electronics Packaging.* New York: Van Nostrand Reinhold, 1989.
17. Hwang, J. S. *Modern Solder Technology for Competitive Electronics Manufacturing.* New York: McGraw-Hill, 1996.
18. Johnson, R. W., R. K. Teng, and J. W. Balde. *Multichip Modules: System Advantages, Major Construction, and Materials Technologies.* New York: IEEE Press, 1991.
19. Sandborn, P. A., and H. Moreno. *Conceptual Design of Multichip Modules and Systems.* New York: Kluwer Academic Publishers, 1994.
20. Nash, F. R., *Estimating Device Reliability: Assessment of Credibility.* New York: Kluwer Academic Publishers, 1993.
21. Gyvez, J. P. *Integrated Circuit Defect-Sensitivity: Theory and Computational Models.* New York: Kluwer Academic Publishers, 1993.
22. Doane, D. A., and P. D. Franzon. *Multichip Module Technologies and Alternatives.* New York: Van Nostrand Reinhold, 1992.
23. Messuer, G., I. Turlik, J. Balde, and P. Garrou. *Thin Film Multichip Modules.* Silver Spring, Md.: International Society for Hybrid Microelectronics, 1992.
24. Manzione, L. T. *Plastic Packaging of Microelectronic Devices.* New York: Van Nostrand Reinhold, 1990.
25. Hymes, L. *Cleaning Printed Wiring Assemblies in Today's Environment.* New York: Van Nostrand Reinhold, 1991.
26. Gilleo, K. *Handbook of Flexible Circuits.* New York: Van Nostrand Reinhold, 1991.
27. Engel, P. A. *Structural Analysis of Printed Circuit Board Systems.* New York: Springer-Verlag, 1993.

28. Suhir, E. *Structural Analysis in Microelectronic and Fiber Optics Systems.* New York: Van Nostrand Reinhold, 1991.
29. Matisoff, B. S. *Handbook of Electronic Packaging Design and Engineering.* New York: Van Nostrand Reinhold, 1989.
30. Prasad, R. P. *Surface Mount Technology.* New York: Van Nostrand Reinhold, 1989.
31. Manko, H. H. *Soldering Handbook for Printed Circuits and Surface Mounting.* New York: Van Nostrand Reinhold, 1986.
32. Morris, J. E. *Electronics Packaging Forum, volume 1.* New York: Van Nostrand Reinhold, 1990.
33. Morris, J. E. *Electronics Packaging Forum, volume 2.* New York: Van Nostrand Reinhold, 1991.
34. Hollomon, J. K., Jr. *Surface-Mount Technology.* Indianapolis, Ind.: Howard W. Sams & Company, 1989.
35. Solberg, V. *Design Guildelines for SMT.* New York: TAB Professional and Reference Books, 1990.
36. Hutchins, C. *SMT: How to Get Start.* Raleigh, N.C.: Hutchins and Associates, 1990.
37. Bar-Cohen, A., and A. D. Kraus. *Advances in Thermal Modeling of Electronic Components and Systems, volume 1.* New York: Hemisphere Publishing Corp., 1988.
38. Bar-Cohen, A., and A. D. Kraus. *Advances in Thermal Modeling of Electronic Components and Systems, volume 2.* New York: ASME Press, 1990.
39. Kraus, A. D., and A. Bar-Cohen. *Thermal Analysis and Control of Electronic Equipment.* New York: Hemisphere Publishing Corp., 1983.
40. Harper, C. A. *Handbook of Microelectronics Packaging.* New York: McGraw-Hill Book Company, 1991.
41. Pecht, M. *Handbook of Electronic Package Design.* New York: Marcel Dekker, 1991.
42. Harman, G. *Wire Bonding in Microelectronics.* Reston, Va.: International Society for Hybrid Microelectronics, 1989.
43. Lea, C. *A Scientific Guide to Surface Mount Technology.* Scotland: Electrochemical Publications, 1988.
44. Lea, C. *After CFCs? Options for Cleaning Electronics Assemblies.* IOM, British Isles: Electrochemical Publications Ltd., 1992.
45. Wassink, R. J. K. *Soldering in Electronics.* Scotland: Electrochemical Publications, 1989.
46. Pawling, J. F. *Surface Mounted Assemblies.* Scotland: Electrochemical Publications, 1987.
47. Ellis, B. N. *Cleaning and Contamination of Electronics Components and Assemblies.* Scotland: Electrochemical Publications, 1986.
48. Sinnadurai, F. N. *Handbook of Microelectronics Packaging and Interconnection Technologies.* Scotland: Electrochemical Publications, 1985.
49. Kasai, J., M. Sato, T. Fujisawa, T. Uno, M. Waki, K. Hayashida, and T. Kawahara. "Low Cost Chip Scale Package for Memory Products," *Proceedings of SMI Conference,* August 1995, pp. 6–17.
50. Murakami, G. "Rationale for Chip Scale Packaging (CSP) Rather Than Multichip Modules (MCM)," *Proceedings of SMI Conference,* August 1995, pp. 1–5.
51. Cha, K., Y. Kim, T. Kang, D. Kang, and S. Back. "Ultra-Thin and Crack-Free Bottom Leaded Plastic (BLP) Package Design," *Proceedings of IEEE Electronic Components & Technology Conference,* May 1995, pp. 224–228.
52. Lee, S., J. Lee, S. Oh, and H. Chung. "Passivation Cracking Mechanism in High Density Memory Devices Assembled in SOJ Packages Adopting LOC Die Attach Technique," *Proceedings of IEEE Electronic Components & Technology Conference,* May 1995, pp. 455–462.
53. Amagai, M. "The Effect of Stress Intensity of Package Cracking in Lead-On-Chip (LOC) Packages," *Proceedings of IEEE Japan International Electronics Manufacturing Technology Symposium,* December 1995, pp. 415–420.
54. Okugawa, Y., T. Yoshida, T. Suzuki, and H. Nakayoshi. "New Tape LOC Adhesive Tapes," *Proceedings of IEEE Electronic Components & Technology Conference,* May 1994, pp. 570–574.

55. Nakayoshi, H., N. Izawa, T. Ishikawa, and T. Suzuki. "Memory Package with LOC Structure Using New Adhesive Material," *Proceedings of IEEE Electronic Components & Technology Conference,* May 1994, pp. 575–579.
56. Kunitomo, Y. "Practical Chip Size Package Realized by Ceramic LGA Substrate and SBB Technology," *Proceedings of SMI Conference,* August 1995, pp. 18–25.
57. Master, R., R. Jackson, S. Ray, and A. Ingraham. "Ceramic Mini-Ball Grid Array Package for high Speed Device," *Proceedings of IEEE Electronic Components & Technology Conference,* May 1995, pp. 46–50.
58. Lall, P., G. Gold, B. Miles, K. Banerji, P. Thompson, C. Koehler, and I. Adhihetty. "Reliability Characterization of the SLICC Package," *Proceedings of IEEE Electronic Components & Technology Conference,* May 1996, pp. 1202–1210.
59. Iwasaki, H. "CSTP: Chip Scale Thin Package," *Proceedings of SEMICON Japan,* November 1994, pp. 488–495.
60. Forman, G., R. Fillion, R. Kole, R. Wojnarowski, and J. Rose. "Development of GE's Plastic Thin-Zero Outline Package (TZOP) Technology," *Proceedings of IEEE Electronic Components & Technology Conference,* May 1995, pp. 664–668.
61. Matsuda, S., K. Kata, and E. Hagimoto. "Simple-Structure, Generally Applicable Chip-Scale Package," *Proceedings of IEEE Electronic Components & Technology Conference,* May 1995, pp. 218–223.
62. Tanigawa, S., K. Igarashi, M. Nagasawa, and N. Yeti. "The Resin Molded Chip Size Package (MCSP)," *Proceedings of IEEE Japan International Electronics Manufacturing Technology Symposium,* December 1995, pp. 410–415.
63. Kayo, T., K. Be, N. Sakaguchi, and S. Wakabayashi. "Reliability of mBGA Mounted on a Printed Circuit Board," *Proceedings of SMI Conference,* August 1995, pp. 43–56.
64. Marcoux, P. P. "Advanced Packaging Approaches to Very Fine-Pitch Surface Mount Interconnections," *Proceedings of NEPCON West,* February 1994, pp. 2056–2059.
65. Badihi, A., and E. Por. "Shellcase—A True Miniature Integrated Circuit Package," *Proceedings of Flip Chip, BGA, TAB, and Advanced Packaging Symposium,* February 1995, pp. 244–252.
66. Yasunaga, M., S. Baba, M. Matsuo, H. Matsushima, S. Nako, and T. Tachikawa. "Chip Scale Package (CSP): A Lightly Dressed LSI Chip," *Proceedings of IEEE Japan International Electronics Manufacturing Technology Symposium,* December 1994, pp. 169–176.
67. Chanchani, R., K. Treece, and P. Dressendorfer. "mini Ball Grid Array (mBGA) Technology," *Proceedings of NEPCON West,* February 1995, pp. 938–945.
68. Lau, J. H., W. Jung, and Y.-H. Pao. "Temperature-Dependent Elasto-Plastic-Creep Analysis of a Chip Scale Package (CSP)'s Solder Joints," *ASME Paper No. 96WA/EEP-10,* November 1996.
69. Lau, J. H. "Solder Joint Reliability of a Low Cost Chip Size Package—NuCSP," *ISHM Proceedings,* October 1997, pp. 691–696.

Chapter

2

Basic Electrical Effects in Electronic Packaging and Interconnects

2.1 Basic Electrical Parameters and Phenomena in Electronic Packaging

Some understanding of the challenges facing electrical designers of packaging systems over the next fifteen years can be gained by examining Figs. 2.1, 2.2, and 2.3. These figures must be considered in the light of the boundary conditions implicit in Figs. 2.4, 2.5, and 2.6. Figure 2.1 is a 1997 update of the well-known Moore's Law graph, which relates functional complexity (e.g., gates/chip) of semiconductor chips to the first year of new product introduction [1]. The figure demonstrates the tremendous progress in functional complexity in the past and shows that today's best thinking projects a continuation and even acceleration of the process through the year 2012. The acceleration evident in the late 1990s is due to an increase in the rate of critical dimension shrinkage due to unexpectedly rapid progress in lithography capability. Figure 2.2 shows that although die size is expected to increase over the next fifteen years, the contribution of die area increase to chip functional complexity is small compared to the expected 50-fold increase in chip complexity over that same period [1].

One of the benefits of the reduction in critical dimension is an increase in the chip processing clock frequency. This is accompanied by an increase in on-chip and off-chip I/O clock frequency, such as is shown in Fig. 2.3. The term *cost-performance* in this figure refers to utilization in products which sell for $3000 maximum and are appropriate for use on a desktop. For these products, cost is generally the primary

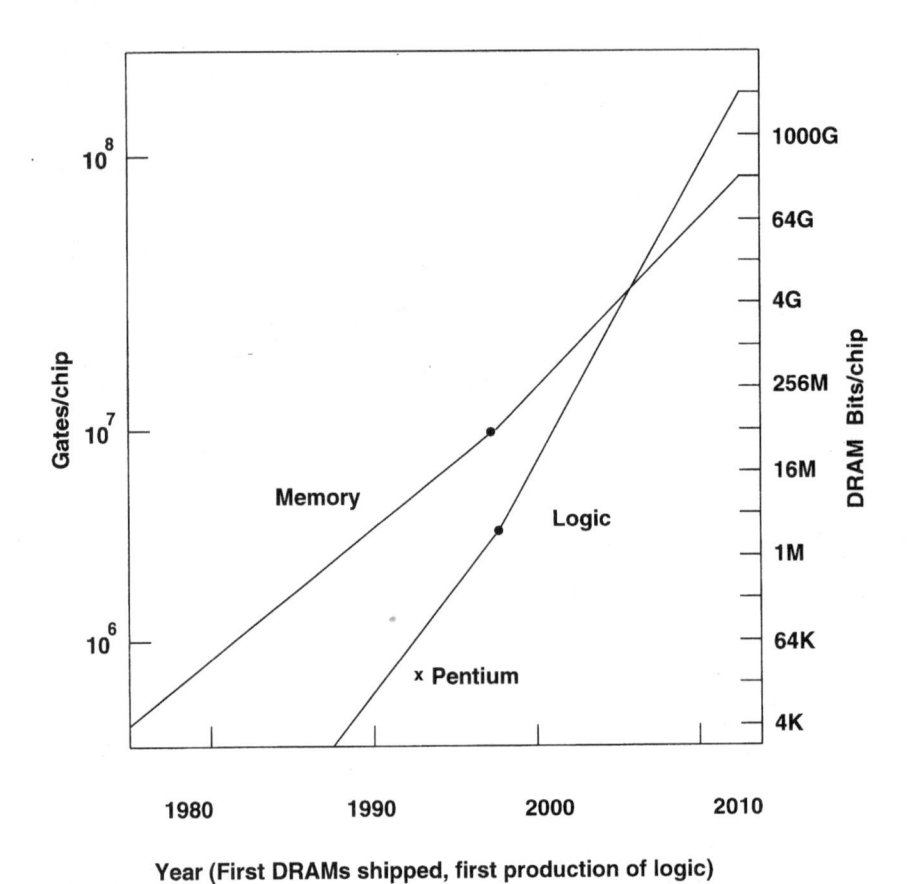

Figure 2.1 Chip functional complexity versus year of first production ("Moore's Law" updated) [1].

figure of merit and performance is a discriminator. *High-performance* refers to usage in products for which performance is the primary figure of merit and cost is a discriminator. High-performance workstations are typical examples of this type of product. The fact that within a decade package and substrate-level clock frequencies can reasonably be expected to lie in the neighborhood of 1–2 GHz for high-volume products has profound implications for package system performance. For example, pulse rise and fall time will be in the neighborhood of 100 ps, and some important frequency components of the pulse signals will extend as high as 10 GHz. Transmission-line design techniques will thus be required for lines with lengths in excess of 0.5 to 1.0 cm. Substrate materials will be required to support microwave frequencies. Radiated noise will become more of a problem as Fourier component frequencies rise and wavelengths decrease. As an example, for a 10-GHz sinusoid, with an effective substrate or package dielectric con-

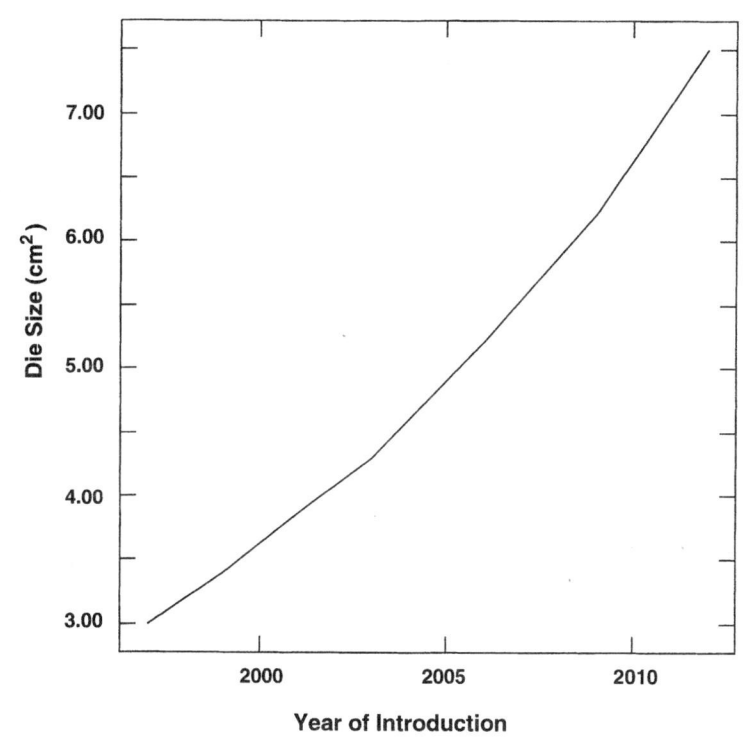

Figure 2.2 Maximum volume production die size versus year of chip introduction [1].

stant of 4.0, the wavelength is 1.5 cm. Since this wavelength is clearly of the order of board and package-level line lengths, both electromagnetic radiation and reception efficiency will rise and will become substantially more of a problem than for current clock frequencies. Noise and extraneous voltages of all sorts will be enhanced by the high edge speed and by the high frequency content of signal pulses.

The ability to design and fabricate package systems that will meet the extreme conditions coming in the near future is constrained by cost considerations. Figure 2.4 shows a sketch of the history of products utilizing Intel microprocessor, over the decade from the early 1980s to the early 1990s [2]. The most important data in this figure are literally the bottom lines. It appears that regardless of the tremendous performance enhancements and the like evident in the figure, cost rules and the selling price remains constant. Figure 2.5 shows cost constraints that must be met at the microprocessor package level, over the next fifteen years. Although there is some elasticity evident in this figure, the challenges of achieving microwave frequency performance at PC-level costs will be substantial. The potential difficulty in achieving the costs

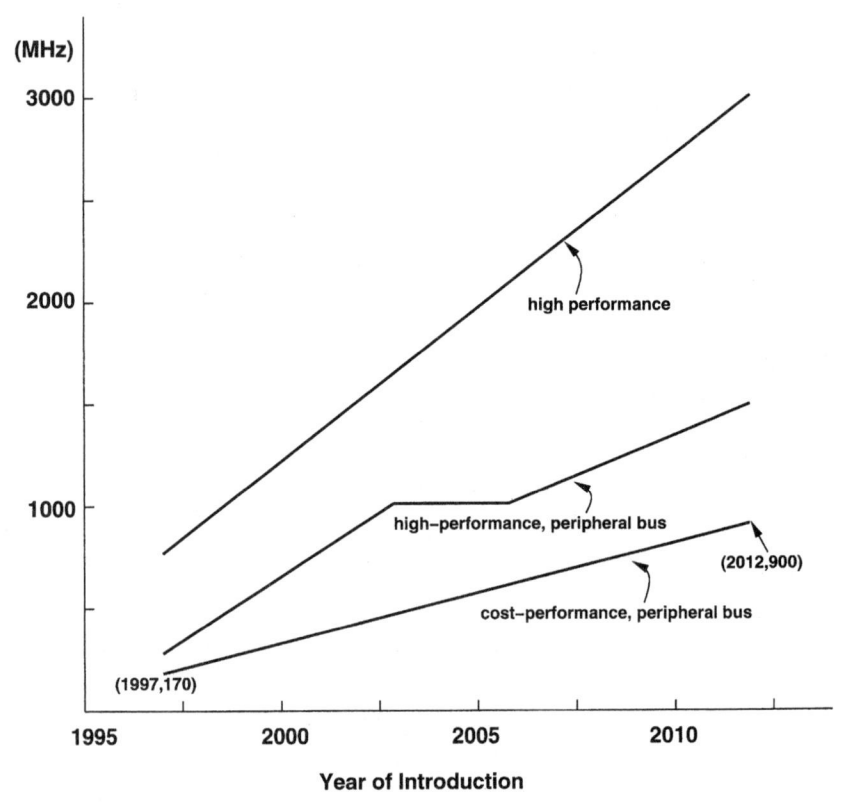

Figure 2.3 Off-chip clock frequency for various busses versus year of introduction [1]. See text for interpretation of "high-performance" and "cost-performance."

projected in Fig. 2.5 can be estimated by examining the Intel microprocessor package cost history of Fig. 2.6 [2]. Clearly package costs rose a factor of seven in approximately one decade of intensive technological progress. High-end package cost as reflected in Fig. 2.5 will be permitted to rise only a factor of two during a period of fifteen years during which time pin count will rise from 500 to 2500 [1] and off-chip operat-

1981	**1987**	**1994**
8088	80386	Pentium
0.3 MIP	3 MIP	66-100 MIP
40 pins	100 pins	400 pins
$ 2.5K	$ 2.5K	$ 2.5 K

Figure 2.4 End product cost history for example cost-performance (PC) systems [2].

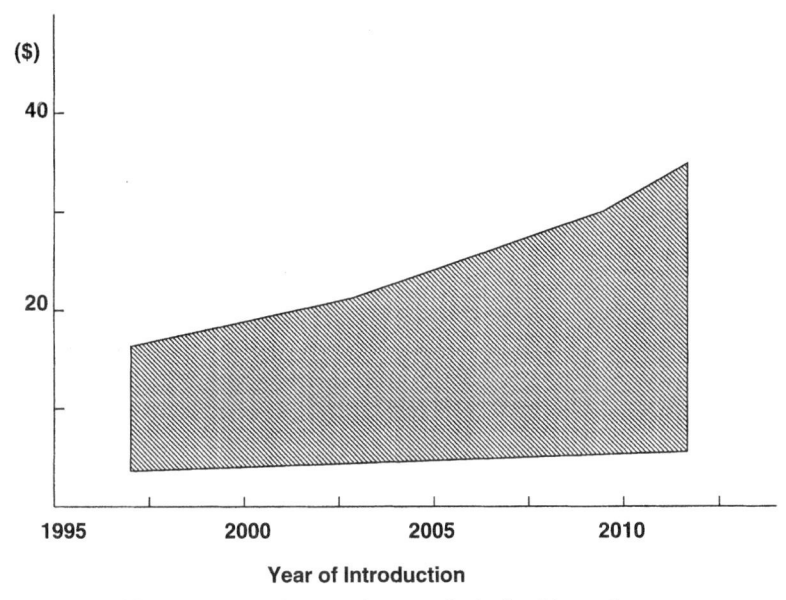

Basic Electrical Effects in Electronic Packaging and Interconnects 21

Figure 2.5 Microprocessor (cost-performance) single-chip package cost versus year of introduction [1].

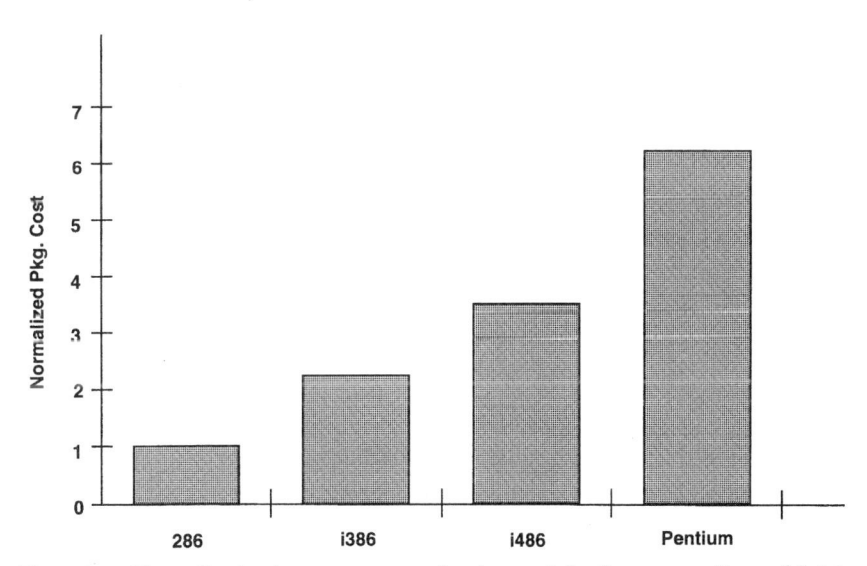

Figure 2.6 Normalized microprocessor packaging cost for four generations of Intel products.

ing frequency will rise a factor of six as shown in Fig. 2.3. This cost limitation has important implications for the electrical design and performance of these packaging structures.

2.1.1 Introduction

From the standpoint of electrical performance, the goal of a packaging structure is to transmit signals (either digital or analog) from point A to point B faithfully and—in the case of digital signals—in minimum time. Also, it is a goal of the packaging structure to supply "clean" power and reference voltage levels to active devices in the system, and at the same time not to introduce radiated electromagnetic energy into the environment nor to receive electromagnetic energy from other radiating structures in the environment [*electromagnetic interference* (EMI) and *electromagnetic compatibility* (EMC)].

To transmit a signal faithfully between two points means to maintain the signal *integrity*, that is, to introduce little or no distortion in the signal [3]. Some types of distortion that may be introduced are amplitude reduction, which may be due to signal reflections or energy losses, and shape distortion, which may be due to signal reflection, energy losses, frequency dependent parameters of the structure, and loading of the signal transmission line by capacitive or inductive elements.

The two types of distortion (generally referred to as "noise," although they are deterministic and not stochastic) may give rise to logic or timing errors, with ultimate failure of the chip or system. Amplitude reduction and shape distortion may, for example (possibly in concert with other circuit or system-level efforts), result in sensing by receivers of false logic levels (a logic "1" is sensed by a receiver as a logic "0" because the logic "1" voltage level is reduced below the input transition voltage level of the receiver, or vice versa for a logic "0" → logic "1" error). They may also both contribute to timing errors, which are caused by signals reaching the receiver input transition point at a time later than (or possibly even earlier than) the time window required for proper chip or system functioning. It is critical to design packaging systems so that pulse distortion efforts of these sorts are predictable, understood, and controlled. First-pass design success through understanding of the nature and magnitude of noise mechanisms and prediction, and control of the effects through analytical and computer-modeling techniques is the goal of electrical design of packaging systems.

To supply "clean" power and reference voltage levels means that dc voltage levels supplied to the chip or chips are not interrupted by transient voltage levels that decrease or increase the dc level. Such transient voltage levels are typically generated when output drivers on a chip switch binary states, and transient currents with large dI/dt flow through inductive paths in the interconnect structure [4].

Concern with supplying clean power and reference voltage levels to the chip comes back to maintaining the integrity of signals. Transient voltage levels superimposed on the dc voltage levels may feed through to chip output pads and either cause sensing of false logic levels at receivers or, at a minimum, cause reduction in the voltage margin between one or more chip output driver voltage levels and the switching threshold of their receivers. It also happens that reduction of net power supply voltage reduces the current driver capability of CMOS gates. This reduction of current driver capability results in a slowing of the charging/discharging rates of capacitive nodes. This slowdown of the performance may result in timing errors at the chip or system level, finally resulting in failure of the system.

For EMI and EMC, typical structures that provide for efficient transmission and reception of electromagnetic energy are either large-area loops in which currents circulate or structures such as floating metal (e.g., heat sinks, etc.), which are not a part of the dc path for power distribution but which are near signal lines.

Concern over EMI and EMC is increasing due to increasing clock frequencies and thus shorter signal rise and fall times (see, e.g., Fig. 2.3). At least two undesirable effects that may occur due to EMI/EMC are failure of the sytem to meet FCC standards, and chip-to-chip or board-to-board radiative noise communication. Use of available, accurate computer modeling and simulation software for both near-field and far-field radiation is a requirement in designing for EMI/EMC control. Such software has only recently become available, albeit in a limited sense.

Examples of these undesirable phenomena that can be introduced by packaging structures are shown below. Figure 2.7a illustrates a pulse at the near (chip) end of a package conductor [V(4)] and the pulse 10 cm away from the chip [V(5)]. This signal conductor is lossy, with conductivity 6.33×10^5 (ohm-cm)$^{-1}$. It is clear that amplitude reduction due to losses is sufficient to render the V(5) signal amplitude too small to be sensed as a logic "1" in a CMOS system. Figure 2.7b shows a pulse output from an off-chip driver and the same pulse at the receiver, which is assumed to behave as a 50 pF capacitor. The rise time of the initial pulse is 50 ps. Clearly the time to rise to 50 percent of the full pulse amplitude is increased by the capacitive loading in Fig. 2.7b. Figure 2.8a shows the voltage coupled into a quiet line by a nearby active line, with the length of coupling equal to 5 cm and the active pulse rise time equal to 0.33 ns. (See also Fig. 2.8b.) The mechanisms for these effects will be discussed in detail in later sections.

In addition to signal integrity, prediction of pulse timing is important in packaging design. In addition to modeling the *time-of flight* (tof) delay (that delay caused by finite velocity of propagation in the dielectric medium), the delay adder caused by pulse distortion must also be modeled and simulated. The *delay adder* is the time in excess of the tof

Lossy Microstrip Line, r11=15.12ohm/cm

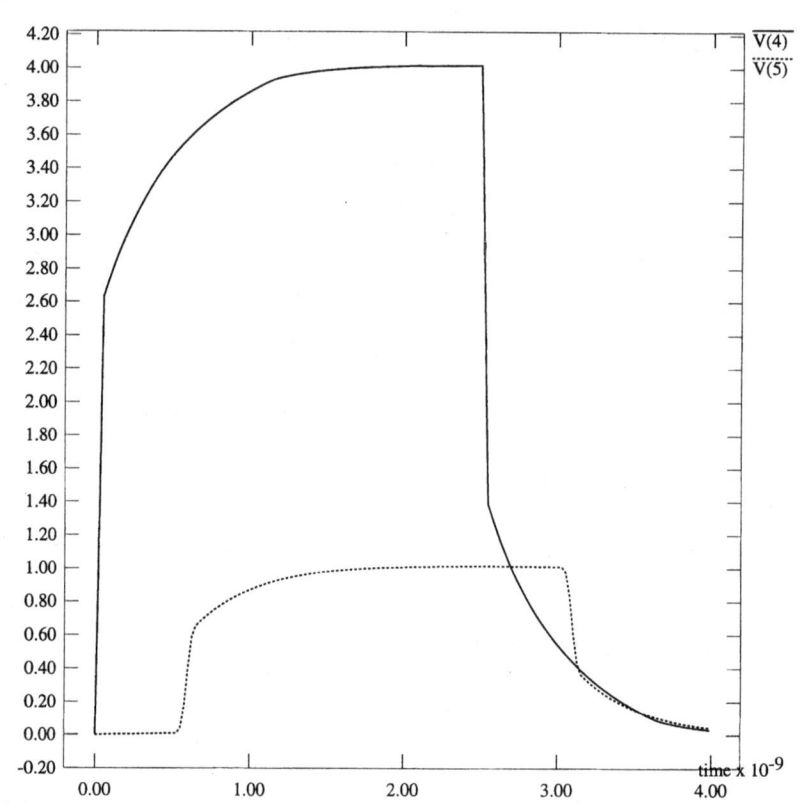

Figure 2.7a Signal pulse distortion due to propagation on a lossy signal trace. V(4) is the near-end pulse voltage and V(5) is the far-end (10-cm) voltage.

delay it takes for the pulse at a receiver to rise to 50 percent of its ideal final value. Both tof delay and extra delay due to a delay adder are evident in Fig. 2.7b.

Two tasks are entailed in predicting the electrical performance of packaging:

1. Package modeling or parameter extraction
2. Package simulation

The two tasks are separable in many classes of problems, but not in all classes of problems. Package modeling can be divided into three types:

1. Lumped-element modeling
2. Distributed-element modeling
3. Non-TEM (very high frequency) modeling

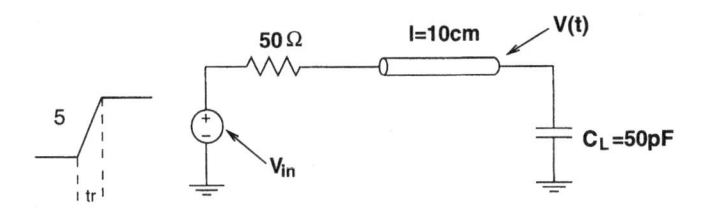

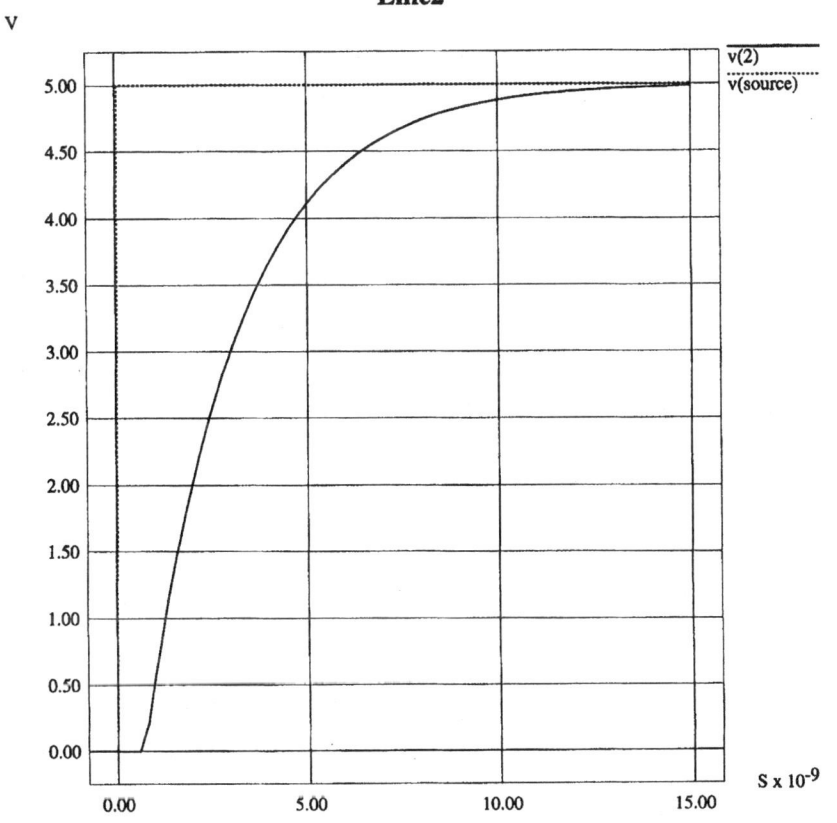

Figure 2.7b Signal pulse distortion due to a capacitive receiver input circuit ($C_L = 50$ pF).

The term *TEM* stands for Transverse ElectroMagnetic, and is discussed below. For both cases 1 and 2, the task is to extract appropriate parameters R, L, C, and G, which can subsequently be used to simulate the electrical performance of the structure in its environment of nearby conductors. For case 3, the task is to extract the propagation constant $\gamma = \alpha + j\beta$, which then is used to simulate the pulse behavior. The three types of package modeling are shown schematically in Fig. 2.9.

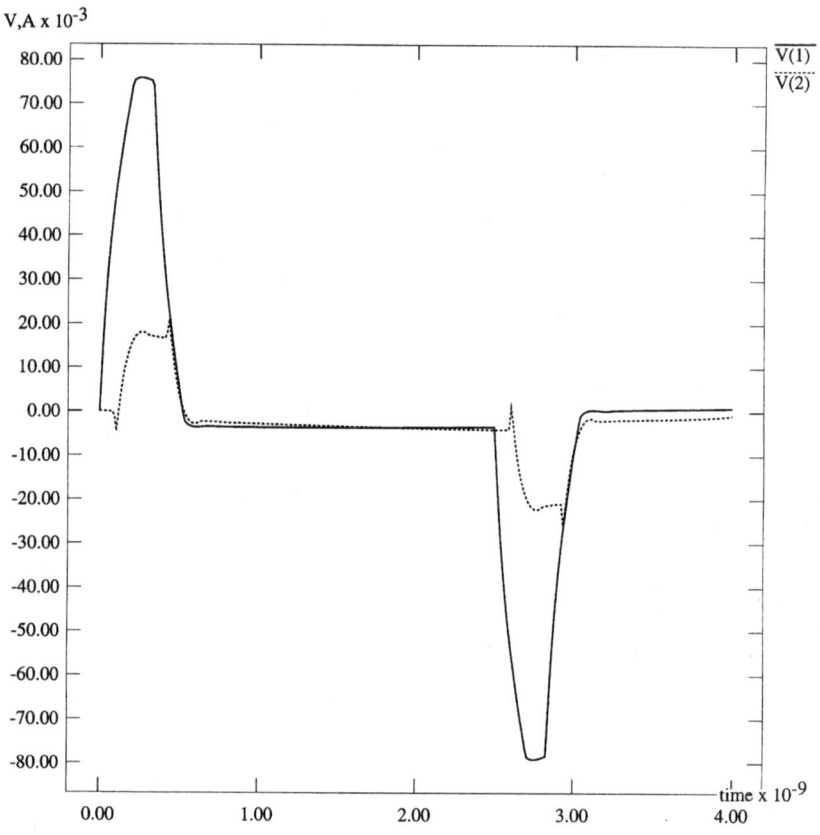

Figure 2.8a Extraneous voltages coupled into a quiet line from an active line. V(1) is the near-end extraneous voltage and V(2) is the far-end extraneous voltage.

Package simulation can be divided into two basic types:

1. TEM-wave simulation, either lumped-element or distributed-element as appropriate
2. Non-TEM wave simulation, generally referred to as full-wave simulation, since the method of simulation is often to simulate the propagation of $\vec{E}$ and $\vec{H}$ waves in the process of a full solution of the governing electromagnetic differential equations

These types and subtypes are arranged in ascending order of computational complexity and run time of simulation software. Since it is a general principle that, beyond run times of one or two minutes, the best approach to a packaging design or analysis problem is the one that gives the shortest run time with acceptable accuracy, it is important to understand this hierarchy of solution methods.

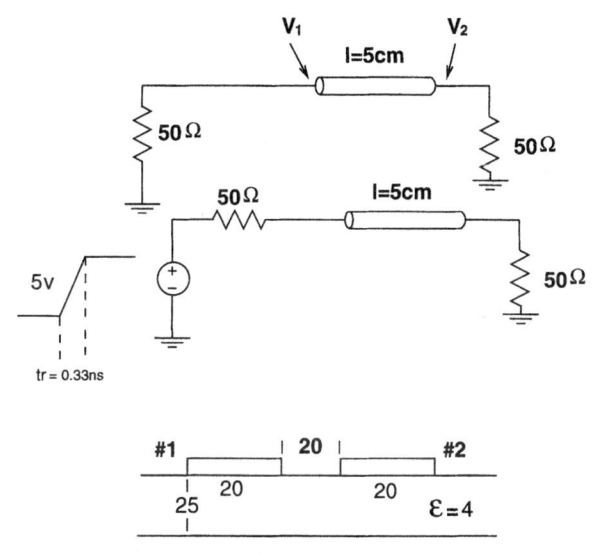

Basic Electrical Effects in Electronic Packaging and Interconnects   27

Figure 2.8b Structure and circuit for Fig. 2.8a.

The term *TEM-wave* is used to refer to a particular type of electromagnetic wave, one in which both the electric and magnetic fields are perpendicular to each other and perpendicular to the direction of propagation of the pulse. In this case, a TEM-wave travels as a plane wave; ordinary definitions of resistance, inductance, capacitance, and

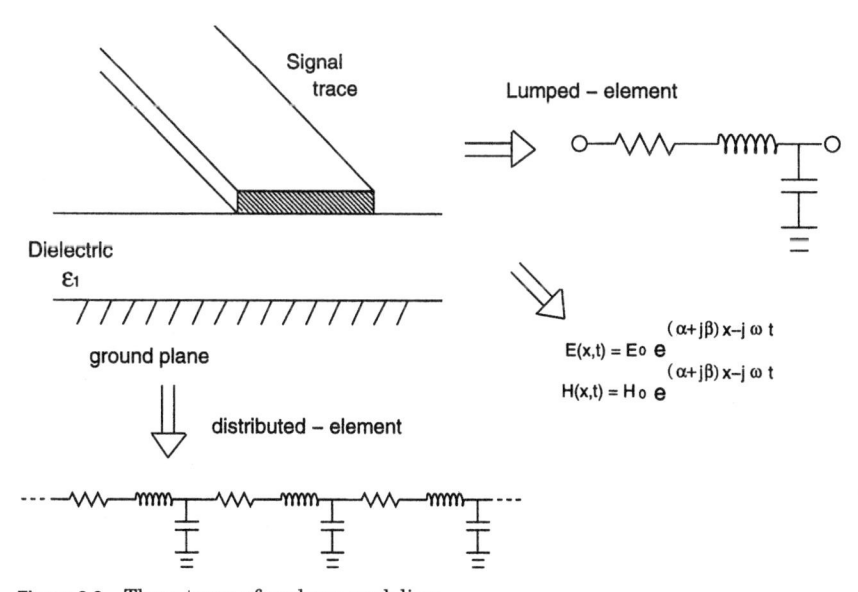

$$E(x,t) = E_0 \, e^{(\alpha+j\beta)x - j\omega t}$$
$$H(x,t) = H_0 \, e^{(\alpha+j\beta)x - j\omega t}$$

Figure 2.9 Three types of package modeling.

conductance can be used in simulating the wave behavior; and single-valued voltages and currents can be extracted. TEM-waves can exist only (1) in a homogeneous medium and (2) if conductors have zero resistance.

Thus, true TEM waves do not exist in a real packaging structure. However, for "low" frequencies (up to a few GHz) and moderate conductor resistance losses, the waves that do propagate are very nearly TEM in nature, and they behave as if they were TEM waves. These are called *quasi-TEM waves* and are treated as if they were true TEM waves. In addition to the earlier restrictions, it is also true that TEM waves, being plane waves, do not turn corners; therefore, the fields in the vicinity of vias, bends, and so forth are not TEM waves. Again for "low" frequencies, the effect of such a discontinuity as a change in direction of a signal conductor, a change in width of the conductor, and the like is taken into account by replacing the discontinuity by a lumped-element circuit. Some equivalent circuits are shown in Fig. 2.10. Parameter values for these equivalent circuits are obtained through quasi-static L-C modeling software tools. Some common conductor configurations are shown in Fig. 2.11.

Full-wave simulation requires solution of the controlling partial differential equations on a discrete grid in space and time [5], with appropriate boundary and initial conditions. Voltages and currents then are obtained through

$$V_{AB} = \int_A^B \vec{E} \cdot d\vec{\ell}$$

and

$$I = \oint \vec{H} \cdot d\vec{\ell}$$

where $\vec{E}$ and $\vec{H}$ are the electric and magnetic fields respectively. In some cases, for non-TEM fields, voltage and current will not be uniquely defined. Full-wave solution gives accurate numerical solutions for both TEM and non-TEM fields. Its drawback is the relatively long solution times even for moderately complex packaging structures.

2.1.2 Intrinsic delay

Information propagates in packaging structures at speeds up to the speed of light in the dielectric medium in the packaging structure. For TEM waves (which, strictly speaking, require homogeneous dielectrics, e.g., stripline in Fig. 2.11) this velocity is

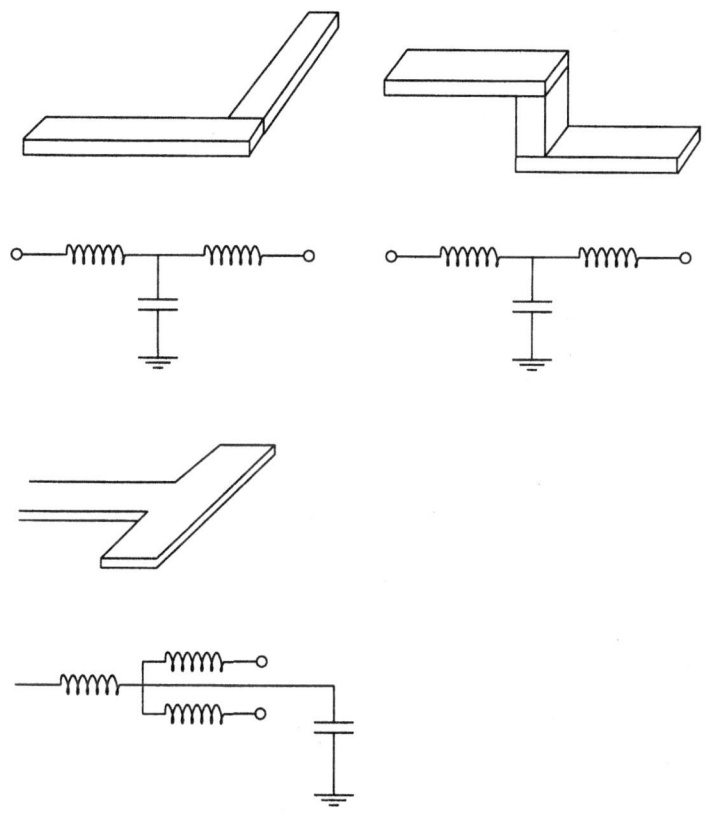

Figure 2.10 Some simple lumped equivalent circuits for common discontinuities.

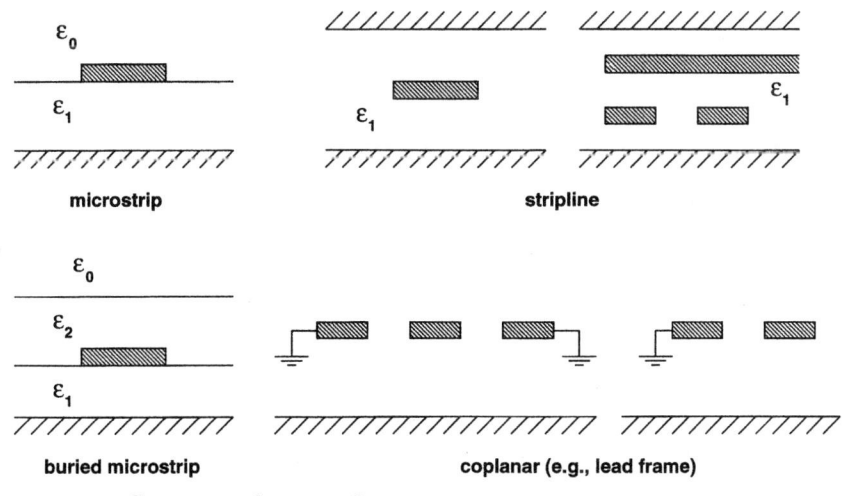

Figure 2.11 Common conductor configurations.

$$v_0 = c(\mu_r \varepsilon_r)^{-1/2} \quad (2.1)$$

where μ_r and ε_r are the relative permeability and permittivity, respectively, of the dielectric medium. The speed of light in a vacuum is

$$c = 3 \times 10^{10} \text{ cm/s}$$

For nonmagnetic dielectric material (the usual case), $\mu_r \cong 1$ and

$$v_0 = c(\varepsilon_r)^{-1/2} \quad (2.2)$$

For air, $\varepsilon_r \cong 1$ and $v_0 \cong c$. Table 2.1 gives ε_r values for some common materials.

For inhomogeneous quasi-TEM structures (e.g., microstrip or buried microstrip in Fig. 2.11), it is common to define an "effective ε," called ε_{eff}, and to use Eqs. 2.1 and 2.2 above with ε_{eff} instead of ε_r. For microstrip, ε_{eff} is usually close to the value of ε_1 but between ε_1 and ε_0. As a rule, the approximation $\varepsilon_{\text{eff}} = \varepsilon_1$ does not lead to gross error when calculating v_0.

A useful parameter related to v_0 is the time delay per unit (p.u.) length of conductor, given by

$$t_d/\ell = 1/v_0 \quad (2.3)$$

For example, t_d/ℓ (glass) = 67 ps/cm and t_d/ℓ (alumina) = 100 ps/cm.

2.1.3 Characteristic impedance

Characteristic impedance Z_0 is simply the ratio $|E|/|H|$ for propagating waves on a single line. Z_0 has no significance in the dc case, where there are no propagating waves. For TEM waves

$$Z_0 \text{ (TEM)} = \sqrt{\frac{\mu}{\varepsilon}} \quad (2.4)$$

TABLE 2.1 Relative Dielectric Constants for Common Packaging Dielectrics

Material	ε_r
Teflon	2.0
Polyimide	3.2
Glass	4.0
PWB (FR-4)	4.8
Glass ceramic	5.2
Alumina ceramic	9.5

Free space values of the permittivity and permeability are

$$\varepsilon_0 = \frac{1}{36\pi} \times 10^{-11} \; F/cm \quad \text{and} \quad \mu_0 = 4\pi \times 10^{-9} \; H/cm$$

As an application example, Z_0 (TEM) $= 120\pi = 377$ ohms for free space. An alternative expression that is useful for quasi-TEM waves is

$$Z_0 = \sqrt{L/C} \quad (2.5)$$

Typical values of Z_0 for packaging structures are in the range 25–100 ohms. A very common value for Z_0 is 50 ohms. This value is usually easy to achieve with typical packaging geometries, and it matches the input and output impedance of measurement instrumentation.

2.1.4 Maxwell's equations, capacitance, inductance, and resistance

2.1.4.1 Maxwell's equations. The governing equations for electric and magnetic fields are Maxwell's equations

$$\nabla \cdot \vec{B} = 0 \quad (2.6a)$$

$$\nabla \cdot \vec{D} = \rho \quad (2.6b)$$

$$\nabla \times \vec{H} = \vec{J} + \partial D/\partial t \quad (2.6c)$$

$$\nabla \times \vec{E} = -\partial \vec{B}/\partial t \quad (2.6d)$$

A general solution to these equations exists only for a few (important) special cases. In fact, full-wave solutions are numerical solutions to Eq. 2.6, and are notably long in computer run time. Some specific solutions are particularly useful in understanding the electrical performance packaging structures, and these are addressed here.

2.1.4.2 Capacitance. Any two conductors with an electric potential difference V will exhibit charge accumulations of $+Q$ and $-Q$, as shown in Fig. 2.12. The capacitance of the system is $C = Q/V$. This can be extended to systems of conductors with arbitrary numbers of conductors. Note that the charge distribution on any conductor is affected by the presence of other conductors, even by conductors that are not connected to a fixed potential (so-called floating conductors). The system of conductors stores an amount of energy

$$U_E = \frac{1}{2} \iiint\limits_{volume} \varepsilon |\vec{E}|^2 \, dV$$

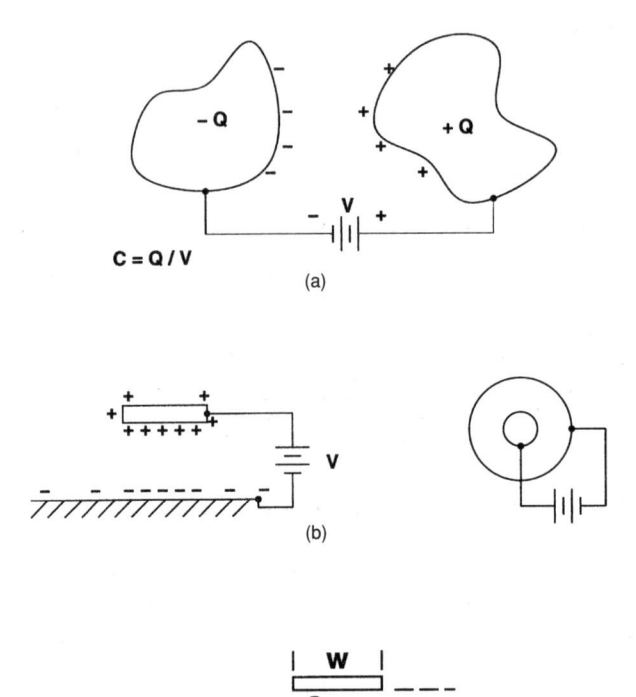

Figure 2.12 System of two conductors with electric potential difference V. (a) general case; (b) and (c) two common specific cases, microstrip and coaxial cable cross sections, respectively.

All charges reside on the surfaces of conductors. A particularly simple expression for the capacitance between two parallel conducting sheets such as those shown in cross section in Fig. 2.12c is

$$c = \varepsilon \frac{A}{h} \qquad (2.7)$$

where A is the area of the conductor with the least area, and h is the separation between conductors. This expression is accurate to within 10 percent of the true value for $w/h \geq 20$.

For a system of more than two conductors (counting the ground conductor as one conductor), a very useful capacitance matrix, called the *Maxwell Matrix,* is extracted by modeling tools that solve the electrostatic equation (Eq. 2.12b), subject to global charge neutrality and a boundary condition on the electric field caused by the applied voltage V. Figure 2.13a shows a three-conductor system with relevant two-

Basic Electrical Effects in Electronic Packaging and Interconnects   33

$$\begin{pmatrix} C_{11} & C_{12} \\ C_{21} & C_{22} \end{pmatrix}$$

$$C_{ij} < 0, \quad i \neq j$$

(a)

$$C_{11} = C_{1g} + C'_{12} \qquad C'_{12} > 0$$

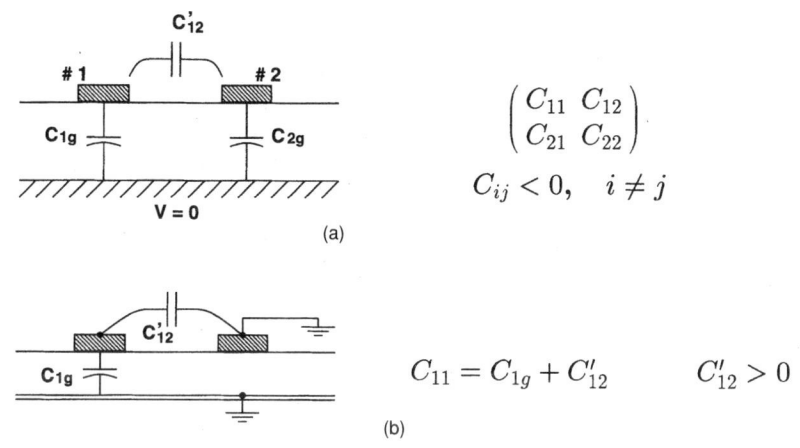

(b)

Figure 2.13 Method of extracting the Maxwell capacitance matrix for three conductors.

terminal capacitance; Fig. 2.13b shows the connection used to determine the Maxwell "self-capacitance" C_{11} per unit length. The elements of the Maxwell Matrix C_{ij} are also called *electrostatic induction coefficients*. In Fig. 2.13b, conductor 1 is put at a potential of +1 volt, all other conductors are put at zero volts, and the total charge on conductors 1 and 2 is determined through a solution of Eq. 2.6b. Note that $Q_1 > 0$ and $Q_2 < 0$ for this case. Then by the definition of the elements of the matrix $C_{11} = Q_1/1 > 0$ and $C_{12} = Q_2/1 < 0$. This process is continued until all non-ground conductors have been held at +1 volt, and all self and mutual elements of the Maxwell Matrix are determined. In the case shown in Fig. 2.13, this involves only putting conductor 2 at +1 volt and solving for $C_{22} = Q_2 > 0$ and $C_{21} = Q_1 < 0$. In this case the entire matrix is now available,

$$C = \begin{pmatrix} C_{11} & C_{12} \\ C_{21} & C_{22} \end{pmatrix}$$

By the principle of reciprocity, symmetrical off-diagonal elements are equal, that is, $C_{ij} = C_{ji}$.

From Fig. 2.13, it is clear that $C_{11} = C_{1g} + C'_{12}$, and $C'_{12} = -C_{12} > 0$. Then $C_{1g} = C_{11} - C'_{12} = C_{11} + C_{12}$. This leads to the general conclusion that for a system of N conductors, which has an $N \times N$ Maxwell Matrix,

$$C_{ii} > 0 \qquad (2.7a)$$

$$C_{ij} < 0 \quad i \neq j \qquad (2.7b)$$

$$C_{ig} = \sum_{j=1}^{N} C_{ij} \qquad (2.7c)$$

$$C_{ij} = C_{ji} \qquad i \neq j \qquad (2.7d)$$

2.1.4.3 Inductance. A magnetic field $\vec{H}$ is associated with any flow of current. The magnetic flux density $\vec{B}$ is obtained through $\vec{B} = \mu \vec{H}$. $\vec{H}$ is obtained through solution of Eqs. 2.6a and 2.6c, subject to appropriate boundary conditions. It is beneficial to understand some simple cases. Figure 2.14 shows the magnetic field caused by an infinitesimally thin filament of current I. For this case

$$\vec{H} = \frac{I}{2\pi r} \hat{\phi} \qquad (2.8)$$

where $\hat{\phi}$ is the unit vector in the ϕ direction. For a round conductor of finite radius a, at low frequency and isolated from the magnetic field of other conductors so that there is no proximity effect and the current is thus uniformly spread across the cross section of the conductor,

$$\vec{H} = \frac{Ir}{2\pi a^2} \hat{\phi} \qquad 0 \leq r \leq a \qquad (2.9a)$$

and

$$\vec{H} = \frac{I}{2\pi r} \hat{\phi} \qquad a \leq r \leq \infty \qquad (2.9b)$$

The energy stored in the magnetic field is

$$U_M = \frac{1}{2} \int_v \mu |\vec{H}|^2 \, dV$$

In both cases it is important to observe that the H-field is radially symmetric and consists only of a component in the $\hat{\phi}$ direction, that is, the magnetic field "curls" around the current element. The direction of $\vec{H}$ (and $\vec{B}$) is clockwise around the current element, if the observer is look-

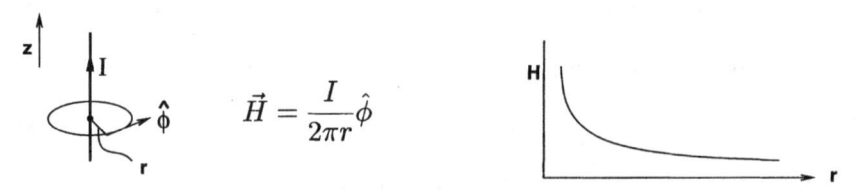

Figure 2.14 External magnetic field caused by an infinitesimally thin current filament.

ing axially in the direction of current flow. It is clear that *if* the spatial distribution of current density is known, it can be approximated by current filaments of varying strengths. The $\vec{H}$ and $\vec{B}$ are then easily obtained by superposition and summation of filamentary contributions using Eq. 2.8.

Inductance is formally defined in terms of a current loop, and is given by

$$L = \frac{1}{I} \int_s \vec{B} \cdot \hat{n}\, dS = \frac{\phi}{I} \qquad (2.10)$$

where S is the surface of the loop and $\hat{n}$ is the normal to that surface. Figure 2.15 shows an arbitrary current loop with a defined direction for I and the resulting flux density B. From Eq. 2.10, the loop inductance is simply the total magnetic flux passing through the area defined by the current loop, divided by the current flowing. This immediately brings to mind a method for reducing inductance: simply reduce the *area* of the current loop, and the loop inductance decreases. This technique is often used in packaging structures. To achieve inductance reduction, signal traces are often placed very near to ground planes that are part of the current loop, or very near to traces carrying ground or power current.

"Partial" inductances are often required for simulation of packaging structures. Figure 2.16 shows a loop decomposed into partial inductances, both self-partial inductances and mutual partial inductances

$$L = \int_s \frac{\vec{B} \cdot \hat{n} ds}{I} = \frac{\Phi}{I}$$

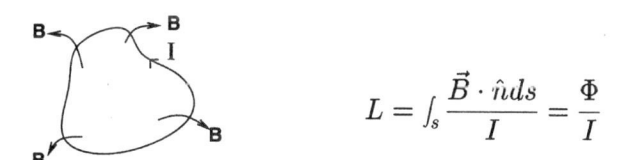

Figure 2.15 An arbitrary current loop showing the direction of I and the resulting flux density B.

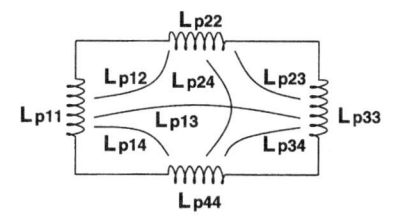

Figure 2.16 Circuit loop inductance decomposed into partial inductances.

(Lp_{ij}, $i \neq j$). The partial self-inductance is simply the total flux linking (encircling) a particular conductor, due to the current flowing in that particular conductor, divided by that current. Thus for a wire of finite radius a and length ℓ and carrying current I, the partial self-inductance per unit length, at low frequency, is

$$L_{p11} = \frac{1}{I} \int_0^a \mu_1 \frac{Ir}{2\pi a^2} dr + \frac{1}{I} \int_a^\infty \mu_2 \frac{I}{2\pi r} dr \qquad (2.11)$$

where μ_1 is the permeability inside the wire and μ_2 is the permeability outside the wire. The second integral is a problem to perform in a straightforward manner but Eq. 2.11 illustrates the principle. Partial mutual inductances are due to the flux from one conductor, say conductor j, which links another conductor, say conductor i, divided by current I_j. Figure 2.17 illustrates the method of calculating L_{pij}.

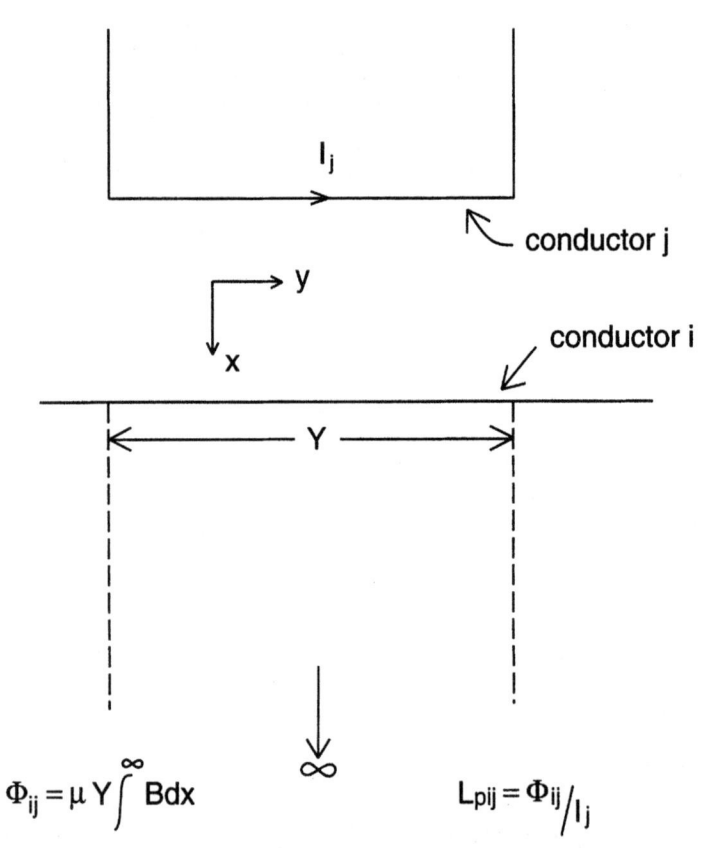

Figure 2.17 Illustration of the method of calculating partial mutual inductances.

Basic Electrical Effects in Electronic Packaging and Interconnects 37

In order to properly assign polarities of voltage drops due to partial mutual inductance, it is necessary to assign a sign to L_{pij}. This can be done using Faraday's law, which is the integral form of Eq. 2.6d [6]. The results from this are shown in Fig. 2.18. Using the information in Fig. 2.18, it is possible to calculate the voltage drop across any of the line segments in Fig. 2.16. For example, if current I flows in the loop

$$V_1 = \left(\sum_{j=1}^{4} L_{p1j} \right) dI/dt \tag{2.12}$$

and, considering that $L_{p12} = L_{p14} = 0$, and $L_{p13} < 0$

$$V_1 = (L_{p11} - |L_{p13}|) \, dI/dt \tag{2.13}$$

It is thus possible to determine the voltage drop across any segment of the loop using partial inductances. The loop inductance is also easily calculable. Noting that $L_{p23} = L_{p34} = 0$ and $L_{p24} < 0$, and simply adding up voltage drops around the loop (see Eq. 2.13),

$$L_{\text{loop}} = L_{p11} + L_{p22} + L_{p33} + L_{p44} - 2|L_{p13}| - 2|L_{p24}| \tag{2.14}$$

From Eqs. 2.9 and 2.11 it is clear that the inductance of a structure can be broken down into the *internal* inductance L_{int} and the *external* inductance L_{ext}. L_{int} is due to the magnetic field *inside* a conductor and corresponds, for example, to the first term in Eq. 2.11. L_{ext} is due to the magnetic field *outside* a conductor and corresponds, for example, to the second term in Eq. 2.11. Some external partial self- and mutual partial inductance equations for rectangular structures are shown in Fig. 2.19. In the equation for M^p, the partial mutual inductance in Fig. 2.19, no explicit parameters for the shape of the square conductors are present. The equation is relatively insensitive to the exact cross-sectional shape of the conductors and therefore the same equation is used for the M^p between circular cross-sectional conductors (e.g., bond wires).

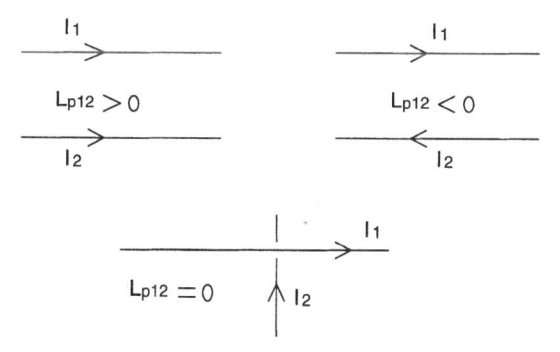

Figure 2.18 Methods of assignment of a sign to L_{pij}.

$$L^p = \frac{\mu l}{2\pi}\left[\ln\left(\frac{2l}{w+t}\right) + \frac{1}{2} + f(t/w)\right] \quad (H)$$

where $f(t/w) \longrightarrow 0$ as $t/w \longrightarrow 0$.

$$M^p = \frac{\mu l}{2\pi}\left[\ln\left(\frac{l}{d} + \sqrt{1 + \frac{l^2}{d^2}}\right) - \sqrt{1 + \frac{d^2}{l^2}} + \frac{d}{l}\right] \quad (nH)$$

Figure 2.19 External partial inductance equations for rectangular structures; all dimensions in cm.

Figure 2.20 shows a bond wire geometry in proximity to a ground plane. For the two cases h finite and h infinite, the external partial self inductance is given by

$$L = \frac{\mu \ell}{2\pi}\left[\ln\left(\frac{4\ell}{d}\right) - 1\right], \quad h = \infty, \quad \frac{\ell}{d} \gg 1 \quad (2.15a)$$

and

$$L = \frac{\mu \ell}{2\pi}\left[\ln\left(\frac{4h}{d}\right) - 2\frac{h}{\ell}\right], \quad h \text{ finite}, \quad \frac{h}{d} \gg 1 \quad (2.15b)$$

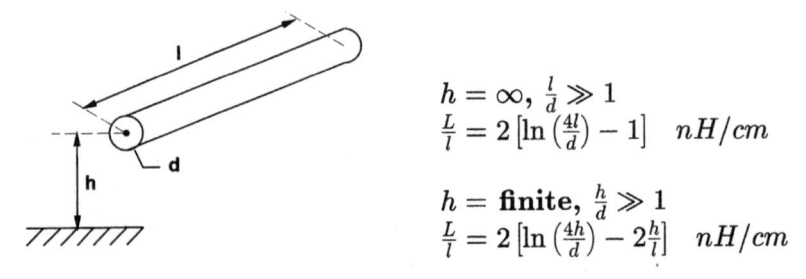

$h = \infty,\ \frac{l}{d} \gg 1$
$\frac{L}{l} = 2\left[\ln\left(\frac{4l}{d}\right) - 1\right]\ nH/cm$

$h = \textbf{finite},\ \frac{h}{d} \gg 1$
$\frac{L}{l} = 2\left[\ln\left(\frac{4h}{d}\right) - 2\frac{h}{l}\right]\ nH/cm$

Figure 2.20 Bond wire geometry for external partial self-inductance calculation.

Again, the external partial mutual inductance between two wires of round cross section is given by the same equation as M^p in Fig. 2.19. As an example, some calculations are shown below for single bond wires and pair of wires.

Example Bond wire external partial inductance calculations

$h = \infty$, $\ell = 0.25$ cm, $d = 25 \times 10^{-4}$ cm, $\mu = \mu_0$

$L_p = 2.62$ nH

$h = 5 \times 10^{-2}$ cm, $\ell = 0.25$ cm, $d = 25 \times 10^{-4}$ cm, $\mu = \mu_0$

$L_p = 1.97$ nH

$h = \infty$, $\ell = 0.25$ cm, separation of wires = 15.2×10^{-3} cm

$M^p = 0.915$ nH

Since bond wire inductance is so important in simultaneous switching output (SSO) noise (discussed in Sec. 2.3), it is worth examining the results of combining multiple bond wires. Figure 2.21 shows schematically two bond wires electrically in parallel. If the partial mutual inductance is zero, then the partial inductance of the combination is $L = L_{p11}/2$, which shows great benefit from combining two bond wires. The true situation, however, is that L_{p12} may be very large for closely spaced bond wires. For the case where $L_{p12} \to L_{p11}$, $L = L_{p11}$ and no benefit in reduction of inductance is seen. One way to reduce inductance is to permit the bond wires to diverge. For two bond wires of length 0.25 cm, diverging at an angle of 11.5°, $L = 0.75 L_{p11}$ from experimental measurements [8]. Figure 2.22 shows loop inductance L and partial voltages V_1 and V_2, caused by partial inductances L_{pij} and by the time-variation of current dI/dt. These connections are directly applicable to SSO noise, which will be discussed in a later chapter.

For combinations of high frequency waves and high conductivity conductors, it is found that the current, and electric and magnetic fields are constrained to a thin layer near the surfaces of the conductor. This

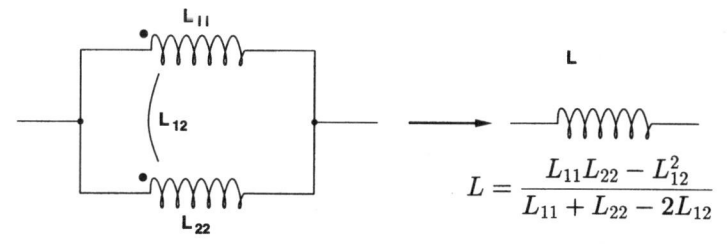

$$L = \frac{L_{11}L_{22} - L_{12}^2}{L_{11} + L_{22} - 2L_{12}}$$

Figure 2.21 Illustration of the effect of using two parallel bond wires on total inductance (all shown are partial inductances).

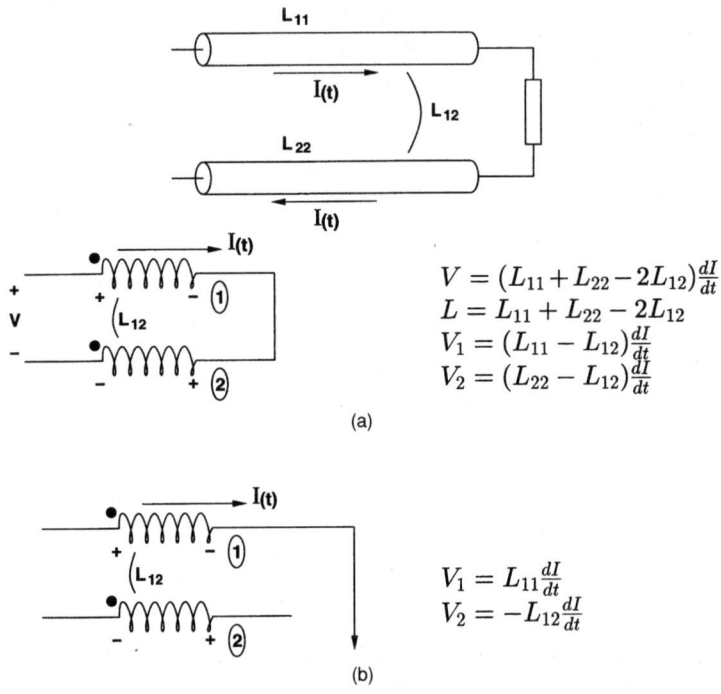

$$V = (L_{11}+L_{22}-2L_{12})\tfrac{dI}{dt}$$
$$L = L_{11} + L_{22} - 2L_{12}$$
$$V_1 = (L_{11} - L_{12})\tfrac{dI}{dt}$$
$$V_2 = (L_{22} - L_{12})\tfrac{dI}{dt}$$

(a)

$$V_1 = L_{11}\tfrac{dI}{dt}$$
$$V_2 = -L_{12}\tfrac{dI}{dt}$$

(b)

Figure 2.22 Loop inductance L and partial voltages V_1 and V_2 due to partial inductances L_{ij} and dI/dt.

is the so-called "skin effect." This causes frequency dependence in both L_{int} and in conductor resistance. The longitudinal (direction of signal propagation) electric field decreases with distance x from the surface as $e^{-x/\delta}$, where δ is the skin depth. The *skin depth* is the distance into the conductor for which the fields and current have decreased by an amount $1/e$ from their surface values. The skin depth δ is given by

$$\delta = (\pi f \mu \sigma)^{-1/2} \tag{2.16}$$

Here σ is the electrical conductivity of the material. Table 2.2 shows values of skin depth for various values of frequency for copper.

TABLE 2.2 Skin Depth for Copper

f (Hz)	δ (μm)
10^4	660
10^6	66
10^8	6.6
10^9	2.1
10^{10}	0.66

The effect on inductance of the skin effect is to decrease the internal inductance as frequency increases. The external inductance is not affected by frequency. The limit as the $f\sigma$ product approaches infinity is $L_{int} = 0$. The importance of the skin effect on inductance can then be estimated for simple cases. For example, the internal partial inductance of a round wire at low frequency is

$$L_{int} = \frac{\mu}{8\pi} \ell \qquad (2.17)$$

For a single 0.25-cm long, 25-micron diameter bond wire isolated from ground, $L_{int} = 0.125$ nH using Eq. 2.17. This can be compared to the previous calculation of $L_{ext} = 2.62$ nH in the last example. Thus the total frequency variation of partial inductance is a decrease of only 5 percent for this wire. Figure 2.23 shows the frequency variation of inductance (2.23a) and resistance (2.23b) for a microstrip geometry such as shown in Fig. 2.24.

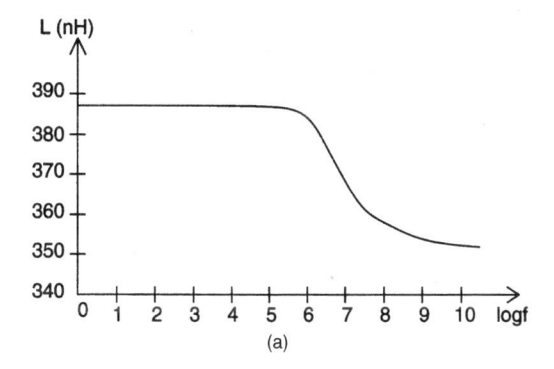

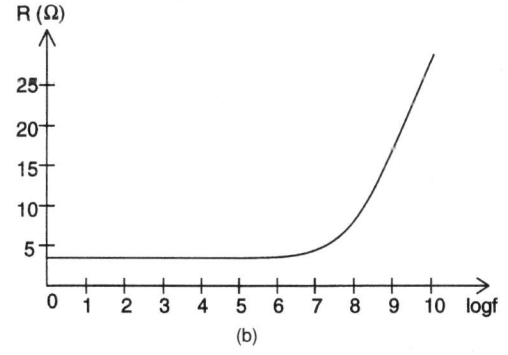

Figure 2.23 Frequency variation of partial inductance and resistance of 1-cm copper microstrip line with dimensions shown in Fig. 2.24.

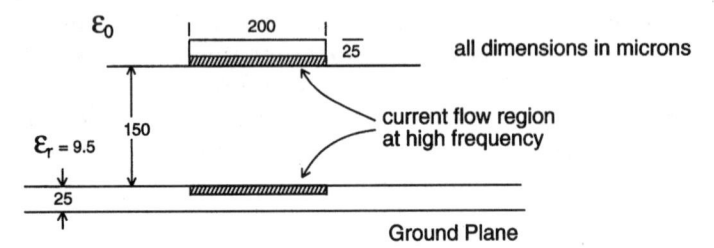

Figure 2.24 Microstrip geometry showing results of skin effect on current distribution.

2.1.4.4 Resistance.

Above a certain frequency, the resistance of a conductor increases with frequency due to the skin effect. This effect is shown in Fig. 2.23 for a microstrip line. For a wire which is round in cross section with radius r, the current-carrying area at low frequency is $A = \pi r^2$ and the resistance p.u. length is $R_{DC} = 1/(\sigma \pi r^2)$. At frequencies high enough such that $\delta \leq r$, the approximation that current is totally constrained to a layer δ in thickness gives

$$R(f) = \frac{1}{\sigma \pi [r^2 - (r-\delta)^2]}, \quad f > \frac{1}{\pi r^2 \mu \sigma} \quad (2.18)$$

The condition on f assures that $R \geq R_{DC}$, which is a physical requirement. As $\delta \to 0$, Eq. 2.18 shows that $R(f)$ increases without limit. It can be shown that for frequencies much above the limit in Eq. 2.18, $R(f)$ takes on the limiting form $R_{HF} = \sqrt{\mu f/\pi\sigma}/2r$ and R_{HF} (the "skin resistance") thus depends on $\sqrt{f}$ in this skin-effect dominant regime. Figure 2.25 shows schematically the frequency dependence of resistance for

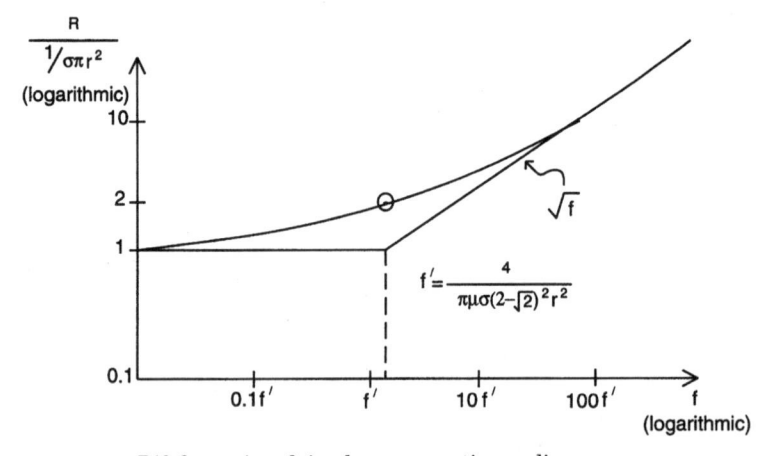

Figure 2.25 $R(f)$ for a wire of circular cross section, radius r.

such a wire. The frequency f' shown in this figure corresponds to the case where $R = 2R_{DC}$, and also corresponds to $\delta = 0.293r$. For a gold bond wire the frequency f' is approximately 100 MHz.

Although much has been made of frequency-dependent resistance (and inductance) in packaging simulations, in fact the skin effect does not lead to major phenomena in many packaging structures. This is because the low frequency resistance is often so low that even increasing it one or two orders of magnitude does not give serious effects. For very advanced structures, with dense and therefore thin wiring layers, the metal thickness is often so small that, although the conductors are very lossy at low frequency, the skin depth at the maximum frequency of interest is larger than one-half the metal thickness. Thus frequency dependence of resistance is not observed for these structures. Referring to Table 2.2, for 5 micron thick metal layers the skin effect would be *dominant* only for $\delta < 0.3 \times 2.5$ microns = 0.75 microns (using an analogy to the skin effect in a round wire). For copper, this requires a frequency of approximately 10 GHz (see Table 2.2). This frequency is the maximum frequency of interest for a clock frequency of 1 to 1.5 GHz. This last can be seen by taking a series of trapezoidal pulses of width $T/2$ and period T, having a t_r equal to $0.1T$. Seven harmonics of the fundamental frequency $1/T$ are required to fairly accurately reproduce the pulse. Four harmonics gives a degraded pulse, while ten harmonics gives a very accurate reproduction. Thus the statement proceeds that a clock waveform of 1 to 1.5 GHz requires a maximum frequency (of harmonics) of 10 GHz (10 GHz = 10×1.0 GHz $\cong 7 \times 1.5$ GHz).

In summary, frequency-dependent losses should be comprehended in a design for safety, but it is not often that they will have to be taken into account in simulation. The same statement applies to frequency-dependent inductance.

2.2 Basic Transmission Line Characteristics

2.2.1 Introduction

Information transfers in a system at or below a maximum velocity $v_0 = 1/\sqrt{\mu\varepsilon}$, the speed of light in the medium in which the information transfers. The actual velocity may be less than v_0 due to various impediments that cause slowing of the wave, such as line resistance or capacitive loading. In a sense all interconnects in a system are transmission lines, although some interconnects may behave as if they were a circuit composed of lumped elements. The question of when an interconnect may be accurately modeled by a small number of lumped elements and when an interconnect must be modeled as a transmission line in order to obtain accurate simulation results will be answered later in this section. Four types of transmission line structures were shown in Fig.

2.11. An important concept concerning transmission lines is that they are wave-guiding structures in which dc connectivity is maintained. Although so-called differential lines are sometimes used, it is common for a two-wire transmission line to be composed of an active line and a "return path" which is at nominal ground potential. Ground potential delivered to the chip by the package is not always equivalent to the system ground potential due to inductive effects. This then gives rise to ground bounce, or power line disturbances, which is a serious problem and is discussed in sec. 2.3.

Transmission lines dealt with in this chapter will primarily be two-line structures, as shown in Fig. 2.11. Of course multiple transmission line (MTL) systems will also be considered when coupled noise is discussed. Generally the lines will be uniform in cross section. These uniform lines are technologically the simplest to fabricate for packaging structures, they have no cutoff frequency (and can thus guide waves of any frequency), and they support the simplest electromagnetic waves which travel with velocity v_0, the speed of light in the medium in which the fields exist. These waves are the TEM waves discussed in the previous section. As discussed there, in nature only quasi-TEM waves exist. However, these are usually very close to TEM waves in behavior and this assumption will be made, that is, quasi-TEM waves behave like TEM waves will be taken as true.

One determinant of the importance of longitudinal fields (pointing in the direction of propagation rather than perpendicular to that direction) is the largest cross-sectional dimension of the line. If the shortest wavelength of electromagnetic wave which propagates on a line has a wavelength which is large compared to the largest cross-sectional dimension, then the longitudinal field components may be small compared to the transverse fields. As an example of this condition, assume that microstrip is used and the maximum cross-sectional dimension is the dielectric thickness, which is 2×10^{-2} cm. Since $\lambda = v_0/f$, this condition for TEM behavior is

$$v_0/f \gg 2 \times 10^{-2} \text{ cm}$$

or

$$f \ll v_0/2 \times 10^{-2} \text{ cm}$$

If the dielectric is alumina ceramic, $\varepsilon \cong 9$ and $v_0 = 10^{10}$ cm/s, which gives

$$f \ll 5 \times 10^{11} \text{ Hz}$$

Thus for waveforms which have a maximum frequency component much less than 500 GHz, TEM modeling is reasonably accurate. Since at least five to ten harmonics of the fundamental frequency of a pulse train are required in order to give a reasonably sharp, high edge-speed pulse, as discussed earlier, the preceding inequality actually leads to a clock frequency limit of 5 to 10 GHz. Thin-film packaging structures may give TEM performance at much higher frequencies because their characteristic cross-sectional dimension is approximately one order of magnitude smaller than alumina ceramic thick film/cofired structures; however, line resistance may be large enough to seriously perturb fields for thin-film structures (see discussion on lossy transmission lines, for example).

Another limit to TEM-like behavior for structures with inhomogeneous dielectrics is geometrical dispersion [9]. A rough estimate of the maximum sinusoidal frequency at which TEM-like behavior is observed is

$$f_{max} = \frac{21.3}{(w + 2h)\sqrt{1 + \varepsilon_r}} \text{ GHz} \quad (2.19)$$

where w and h are in mm and are defined in Fig. 2.26. As a simple example, take cofired ceramic with $\varepsilon_r = 9$, $w = 0.2$ mm, $h = 0.2$ mm; then $f_{max} = 11.2$ GHz. For a thin-film structure, $w = 0.01$ mm, $h = 0.01$ mm, and $\varepsilon_r \cong 4$; then $f_{max} = 318$ GHz. These limits are to be taken as guidelines rather than hard limits.

2.2.2 Distributed versus lumped-element simulation for transmission lines

Figure 2.27 shows schematically the choice of analysis modes for interconnects. Depending on line delay and input signal rise time, a lumped element equivalent circuit may be used for simplicity of analysis (assuming a circuit analysis computer program is used). For sufficiently short rise times relative to the line delay, a distributed element analysis ("transmission line" analysis) is required for accuracy.

The general rule of thumb for transmission line behavior is that a line behaves as a transmission line if the rise time of the driving signal

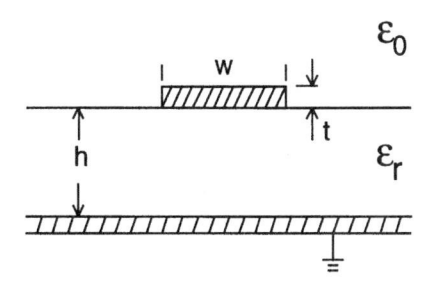

Figure 2.26 Dimensioned microstrip line.

- neglect losses

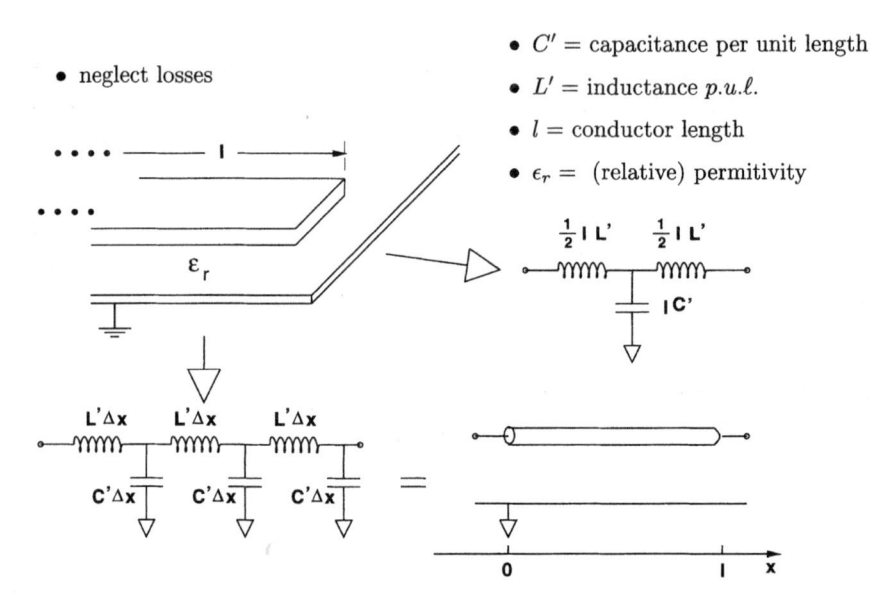

- C' = capacitance per unit length
- L' = inductance p.u.ℓ.
- l = conductor length
- ϵ_r = (relative) permitivity

Figure 2.27 Distributed and lumped models for a transmission line structure.

voltage waveform (t_r) is completed before any reflections from the far end of the line arrives back at the source end (near end) of the line. Taking the line delay T_d to be

$$T_d = \ell/v_0 \tag{2.20}$$

where ℓ is the length of the line, this condition requires

$$t_r < 2T_d \tag{2.21}$$

where $2T_d$ is the round-trip delay of the line. This can be understood by taking the case where the signal rise time t_r is much longer than $2T_d$. In this case, many reflections from the far end of the line arrive back at the source during the input rise time, and the near-end voltage is influenced by the far-end condition, that is, the load. The source end and load ends of the line thus affect each other during the signal rise and the role of the line can be modeled using lumped elements. For the case where the rise time of the near-end voltage is less than $2T_d$, the input signal rises to its maximum value before any reflections arrive back from the far end. During the time $2T_d$, the near-end voltage is unaffected by the far-end conditions and the distributed transmission line model must be used to obtain this temporal isolation of near and far ends. Note that distributed-element analysis always gives the ideal, correct answer; lumped-element analysis may under certain circumstances give a good approximation to the transmission line analysis and may be much simpler and faster to perform.

For comparison of lumped-element and transmission line analysis results, the circuit of Fig. 2.28 was used to simulate performance for a range of input voltage rise times which ranged from $t_r = 4T_d$ to $t_r = T_d/4$. In the figure, solid lines are transmission line analysis results, and dotted lines are lumped-element results. Based on the results shown in these figures, it is clear that for accurate results distributed element modeling must be used for $t_r < 2T_d$ and lumped-element modeling may be used for $t_r > 4T_d$. In the range $2T_d < t_r < 4T_d$, lumped-element modeling may be used, with increased error. These are not hard conditions, and if it is possible to accept less accuracy, then the use of lumped-element modeling may be extended to values of t_r somewhat less than $2T_d$.

Example Given a trace of length $\ell = 2.5$ cm in an Al_2O_3 package with $\varepsilon_r = 9.6$. What is the minimum t_r for which the trace does not exhibit transmission line characteristics?

$$v_0 = c/\sqrt{\varepsilon_r} = 3 \times 10^{10}/\sqrt{9.6} \text{ cm/s} = 9.7 \times 10^9 \text{ cm/s}$$

$$T_d = \ell/v_0 = 0.262 \text{ ns}$$

$$4T_d = 1.05 \text{ ns}$$

For $t_r \geq 1$ ns the trace exhibits lumped element behavior.

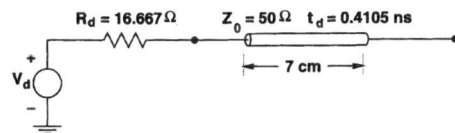

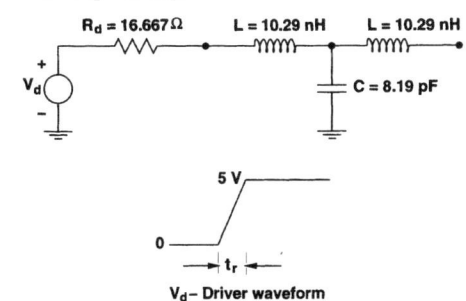

Figure 2.28 Comparisons of results from distributed and lumped models for the circuit shown at the top.

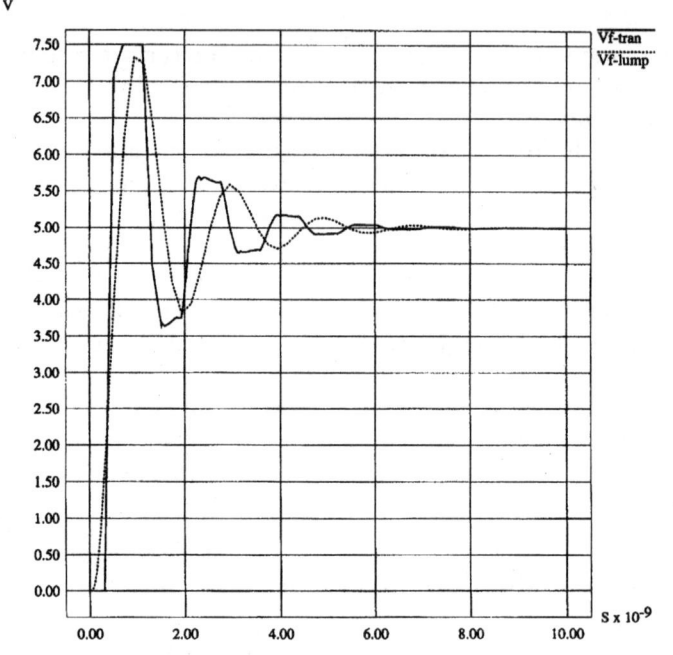

Figure 2.28 (*Continued*) Far-end waveforms, rise time $T_d/4$.

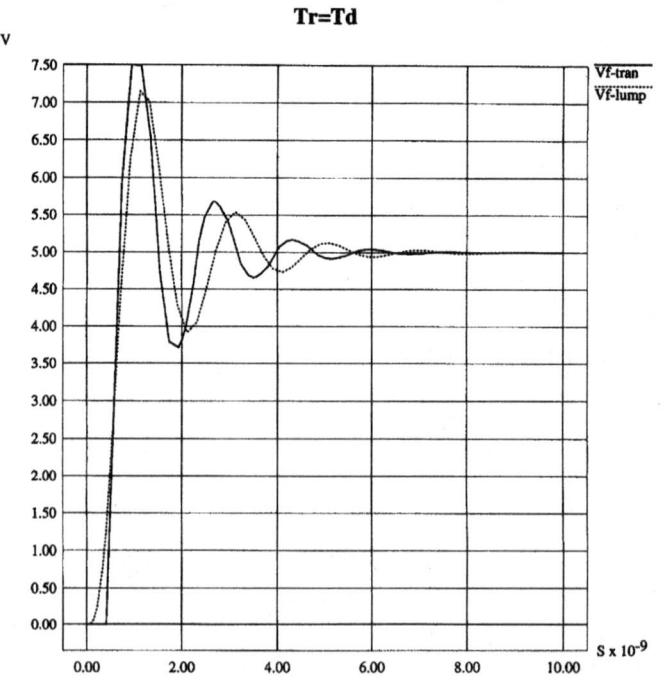

Figure 2.28 (*Continued*) Far-end waveforms, rise time T_d.

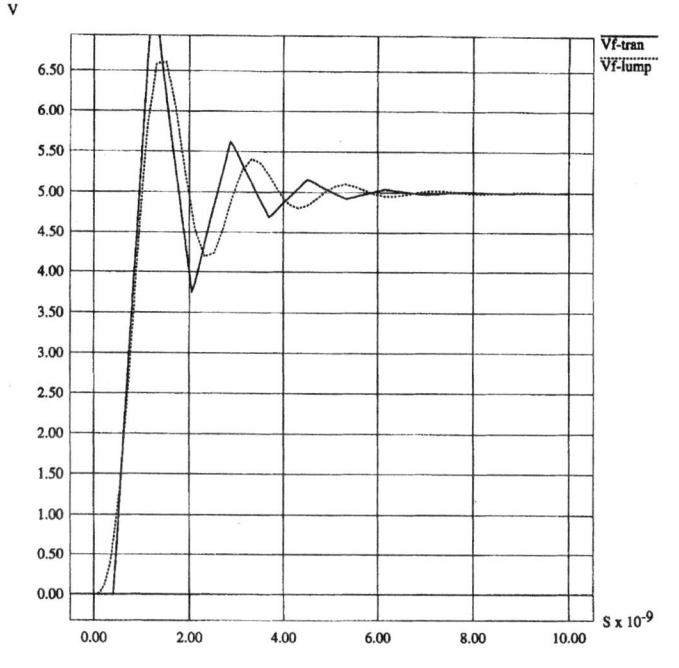

Figure 2.28 (*Continued*) Far-end waveforms, rise time $2T_d$.

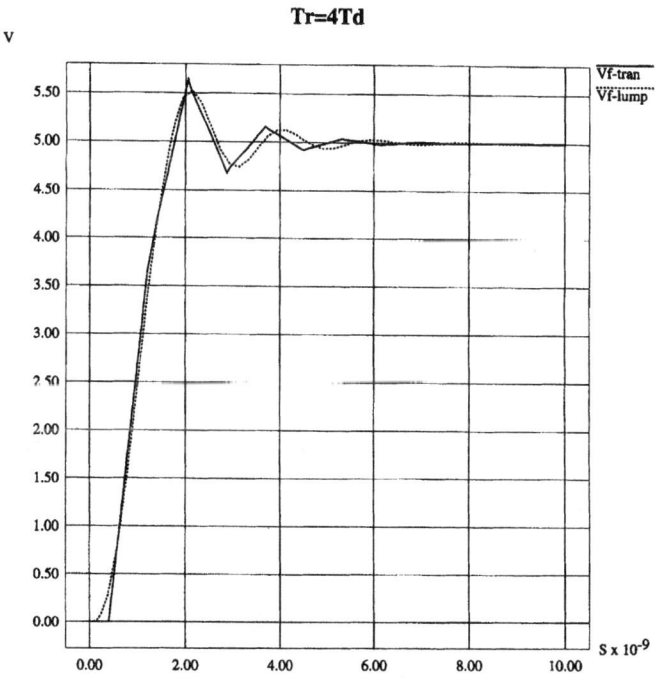

Figure 2.28 (*Continued*) Far-end waveforms, rise time $4T_d$.

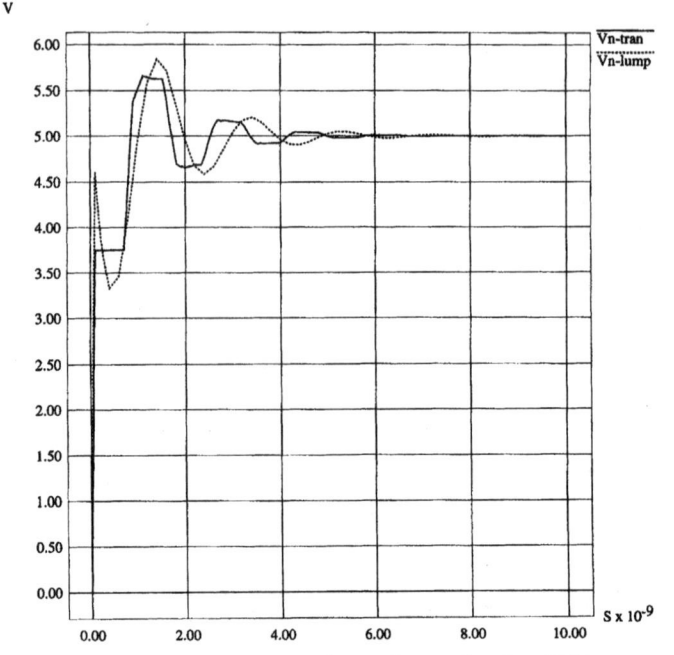

Figure 2.28 (*Continued*) Near-end waveforms, rise time $T_d/4$.

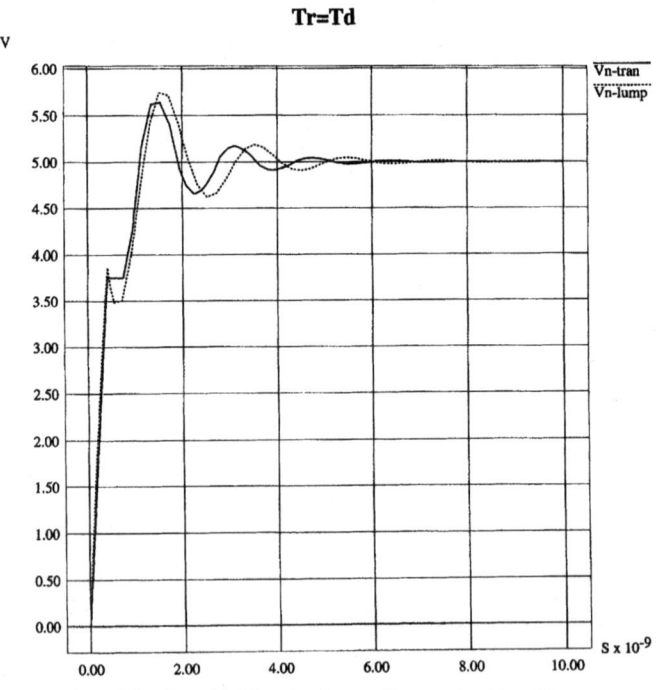

Figure 2.28 (*Continued*) Near-end waveforms, rise time T_d.

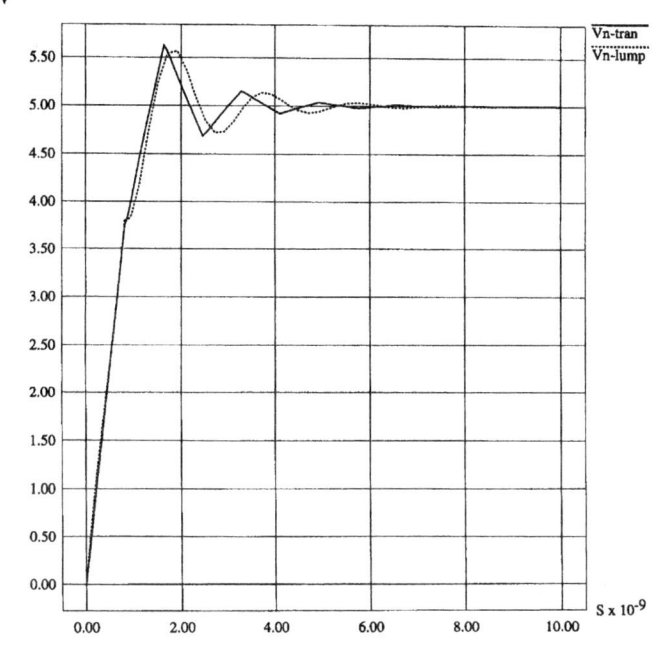

Figure 2.28 (*Continued*) Near-end waveforms, rise time $2T_d$.

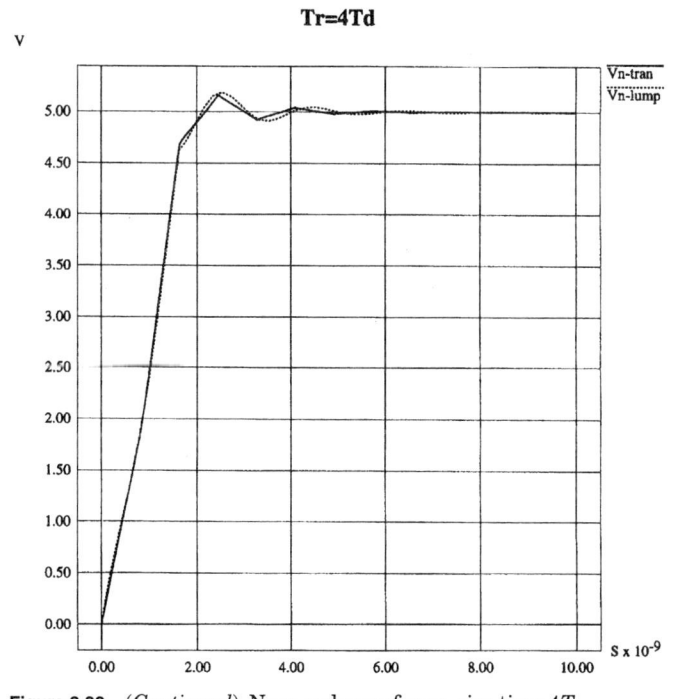

Figure 2.28 (*Continued*) Near-end waveforms, rise time $4T_d$.

2.2.3 Lossless transmission line equations

2.2.3.1 Solutions to the Telegraphers' Equations.
Figure 2.29 shows a discretized section of a transmission line. L' and C' are p.u. length quantities. Remembering that in general $V_L = L\, dI/dt$ and $I_C = C\, dV/dt$, and neglecting a term $d\Delta V/dt$, it is possible to write, directly from Fig. 2.29,

$$V = (L'\, \Delta x)\, dI/dt + V + \Delta V \qquad (2.22a)$$

and

$$I = (C'\, \Delta x)\, dV/dt + I + \Delta I \qquad (2.22b)$$

Simplifying and taking the limit as $\Delta x \to 0$,

$$\partial V/\partial x = -L'\, \partial I/\partial t \qquad (2.23a)$$

and

$$\partial I/\partial x = -C'\, \partial V/\partial t \qquad (2.23b)$$

These are called the Telegraphers' Equations.

They can be combined by taking $\partial/\partial x$ of Eq. 2.23a and $\partial/\partial t$ of Eq. 2.23b, and adding the results to give

$$\partial^2 V/\partial t^2 = +\frac{1}{L'C'}\, \partial^2 V/\partial x^2 \qquad (2.24)$$

This wave equation (second order in space and time) has solutions

$$V(x,t) = f_1^{\pm}(u) \qquad (2.25)$$

and

$$I(x,t) = f_2^{\pm}(u)$$

where $u^{\mp} = x \pm v_0 t$, $f_1(u)$ is any single-valued function of u, and $f_2(u) = f_1(u)/Z_0$, as shown in the following paragraphs. The propagation velocity v_0 is given by

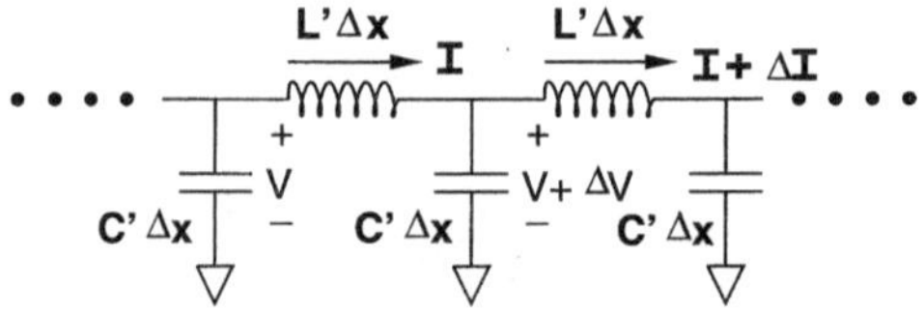

Figure 2.29 Discretized equivalent circuit of a section of a lossless transmission line.

Basic Electrical Effects in Electronic Packaging and Interconnects 53

$$v_0 = 1/\sqrt{L'C'} \tag{2.26}$$

The solutions $f_1^\pm(u)$, $f_2^\pm(u)$ are voltage and related current waveforms that propagate in the $+x$ or $-x$ direction with velocity v_0, without changing shape. The function $f_1(x - v_0 t)$ propagates in the $+x$ direction with velocity v_0. This is easily seen by following in time and space a particular point on the waveform $f_1^+(u)$, say the point $f_1^+(u_a)$, where $u_a^+ = x_a - v_0 t_a$. If t_a increases by an amount Δt, x_a must likewise increase by $\Delta x = v_0 \Delta t$ in order for u_a^+ to be constant. Thus

$$v_x = \frac{\Delta x}{\Delta t} = v_0 \tag{2.27}$$

Likewise the function $f_1(x + v_0 t)$ propagates in the $-x$ direction with velocity v_0. In general, the voltage on the line is a superposition of negatively and positively traveling waves,

$$V(x,t) = g_1^+(x - v_0 t) + h_1^-(x + v_0 t) + K \tag{2.28}$$

where K is a constant. The line current I is obtained from V.

A brief discussion of the time and position dependence of voltage and current on the line is beneficial. Figure 2.30 shows a positively

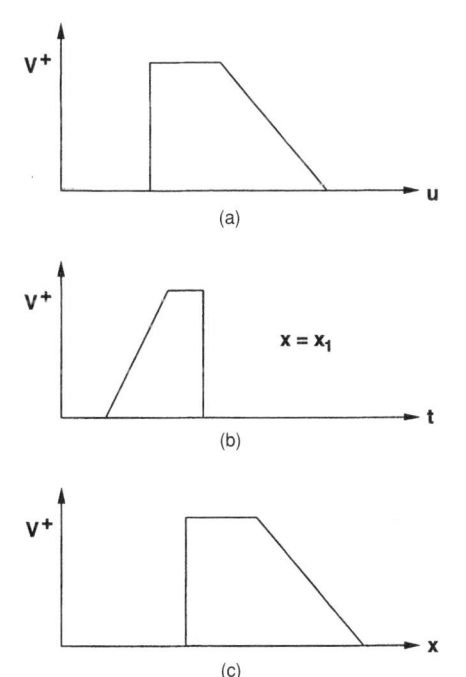

Figure 2.30 Time and space behavior of voltages and currents on a transmission line.

propagating voltage waveform on a transmission line, $V^+(u) = f(x - v_0 t)$. The behavior of $f^+(u)$ is shown in Fig. 2.24a, versus u. At an arbitrary point x_1 on the line, the time dependence of $V^+(x_1,t)$ is shown in Fig. 2.30b. This waveform is obtained by observing that for t large, $u^+ = x_1 - v_0 t$ is small and for t small, u^+ is large. Thus there is an inversion and change of scale in the dependence of $V^+(x_1,t)$ on t as compared to the dependence on u. At fixed time t_1, $u = x - v_0 t$ is large for x large and is small for x small. Thus there is no inversion (although generally there is a change of scale) in the dependence of $V^+(x,t_1)$ on x as compared to the dependence on u. This explains the shapes of $V^+(x_1,t)$ and $V^+(x,t_1)$.

The relationship between current and voltage propagating on a transmission line depends on the direction of propagation of the waveform, and it applies only to the propagating pulse, not to preexisting voltage and current values. To obtain the V-I relationship between the positively propagating current and voltage waveforms, note that from Eq. 2.23a

$$\partial V/\partial x = -L' \, \partial I/\partial t$$

and by the chain rule of differentiation

$$\frac{\partial V^+}{\partial x} = \frac{\partial V^+}{\partial u}$$

and

$$\frac{\partial I^+}{\partial t} = -v_0 \frac{\partial I^+}{\partial u}$$

Then
$$\frac{V^+}{I^+} = L'v_0 = \sqrt{L'/C'} = Z_0 \qquad (2.29)$$

It is easily possible in the same way to show that

$$\frac{V^-}{I^-} = -\sqrt{L'/C'} = -Z_0 \qquad (2.30)$$

Here Z_0 is called the characteristic impedance, and usually ranges from 20Ω to 100Ω for normal packaging structures. It must be reemphasized that Z_0 only relates the traveling voltage and current pulse on the transmission line. For digital systems, after switching transients have died out, Z_0 has no significance since there are no traveling pulses on the line.

Example What is Z_0 for a microstrip line such as that shown in Fig. 2.26, with $\varepsilon_r = 5$, $w = 25$, $t = 5$, and $h = 15$ (notice that the units are arbitrary; for lossless lines the value of L' and C' depends on proportions, not on actual units). Using a simple computer modeling tool, it is found that $L' = 2.95$ nH/cm and $C' = 1.17$ pF/cm. Then

$$Z_0 = \sqrt{L'/C'} = 50\,\Omega$$

Incidentally, $v_0 = 1/\sqrt{L'C'} = 1.7 \times 10^{10}$ cm/s and the delay per unit length of line is 59 ps/cm.

There is a particularly simple case in which Z_0 depends only on C'. If a transmission line is in a homogeneous medium of relative dielectric constant ε_r, then v_0 is known, $v_0 = c/\sqrt{\varepsilon_r}$. Since $v_0 = 1/\sqrt{L'C'}$,

$$L' = \frac{1}{v_0^2 C'} \tag{2.31}$$

and

$$Z_0 = \frac{1}{v_0 C'} \tag{2.32}$$

Thus it is easy to see the behavior of Z_0 as a trace is widened or narrowed, or as a trace is moved relative to a ground plane. For example, a wider trace increases C' as indicated by the parallel plate capacitance equation, Eq. 2.7. Then Z_0 and L' are reduced as indicated in Eqs. 2.31 and 2.32. The error introduced by non-TEM structures such as microstrip comes about because it is not exactly true that $v_0 = 1/\sqrt{\mu_r \varepsilon_r}$, so the value of v_0 used in Eqs. 2.31 and 2.32 will be in error. To see the approximate size of the error, take the case discussed in the previous example. Using $\varepsilon_r = 5$, v_0 is estimated to be 1.34×10^{10} cm/s, using $v_0 = 1/\sqrt{\varepsilon_r \mu_r}$, while the true value is 1.7×10^{10} cm/s. This gives an error in Z_0 of +25 percent and an error in L' of +61 percent. However, the trends are clearly shown by the equations even for this inhomogeneous case.

For the case of inhomogeneous dielectrics (e.g., microstrip line) there is an approach to calculating Z_0 exactly, which requires only calculation of capacitance and does not require knowledge of v_0. The algorithm is:

1. Calculate the capacitance p.u. length of the structure with all dielectrics replaced with air; call this C_0'
2. For the case in (1), $v_0 = c$, $\therefore L' = \dfrac{1}{c^2} C_0'$
3. For the actual structure, $v_0 = 1/\sqrt{L'C'} = c\sqrt{C_0'/C'}$
4. For the actual structure, $Z_0 = \sqrt{L'/C'} = [C(C_0'C')^{1/2}]^{-1}$

If approximate expressions for C_0' and C' are used, this calculation reduces to a "back-of-the-envelope" calculation.

2.2.3.2 Response of transmission lines to ramp and step inputs.

In general driving voltage (source voltage) waveforms can be approximated as a combination of ramp and step voltages. Figure 2.31 shows a schematic with a transmission line of length ℓ and propagation velocity v_0. Also shown is a driving waveform V_s. Here $u(t)$, the unit step function, is not to be confused with the dummy variable used in solving the Telegraphers' Equations. In analyzing the behavior of the near-end and far-end voltages and currents, a pulse propagating in the $+x$ direction is launched on the line at the point $x = 0$ and $t = 0$. To determine $V^+(0,0)$, the line is replaced by its electrical equivalent for the transient case, Z_0. This is shown in Fig. 2.32a. From this figure the voltage impressed on the line at $(x,t) = (0,t)$, $t < t_d$, is

$$V^+(0,t) = \frac{Z_0}{Z_d + Z_0} V_s(t) \tag{2.33}$$

As this pulse develops in time, the front edge of the pulse propagates toward the far end with velocity v_0. The leading edge of the pulse will arrive there at $t_d = \ell/v_0$. Some of the pulse may be reflected at $x = \ell$, and arrive back at the near end at $t = 2t_d$ as a negatively propagating pulse. Note that for $0 < t < 2t_d$, the near-end voltage is unaffected by conditions at the far end of the line because the reflected pulse (if any) has not yet arrived at $x = 0$. If $t_r < 2t_d$, the near-end voltage reaches its maximum value before the reflected pulse arrives at $x = 0$ and transmission line effects will be evident. For $t_r > 4 t_d$, at least two reflections will arrive at $x = 0$ before the rise of the driving volt-

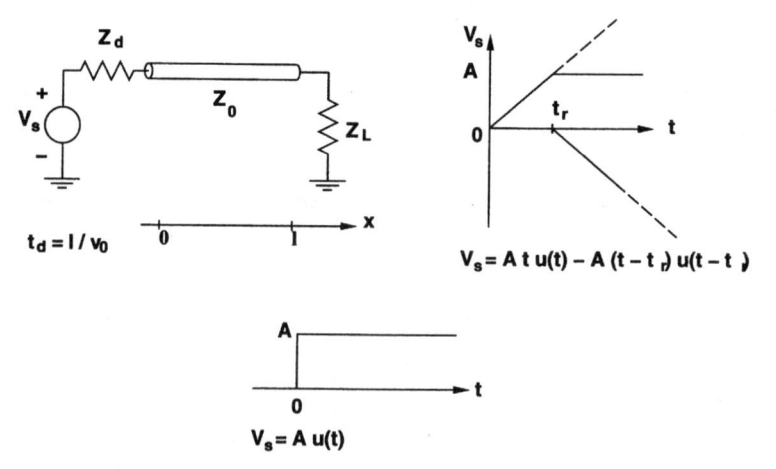

Figure 2.31 Transmission line system with ramp and step source voltage.

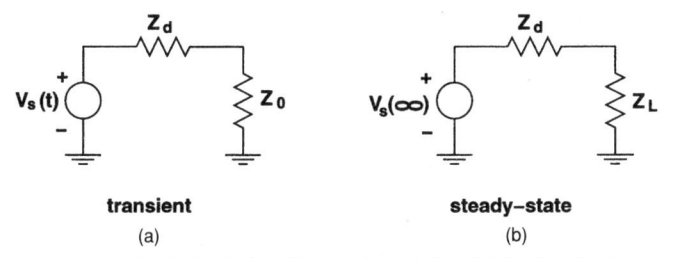

Figure 2.32 Equivalent circuits seen by a driver (a) during the transient phase and (b) at $t = \infty$, when all transients have died out.

age is complete, and the near-end voltage is thus affected by load conditions during the rise of $V^+(0,t)$. Transmission line effects will be less evident (or practically absent) for this condition. Finally, Fig. 2.32b shows the circuit for the steady state ($t = \infty$) case. In that case, the transmission line acts as a dc short circuit connecting near and far ends of the line.

Reflection of propagating waveforms occurs when a discontinuity is encountered. A discontinuity is any condition for which the line does not see $Z = Z_0$. Thus a terminating impedance $Z_L = Z_0$ is not a discontinuity, and no reflections will be generated at $x = \ell$ if $Z_L = Z_0$. For any other Z_L, including inductive or capacitive terminations, reflections will occur. In order to quantify reflections at the load, take the circuit of Fig. 2.33. Waveforms V^+ and I^+ are incident on $x = \ell$ at $t = t_d$. Waveforms V^- and I^- are reflected waveforms. It must be true that the total voltage at (ℓ, t_d) is

$$V(\ell,t_d) = V^+(\ell,t_d) + V^-(\ell,t_d) \tag{2.34}$$

and that likewise the total current at (ℓ, t_d) is

$$I(\ell,t_d) = I^+(\ell,t_d) + I^-(\ell,t_d) \tag{2.35}$$

Assuming that $Z_L = R_L$,

$$V(\ell,t_d) = R_L I(\ell,t_d) \tag{2.36}$$

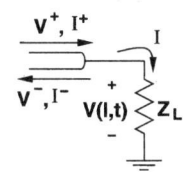

Figure 2.33 Voltage and current conditions at the far end of a transmission line.

and, referring back to Eqs. 2.29 and 2.30,

$$V^+/I^+ = Z_0$$

and

$$V^-/I^- = -Z_0$$

Solving the system of five equations in six unknowns,

$$V^-/V^+ \big|_{x=\ell} = \frac{R_L - Z_0}{R_L + Z_0} = \Gamma_L \qquad (2.37)$$

where Γ_L is called the reflection coefficient at the load.

Reflections may occur in any discontinuity location. The near end may reflect waveforms which are propagating back to $x = 0$ after reflection at $x = \ell$. Thus

$$V^+/V^- \big|_{x=0} = \frac{R_d - Z_0}{R_d + Z_0} = \Gamma_d \qquad (2.38)$$

where it is assumed that the driver impedance is resistive, $Z_d = R_d$.

Example What is the reflected voltage wave at $x = \ell$ if $R_L = Z_0$?

$$\Gamma_L = 0, \qquad V^-(\ell, t_d) = \Gamma_L V^+(\ell, t_d) = 0$$

What is the reflected voltage wave at $x = \ell$ if $R_L = 0$?

$$\Gamma_L = -1, \qquad V^-(\ell, t_d) = -V^+(\ell, t_d)$$

What is the reflected current wave at $x = \ell$ if $R_L = 0$?

$$I^+ = V^+/Z_0, \qquad I^- = -V^-/Z_0 = V^+/Z_0 = -I^+$$

What is the total current at $x = \ell$, $t = t_d$ if $R_L = 0$?

$$I(\ell, t_d) = I^+(\ell, t_d) + I^-(\ell, t_d) = 2I^+(\ell, t_d)$$

current doubling occurs

What is the reflected voltage wave at $x = \ell$ if $R_L = \infty$?

$$\Gamma_L = +1, \qquad V^-(\ell, t_d) = \Gamma_L V^+(\ell, t_d) = V^+(\ell, t_d)$$

voltage doubling occurs

What is the reflected current wave at $x = \ell$ if $R_L = \infty$?

$$I^+ = V^+/Z_0, \qquad I^- = -V^-/Z_0 = -V^+/Z_0 = -I^+$$

What is the total current at $x = \ell$, $t = t_d$ if $R_L = \infty$

$$I(\ell, t_d) = I^+(\ell, t_d) + I^-(\ell, t_d) = 0 \qquad \text{no current can flow into an open circuit}$$

It may be difficult to visualize the development of a reflection in time if the traveling waveform is complex. There is, however, a particularly simple and elegant technique to aid this [10]. Figure 2.34a shows a $V^+(x,t)$, and an "imaginary" $V^-(x_1,t) = \Gamma_L V^+(x,t)$ located at $x = \ell + x_1$. Figure 2.34b shows the two pulses after the leading edge of V^+ has arrived at $x = \ell$, but before the reflection is completed. Figure 2.34c shows the total voltage $V(x,t)$ corresponding to the condition of Fig. 2.34b. Finally, Fig. 2.34d shows the reflected pulse on the line after reflection is completed.

In order to keep track of the voltages on the transmission line and the traveling pulses, a "bookkeeping" tool known as the lattice diagram is useful. Figure 2.35 shows such a diagram. The horizontal scale is distance down the line. Note that only two points are selected for computation of total line voltage. These points are $x = 0$ and $x = \ell$. The vertical scale is a time scale, with gradations in integral multiples of t_d. Note that the total line voltage at a point x_0 is the sum of the preexisting

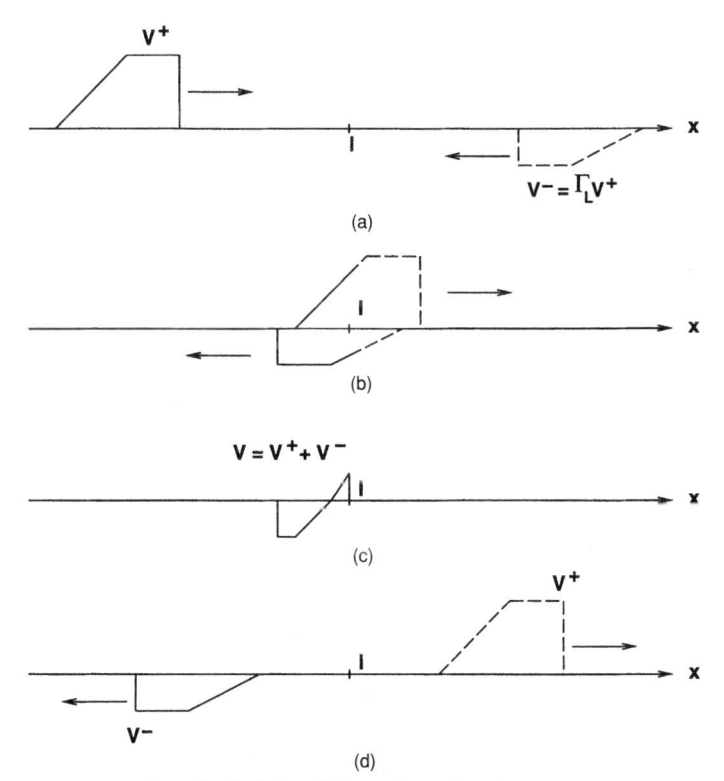

Figure 2.34 A method of visualizing pulse reflection.

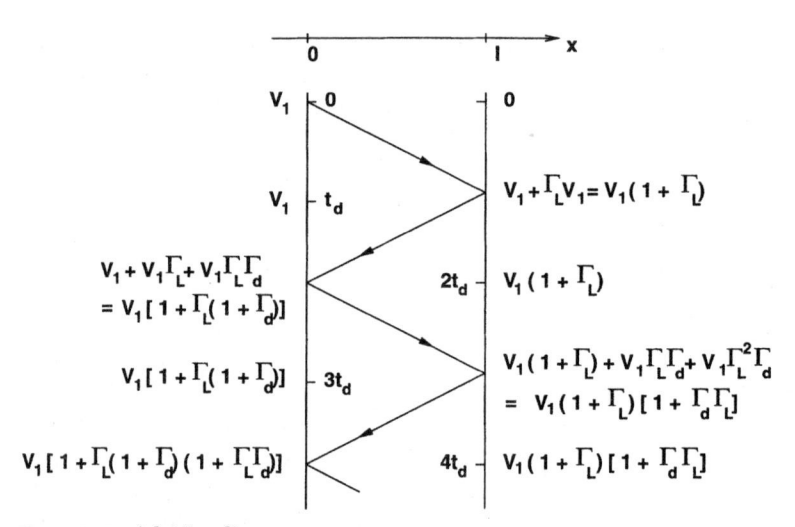

Figure 2.35 A lattice diagram.

voltage V_0 and the voltage due to propagating pulses $V^+(x_0,t_0)$ and $V^-(x_0,t_0)$. To use the lattice diagram for a case where the initial line voltage is zero, first find the near-end voltage V_N at $t = 0$,

$$V_N(0,0) = \frac{Z_0}{R_d + Z_0} V_S = V_1 \tag{2.39}$$

where V_S is the open circuit source voltage. At $t = t_d$, this pulse arrives at $x = \ell$, and the reflected pulse is $V^- = \Gamma_L V_1$. Since the preexisting voltage at $x = \ell$ is zero, the far-end voltage at t_d is

$$V_F(t_d) = V(\ell,t_d) = V^+(\ell,t_d) + V^-(\ell,t_d) = (1 + \Gamma_L)V_1 \tag{2.40}$$

At $t = 2t_d$ the reflected pulse $\Gamma_L V_1$ arrives back at $x = 0$ and is subject to reflection at the driver, giving a traveling pulse $V^+ = \Gamma_d \Gamma_L V_1$. At the near end

$$V_N(2t_d) = V(0,2t_d) = V_1 + \Gamma_L V_1 + \Gamma_d \Gamma_L V_1$$

or

$$V_N(2t_d) = V_1(1 + \Gamma_L + \Gamma_L \Gamma_d) \tag{2.41}$$

At this time $2t_d$ the line voltage at the far end remains at $V_1(1 + \Gamma_L)$ because no pulses are arriving at $x = \ell$ at that time. By carrying this procedure further any number of reflections can be taken into account. Since $-1 \leq \Gamma \leq 1$, some or all of the reflections may entail inversions of the waveform.

Some examples will show the usefulness of the lattice diagram. Figure 2.36 shows a simple transmission line system. Figure 2.37 shows voltage waveforms for V_N and V_F, the lattice diagram, and equivalent circuits at $t = 0$ and $t = \infty$ for the specific case $R_s = Z_0/2$, $R_L = Z_0$, $C = 0$, $V_{in} = 5u(t)$. The preexisting line voltage is 0. The necessary initial calculations are

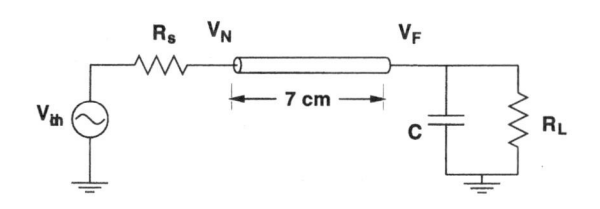

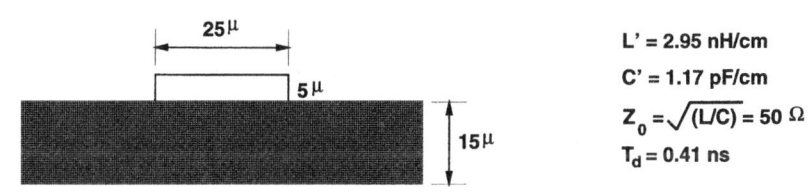

Figure 2.36 A single transmission line system.

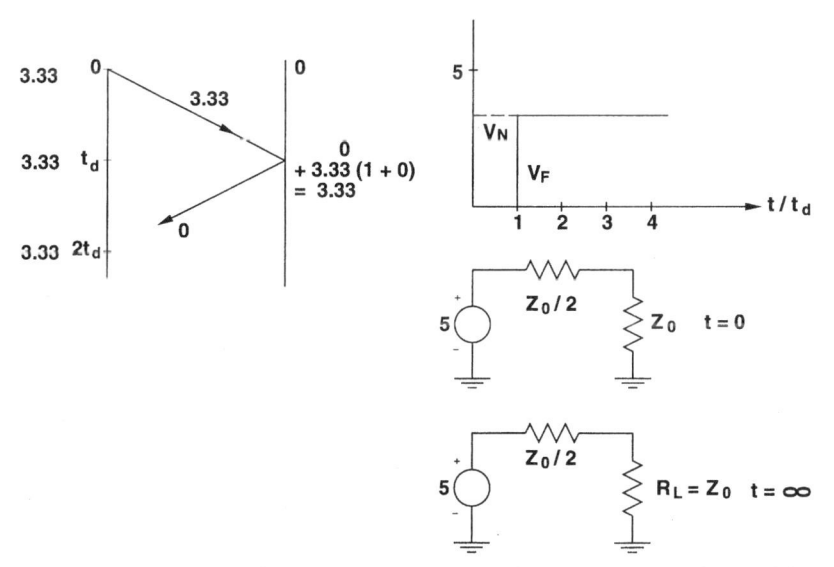

Figure 2.37 Voltage waveforms for the transmission line of Fig. 2.36, with $R_S = Z_0/2$, $R_L = Z_0$, $V_{in} = 5u(t)$ volts.

62 Chapter Two

$$V_1 = \frac{Z_0}{R_s + Z_0} V_{in} = 3.33u(t)V$$

$$\Gamma_L = 0, \quad \Gamma_d = -\tfrac{1}{3}$$

There are no reflections, since the load is matched to the line and $\Gamma_L = 0$. After t_d, all transients have died out and dc conditions prevail.

Figure 2.38 shows line voltages for the case $R_S = Z_0/2$, $R_L = \infty$, $C = 0$, $V_{in} = 5u(t)$. The necessary initial calculations are

$$V_1 = \frac{Z_0}{R_S + Z_0} V_{in} = 3.33u(t)V$$

$$\Gamma_L = 1, \quad \Gamma_d = -\tfrac{1}{3}$$

Under dc conditions, the far end of the line is an open circuit and the V_N and V_F should be 5V. From Fig. 2.38 it does appear that V_N and V_F are approaching 5V as a limit. With voltage doubling at the load ($\Gamma_L = +1$), the initially low voltage applied to the line (3.33V) is doubled to 6.66V at first incidence (i.e., when the input waveform first arrives at $x = \ell$). The ringing or oscillation observed for this case is a consequence of having one reflection coefficient positive and one coefficient negative. Figure 2.38 also shows the spatial variation of line voltage at a time $t = 1.5t_d$, when the first reflected pulse has arrived at the midpoint of the line.

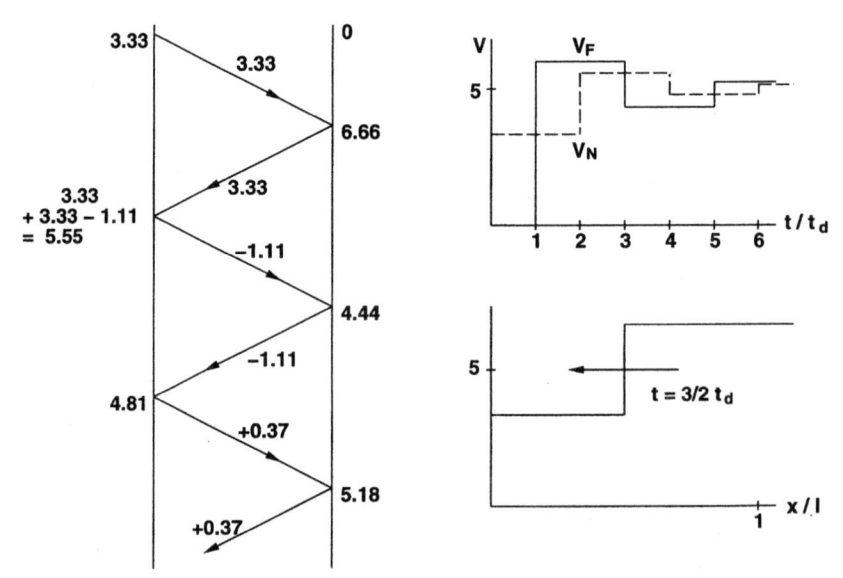

Figure 2.38 Voltage waveforms for the transmission line of Fig. 2.36, with $R_S = Z_0/2$, $R_L = \infty$, $C = 0$, $V_{in} = 5u(t)$ volts.

Once the voltage-time characteristic is obtained for a positive-going step voltage, it is simple to find the behavior of the line driver by a complete pulse, for example $5[u(t) - u(t - 3.5t_d)]$V. Figure 2.39 shows the response of the line to a waveform $5u(t)$V and a waveform $-5u(t - 3.5t_d)$V. Notice that the response to the second waveform is just the response to the first waveform inverted and shifted to the right by an amount $3.5t_d$. Figure 2.40 shows the superposition of the two waveforms in Fig. 2.39 to form the response to the complete waveform.

2.2.3.3 CMOS driver resistance limits for reliable switching. Voltage requirements on transmission lines for CMOS systems are generally as follows:

1. There must be first-incidence switching of receivers, that is, the switching threshold voltage + noise margin must be exceeded by the far-end voltage when the front edge of the pulse arrives, $V_F(t_d)$; thus $V_F(t_d) > V_{IH}$ (min).

2. Once the V_{IH} (min) is exceeded, V_F cannot fall below V_{IH} (min) due to subsequent reflections.

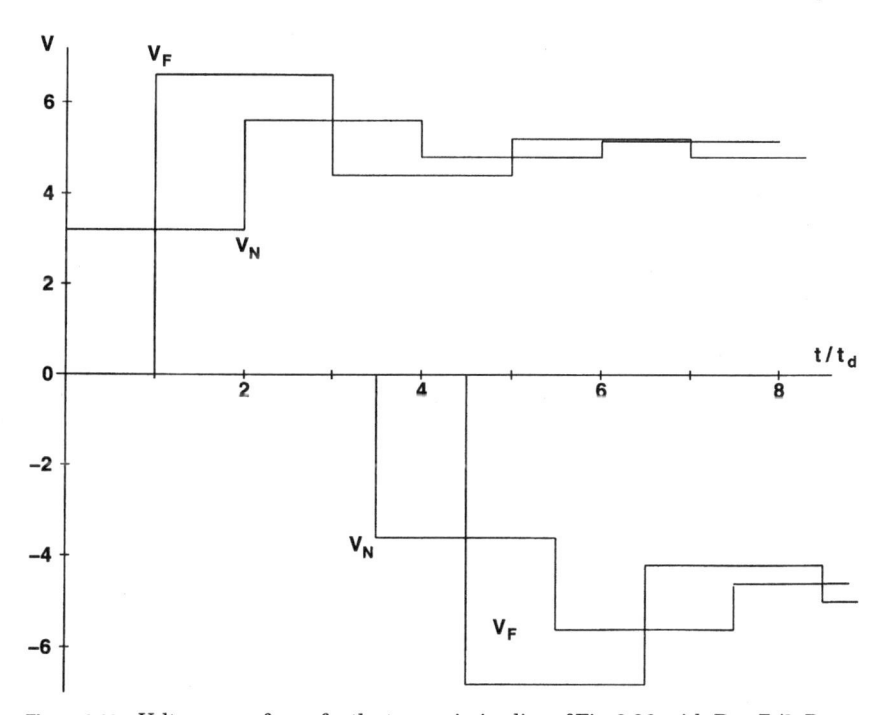

Figure 2.39 Voltage waveforms for the transmission line of Fig. 2.36, with $R_S = Z_0/2$, $R_L = \infty$, $C = 0$, $V_{in} = 5u(t)$, and $V_{in} = -5u(t - 3.5t_d)$ volts.

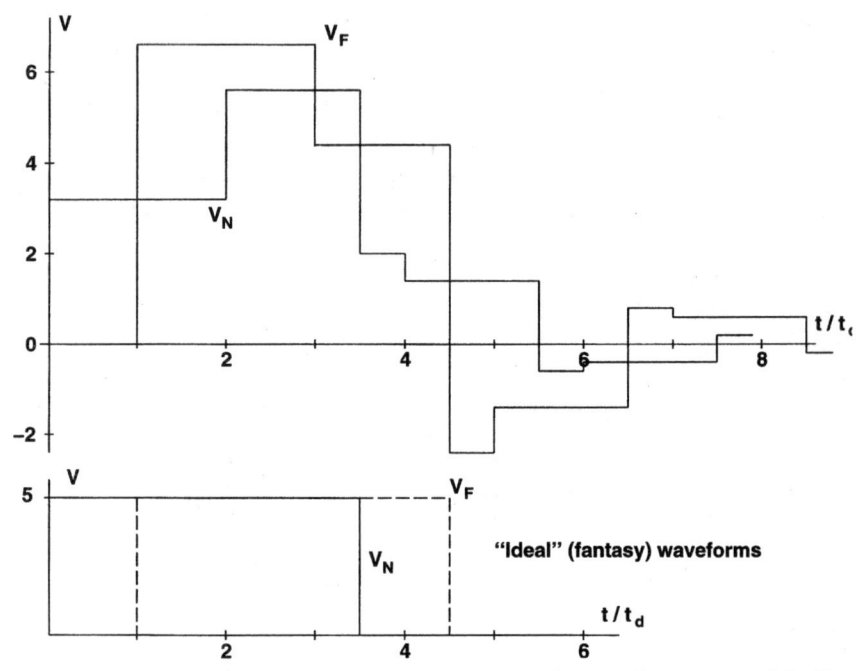

Figure 2.40 Sum of two voltage waveforms in Fig. 2.39, showing the response of the line to $V_{in} = 5[u(t) - u(t - 3.5t_d)]$ volts.

3. The maximum and minimum line voltages cannot exceed those values set by reliability or other requirements; thus $V_F < V_{max}$.

For CMOS systems, reasonable nominal values for these parameters are V_{IH} (min) $= 0.75V_{dd}$ and $V_{max} = 1.5V_{dd}$.

To ensure that the preceding requirements are met for all combinations of temperature, supply voltage, switching pattern, and so forth, is difficult for an arbitrary wiring pattern. Even for a multidrop bus such as is shown in Fig. 2.41, prediction can be complicated. However, for point-to-point wiring as is shown in that figure, design is easier. Taking only one receiver with $R_L = \infty$, which is the normal case for CMOS inputs, the limits on R_d that ensure that the preceding conditions (1) to (3) are met are

$$\tfrac{1}{3}Z_0 \leq R_d \leq \tfrac{5}{3}Z_0 \qquad (2.42)$$

The behavior of V_N and V_F for the two limits is shown in Fig. 2.42. The requirements on V_F are met. It should be clear that the voltage doubling at the load end, which is typical of wiring in CMOS systems, is a very useful characteristic for CMOS systems if used with care. It permits weaker driver circuits to drive transmission lines satisfactorily.

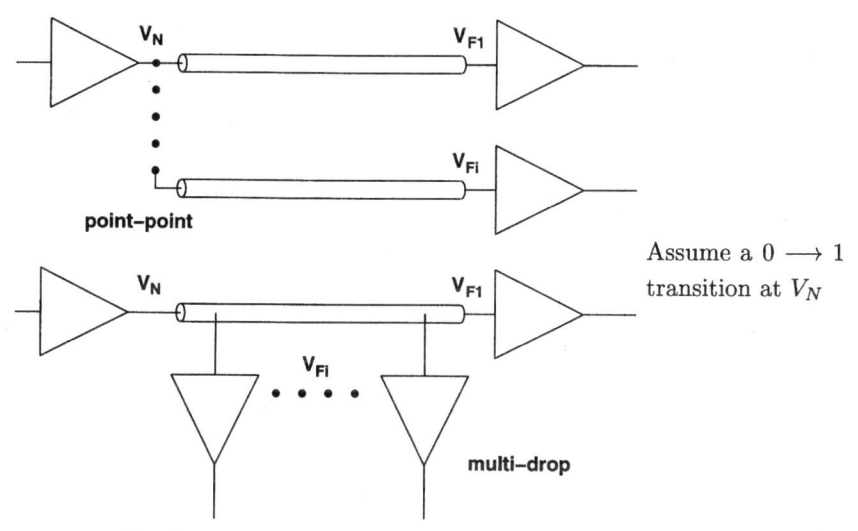

Basic Electrical Effects in Electronic Packaging and Interconnects 65

Figure 2.41 Two basic types of CMOS system wiring.

2.2.3.4 Step response of capacitive and inductive discontinuities and terminations. The step response of a capacitively terminated line gives insight into the time delay penalty paid for this type of load. Figure 2.43 shows such a transmission line, driven by a voltage step of amplitude $V_s = V_d$ at $t = t'$, with $R_d = Z_0$. This results in a voltage step applied to the near end

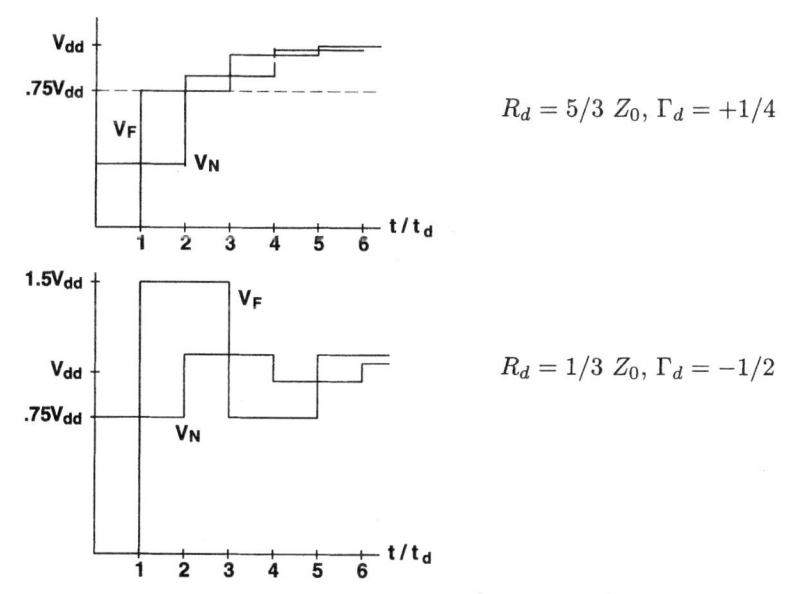

$$R_d = 5/3\ Z_0,\ \Gamma_d = +1/4$$

$$R_d = 1/3\ Z_0,\ \Gamma_d = -1/2$$

Figure 2.42 Transmission line voltages for $R_d = \tfrac{5}{3}Z_0$ and $R_d = \tfrac{1}{3}Z_0$, showing first incidence switching.

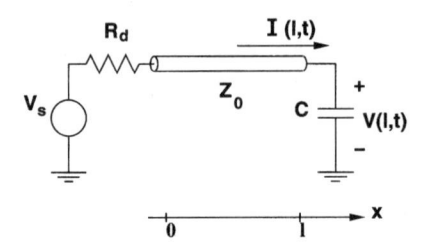

Figure 2.43 Capacitively terminated transmission line.

$$V^+(0,t') = \frac{Z_0 V_S}{R_d + Z_0} u(t') = Au(t) \qquad (2.43)$$

where A is $V_s/2$. The leading edge of the step arrives at $x = \ell$ at a time $t = 0$. Solving for $V(\ell,t)$, utilizing Eqs. 2.29, 2.30, 2.34, and 2.35 and $I(\ell,t) = C \, dV(\ell,t)/dt$,

$$V(\ell,t) = 2A(1 - e^{-t/\tau}) \qquad (2.44)$$

where

$$\tau = CZ_0 \qquad (2.45)$$

Note that

$$V(\ell,\infty) = 2A \qquad (2.46)$$

so that in the limit of long times the expected voltage doubling occurs (the capacitive load gives effectively $R_L = \infty$ at long times).

Figure 2.44 shows the normalized time behavior of the load voltage, along with the idealized load voltage in dotted lines. The time for $V(\ell,t)$ to rise to 50 percent of its final value is $0.693\tau \cong 0.7\tau$. This is the delay adder due to the capacitive load. Some simple calculations will put this delay adder into perspective. For $Z_0 = 50$ ohms, $\tau = 50$ ps for $C = 1$ pF and $\tau = 2.5$ ns for $C = 50$ pF. There are some systems for which a 50-ps delay adder is substantial. There are many more systems for which a delay adder of 2.5 ns is substantial.

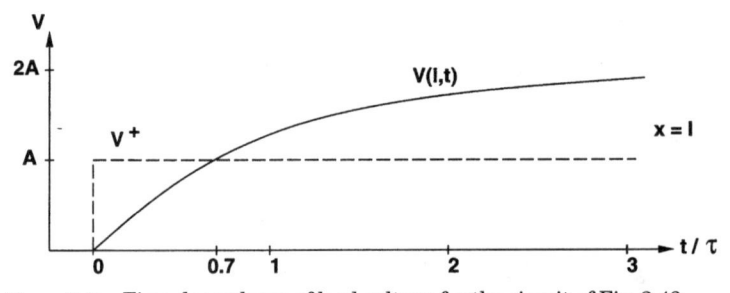

Figure 2.44 Time dependence of load voltage for the circuit of Fig. 2.43.

It is important to understand the nature of the reflected pulse and the behavior of the line voltage. Figure 2.45a shows the reflected pulse at $x = \ell$. It is obtained from

$$V^-(\ell,t) = V(\ell,t) - V^+(\ell,t) \tag{2.47}$$

and is

$$V^-(\ell,t) = A(1 - 2e^{-t/\tau}) \tag{2.48}$$

The voltage on the line for $0 < t < t_d$ is simply the superposition of the preexisting line voltage A and the reflected pulse $V^-(x,t)$. It is easy to understand the general nature of $V^-(x,t)$. The sharp corners of $V^+(x,t)$ are due to very high frequency Fourier components of the waveform. At very high frequencies, a capacitor behaves as if it were a short circuit, $|Z_c| = (2\pi f C)^{-1} \cong 0$, and for a short circuit

$$\Gamma_L \text{ (short circuit)} = -1 \tag{2.49}$$

- the reflected pulse contains most of the high frequencies in V^+

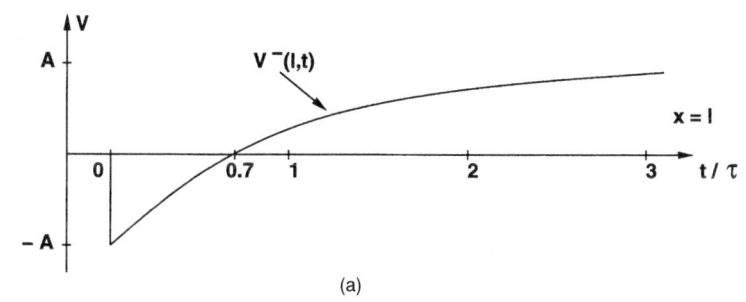

- the voltage on the line for $0 < t < t_d$ is the superposition of $V^+ = A$ and $V^-(l - v_0 t, t)$

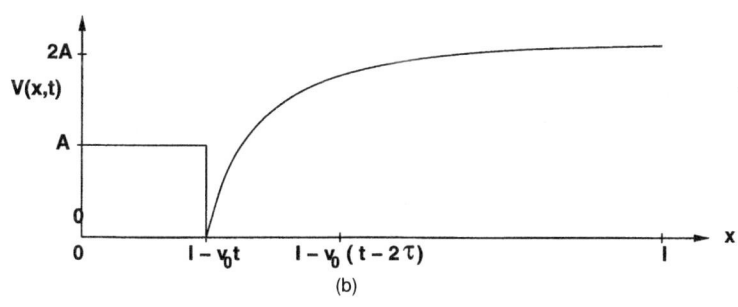

Figure 2.45 Reflected pulse (a) and voltage on the transmission line for $0 < t < t_d$ (b), circuit of Fig. 2.37.

Thus the high frequency components are inverted and reflected back and are seen as the negative going "spike" in Fig. 2.45a. The full width at half maximum of this spike is 0.7τ, which as has been demonstrated can reasonably be in excess of 1 ns. The spike will be seen by any circuits distributed along the line, and those circuits which are fast enough to respond may give logic errors due to the spike. Note that the actual waveform at the receiver has lost all its high frequency components and thus is not sharp, but rounded (see Fig. 2.45b). For the case of a driving pulse with nonzero rise time (i.e., ramp), the response at the load is a delayed ramp if $τ/t_r \ll 1$. The delay adder is τ in this case.

Discontinuities in the line can be caused by geometrical or material inhomogeneities, such as those shown in Fig. 2.46, or by point loading caused by connections to receivers, and so forth. Some simple discontinuities can be examined analytically, but many involve complex equivalent circuits and can be handled only by computer simulation. Two discontinuities which are simple and important are capacitive and inductive discontinuities.

Figure 2.47 shows a capacitive discontinuity with incident, reflected and transmitted pulses, and capacitor charging current. Take the case for an incident voltage pulse V^+ of amplitude A arriving at $x = 0, z = 0$ at $t = 0$, with no preexisting voltage on capacitor C. Then

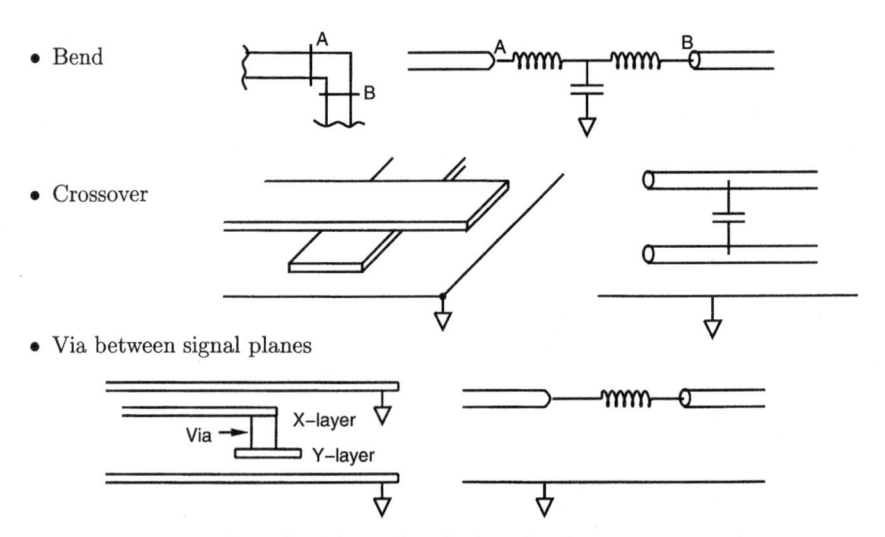

$$V(0,t) = A(1 - e^{-t/τ}) \qquad (2.50)$$

- Bend

- Crossover

- Via between signal planes

Figure 2.46 Various discontinuities and equivalent circuits.

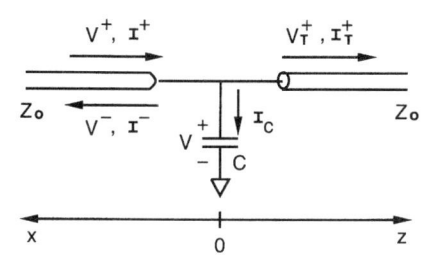

Basic Electrical Effects in Electronic Packaging and Interconnects 69

Figure 2.47 Diagram of currents and voltages at a capacitive discontinuity.

where

$$\tau = \tfrac{1}{2}Z_0 C$$

$V(0,t)$ is shown in Fig. 2.48. The reflected pulse at $x = 0$ is

$$V^-(0,t) = A e^{-t/\tau} \tag{2.51}$$

and is shown in Fig. 2.49. This waveform propagates back toward the source, as shown in Fig. 2.50. The result of this reflected pulse is that there is a negative-going spike, to zero volts, traveling back up the line. This spike can cause logic errors if it is wide enough to be sensed by other gates connected to this line. The full width at half-maximum of the pulse is 0.693τ. Since $\tau = \tfrac{1}{2}Z_0 C$, $\tau = 125$ ps for $Z_0 = 50$ ohms, $C = 5$ pF. This is wide enough to be sensed by fast gates.

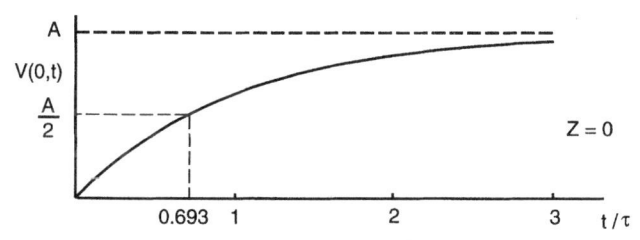

Figure 2.48 Time dependence of line voltage at a capacitive discontinuity, step voltage input.

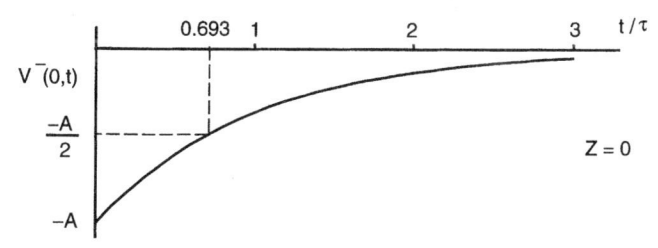

Figure 2.49 Time dependence of the reflected pulse at a capacitive discontinuity, step voltage input.

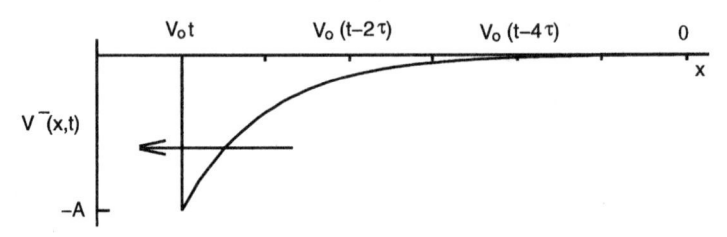

Figure 2.50 Reflected pulse propagating toward the source, capacitive discontinuity with step input.

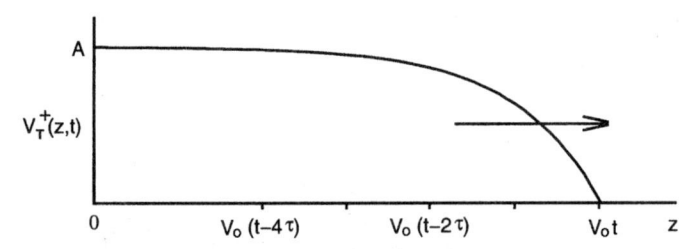

Figure 2.51 Transmitted pulse past a capacitive discontinuity step voltage input.

Figure 2.51 shows the transmitted voltage $V_T^+(Z,t)$ propagating with velocity v_0. Note that the pulse has lost most of its high frequency components and is quite rounded. Of course, the high frequency components were reflected, and are clearly seen in Figs. 2.49 and 2.50. The effect of the reflection is to introduce a delay adder of 0.7τ to the original delay. This delay adder is cumulative; that is, if a line has several capacitive discontinuities the total delay adder is the sum of the individual delay adders. In this way, relatively innocuous individual delay adders can add up to trouble in a system.

The step response of an inductive discontinuity is obtained from the diagram of Fig. 2.52. Assume that an incident step voltage of amplitude A arrives at $x = 0$ at $t = 0$, and that there is no current through L prior to $t = 0$. Further assume that for practical purposes $x = 0$ and $z = 0$ are coincident, that is, that L is very small physically compared to the minimum wavelength harmonic which is of any significance. In this case, propagation delay through L does not have to be taken into account. Then

$$V(0,t) = A(1 + e^{-t/\tau}), \tag{2.52}$$

$$V^-(0,t) = Ae^{-t/\tau} \tag{2.53}$$

and

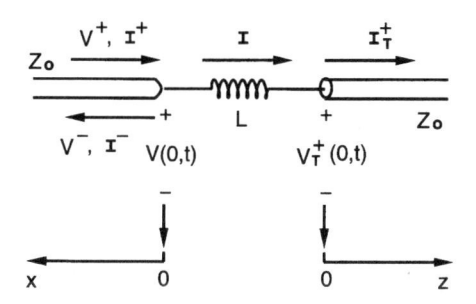

Basic Electrical Effects in Electronic Packaging and Interconnects 71

Figure 2.52 Diagram of currents and voltages at an inductive discontinuity.

$$V_T^+(0,t) = A(1 - e^{-t/\tau}) \quad (2.54)$$

where $\tau = L/2Z_0$. Figure 2.53 shows the behavior of $V(0,t)$ and $V_T^+(0,t)$. The time dependence of $V(0,t)$ can be inferred from the fact that for very high frequencies, the inductive impedance is $|Z_L| = 2\pi fL \gg Z_0$, and so the inductor behaves instantaneously, on arrival of the sharp edge of V^+, as an open circuit. Thus $\Gamma = +1$ and voltage doubling occurs, giving an instantaneous peak voltage of $2A$, as shown in Fig. 2.53. The voltage doubling results in the initial current through the inductor being zero, similar to the current through an $R_L = \infty$.

As expected, the reflected high frequencies are contained in the reflected pulse, as shown in Fig. 2.54. The reflected waveform is a "spike" of full width at half maximum equal to 0.7τ. The transmitted voltage pulse, again as expected, is a square pulse that has lost most of its high-frequency content; that is, it is a rounded pulse such as that shown in Fig. 2.55. The loss of the high frequency content, and the rounding of the pulse due to that, results in a delay adder of 0.7τ, as shown in Fig. 2.55.

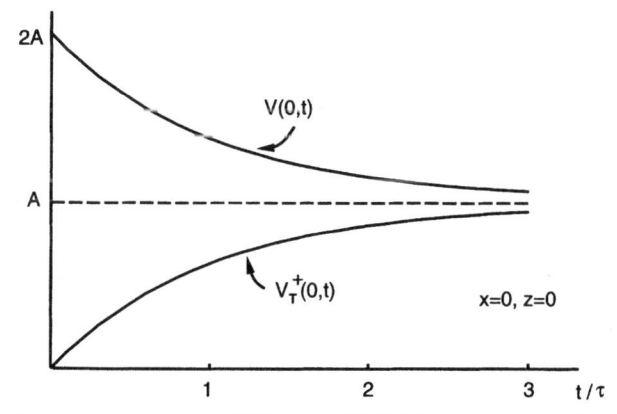

Figure 2.53 Behavior of $V(0,t)$ and $V_0^+(0,t)$ for an inductive discontinuity.

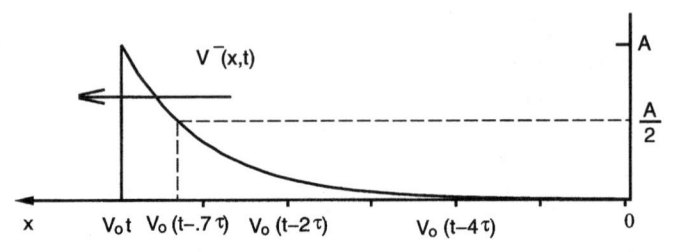

Figure 2.54 Reflected pulse propagating toward the source, inductive discontinuity with step input.

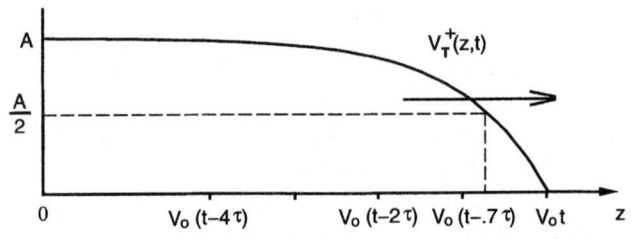

Figure 2.55 Transmitted pulse past an inductive discontinuity, step voltage input.

Example What is the delay adder due to a 0.025-mm diameter via which is 0.25 mm in length if $Z_0 = 50\Omega$? Assume $L_p = 0.26$ nH

$$0.7\tau = 0.7(0.26 \times 10^{-9}/2 \times 50) = 1.82 \text{ ps}$$

2.2.3.5 Modeling of delay adders due to capacitive and inductive discontinuities. If T_d is the total delay to the 50 percent point of the maximum voltage at $x = \ell$, for the circuit of Fig. 2.56, the delay adder is 0.7τ and

$$T_d = t_d + 0.7\tau \tag{2.55}$$

which can be put in the form

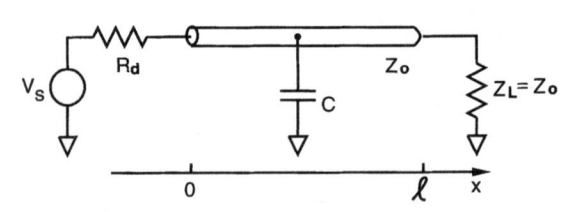

Figure 2.56 Transmission line with single point capacitive discontinuity.

Basic Electrical Effects in Electronic Packaging and Interconnects 73

$$T_d = t_d\left[1 + 0.35 \frac{C}{\ell C'}\right] \quad (2.56)$$

where the identity

$$CZ_0 = t_d(C/\ell C') \quad (2.57)$$

has been used. The total line capacitance, with $C = 0$, is $\ell C'$.

Another approach to estimating T_d is to average the capacitive loading over the entire line length and define a total capacitance p.u. length for the line

$$C'_T = C' + C/\ell \quad (2.58)$$

This can be used to determine an effective delay time for a pulse,

$$T_d^e = \frac{\ell}{v_0} = \ell\sqrt{L'C'_T} = t_d\sqrt{1 + (C/\ell C')} \quad (2.59)$$

This approach gives no information about the shape of the transmitted pulse or the reflected pulse; it effectively treats the transmitted pulse as a delayed step. For $C/\ell C' \leq 0.8$,

$$t_d\left(1 + 0.5 \frac{C}{\ell C'}\right) \leq T_d^e \leq 1.4 t_d \quad (2.60)$$

while for $C/\ell C' \gg 1$,

$$T_d^e = t_d \sqrt{C/\ell' C} \quad (2.61)$$

It is clear that the use of T_d^e is relatively accurate only when $C/\ell C' \ll 1$, since for this case $T_d^e = t_d(1 + 0.5C/\ell C')$ which is approximately equal to Eq. 2.56.

The usefulness of this approach becomes apparent only when the case of multiple capacitive discontinuities distributed along a line, such as shown in Fig. 2.57, is considered. For a line loaded with N point loads of C each, an average total capacitance p.u. length

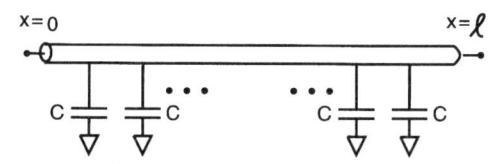

Figure 2.57 Transmission line with distributed capacitive loading.

$$C'_T = C' + NC/\ell \tag{2.62}$$

can be defined. This results in reduced Z_0, since the effective Z_0 is now

$$Z_0^e = \sqrt{L'/C'_T} = Z_0(1 + NC/\ell C')^{-1/2} \tag{2.63}$$

The effective velocity of propagation is also decreased due to the increased capacitance,

$$v_0^e = v_0(1 + NC/\ell C')^{-1/2} \tag{2.64}$$

Thus

$$T_d^e = \ell/v_0^e \tag{2.65}$$

or

$$T_d^e = t_d(1 + NC/\ell C')^{1/2} \tag{2.66}$$

The effect of distributed capacitive loading is to:

1. Decrease Z_0
2. Decrease v_0
3. Increase T_d by a delay adder $t_d\left[(1 + NC/C')^{1/2} - 1\right]$

Alternatively, the effect of distributed (series) inductance loading is to increase Z_0, decrease v_0, and introduce a delay adder $t_d[(1 + NL/\ell L')^{1/2} - 1]$. In either case, the reflected pulses are a stream of spikes, positive- or negative-going as the discontinuities are inductive or capacitive, respectively, propagating back down the line toward the source. Each spike has successively lower high-frequency content compared to the preceding spike.

The effect of multiple capacitive or inductive discontinuities on delay can also be obtained by assuming that each discontinuity contributes independently a delay adder of $0.35t_dC/\ell C'$ or $0.35t_dL/\ell L'$. Then for N identical discontinuities, as mentioned earlier,

$$T_d = t_d[1 + 0.35NC/\ell C'] \tag{2.67}$$

or

$$T_d = t_d[1 + 0.35NL/\ell L'] \tag{2.68}$$

One observation on the effect of discontinuities on signals is important. Quite often, the additional delay caused by discontinuities is unimportant [11]. This is particularly true if the discontinuities are line corners, bends, or vias. From the design standpoint, this is fortunate, because if the discontinuities are unimportant then considerable three-dimensional modeling is avoided.

2.2.4 Lossy transmission lines

Losses in transmission line structures come about through conductor resistance, dielectric losses which are modeled as a parallel conductance p.u. length G', and conversion of energy to other traveling wave modes, for example, radiation (EMI). In this section, losses due to conductor resistance are addressed.

Two different approaches can be taken to understanding the behavior of lossy transmission lines, a time domain approach and a frequency domain approach. In the time domain approach, assuming that $V^+(0,t) = Au(t)$ and $G' = 0$, and assuming no reflections occur, analysis shows [12]

$$V(\ell,t_d) = Ae^{-\alpha\ell}[1 + f(\ell,v_0,t)]u(t - \ell/v_0) \quad (2.69)$$

Here

$$\alpha = R'/2Z_0 \quad (2.70)$$

and $f(\ell,v_0,t)$ is a slowly monotonically increasing function which has final value $e^{\alpha\ell} - 1$.

Thus, the far-end waveform consists of a step of amplitude $Ae^{-\alpha\ell}$ followed by a slow rise to full amplitude A, which is similar to an RC charging time constant. Figure 2.58 shows this time behavior. For first-incidence switching at the far end, $V(\ell,t_d) \geq 0.75 \times 2A$, if it is assumed that voltage doubling occurs at the load end and that the line is matched to the driver. Thus, the step in voltage at (ℓ,t_d) must equal $0.75 \times 2A$. This requires

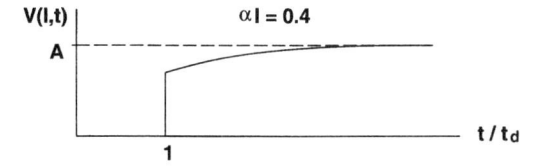

$$-\alpha\ell = \ln 0.75 = -0.288$$

Figure 2.58 Far-end voltage for a lossy transmission line with $\alpha\ell = 0.4$.

or

$$R'\ell \le 0.58 Z_0 \tag{2.71}$$

Therefore for a transmission line for which the total line resistance is less than about half of Z_0, no significant delay will be caused by lossy-line effects in CMOS-based systems.

Example For what length of line will lossy-line effects begin to *dominate* the far-end voltage of a microstrip line, if $\varepsilon_r = 7$, $w = 25$ µm, $t = 5$ µm, $h = 25$ µm, and $\rho = 1.58 \times 10^{-6}$ Ω/cm(Cu)?

If $\alpha\ell \ge 0.7$, $e^{-\alpha\ell} \le 0.497$ and the step in voltage at $x = \ell$ will be less than half the lossless-line value. Thus the lossy-line effects will be dominant. By calculation, $R' = 1.26$Ω/cm and $Z_0 = 56.3$Ω. Thus

$$\frac{1}{2}\frac{R'\ell}{Z_0} \ge 0.7$$

requires

$$\ell \ge 6.26 \text{ cm}$$

If w and t are changed to 0.5 µm and 1.0 µm, respectively, then $\ell \ge 0.25$ cm.

Frequency domain analysis provides insight but does not provide quantitative requirements on $R'\ell$. Figure 2.59 shows a section of a lossy transmission line. Intuitively it is clear that if

$$R' \ll \omega L' \tag{2.72}$$

(for example, if $R' \le 0.1\omega L'$), then the effects of loss will be negligible at that frequency. The question is, what frequency shall be used in Eq. 2.72? If ω is taken to be the maximum frequency of interest and that is taken to be $1/t_r$, it can be shown that

$$\frac{R'}{\omega L'} = \frac{1}{\pi}\frac{t_r}{t_d}\alpha\ell \tag{2.73}$$

If $t_r/t_d \cong 1$ (a questionable generalization) then the condition

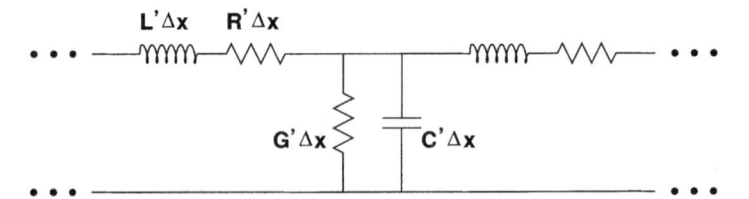

Figure 2.59 Section of a lossy transmission line.

Basic Electrical Effects in Electronic Packaging and Interconnects 77

$$\frac{R'}{\omega L'} \leq 0.1 \tag{2.74}$$

reduces to

$$\alpha \ell \leq 0.1\pi \cong 0.3 \tag{2.75}$$

or

$$R'\ell \leq 0.6 Z_0 \tag{2.76}$$

This is the same condition arrived at in the previous time-domain derivation. However, in this case the condition was arrived at using qualitative and intuitive arguments. Nevertheless the principles of this frequency domain argument are sound and give insight into lossy lines.

2.2.5 Coupled noise (cross talk)

Coupled noise is due to energy coupled from an active (driven) line to another line—at worst case, a quiet (inactive) line. The coupling is non-radiative and is due to mutual capacitance and inductance. The coupled voltages and currents are not technically noise, since they are not random but are predictable given driving voltage waveforms. However, they are referred to as noise. The coupling may be primarily capacitive or inductive. The coupled noise may, in concert with other effects, cause false switching.

The case of weak coupling will be analyzed below. Weak coupling means that the active-line voltage waveform is practically unperturbed by the energy coupled from it, and unperturbed by the energy recoupled back from the quiet line. Figure 2.60 shows an active line which couples energy into two lines, line 1 and line 2. The length of line over which coupling occurs is ℓ_1 and ℓ_2, as indicated in the figure. There are two types of noise coupled, v_f, a forward cross-talk voltage that propagates in the direction of propagation of the pulse on the active line, and v_b, a backwards cross-talk voltage that propagates in the opposite direction. For line 1, receiver 1 sees v_f before any reflections occur. For line 2, receiver 2 sees v_b before any reflections occur. Figure 2.61 shows the shapes of the v_f and v_b assuming a trapezoidal driving pulse on the active line, of width T, rise and fall time t_r and t_f and amplitude A. In Fig. 2.61, it is assumed that no reflections take place at near and far ends of the line, that is, the line has a matched termination. As is determined later, the sequence of polarities shown for v_f can be reversed. Also, the amplitude of v_b is at most a value $v_b(\text{sat})$. If the amplitude is reduced below v_b (sat), the timing changes as shown at the bottom of Fig. 2.61.

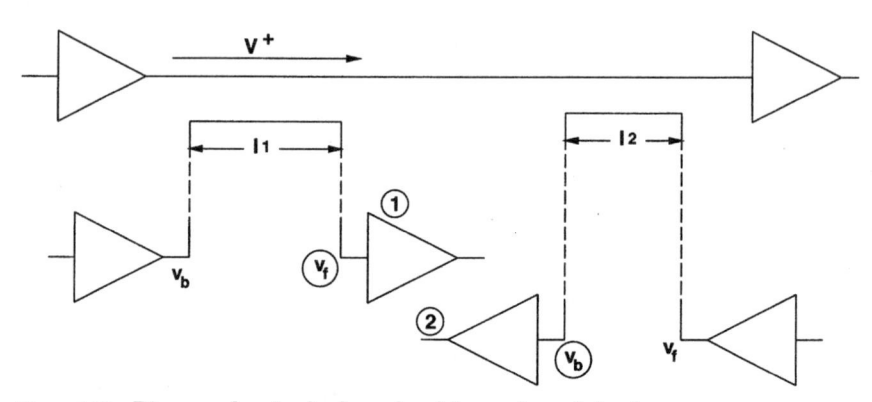

Figure 2.60 Diagram showing backward and forward coupled noise.

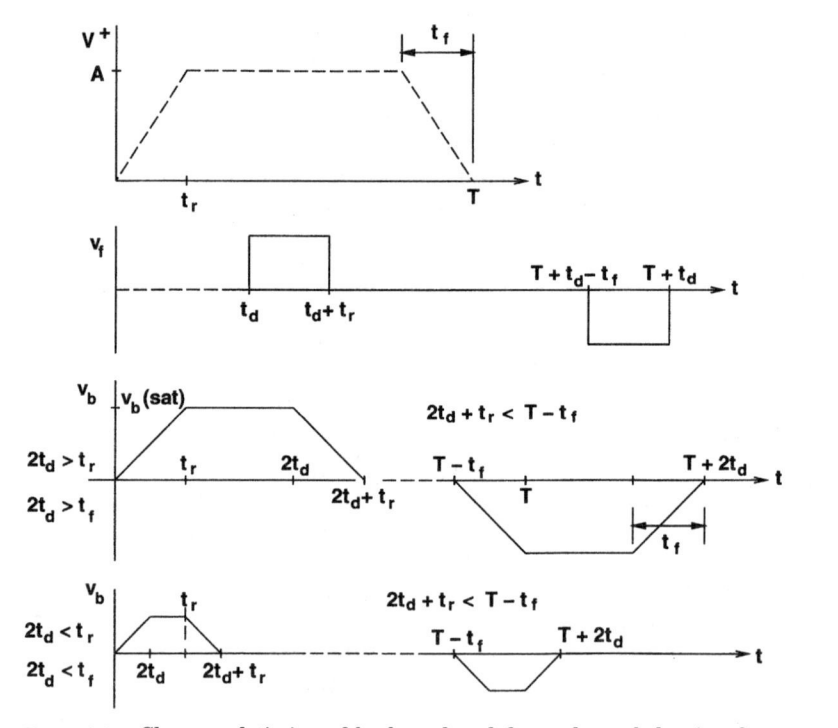

Figure 2.61 Shape and timing of backward and forward coupled noise, for a trapezoidal pulse on the active line.

Basic Electrical Effects in Electronic Packaging and Interconnects 79

Figure 2.62 shows the schematic diagram that is used to analyze v_b and v_f. It consists of a section of width Δx taken out of the middle of the active and quiet lines, connected to the external elements such as R_b, R_f, and so forth. Assume the voltages on line 2 are much less than the propagating pulse on line 1, $V_2 \ll V_1^+$. Then since $I_c = C\, dV_c/dt$ and $V_L = L\, dI_L/dt$,

$$\Delta i_b + \Delta i_f = (C'_{12}\, \Delta x)\, \frac{dV_1^+}{dt} \tag{2.77}$$

and

$$\Delta v_b - \Delta v_f = (L'_{12}\, \Delta_x)\, \frac{dI_1^+}{dt} \tag{2.78}$$

Here Δi_b is a backwards coupled wave and is flowing in the direction of propagation of the wave, so $\Delta i_b = \Delta v_b/z_0$. Likewise $\Delta i_f = \Delta v_f/z_0$. Using $Z_0 = \sqrt{L'_{11}/C'_{11}}$, which is not quite rigorously true, and using $v_0 \cong 1/\sqrt{L'_{11}C'_{11}}$, Eqs. 2.77 and 2.78 give (in the limit of infinitesimal Δx)

$$\frac{dv_b}{dx} = \frac{1}{2v_0}\left[\frac{L'_{12}}{L'_{11}} + \frac{C'_{12}}{C'_{11}}\right]\frac{dV^+}{dt} \tag{2.79}$$

and

$$\frac{dv_f}{dx} = \frac{1}{2v_0}\left[-\frac{L'_{12}}{L'_{11}} + \frac{C'_{12}}{C'_{11}}\right]\frac{dV_1^+}{dt} \tag{2.80}$$

For the V_1^+ shown in Fig. 2.61,

$$\frac{dV_1^+}{dt} = +\frac{A}{t_r}, \qquad 0 < t < t_r \tag{2.81}$$

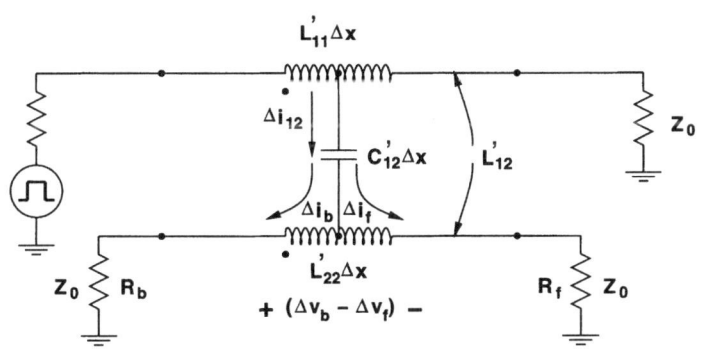

Figure 2.62 Equivalent circuit for derivation of backward and forward cross-talk waveforms.

and

$$\frac{dV_1^+}{dt} = -\frac{A}{t_f}, \quad T - t_f < t < T \quad (2.82)$$

Substituting Eqs. 2.81 and 2.82 into 2.80, the forward-coupled voltage at $x = \ell$ will be

$$v_f(\ell) = \frac{1}{2}\left[\frac{L'_{12}}{L'_{11}} + \frac{C'_{12}}{C'_{11}}\right]\frac{t_d}{t_r}A \quad (2.83)$$

$$t_d < t < t_d + t_r$$

and

$$v_f(\ell) = -\frac{1}{2}\left[-\frac{L'_{12}}{L'_{11}} + \frac{C'_{12}}{C'_{11}}\right]\frac{t_d}{t_f}A \quad (2.84)$$

$$T - t_f + t_d < t < T + t_d$$

Note that the bracketed term in Eqs. 2.83 and 2.84 may be either positive or negative. The negative sign ensures that for the two pulses there is one inversion, so that there is always a negative pulse and a positive pulse.

It can be shown that for a homogeneous structure such as a stripline, where TEM waves propagate, $L'_{12}/C'_{12} = L'_{11}/C'_{11}$. Thus, for such structures, the forward cross talk is zero.

It is worth noting that $v_f(\ell)$ is directly proportional to ℓ since $t_d = \ell/v_0$. Also, the v_f pulses are narrow, of width t_r and t_f. The strength of the pulses is independent of t_r and t_f, since $v_f(\ell) \times t_f$ and $v_r(\ell) \times t_r$ are independent of t_r and t_f. It is nevertheless possible to have relatively high-amplitude pulses that are so narrow that they cannot be sensed by gate inputs.

To calculate $v_b(t)$ requires a little more subtlety, since as V_1^+ propagates toward $x = \ell$ it continuously couples a voltage into line 2, which propagates backwards toward $x = 0$. Thus, the coupled voltage waveform is stretched out in time (sort of a Doppler effect, but not exactly). If $2t_d > t_r$, the result is

$$V_b(0,t) = \frac{1}{4}\left[\frac{L'_{12}}{L'_{11}} + \frac{C'_{12}}{C'_{11}}\right][V_1^+(0,t)u(t) - V_1^+(0,t - 2t_d)u(t - 2t_d)] \quad (2.85)$$

$$= v_b\,(\text{sat})\left[\frac{V_1^+}{A}(0,t)u(t) - \frac{V_1^+}{A}(0,t - 2t_d)u(t - 2t_d)\right]$$

where

Basic Electrical Effects in Electronic Packaging and Interconnects   81

$$v_b \text{ (sat)} = \frac{A}{4}\left[\frac{L'_{12}}{L'_{11}} + \frac{C'_{12}}{C'_{11}}\right] \qquad (2.86)$$

Figure 2.63 shows the waveform of Eq. 2.85, under the further condition that $T - t_f > 2t_d + t_r$, as the first trapezoidal waveform. This first waveform can be understood by first observing that it is entirely due to a time-dependent part of $V_1^+(x,t)$, specifically the front edge of $V_1^+(x,t)$. As $V_1^+(0,t)$ rises to amplitude A with rise time t_r, a fraction $\frac{1}{4}[(L_{12}/L_{11}) + (C_{12}/C_{11})]$ is coupled into line 2. This gives the waveform for $0 < t \le t_r^+$. As the front edge of $V_1^+(x,t)$ propagates toward $x = \ell$, it couples energy which propagates back to $x = 0$, maintaining the voltage $V_2^-(0,t)$ at $\frac{1}{4}[(L_{12}/L_{11}) + (C_{12}/C_{11})]A$. When the front edge of $V_1^+(x,t)$ arrives at $x = \ell$, that is, when $t = t_d$, the energy coupled back toward $x = 0$ begins to decrease. This start of the decrease of $v_b(0,t)$ only begins at $x = 0$ when enough time has elapsed to allow that part of the waveform to propagate back from $x = \ell$. Thus the decrease in $v_b(0,t)$ begins at $t = 2t_d$ and ends at $t = 2t_d + t_r$, as shown in Fig. 2.63. The rightmost waveform in Fig. 2.63 is the $v_b(0,t)$ waveform caused by the falling edge of $v_1^+(x,t)$. Identical reasoning to that above will justify the shape and timing of this part of the waveform.

For the case $2t_d < t_r$ a backward coupled noise waveform such as that shown in Fig. 2.64 will be obtained. In this case, the decrease in $V_2^-(0,t)$ that starts at t_d begins before the rise in $V_1^+(0,t)$ stops at $t = t_r$. This results in flattening of the rise at a value

$$v_b(0, 2t_d) = \frac{At_r}{2t_d} \times \frac{1}{4}\left[\frac{L'_{12}}{L'_{11}} + \frac{C'_{12}}{C'_{11}}\right] \qquad (2.87)$$

$$v_b(0, 2t_d) < v_b \text{ (sat)}$$

The shape of the remainder of the waveform can be obtained through reasoning such as that used to explain the waveform in Fig. 2.63.

From Eq. 2.86 it is clear that v_b (sat) itself is not a function of t, ℓ, t_r, or t_f. However, the line length at which saturation of $v_b(0,t)$ occurs is a function of ε and is a function of t_r through the saturation condition

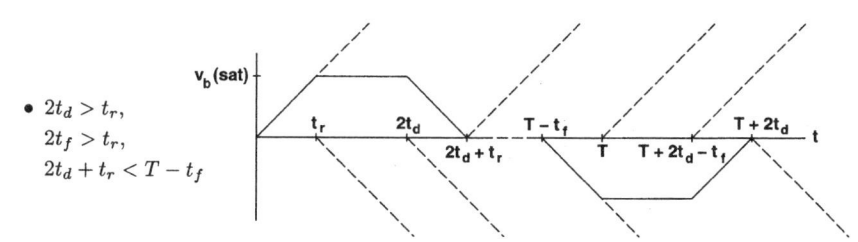

- $2t_d > t_r$,
  $2t_f > t_r$,
  $2t_d + t_r < T - t_f$

Figure 2.63   Saturated backward cross-talk voltage for trapezoidal active-line waveform.

- $2t_d < t_r,$
- $2t_f < t_r,$
- $2t_d + t_r < T - t_f$

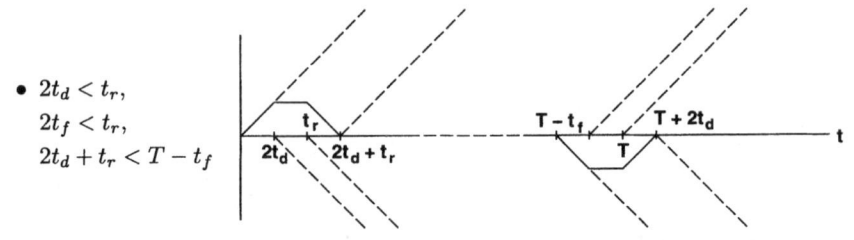

Figure 2.64 Nonsaturated backward cross-talk voltage for trapezoidal active-line waveform.

$$2t_d > t_r \tag{2.88}$$

The coupled length at which saturation occurs is

$$\ell_{(\text{sat})} = \frac{v_0 t_r}{2} = f(\varepsilon, t_r) \tag{2.89}$$

where v_0 is the propagation velocity. Backward cross talk is particularly dangerous to a system because, if design rules are not properly applied, very wide $v_b(0,t)$ waveforms are obtained. Thus the length of the saturated $v_b(0,t)$ is $2t_d + t_r$, which can be quite long, depending on t_d. Wide waveforms are easily sensed by other gates connected to a transmission line. Control of $v_b(0,t)$ is through control of ℓ and through control of L'_{12} and C'_{12}. The most direct control is to control the amplitude of $v_b(0,t)$ through limiting the coupled length ℓ, thereby decreasing $v_b(0,t)$ below v_b (sat). Rerouting to control ℓ will usually have the effect of decreasing both L'_{12} and C'_{12}. Given the permitted value of $v_b(0,t)$, Eq. 2.87 gives the value of permitted ℓ through knowledge of $t_d = \ell/v_0$.

Example Given the two-line transmission line system of Fig. 2.65, with air covering the lines, calculate $v_f(\ell,t)$ and $v_b(0,t)$ if $\ell = 5$ cm, $t_r = 0.5$ ns, and the $V_i^+(x,t)$ amplitude is A. Assume that no reflection occurs.

From two-dimensional L,C modeling tools,

$$L'_{11} = 5.48 \text{ nH/cm}, \quad L'_{12} = 1.42 \text{ nH/cm}$$

$$C'_{11} = 0.557 \text{ pf/cm}, \quad C'_{12} = 0.096 \text{ pf/cm}$$

$$v_0 \cong (L_{11}C_{11})^{-1/2} = 1.8 \times 10^{10} \text{ cm/s}$$

$$t_d = \ell/v_0 = 0.276 \text{ ns}$$

$$2t_d = 0.552 \text{ ns} > t_r; V_b \text{ is saturated}$$

$$v_f(\max) = \frac{1}{2}\left[-\frac{L'_{12}}{L'_{11}} + \frac{C'_{12}}{C'_{11}}\right]\frac{t_d}{t_r}A = -0.044A$$

$$v_b(\text{sat}) = \frac{1}{4}\left[\frac{L'_{12}}{L'_{11}} + \frac{C'_{12}}{C'_{11}}\right]A = 0.11A$$

The first waveform for $v_b(0,t)$ is shown in Fig. 2.66.

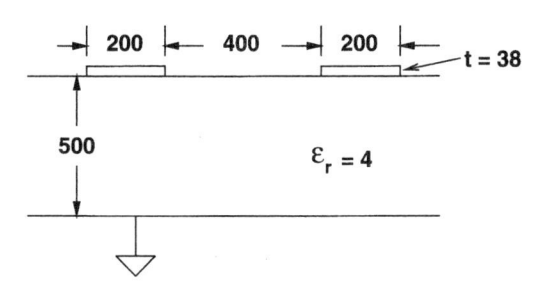

Figure 2.65 Two-conductor transmission line system for cross-talk example; dimensions in microns.

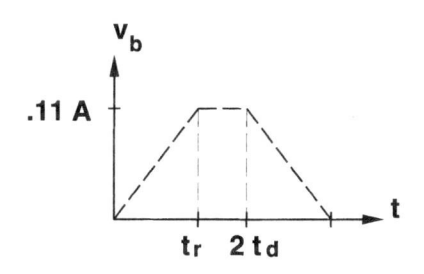

Figure 2.66 Backward cross-talk waveform for cross-talk example.

2.3 Power and Ground Bus Disturbances

Power and ground voltage disturbances are the system-limiting noise sources for most single-chip packages and most systems. One of the factors driving this noise source is the number of output pads per bus per chip. This factor is discussed later in this chapter. Another factor is the total number of I/Os per chip and the number of busses per chip. Figures 2.67a and 2.67b show projections [1] of total single-chip package "pins" (or balls) for microprocessor and ASIC packages, respectively. The time scale for these figures can be obtained using Fig. 2.1. The relationship between I/O count and gate count for a system of any level of complexity is usually taken to be a form of "Rent's Rule." In Rent's Rule, developed in the 1960s for IBM mainframe architectures, the number of pins N_p is related to the number of gates N_g through

$$N_p = K N_g^\beta$$

where K and β are system- and architecture-dependent constants. A wide selection of values for K and β can be gotten through examination of recent literature. Some values which may be used [7] are:

	K	β
Microprocessor	0.82	0.45
ASIC	1.9	0.50

It is interesting to compare these values to Rent's original values ($K = 2.5$, $\beta = 0.6$), which were extracted from architectures that have

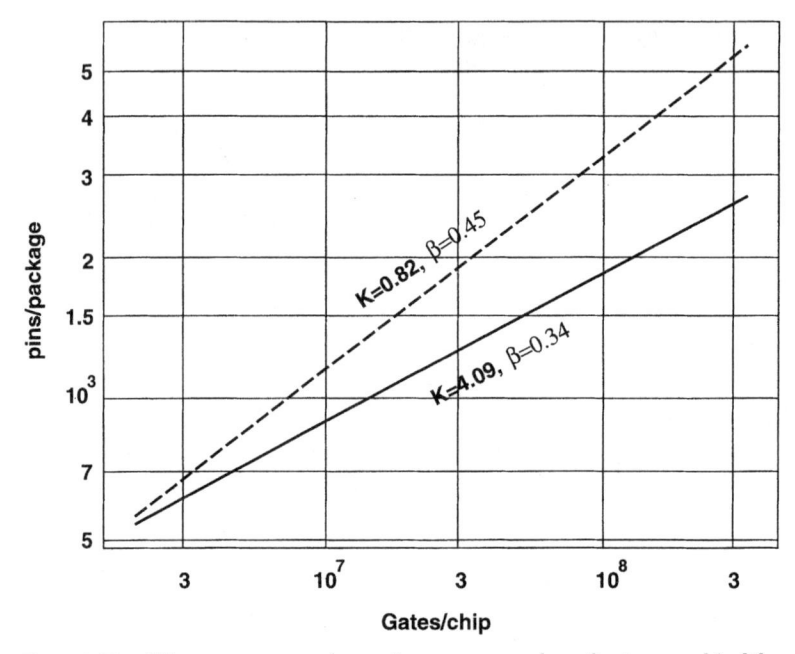

Figure 2.67a Microprocessor package pins versus number of gates per chip [1].

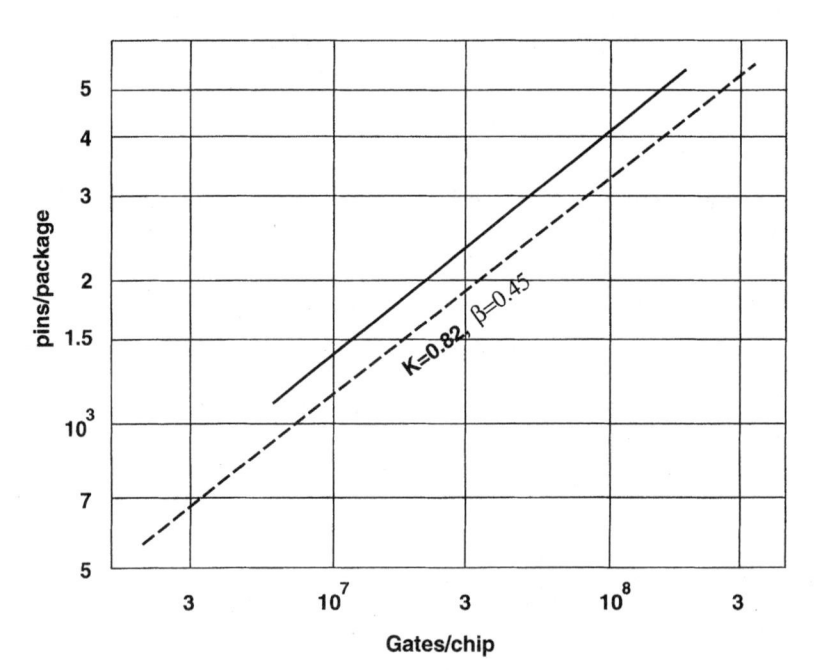

Figure 2.67b ASIC package pins versus number of gates per chip [1].

Basic Electrical Effects in Electronic Packaging and Interconnects   85

been dead for decades! In Figs. 2.67a and 2.67b the models are plotted as dotted lines. The projections are plotted as solid lines and have best-fit values for K and β. The depression in the value of β for microprocessors, which is evident in Fig. 2.67a, is partly due to the inclusion of significant amounts of memory on-chip.

Recent history shows that the number of power plus ground pins is around 25 percent of the total number of pins. Thus for advanced microprocessors, the total number of signal I/Os is projected to exceed 2000 in the year 2012, and for leading-edge ASIC packages the signal I/O count will probably exceed 4000 in 2012. Implications of this large number of I/Os can be seen in subsequent paragraphs in this section. One factor that adds to the complexity of the problem is the expected decrease of chip core voltage, shown in Fig. 2.68 [1]. Some types of noise, for example cross-talk noise, scale as power supply voltages are scaled. Power and ground voltage disturbances, since they are usually due to inductive effects, are independent of supply voltage but are sensitive to pulse-edge speeds (which are themselves influenced by clock

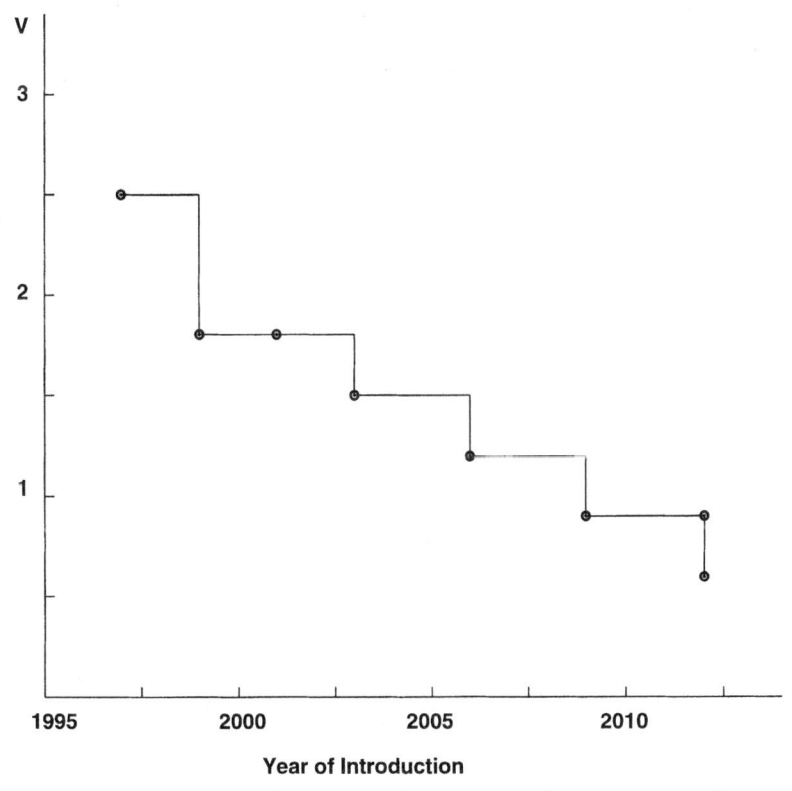

Figure 2.68   Core supply voltage versus first year of product introduction [1].

frequencies). The significance of noise for CMOS systems is gauged by its size relative to V_{dd}. Thus there are three critical factors that act together to cause power and ground noise to be of increasing concern over time: (1) the strong increase in package pins with passage of time; (2) the accompanying strong increase in off-chip clock frequency; and (3) the projected strong decrease in chip core power supply voltage. With these factors all driving in the same direction (increasing noise problems), problems with the power and ground noise component can be expected to worsen with time. This situation only increases the emphasis on understanding the noise mechanism so that it can be controlled or designed around.

2.3.1 Introduction

Power and ground bus disturbances consist of transient voltages on the chip ground or power bus, which cause the voltage on the bus to differ from $0V$ or V_{dd} V. The disturbances are caused by voltage drops across parasitic impedances, such as the voltage drops shown across Z_{vdd} and Z_{vss} in Fig. 2.69. The parasitic impedances (Z_{vdd} and Z_{vss}) may be primarily resistive, but are usually primarily inductive if the bus disturbances due to off-chip drivers are considered. If disturbances due to switching of the chip core logic are examined, global disturbances are primarily due to resistive parasitics, since switching of the core results in a time-averaged current draw that is relatively constant. Switching of the off-chip drivers, however, results in identifiable current pulses

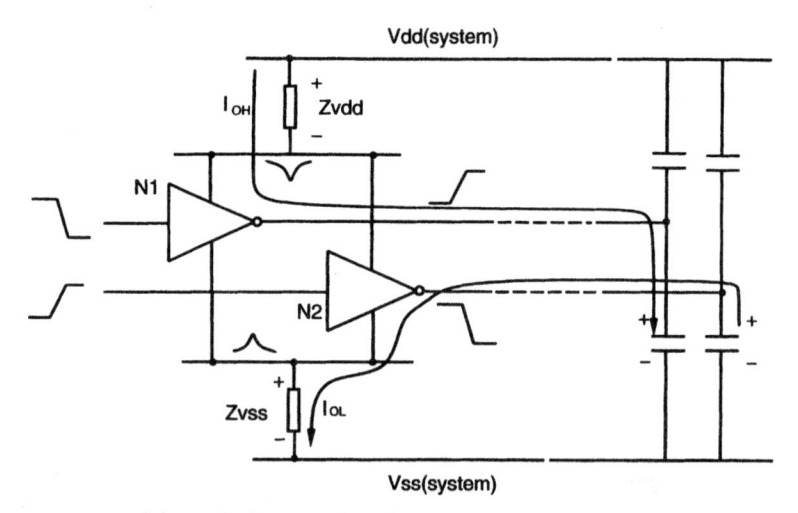

Figure 2.69 Schematic diagram of N1 drivers switching 0→1 at the output and N2 drivers switching 1←0 at the output, showing voltage disturbances on the chip ground and power busses.

drawn from or sunk into the busses, and the voltage noise generated is primarily due to $L\,dI/dt$ effects and not due to IR effects. These inductive effects are called Simultaneously Switching Output (SSO) Noise, or simply SSN.

Examination of Fig. 2.69 will aid in understanding SSO noise. Take the N1 off-chip drivers, which are switching output Low to High. Assume that Z_{vdd} is due to an inductance L_{vdd}. The drivers act as switches connecting the load capacitance to the chip V_{dd} bus. Current is sourced out of the system supply voltage V_{dd}, through L_{vdd}, to charge the load capacitance. This causes a positive voltage drop across L_{vdd} as the current through the gate rises to a maximum value, causing the chip V_{dd} bus voltage to drop below V_{dd}. As the capacitor charges, the output current decreases and the voltage across L_{vdd} becomes negative (and smaller in magnitude than the positive portion of the pulse). The chip V_{dd} bus voltage therefore rises slightly above V_{dd}. Thus the effective supply voltage $V_{dd} = V_{ss}$ is temporarily decreased due to the positive pulse. This results in reduction of noise margins and temporary slowing of the off-chip drivers due to reduction of the power supply voltage. When outputs switch High to Low (e.g., the N2 off-chip drivers in Fig. 2.61), the same mechanisms give rise to a transient increase in the V_{ss} (ground) voltage on the chip.

If the N1 drivers shown in Fig. 2.69 are quiet with output held low, and the N2 drivers are switching as shown, then the noisy chip ground voltage is connected through the N1 gates to their outputs. This is more clearly shown in Fig. 2.70 for a pair of gates, one gate quiet. In this way, SSO noise can propagate to receivers connected to quiet drivers and decrease the available noise margin or even independently cause logic errors in extreme cases.

Figure 2.71 shows a primitive driver with various parasitics connected. As V_{in} rises toward V_{dd}, the current through the N-channel

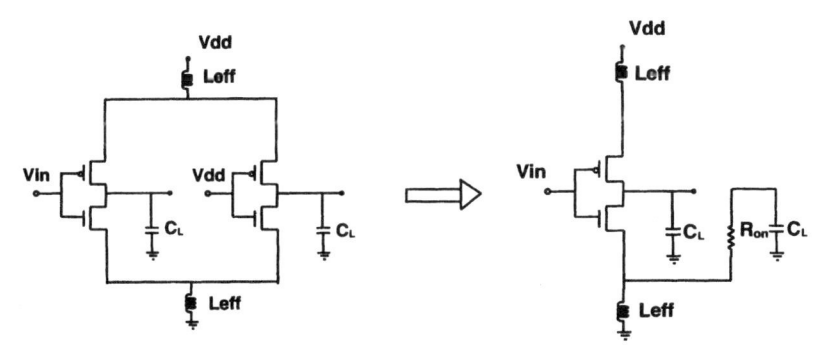

Figure 2.70 Diagram showing mechanism of coupling SSO noise to the receiver of a quiet off-chip driver.

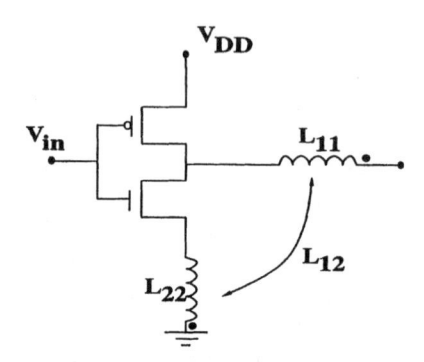

Figure 2.71 Primitive off-chip driver with various parasitics connected.

MOSFET and L_{22} rises and V_n, the noise voltage at the source of the N-channel MOSFET, is positive. This corresponds to the initially rising portion of the current waveform of Fig. 2.72. The NMOS current reaches a maximum value I_{max} and eventually begins to decrease as the load capacitance discharges and the output voltage drops to the point that the NMOS comes out of saturation. Examining Fig. 2.72 for dI/dt, it is clear that V_n first goes positive, then goes to zero before it goes negative. The most dangerous part of the waveform is the positive V_n pulse, which is due to the leading edge of the current waveform.

2.3.2 Present and future concerns

A straightforward analysis will shed some light on SSO noise phenomena and demonstrate the reasons for concern with this sort of noise. Take a single off-chip driver that sinks or sources current according to the $I^* - t$ characteristics of Fig. 2.72. The time t^* is approximately the risetime of input waveform to the final driver stage shown in Fig. 2.71. Take the case of an output 1→0 transition. The current that is discharging the load capacitance C_L is approximately I_{max}. Since $I = C\, dV/dt$ for a capacitor, the I_{max} required to discharge a C_L of 10 pf from 3.3V to 0V in 10^{-9} s is $I_{max} = 33$ mA. This can easily be generalized. For a clock frequency of f_c, the period of the clock waveform is $T = 1/f_c$. A well-shaped pulse has rise and fall time of about 10 percent of the

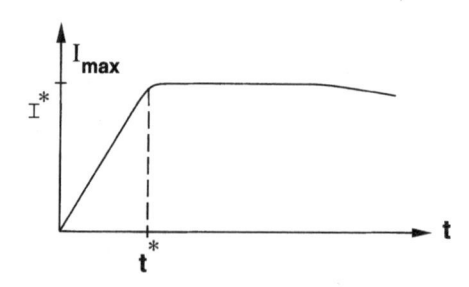

Figure 2.72 I-t waveform for discharge of load capacitors in a 1→0 output transition.

period. Thus, the rise and fall times of the output voltage are $dt = 0.1/f_c$, and

$$I_{max} = 10 f_c C_L \, dV \tag{2.90}$$

For $dV = 3.3V$ and $C_L = 10$ pf, $I_{max} = 3.3 \times 10^{-10} f_c$, which is plotted in Fig. 2.73. Clearly I_{max} becomes very large for high clock frequency.

The noise voltage generated by n outputs switching is, if the negative feedback effect [13] is neglected,

$$V_n = nL \frac{dI}{dt} \cong nL \frac{I_{max}}{t^*} \tag{2.91}$$

Generally $t^* \geq t_r$ of the final stage input waveform. Thus

$$V_n \leq n \cdot 10 f_c L I_{max} \tag{2.92}$$

or

$$V_n \leq 3.3 \times 10^{-9} \, nL f_c^2 \tag{2.93}$$

For a specified maximum V_n, Eq. 2.93 allows determination of L required, given n and f_c. Bus widths for cost-performance microprocessors are forecasted to move from 64 bits (1997) to 256 (2006), as shown in Fig. 2.74 [1]. Maximum off-chip clock frequencies will likely move from 170 MHz to 600+ MHz in that same time period (see Fig. 2.3). Then for $V_n(max) = 0.2V$, $n = 64$ and $f_c = 170$ MHz,

$$L \cong 0.07 \text{ nH}$$

and

$$nI_{max} = 2.1 A$$

while for $n = 256$ and $f_c = 600$ MHz

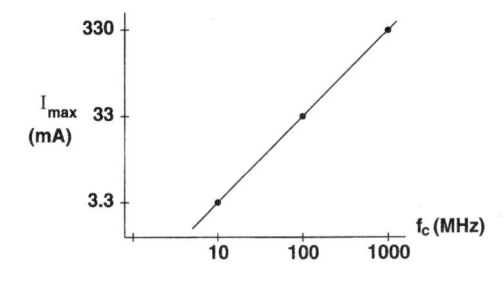

$$L \cong 0.7 \text{ pH}$$

Figure 2.73 I_{max} versus clock frequency for $V_{dd} = 3.3V$, $C_L = 10$ pf.

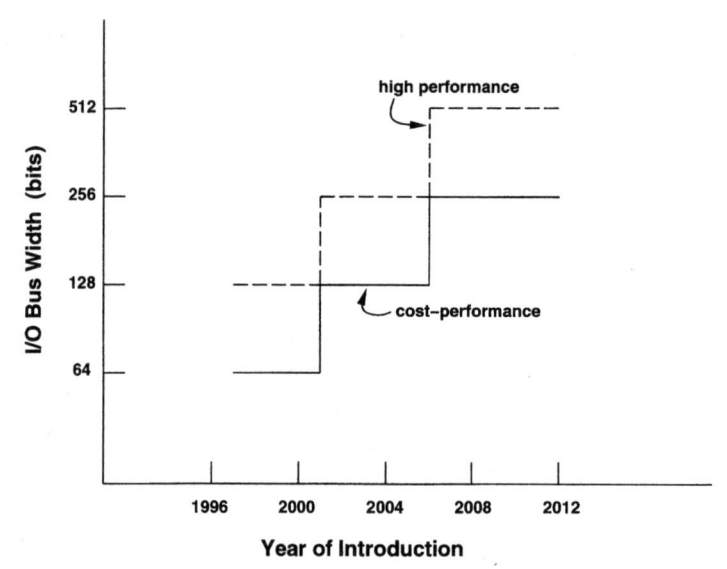

Figure 2.74 I/O bus width versus year of introduction for cost-performance and high-performance products [1].

and

$$nI_{max} = 42.2A$$

The first inductance value is achievable for the proper technology and design. The second value is most likely not achievable. Therein lies a problem, which is directly related to the very high current peak for the switching drivers.

2.3.3 SSO noise modeling

There are two possible approaches to modeling and simulating SSO noise. The first is full SPICE simulation without analytic modeling; the second is analytic modeling followed by SPICE simulation. The second approach is taken here. This is a design approach rather than a solely analytical approach.

The following is limited to SSO noise on the chip ground bus. The behavior of SSO noise on the chip V_{dd} bus is entirely analogous. Since the entire SSO noise is due to dI/dt, the time dependence of current through the N-channel MOSFET must be understood. Figure 2.75 shows an off-chip driver and the N-channel MOSFET current I_n versus time, for different values of load capacitance C_L. It is assumed that the intrinsic response time of the driver is less than the rise time of the input waveform, and that the MOSFETs behave as long-channel devices such that

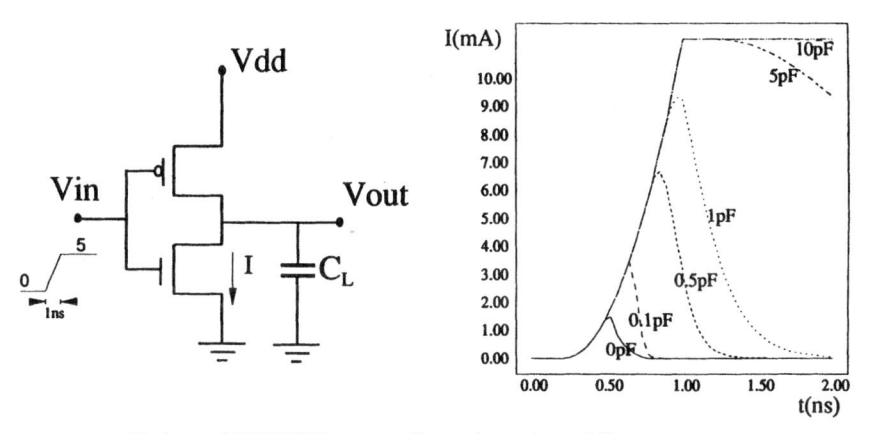

Figure 2.75 N-channel MOSFET current for various sizes of C_L.

$$I_n = \frac{K}{2}(V_{in} - V_T)^2 \quad (2.94)$$

where $K = \mu C_{ox} W/L_c$, L_c = channel length, W = channel width, μ = electron surface mobility, C_{ox} = gate oxide capacitance per unit area, and V_T = threshold voltage. From Fig. 2.75, it is clear that the maximum I_n current and the maximum dI/dt are obtained for a C_L in excess of a critical value, which is somewhat less than 5 pf for the case shown in the figure. That case corresponds to $K = 1.6 \times 10^{-3}$ A/V^2 and $V_T = 1V$. The principle behind the achievement of maximum I and dI/dt is that for C_L larger than a critical value C_{crit}, the N-channel device remains in saturation throughout the rise of V_{in}, and therefore Eq. 2.94 predicts the maximum I_n,

$$I_n(\max) = \frac{K}{2}(V_{in}(\max) - V_T)^2 \quad (2.95)$$

Since $V_{in} = V_{in}(\max)t/t_r$, Eq. 2.94 also gives the maximum dI/dt,

$$\frac{dI}{dt}(\max) = \frac{KV_{in}(\max)}{t_r}(V_{in}(\max) - V_T) \quad (2.96)$$

Figure 2.76 shows $dI/dt(\max)$ for values of C_L up to C_{crit}, demonstrating that the absolute maximum is obtained for $C_L = C_{crit}$. The C_{crit} can be calculated through [14]

$$C_{crit} = \frac{Kt_r}{6V_{dd}V_T}[V_{dd} - V_T]^3 \quad (2.97)$$

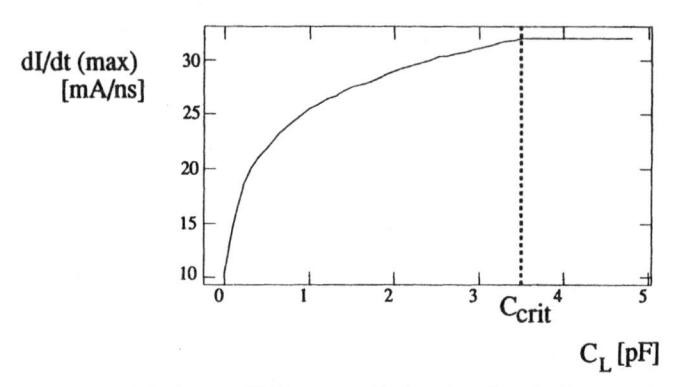

Figure 2.76 Maximum dI/dt versus C_L showing the absolute maximum occurring for $C_L \geq C_{crit}$.

For a 2.5-mA driver ($K = 1.6 \times 10^{-3}\ A/V^2$) with $t_r = 10^{-9}$ S, $V_{dd} = 5V$ and $V_T = 1V$, $C_{crit} = 3.4$ pf. For a 25-mA driver ($K = 16 \times 10^{-3}\ A/V^2$), $C_{crit} = 34$ pf. If $C_L < C_{crit}$, the peak SSO noise will be reduced.

As a first approximation to modeling SSO noise for long-channel devices the $I(t)$ characteristic is approximated by a linear rise to I_{max} in time t_r such as is shown in Fig. 2.77 for a total driver system current I_{tot}. Also shown in the figure is the resultant SSO noise, a positive square pulse of amplitude V_n^+ followed by a lower amplitude negative-going square pulse. Under this model

$$\frac{dI_{tot}}{dt} = \frac{I_{tot}(\max)}{t_r}, \quad t \leq t_r \quad (2.98)$$

where

$$I_{tot}(\max) = NI_n(\max)$$

N is the number of simultaneously switching outputs and t_r is the rise time of the input to the drivers. Then

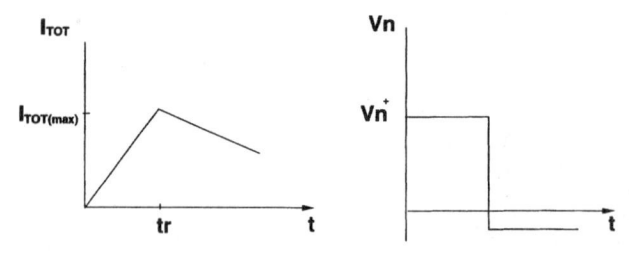

$$V_n = L\frac{I_{tot}(\max)}{t_r} \quad (2.99)$$

Figure 2.77 First approximation of $I(t)$ for SSO noise modeling.

Basic Electrical Effects in Electronic Packaging and Interconnects 93

and, combining with Eq. 2.95,

$$V_n = V_n^+ = \frac{t_r}{LNK}\left[1 - \sqrt{1 + 2V_K \frac{LNK}{t_r}}\right] \qquad (2.100)$$

where $V_K = V_{dd} - V_T$. In the limit of small N, Eq. 2.100 has the limit

$$V_n = \frac{1}{2} KV_K^2 \frac{NL}{t_r} \qquad (2.101)$$

while for large N,

$$V_n \to V_K \qquad (2.102)$$

Figure 2.78 shows a plot of $V_N(\text{max}) = V_n^+$ versus N, which shows the saturation of $V_n(N)$ at large N. This is due to the so-called negative feedback effect, and is important only when V_n approaches or exceeds approximately 10 percent of the supply voltage V_{dd}. It is very important to note that for small values of V_n^+, V_n^+ is linearly proportional to N (see Eq. 2.101).

A more rigorous analysis for long-channel devices takes into account the actual N-channel MOSFET current characteristic given in Eq. 2.95. The result of this is

$$V_n(\text{max}) = V_K^* + \frac{V_K^* t_r}{2V_{dd}NLK}\left[1 - \sqrt{1 + \frac{4V_{dd}NLK}{t_r}}\right] \qquad (2.103)$$

where $V_K^* = (V_{dd}/t_r)\, T - V_T$, and T is the time at which the N-channel MOSFET comes out of saturation [14]. For $C_L \geq C_{\text{crit}}$, $T = t_r$ and $V_K^* = V_K$. Figure 2.79 shows the $V_n(t)$ characteristic from SPICE simulation and can be compared to the shape predicted by Eq. 2.100 (see Fig. 2.77). This shape is accurately predicted by Eq. 2.103 for $t \leq t_r$, as can be seen from taking the time derivatives of Eq. 2.94 with $V_{\text{in}}(t) = V_{dd}t/t_r$. Since

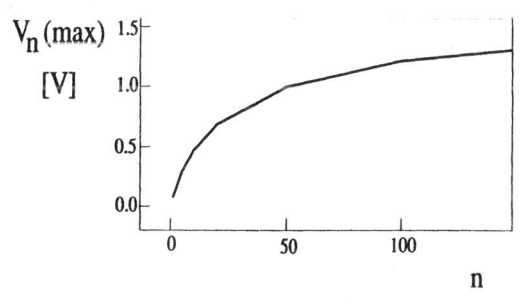

Figure 2.78 Typical $V_n(\text{max})$ versus the number of simultaneous switching drivers, showing the negative feedback effect, long channel drivers.

94 Chapter Two

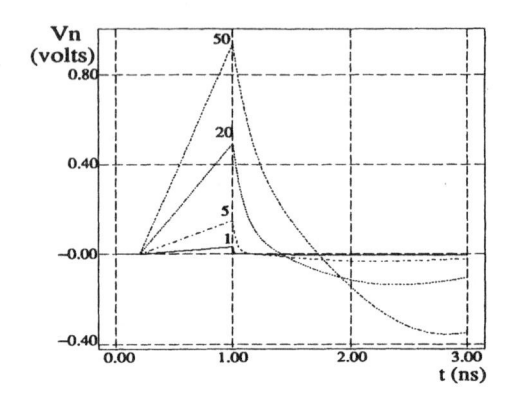

Figure 2.79 $V_n(t)$ using SPICE simulation, various numbers of long channel drivers, $C_L > C_{crit}$, $V_{dd} = 5V$, $t_r = 1$ ns, $L = 1$ nH, $K = 1.6 \times 10^{-3} A/V^2$, $V_T = 1V$.

the equation is quadratic in time, the derivative dI/dt is linear in time and gives the linear rise of $V_n(t)$ up to $V_n(\text{max})$. The negative feedback effect is evident in Fig. 2.79, since for $N = 20$ and $N = 50$ the values of $V_n(\text{max})$ are less than would be predicted by a linear extrapolation of the value for $V_n(\text{max})$ obtained for $N = 5$.

Up to this point, two analytic models have been developed which can be used either in design mode or analysis mode to apply to SSO noise problems. Much can be learned about SSO noise from the simple models, particularly after models for the package inductances are developed. However, there is a fundamental shortcoming of both models: long-channel device models were used in developing the noise models. Driver transistors are now characterized by short-channel $I_n(V_{in})$ behavior. The difference between short-channel and long-channel behavior is that for the same value of K, the short-channel device will have much less drive current $I_n(V_{in})$ for the same value of V_{in}. This is primarily due to saturation of the electron velocity in the MOSFET channel. Figure 2.80 shows two NMOS $I_n(V_{in})$ characteristics, one long-channel and one short-channel. The two devices have been matched so that the current of each is equal at $V_{gs} = 2.2V$. The required value of $K_{\text{short channel}}$ is nearly ×7 the $K_{\text{long channel}}$ in order for this match to take place.

To address the modeling of SSO noise for short-channel device drivers, a simple model taking into account only velocity saturation is used [15]. For this case the N-channel MOSFET current is

$$I_n = \frac{1}{2} KV_C(V_{GS} - V_T - V_C) \quad (2.104)$$

in the saturation region. Here $K = \mu_0 C_{ox} W/L_{\text{eff}}$ as before, where μ_0 is the low-field value of electron surface mobility. V_C is the drain voltage at which velocity saturation occurs, $V_C = v_{\text{max}} L_{\text{eff}}/\mu_s$, where μ_s is the high-field value of electron surface mobility, and L_{eff} is the effective NMOS

Basic Electrical Effects in Electronic Packaging and Interconnects 95

Long channel

$K = 21.4 \times 10^{-3}$ A/V^2

Short channel

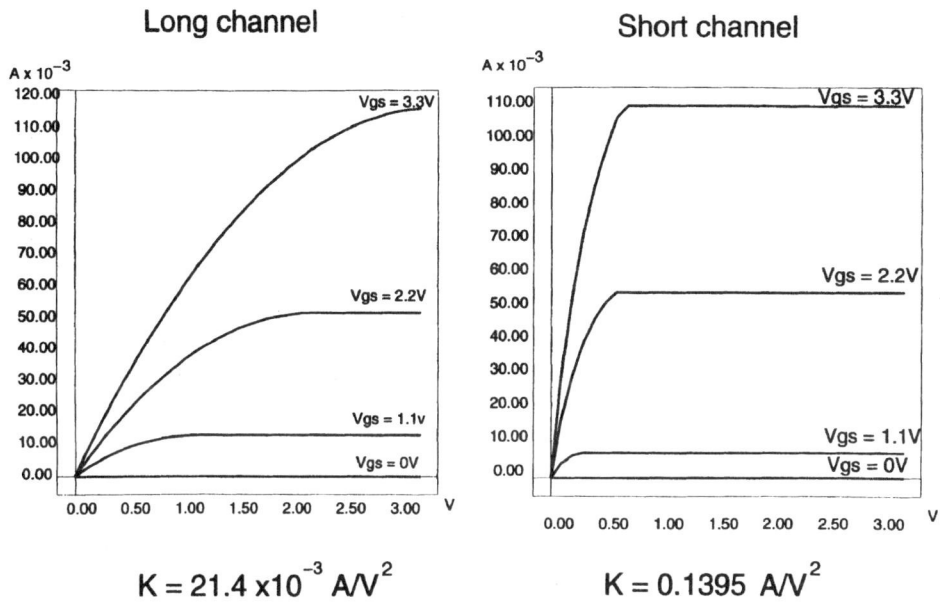

$K = 0.1395$ A/V^2

Figure 2.80 $I_d - V_d$ characteristics of long-channel and short-channel NMOS Transistors, $V_{dd} = 3.3V$. The saturation currents are approximately matched at $V_{GS} = 2.2V$.

channel length. In this work, the electron saturation velocity is taken to be $v_{max} = 7 \times 10^6$ cm/s. Analysis shows [15] that

$$V_n(\max) = V_{dd} \frac{t}{t_r}\left(1 - e^{-(t_r - t_t)/t}\right) \quad (2.106)$$

where $t_t = t_r V_T/V_{dd}$ and

$$\tau = \tfrac{1}{2} NLKV_C \quad (2.107)$$

Remember the L in Eq. 2.107 is the system inductance. The bracketed term in Eq. 2.106 contains the negative feedback effect; for N small the exponential is small and

$$V_n(\max) \simeq V_{dd} \frac{\tau}{t_r}, \quad N \text{ small} \quad (2.108)$$

Thus for small N, $V_n(\max) \propto N$, while for large N, saturation of $V_n(N)$ sets in just as for SSO noise for long-channel devices.

Example Calculate $V_n(\max)$ using the best long-channel model, for $K = 2.14 \times 10^{-3}$ A/V^2, $V_{dd} = 3.3V$, $V_T = 0.7V$, $t_r = 1$ ns, $L = 1$ nH, using $N = 1$.

$$V_n(\max) = V_K\left[1 + \frac{t_r}{2NKLV_{dd}}\left(1 - \sqrt{1 + 4V_{dd}\frac{NLK}{t_r}}\right)\right]$$

$$V_n(\max) = 0.161V$$

Calculate $V_n(\text{max})$ for $N = 1$ using the short-channel model and using the preceding parameters; in addition, use $V_C = 0.76V$.

$$V_n(\text{max}) = V_{dd}\frac{\tau}{t_r}\left[1 - e^{(t_r - t_0)/\tau}\right]$$

$$V_n(\text{max}) = 0.0268V$$

Calculate $V_n(\text{max})$ for $N = 1$ using the short-channel model and using the preceding parameters, except use $K = 139.5 \times 10^{-3} A/V^2$ (the value required to roughly match I_d at $V_{GS} = 2.2V$ for the two MOSFET models; see Fig. 2.80).

$$V_n(\text{max}) = 0.175V$$

From the preceding example, it is clear that for short-channel MOSFETs of similar drive capability to long-channel MOSFETs, the predicted SSO noise is similar. This is as expected, since "similar drive capability" requires the ability to sink or source similar amounts of current and implies similar dI/dt and thus $V_n(\text{max})$ values.

For a comparison of the predicted $V_n(\text{max})$ values versus results from SPICE simulation, Figs. 2.81 and 2.82 should be consulted. In Fig. 2.81, SPICE output shows $V_n(t)$, whereas the horizontal dotted lines correspond to predicted long-channel $V_n(\text{max})$ values from Eq. 2.103. Different values of inductance L are associated with the number of ground pins in this figure. It seems clear that the analysis model does show excellent agreement with the SPICE simulation results. Figure 2.82 shows similar results using a short-channel MOSFET model. Good agreement between the analytic model and SPICE simulation results are also shown in this figure.

It is plain, after the foregoing discussion, that the partial inductance of the package is required for calculation of SSO noise. Figure 2.83 shows, for one driver, the partial inductance of the signal trace and the

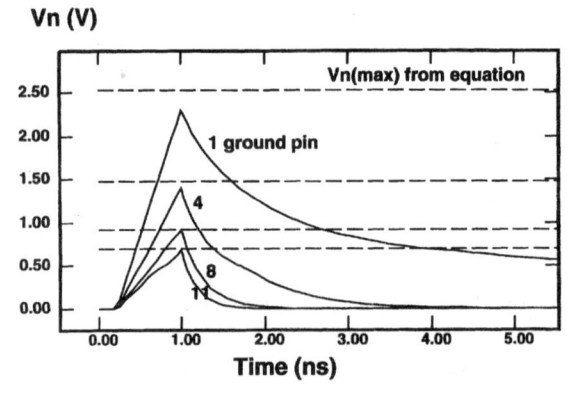

Figure 2.81 SPICE simulation results and predicted V_n (max) values (dotted lines) from Eq. 2.103, long-channel model.

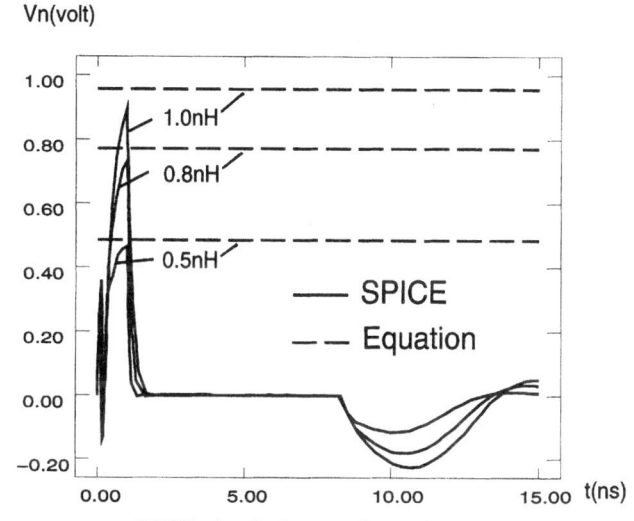

Basic Electrical Effects in Electronic Packaging and Interconnects   97

Figure 2.82 SPICE simulation results and predicted V_n (max) values (dotted lines) from Eq. 2.106, short-channel model.

ground returns, and the partial mutual between the two. The partial voltage V_n is given by

$$V_n = (L_{p22} - L_{p12}) \frac{dI}{dt} \qquad (2.109)$$

where L_{p22} is the partial inductance of the ground plane and the current I is the signal trace current that is sunk into the ground return [16].

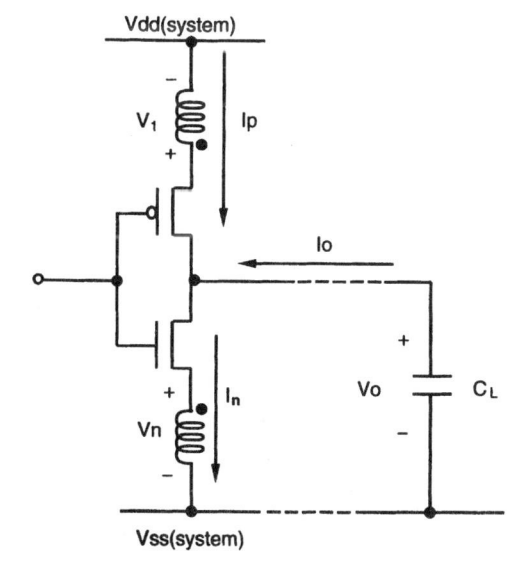

Figure 2.83 Driver circuit for the calculation of the mutual effect between L_{vss} and L_{vdd} on V_n.

Other partial mutual inductances exist, such as the mutual between L_{vss} and L_{vdd} in Fig. 2.82. In this case, taking L_m into account,

$$V_n = \left(L_{p22} - L_{p12}\frac{dI}{dt}\right)\frac{dI_0}{dt} - L_m\frac{dI_p}{dt}$$

where L_m is the mutual between L_{vss} and L_{vdd}. However, except for a possible very narrow pulse, the P-channel current $I_p \cong 0$ and the effect of L_m on V_n is small.

The approach to extracting inductances for SSO noise is simpler for lead-frame packages than for packages with ground planes. Figure 2.84 shows an incomplete (for clarity) schematic diagram of a leadframe package with m drivers and two power and two ground connections. Some inductances have been omitted for simplicity. The partial self and mutual inductances can be calculated using equations for partial inductances of rectangular structures given in Sec. 2.1 [17]. An idea of the importance of mutual interactions between signal and ground/power traces can be gotten by examining Figs. 2.85 and 2.86. The figures show a 1-cm chip bonded out in a 208-pin PQFP, with 32 off-chip drivers and two power/ground pin pairs. In Fig. 2.85 the power/ground pin pairs are optimally located to minimize the package partial ground/power inductance. In Fig. 2.86 the pairs are located such that mutual interactions with signal traces will be small, and effective package inductance will not be minimized. The result of this is shown in Fig. 2.87 for various values of N. It is plain to see that maximizing the mutual inductance interactions between ground/power traces and signal traces results in lower package L_{eff} (effective inductance) and lower V_n(max). Please note that L_{eff} refers to package inductance, and not effective channel length of a short-channel MOSFET.

To extract the package effective inductance from a leadframe package is possible. The algorithm is given in [17]. The inductance can then be used with Eq. 2.106, for V_n(max). Alternatively, the equivalent circuit as shown in Fig. 2.84 may be used for a straightforward SPICE simulation.

Modeling and simulating SSO noise in packages with ground planes is conceptually simple and similar to that of leadframe packages. Figure 2.88a shows a signal trace with a ground plane underneath and a ground contact near the end of the signal trace. The cross section of the signal trace and the ground plane are shown discretized into rectangular elements. The discretization width must be adjusted considering skin depths and metal thicknesses. The discretization limits for the ground plane can be set taking into account the extent of return current spreading under the signal trace. Figure 2.88b shows current density as a function of position for a specific microstrip structure. The

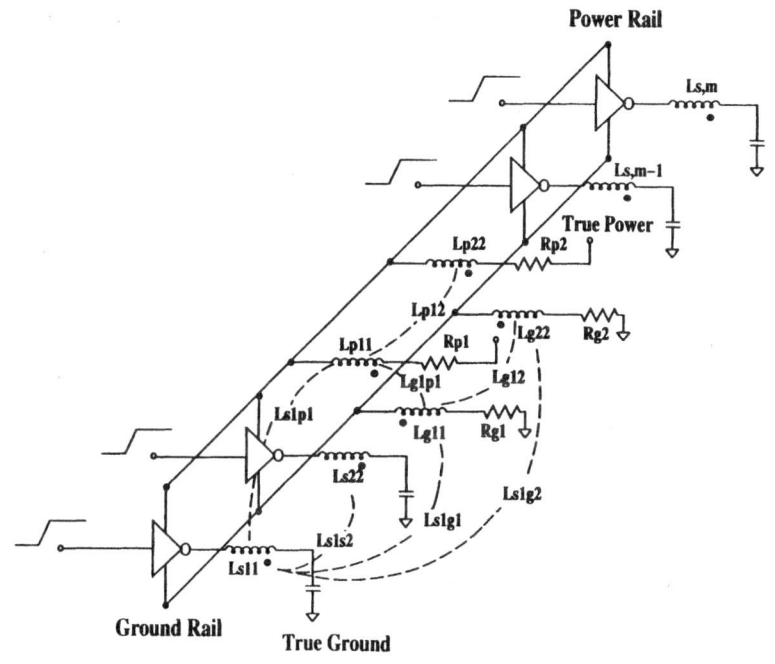

Basic Electrical Effects in Electronic Packaging and Interconnects   99

Figure 2.84 Schematic diagram of m drivers in a leadframe package with two power and two ground connections. Some inductances omitted for clarity.

sharp peaking of the current at high frequencies (frequencies for which dI/dt is large) underneath the signal trace indicates that the ground plane can be terminated (in software) and accurate modeling results still obtained. In this particular case, for example, termination on the left at $x = 0.20$ µm and on the right at $x = 0.80$ µm will introduce very little inaccuracy.

It may happen that for a particular packaging structure the contacts from the chip to the ground plane are not very close to the near end of the signal trace and/or the ground contacts (to package pins) at the far end of the signal trace are not very near the far end. In that case, so-called excess inductances must be added in series with the normal package inductance components. Figure 2.89 [16] shows a prototype package with four pins at corners of the package (closed circles), four chip ground bus contacts near the corners of the chip (open circles), and a signal trace between A-A'. Since there is no package pin near to the point A', current must flow laterally in the ground plane (for example, between A' and B') in order to leave the package or enter the package, depending of course on current direction. Figure 2.90 [16] shows an expanded view of the excess inductance paths through which the ground return current flows. Since the excess inductance has minimal

BODY SIZE: 28 X 28 mm LEAD COUNT: 208

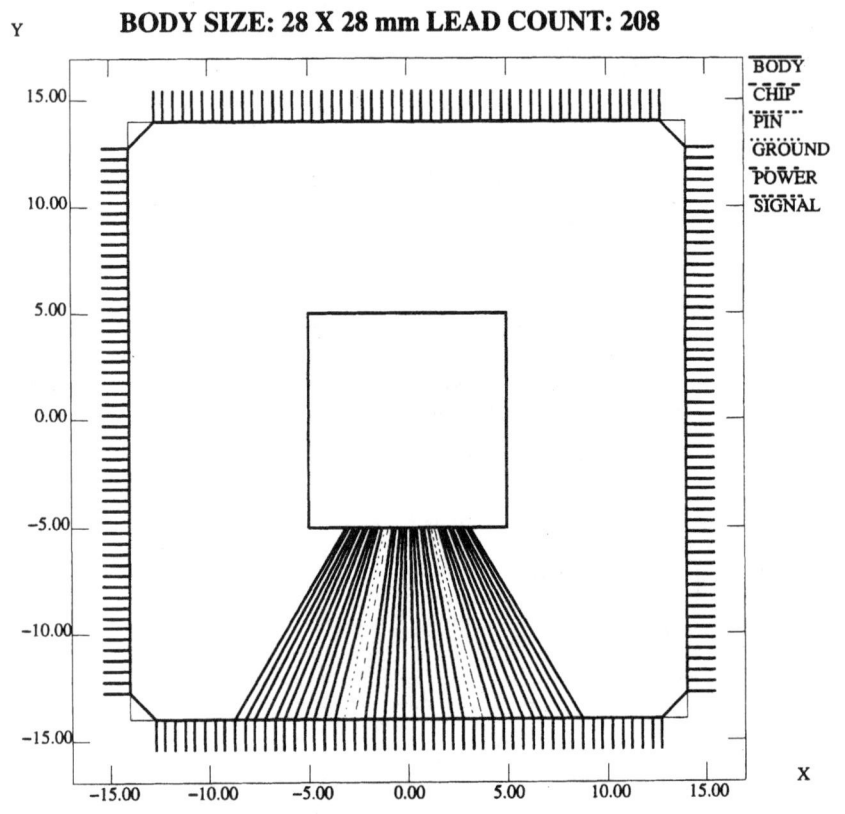

Figure 2.85 Diagram of PQFP with 32 drivers and two power/ground pin pairs in optimal position for SSO noise, Case A.

mutual magnetic interactions with the signal current due to distance and current orientation, even a short excess inductance path can add appreciable inductance to the overall effective inductance.

An exercise which clearly shows the effect of location of ground contacts relative to signal trace position is shown in Fig. 2.91. This case concerns a hypothetical package with up to four drivers, having fixed locations of contacts to the ground plane and package pin locations. The figure shows combinations which were modeled and then simulated using SPICE. Table 2.3 shows resulting SSO noise values. Absolute values here are not important, but relative values are important. Observe that if coupling between signal traces and ground plane are ignored, large values of $V_n(\text{max})$ are obtained. Also observe that for unfavorable placement of signal traces, in locations where the excess inductance is high, relatively large V_n is obtained. For optimal placement of signal traces, minimum values of V_n are obtained as can be

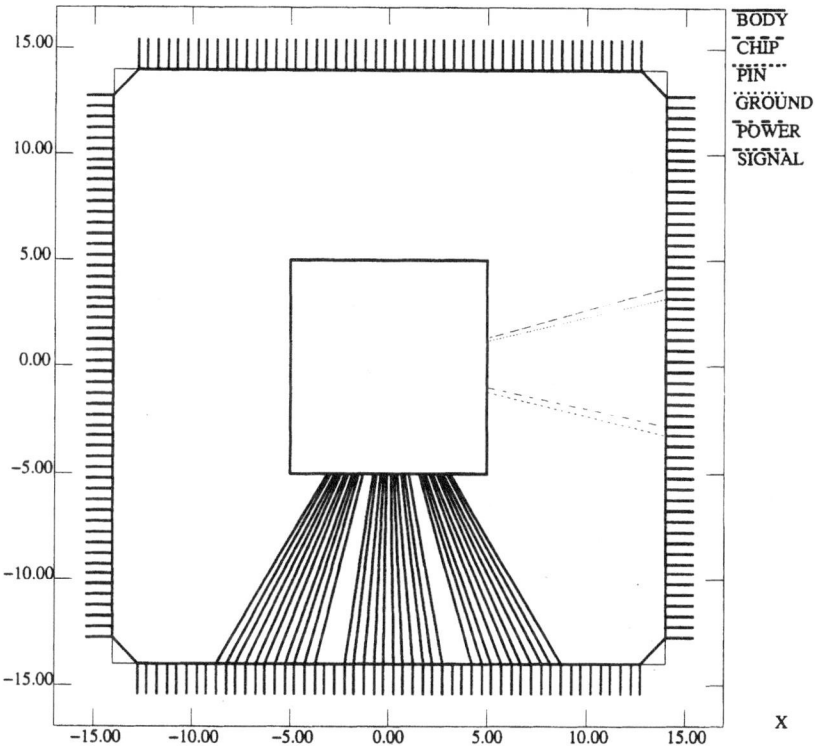

Basic Electrical Effects in Electronic Packaging and Interconnects 101

Figure 2.86 Diagram of PQFP with 32 drivers and two power/ground pin pairs in non-preferred position for SSO noise, Case B.

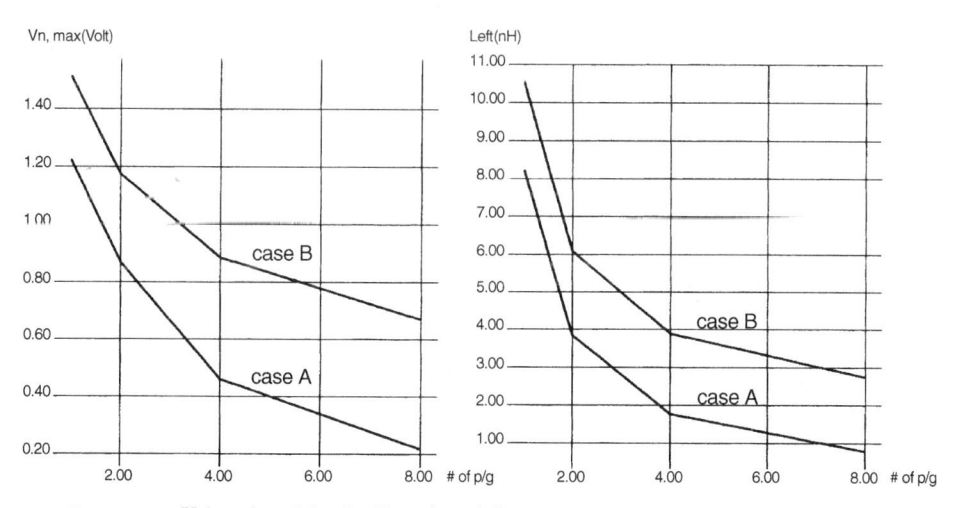

Figure 2.87 V_n(max) and L_{eff} for Case A and Case B (see Figs. 2.85 and 2.86) versus N, 0.5 μm driver, $W/L_{\text{eff}} = 320$, $V_T = 0.7V$, $\mu_n = 800$ $V^{-1}s^{-1}\sqrt{1}$ s^{-1}, $v_{\max} = 7 \times 10^6$ cm/s, $t_r = 1$ ns.

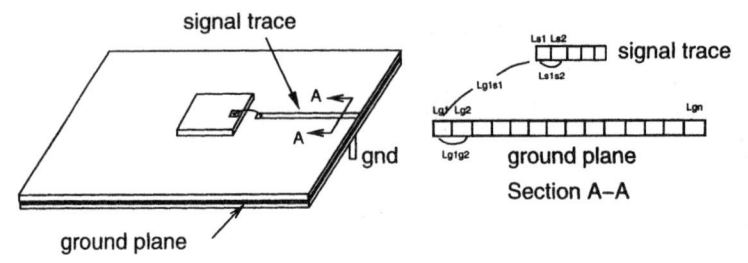

Figure 2.88a Discretization of ground plane and signal trace.

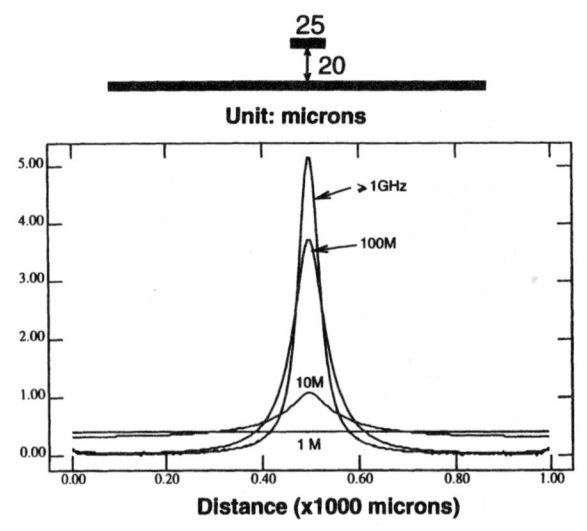

Figure 2.88b Current density spatial distribution as a function of frequency.

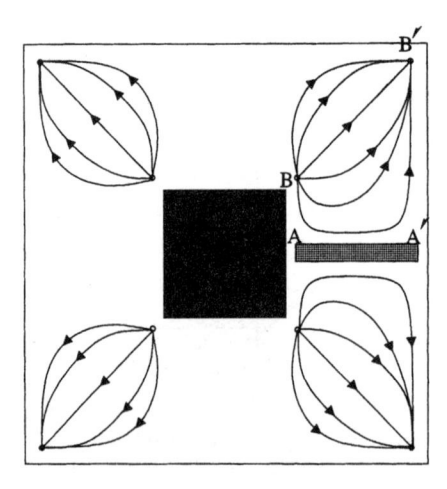

Figure 2.89 Sketch of the current flow distribution on a ground plane for a particular combination of contacts.

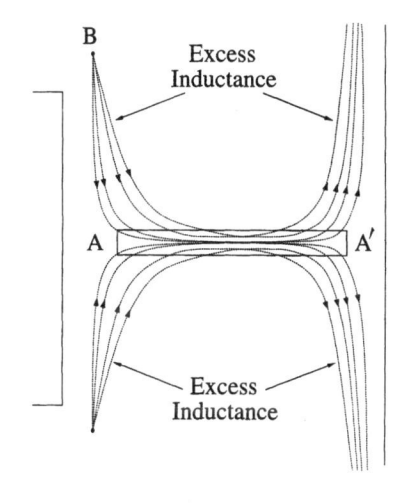

Figure 2.90 Expanded view of the current paths below and near the signal conductor of Fig. 2.89, showing excess inductance paths.

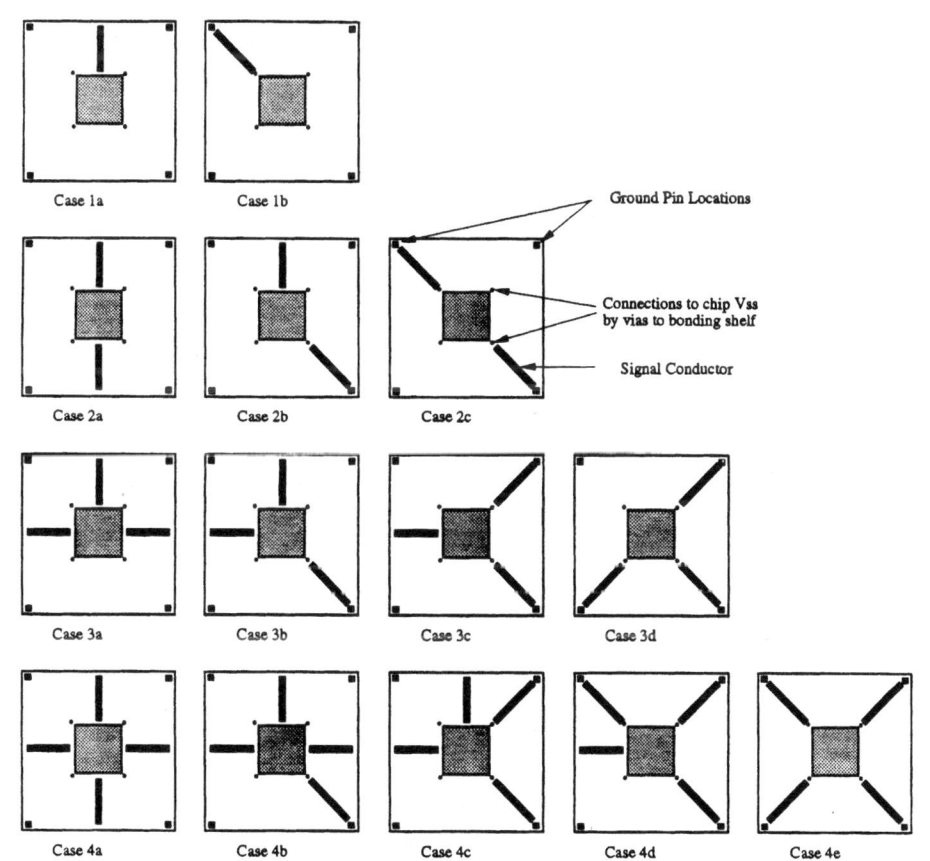

Figure 2.91 Sketches representing various positions of signal traces relative to connections to chip ground and ground pin locations.

seen from the last values in each column in Table 3.1 [16]. In fact, from Table 3.1, four drivers optimally located give less noise than two drivers improperly located (cases 4e versus 2a).

Although this section has concentrated on SSO ground noise, it must be evident that power bus noise is approached in the same manner, and that by inserting "P-channel MOSFET" for N-channel MOSFET and performing a few other housekeeping tasks, all conclusions and equations have an obvious analog for power bus noise.

2.3.4 Effects of decoupling capacitors

In order to control the decrease of the effective power supply voltage at the chip, decoupling capacitors are sometimes used. The benefit of using decoupling capacitors is highly dependent on the physical location and electrical connections, and on the quality of the capacitor itself. Figure 2.92 shows a single off-chip driver with a decoupling capacitor C_D located and connected as if it were in or was a part of the package. Other possible locations for C_D are: on the chip itself (effective connections are at the sources of the two MOSFETs), and outside the package altogether (effective connections are at V_{dd} and ground). In examining Fig. 2.92, it should become clear that voltage disturbances on the chip power and ground busses do not go away, although they may decrease in amplitude. For example, during most of the output 0→1 switching cycle for the driver, the P-channel MOSFET is off and the current through the N-channel device is that current which discharges C_L. Assume that C_D is a very large value capacitor; then for all frequencies above zero it will be practically a short circuit. The time-varying current I_N thus sees two paths to ground, one through the lower L_{bond} and $L_{pin} + L_{plane}$, and one through the upper $L_{pin} + L_{plane}$ and through the V_{dd} supply to ground. The equivalent circuit thus becomes that of Fig. 2.93. It can be seen clearly that the SSO noise on the chip ground bus is not zero, since there are voltage drops across L_{bond} and

TABLE 2.3 **Switching Noise Values for Different Number and Positions of Conductors (Cases Refer to Fig. 2.91)**

1 Driver		2 Drivers		3 Drivers		4 Drivers	
Case	Noise (mV)	Case	Noise (mV)	Case	Noise (mV)	Case	Noise (mV)
No coupling	290	No coupling	526	No coupling	741	No coupling	896
1a	250	2a	401	3a	526	4a	630
1b	136	2b	318	3b	470	4b	585
		2c	225	3c	388	4c	526
				3d	300	4d	463
						4e	365

$L_{\text{pin}} + L_{\text{plane}}$. The SSO ground noise amplitude will be reduced somewhat, due to the two $L_{\text{pin}} + L_{\text{plane}}$ being in parallel. However, the mutual inductance between the two is positive, which acts to reduce the benefit of paralleling. Also, the L_{bond} is normally the same size as or larger than L_{pin} and L_{plane}. The conclusion is that SSO ground noise will be reduced, but probably not significantly ($\approx$10 to 20 percent, perhaps). The same conclusions apply to V_{dd} noise, also.

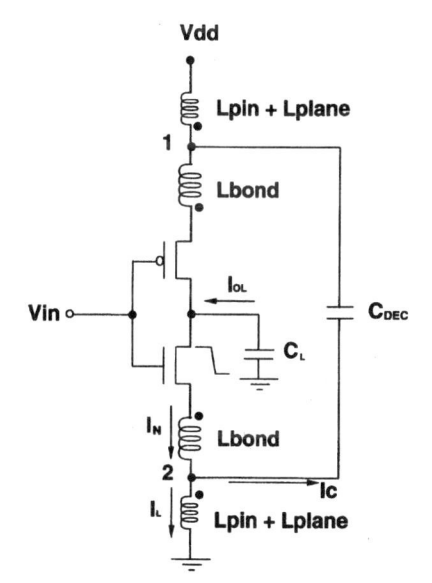

Figure 2.92 Schematic diagram of driver with decoupling capacitor connected.

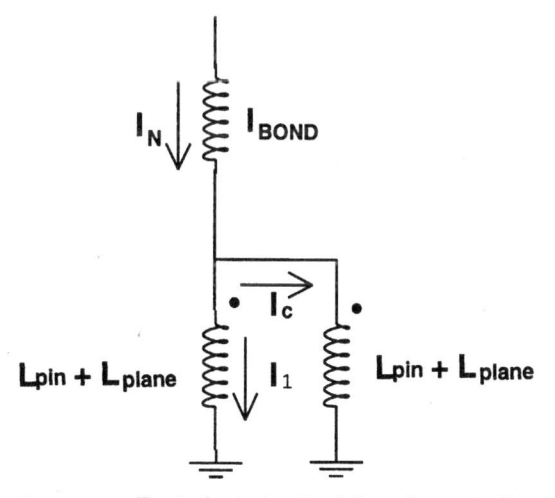

Figure 2.93 Equivalent circuit of the schematic diagram of Fig. 2.92, at high frequencies.

From the previous discussion, it is clear that decoupling capacitors are not the solution to SSO noise for off-chip drivers, since the noise waveform still exists at the sources of the N-channel and P-channel MOSFETs. It may be reduced somewhat due to C_D, as discussed earlier. However, decoupling capacitors can be used to prevent power supply collapse. Here *power supply collapse* means a temporary decrease in the power minus ground voltage supplied to the chip busses. This collapse occurs when noise appears on either or both of the N-channel and P-channel sources.

Decoupling for core (internal) gates is shown in Fig. 2.94. In this figure, C_2 is the on-chip capacitance charged or discharged in one clock cycle and C is the on-chip capacitance of quiet gates. The decoupling capacitor C_D is on-chip. During switching, it is assumed that V_{dd} and V_{ss} are effectively disconnected from the circuit due to $L\,dI/dt$ voltage drops across package inductances. When switching occurs, C_D and C_1 are already precharged to V_{dd}. They act as the chip power source for a few nanoseconds. They share charge with C_2 and in doing so allow C_2 to charge to a voltage V. Since charge must be conserved (equal before and after switching)

$$(C_1 + C_D)V_{dd} = (C_1 + C_2 + V_{dd})V \qquad (2.110)$$

This gives

$$V_{dd} - V = \frac{C_2}{C_1 + C_2 + C_D} V_{dd} \qquad (2.111)$$

which shows the power supply voltage drop $V_{dd} - V$.

Example Calculate the on-chip decoupling capacitance C_D required if $\Delta V = 0.25V$ is permissible for the power supply, $V_{dd} = 5V$, and $C_1 = C_2 = 15 \times 10^{-9}$ F.

From Eq. 2.111

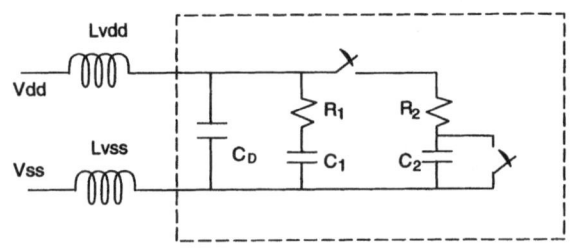

Figure 2.94 Equivalent circuit for on-chip decoupling capacitor and chip core logic.

$$C_D = C_2 \frac{V_{dd}}{\Delta V} - (C_1 + C_2)$$

$$C_D = 270 \times 10^{-9} \text{ F}$$

The effect of the partial mutual inductance between L_{VSS} and L_{VDD} is to partially counteract power supply collapse. Take the driver shown in Fig. 2.95a. The dots on the inductors define the sign of L_{12}. For current entering the lower inductor at the dotted end, a positive voltage is induced at the dotted end of the top inductor with respect to the other end of the top conductor. The upper inductor thus behaves as a voltage source in series with the V_{dd} supply, during switching. For the case where L_{12} is zero (Fig. 2.95a), very little voltage disturbance is seen on the V_{dd} bus, even though the ground noise is substantial. For the case where a healthy L_{12} exists (Fig. 2.95b), the voltage induced

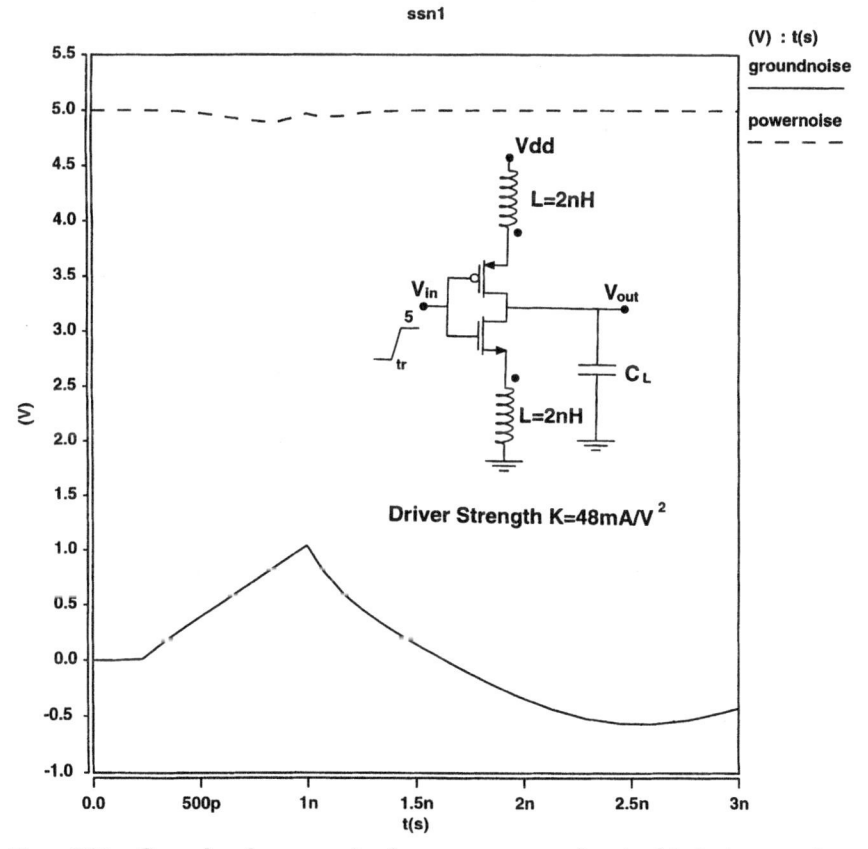

Figure 2.95a Ground and power noise for no power-ground mutual inductance and no decoupling capacitor.

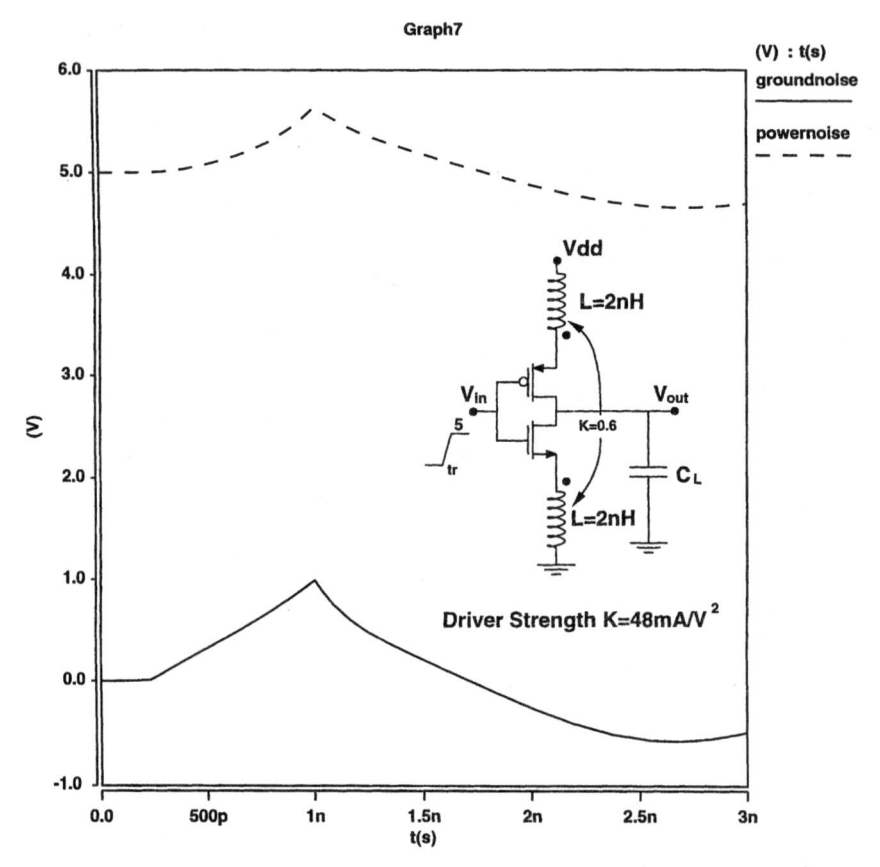

Figure 2.95b Ground and power noise with power-ground mutual inductance, no decoupling capacitor.

across L_{VDD} causes the chip V_{dd} bus to "bounce up" similarly to the ground bus, and less net power supply collapse is seen. Finally, for the case where an on-chip decoupling capacitor and a good-sized L_{12} exist (Fig. 2.95c), power supply collapse may be avoided entirely. Note that for the case of a package-level C_D, which is necessarily connected outside the bond wires (see Fig. 2.92), more supply collapse would occur due to inductive voltages across the bond wire inductances. Finally, for board-level decoupling, the C_D would be connected outside the package entirely. This would do no good in decoupling the chip from the package-level inductive voltage drops, and is of questionable usefulness as a decoupling capacitor although it could have other unrelated circuit benefits.

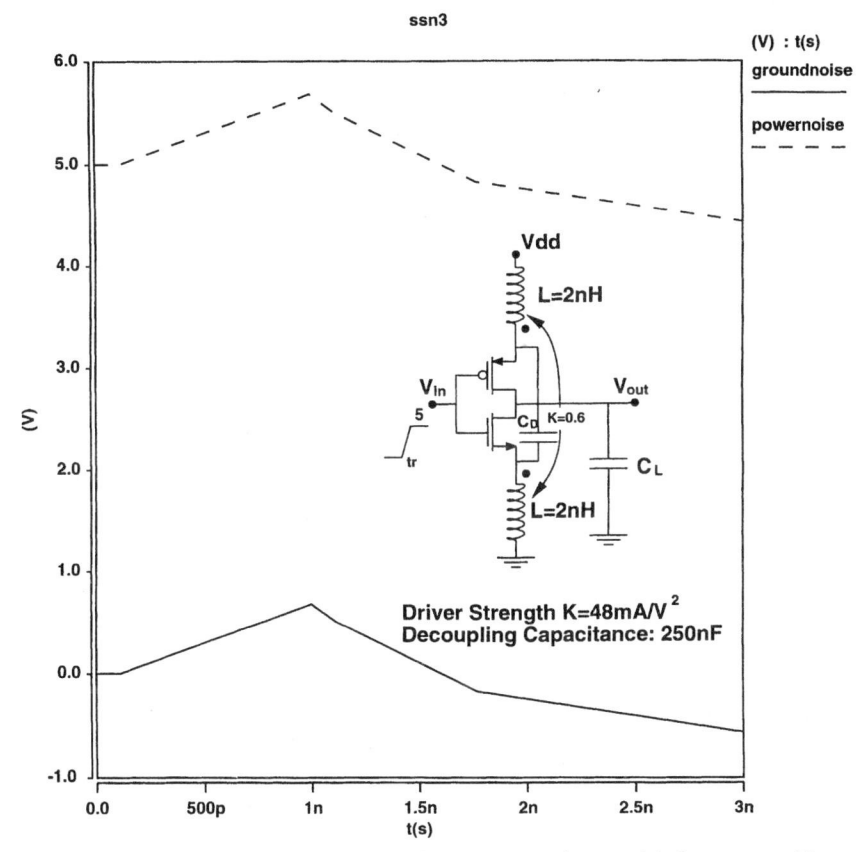

Basic Electrical Effects in Electronic Packaging and Interconnects 109

Figure 2.95c Ground and power noise with power-ground mutual inductance and large value C_D.

2.4 References

1. *The National Technology Roadmap for Semiconductors*, San Jose, Calif.: Semiconductor Industry Association, 1997.
2. Siu, B. Intel Corporation, personal communication.
3. Johnson, H. W., and M. Graham. *High Speed Digital Design*. Englewood Cliffs, N.J.: PTR Prentice Hall, 1993.
4. Senthinathan, R., and J. L. Prince. *Simultaneous Switching Noise of CMOS Devices and Systems*. Boston: Kluwer Academic Publishers, 1994.
5. Yee, K. S. "Numerical Solution of Initial Boundary Value Problems Involving Maxwell's Equations," *IEEE Trans. Antennas Propag., AP-14,* May 1966, pp. 302–307.
6. Grover, F. W. *Inductance Calculations: Working Formulas and Tables*. New York: Dover, 1973.
7. Bakoglu, H. B. *Circuits, Interconnections and Packaging for VLSI*. Reading, Mass.: Addison-Wesley Publishing Co., 1990.
8. Lamson, M. Texas Instruments, Inc., personal communication.

9. Edwards, T. C. *Foundations for Microstrip Circuit Design.* New York: John Wiley and Sons, 1981.
10. Seraphim, D. P., R. Lasky, and C.-Y. Li. *Principles of Electronic Packaging.* New York: McGraw-Hill, 1989.
11. Ventakaraman, S. "Development of Design Rules in a Simulation Environment for Packaging Applications," M.S. Thesis, The University of Arizona, 1994.
12. Ho, C. W., D. A. Chance, C. H. Bajorek, and R. E. Acosta. "The Thin Film Module as a High-Performance Semiconductor Package," *IBM J. Res. Dev., 26,* May 1982, pp. 286–296.
13. Senthinathan, R., and J. L. Prince. "Simultaneous Switching Ground Noise for Packages CMOS Devices," *IEEE J. Solid-State Circuits, 11,* November 1991, pp. 1724–1728.
14. Vaidyanath, A., B. Thoroddsen, and J. L. Prince. "Effect of CMOS Driver Loading Conditions on Simultaneous Switching Noise," *IEEE Trans. Components, Packaging and Manuf. Technology—B: Advanced Packaging, 17,* November 1994, pp. 480–485.
15. Yang, Y., and J. R. Brews. "Design for Velocity-Saturated, Short-Channel CMOS Drivers with Simultaneous Switching Noise and Switching Time Consideration," *J. Solid-State Circuits, 31,* Sept. 1996, pp. 1357–1360.
16. Thoroddsen, B. "Modeling of Simultaneous Switching Noise for Packages with Power Distribution Planes," M.S. Thesis, The University of Arizona, 1994.
17. Huang, C., Y. Yang, and J. L. Prince. "A Simultaneous Switching Noise Design Algorithm for Leadframe Packages With or Without Ground Plane," *IEEE Trans. Components, Packaging and Manuf. Technology—B: Advanced Packaging, 19,* February 1996, pp. 15–22.

Chapter 3

Thermal Design, Analysis, and Measurement of Electronic Packaging

3.1 Introduction

The performance of the electronic system deteriorates precipitously when the temperature of the electronic devices trips beyond a certain bound. The temperature also determines the service life of electronic equipment. Excessively high temperature degrades the chemical and structural integrity of various materials used in the equipment. Large fluctuations of temperature as well as large spatial variations of temperature in the equipment become responsible for malfunction and eventual breakdown of the equipment. The purpose of thermal design is to create and maintain throughout the equipment a temperature distribution having limited variations around a moderate level. This requires strategic location of heat-generating components and the measures to dispose the heat to the exterior of the equipment.

The task for the packaging designer is becoming increasingly challenging, due on one hand to the rapid increase of heat dissipation from logic modules, and on the other to the shrinking system's physical size. The margin for redundancy in cooling design is shrinking. Miscalculation may lead to increasingly grave results. Besides, the time to be spent on design work is shrinking due to accelerating product development cycles. This is obviously a new environment for the designer, unseen in the 1970s and 1980s when thermal problems were largely associated with big systems.

Heat transfer in electronic equipment is a complex process. Writing about it in a limited space requires a certain standpoint. This chapter is an attempt to convey rudimentary concepts about thermal management. Detailed derivations of the formulas and discussions about the empirical correlations and the data are left to the heat transfer textbooks and the original literature as much as possible. Hopefully, this way of writing may meet the need to learn the concept of thermal management in a reasonably short period of time.

The complexity of heat transfer in electronic equipment is attributed to the following facts.

- *Multiple heat sources and heat flow paths.* Heat generated by chips and other components ultimately finds its way to the atmospheric air. Between the heat sources and the atmosphere there exist multiple heat flow paths, and those paths merge and branch at a number of locations in the components and the equipment housing.

- *Involvement of wide spectra of length and time scales.* The electronic equipment is a structure built with components of different materials having different sizes; there are chips, solder bumps, wires, pins, modules, wiring boards, coolant paths, and other structures. Heat diffuses through such composite structures containing a wide spectrum of length scales. Also, those components have different heat capacitance, so that their temperatures change with different time scales.

The complex nature of heat transfer requires an art of modeling in the thermal design of electronic equipment. The modeling is an art where details are even smeared and replaced by equivalent entities or properties. In an extreme example, the entire electronic equipment is viewed as a solid lump having a certain heat capacitance. With such a model, we find from a very simple equation the thermal transient of the equipment, for instance, after power-on. In another example, we narrow our vision to a chip and replace the details beyond the chip by certain "boundary conditions." As such, the modeling is coupled with the range of our vision. The complexity of heat transfer problems dictates that we start with a system-level view, then narrow our vision in stepwise fashion, and finally find a "junction temperature," which belongs to a very small zone on the chip.

This chapter is written with a conscious effort to shed light on the modeling. In Sec. 3.2, we trace heat flows from the transistor junctions to the atmospheric air. Along the way, the principles of heat transfer are explained. Section 3.3 describes thermal analysis, which has to proceed following a way back from the system-level view to smaller zones. Section 3.4 is a guide to the hardware components that are needed to implement thermal designs. Section 3.5 describes the measurements of temperature and flow fields. Section 3.6 summarizes the contents of the chapter.

The literature cited are limited to those papers and books which were used in writing the text and in making figures and tables [1–21]. The Reference section also includes additional information: the annual volume series dedicated to thermal management issues [22], the review articles published by the author of this chapter [23–26], the archival textbooks [27–31], the journals where thermal management papers can be found [32–36], and the conferences devoted to thermal management issues [37–39].

3.2 Heat Diffusion from Junctions to Systems

3.2.1 Definition of thermal resistance

As a preliminary prior to our tracking of heat flows in electronic equipment, we define the thermal resistance and see its basic forms. Throughout Secs. 3.2 and 3.3 the explanations will evolve on the concept of thermal resistance.

Where a temperature difference, $(T_1 - T_2)$ [K], drives a heat flow, Q [W], we write

$$T_1 - T_2 = RQ \qquad (3.1)$$

The R is the thermal resistance in [K/W]. The analogy to Ohm's law for electric circuit is implied (Fig. 3.1a), although the real analogy between electric current and heat flow exists only in limited circumstances.

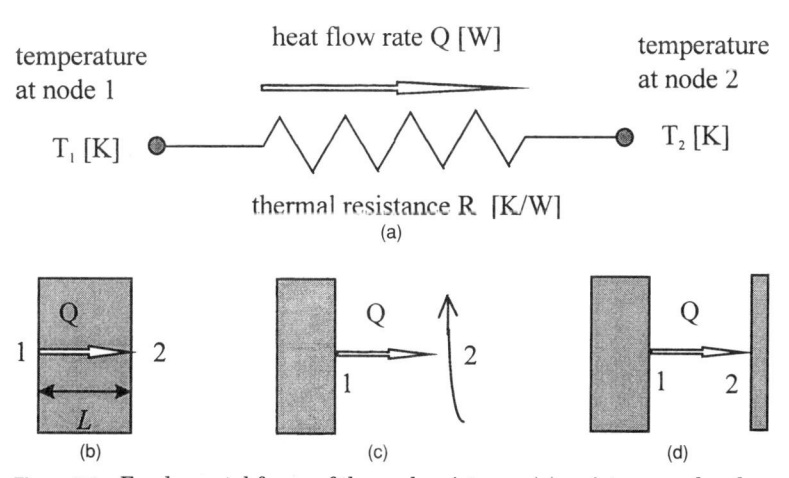

Figure 3.1 Fundamental forms of thermal resistance; (a) resistance and nodes; (b) heat conduction through medium; (c) heat flow from surface to fluid; (d) radiative heat flow from surface to surface.

What the nodal points in the resistance network (1 and 2 in Fig. 3.1) represent depends on our viewpoint. Node 1 can be a representation of the entire electronic equipment and node 2 is the atmospheric air. In a narrower scope, node 1 may represent a module and node 2 is in the coolant over the module. Narrowing further the range of our view, we reach a certain stage where the nodes represent elements of basic geometry. The three sketches in Fig. 3.1 illustrate such basic configurations and the modes of heat transfer associated with them. The thermal resistance defined on basic configurations serves as building blocks for a more complex model.

Figure 3.1*b* shows heat conduction flow Q [W] through a medium of thickness L [m]; node 1 represents one side of the medium and node 2 is on the other side. The thermal resistance is given by

$$R = \frac{L}{\lambda A} \tag{3.2}$$

where A [m²] is the cross-sectional area normal to the heat flow, and λ [W/m K] is the thermal conductivity of the medium.

Figure 3.1*c* shows heat transfer from the surface (node 1) having the area A [m²] to the mass of coolant (node 2). The thermal resistance is given by

$$R = \frac{1}{hA} \tag{3.3}$$

where h [W/m² K] is the heat transfer coefficient.

Heat transfer by radiation plays important roles in many circumstances. Where two surfaces, 1 and 2, having the area A [m²], face each other (Fig. 3.1*d*), the thermal resistance is given by

$$R = \frac{1}{h_R A} \tag{3.4}$$

where h_R [W/m² K] is the radiative heat transfer coefficient.

To model transient heat transfer we need to take into account the heat capacitance of the medium as depicted in Fig. 3.2. In Fig. 3.2*a* node 2 has a heat capacitance C [J/K]. Figure 3.2*b* depicts a situation where node 1 is a strip of heat source dissipating a pulse of heat, P [J], node 2 is a medium having the thickness L [m], and node 3 is the other surface of the medium. To find the response of the temperature at node 2 to the pulse heat input from node 1, we write the network composed of two equal resistances and a capacitance. Where the thermal conductivity and the specific heat of the medium are λ [W/m K] and c [J/kg K], respectively, the thermal resistance and the heat capacitance are written as

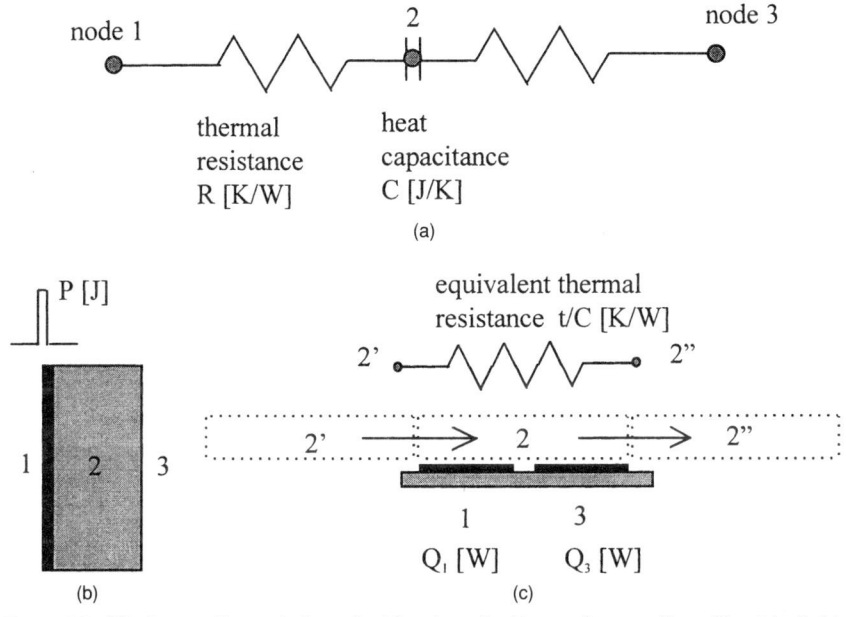

Thermal Design, Analysis, and Measurement of Electronics Packaging 115

Figure 3.2 Heat capacitance in transient heat conduction and convection of heat in fluid; (a) capacitance, resistance, and nodes; (b) heat pulse from strip 1; (c) mass of coolant 2 picking up heat from sources 1 and 3 in time t.

$$R = \frac{L}{2\lambda A} \quad (3.5)$$

$$C = c\rho L A \quad (3.6)$$

where ρ [kg/m³] is the density of the medium, and A [m²] is the area normal to the heat flow.

Figure 3.2c depicts a situation where node 2 represents a mass of coolant (M [kg]) flowing over the heat sources (nodes 1 and 3). The thermal resistance from node 1 to 2, and that from node 3 to 2, can be found by applying Eq. 3.3. We define that nodes 2' and 2" represent the state of the coolant before and after the coolant passes over the heat sources, respectively. Heat transfer from the heat sources (Q [W]) raises the coolant temperature by

$$\Delta T = \frac{t}{C} Q \quad (3.7)$$

where t [s] is the time for the coolant mass to pass over the heat sources, and C [J/K] is the heat capacitance of the coolant mass. Comparison of Eqs. 3.1 and 3.7 points to the equivalence of the term t/C to the thermal resistance, which links nodes 2' and 2". Using the specific heat of the coolant, c [J/kg K], and the mass flow rate, M [kg/s], the equivalent thermal resistance t/C can be written as

$$\frac{t}{C} = \frac{1}{cM} \tag{3.8}$$

3.2.2 Heat diffusion from transistor junctions

The purpose of thermal design is to contain the temperature of transistor junctions, called *junction temperature,* in a certain range. However, the real meaning of junction temperature is usually not questioned in thermal design. The average temperature of the chip is considered synonymous with the junction temperature. Nevertheless, this section considers heat diffusion from the transistor junctions. The subject serves as a convenient example on which some modeling methods and concepts will be explained.

One of the useful concepts in the analysis of complex physical systems is the superposition of solutions. In the temperature range where the electronic equipment normally operates, the conduction of heat can be regarded as a linear process; namely, the thermal resistance in Eq. 3.2 is independent of the temperature. On the linear system the method of superposition can be applied, which helps us to sort out interactions from various heat sources and influences from complex boundary conditions.

The superposition method is now illustrated using an example of a point heat source on the chip's surface. The heat source is assumed to dissipate heat in a train of pulses, each pulse instantaneously dissipating P [J], with the pulsing interval Δt [s]. The heat source fired the first pulse at $t = 0$, the second pulse Δt later, and so on. By solving a heat conduction equation we find a temperature distribution created by each pulse. The temperature distribution created by the jth pulse is written as

$$u_j = \frac{A}{\tau_j^{3/2}} \exp\left(-\frac{r^2}{B\tau_j}\right) \tag{3.9}$$

where $\tau_j = t - (j-1)\Delta t$,

$$A = \frac{P/\rho c}{4(\pi\kappa)^{3/2}}, \quad B = 4\kappa$$

ρ [kg/m^3] is the density, c [J/kg K] is the specific heat, and κ [m^2/s] is the thermal diffusivity of the chip material, respectively.

Superposition of u_j's gives the temperature attributed to all J pulses,

$$T_d = \sum_{j=1}^{J} u_j \tag{3.10}$$

where the suffix d means the discrete-pulse solution.

Meanwhile, the waves of heat originated by heat pulses are diffused as they travel in the chip; at a certain distance from the heat source and at a sufficiently long period of time after the start of the pulse train, the waves from successive pulses become indistinguishable from each other. Hence, we may approximate the train of pulses by continuous heat generation at the point source with an equivalent intensity $q = P/\Delta t$ [W]. Then, the temperature at r is given by

$$T_c = \frac{A'}{r} \operatorname{erfc}\left(\frac{r}{\sqrt{Bt}}\right) \quad (3.11)$$

where the suffix c means the continuous-source solution, erfc (x) is the complementary error function, and

$$A' = \frac{q/\rho c}{2\pi\kappa}.$$

The ratio T_d/T_c is shown in Fig. 3.3 for the case of $\Delta t = 10^{-8}$ [s] (= 10 ns) and $\kappa = 89.2 \times 10^{-6}$ [m²/s] (silicon). The curves in Fig. 3.3 show two distinct features. First, beyond a certain radius, about 2 µm in this example, T_c becomes nearly equal to T_d. We write this radius as r^*. Substituting 10^{-8} [s] (= Δt) for τ_j and $4\kappa = 3.57 \times 10^{-4}$ [m²/s] for B in the exponential function on the right-hand side of Eq. 3.9, we find that for

$$r > r^* \equiv \sqrt{B\,\Delta t} \approx 2\ \mu m \quad (3.12)$$

the temperature wave from an individual heat pulse is diffused over and loses its identity.

Figure 3.3 shows another feature; the curves for $t = 500$ and 1000 ns are indistinguishable, while that for $t = 100$ ns shows deviation. This means that the continuous-heat approximation does well after about 50 pulses are fired.

The continuous-heat approximation gives a concise form of thermal resistance for heat spreading from the heat source. In Eq. 3.11, as t approaches infinity, the complementary error function approaches unity, hence,

$$(T_c)_{l \to \infty} = \frac{A'}{r} \quad (3.13)$$

Equation 3.13 actually gives the temperature rise (ΔT) on the spherical surface having the radius r from the temperature of the distant surrounding. Let us consider a hemispherical surface (S) having the radius r_c ($>r^*$ of Eq. 3.12) around the heat source. The ratio $q/\Delta T$ gives the thermal resistance for heat spreading (R_{sp}) from S to the surrounding environment.

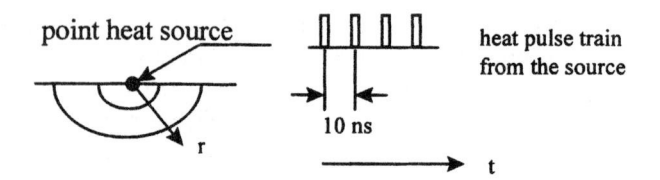

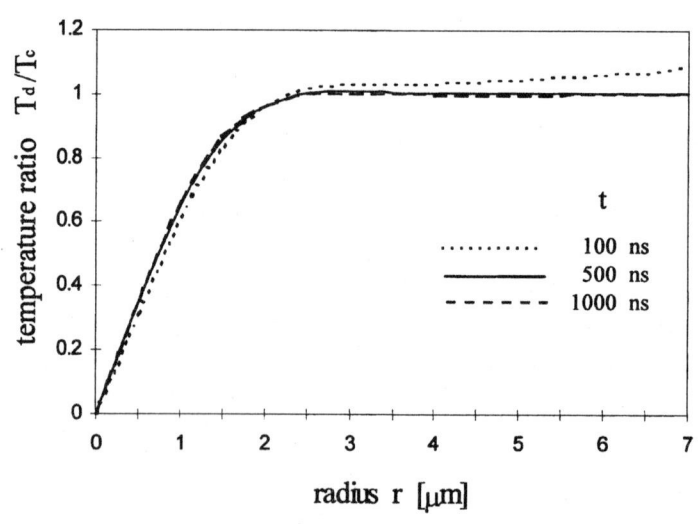

Figure 3.3 Ratio of discrete-pulse solution (T_d) to approximate continuous-heat solution (T_c).

$$R_{sp} = \frac{1}{2\pi r_c \lambda} \quad (3.14)$$

where λ [W/m K] is the thermal conductivity, and the relationship $\kappa = \lambda/\rho c$ was used.

Now we consider superposition of the steady-state solutions associated with the point heat sources A and B, separated by a 2-s distance, each dissipating q [W]. When viewed from a sufficiently long inward distance from the chip's surface, the two heat sources become indistinguishable; hence, they can be replaced by a single heat source (C) having the heat dissipation strength $2q$ [W]. After some manipulation we find a radial distance r^* beyond which the single-source solution approximates the superposed solution within the error ε.

$$r^* \approx \frac{s}{\sqrt{2\varepsilon}} \quad (3.15)$$

Suppose that our point sources A and B are located at the extreme ends of the actual source, for example, source A on one corner of the

FET channel and source B on an opposite end of the channel. Where the solutions associated with A and B can be approximated by the point source C, then, the whole region of the FET channel can also be represented by C. Assuming a typical feature size of the heat source on the chip as 1 μm (= 2 s) we find from Eq. 3.15 that the previous example $r^* = 2$ μm (Eq. 3.12) corresponds to $\varepsilon = 0.03$. This error is sufficiently small compared to possible errors attributed to other causes such as the uncertainty in thermal diffusivity.

Summarizing this section we define the junction temperature, T_J, which is a measure of ultimate importance for thermal management of electronic equipment. The T_J represents the temperature at a certain depth from the chip's surface where temporal variations of temperature caused by switching operation of transistors and spatial variations of temperature due to geometric features of electronic circuits are diffused over to create a nearly steady thermal field. The measure of such depth is given by $\sqrt{4\kappa/\Delta t}$ (Eq. 3.12), where Δt is the switching interval. This depth is much less than the chip thickness; hence, we assume that the heat is generated on the chip's surface.

3.2.3 Heat conduction from integrated circuits

From the distance of $\sqrt{4\kappa \Delta t}$ the integrated circuits can be viewed as an aggregate of point heat sources spread in sheet form over the chip's surface. The heat generated from the integrated circuits is conducted away through two paths, one crossing the chip thickness and another crossing the material in contact with the circuit side of the chip. Figure 3.4 illustrates a situation where the chip is flip-bonded to the substrate, and the heat generation (Q) is divided to a heat flow across the chip (Q_u) and that to the substrate (Q_l). The thermal resistance for Q_u is written as

$$R_u = \frac{\delta_c}{\lambda A} \tag{3.16}$$

where δ_c is the thickness of the chip, A is the area of the chip, and λ is the thermal conductivity of chip material.

The thermal resistance for Q_l is composed of thermal resistance through the solder bumps and that through the medium filling the free space, and given as

$$R_l = \frac{1}{(\lambda_s A_b/\delta_b) + (\lambda_i A_i/\delta_b)} = \frac{\delta_b}{\lambda_s A_b + \lambda_i A_i} \tag{3.17}$$

where λ_s and λ_i are the thermal conductivity of the solder and that of the filler medium, respectively. We model the individual solder bump

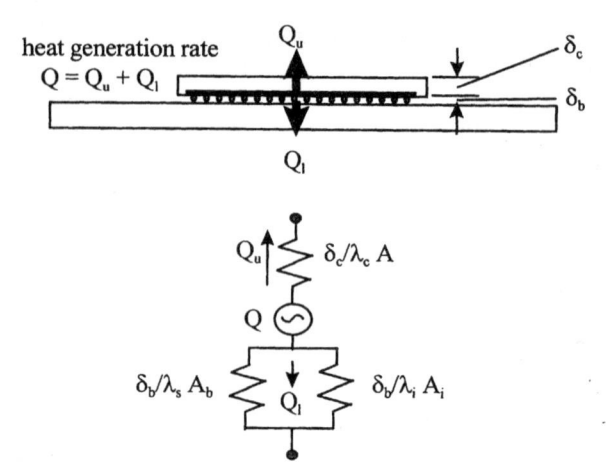

Figure 3.4 Heat conduction from integrated circuits across the chip (Q_u) and to the substrate (Q_l), and thermal resistance network.

by a cylindrical column having the height δ_b and the area a_b. The A_b is the sum of a_b, namely, $A_b = N_b a_b$, where N_b is the number of bumps, and $A_i = A - A_b$.

Typical examples will show the order of magnitude of thermal resistance. The thermal conductivity data are taken from Tables 3.1 and 3.2. The chip is assumed to have a size $12 \times 12 \times 0.3$ mm and 900 solder bumps. Each bump has a 100-µm diameter and a 100-µm height. Where the free space between the chip and the substrate is left to air,

$$R_l = 0.39 \text{ [K/W]}$$

The thermal resistance for Q_u is

$$R_u = 0.014 \text{ [K/W]}$$

These numerical examples suggest that the temperature drop across the chip may be much less than that across the solder bumps. The temperature drop in the chip becomes very small, where there is a large thermal resistance beyond R_u, which reduces the heat flow component Q_u. This is the case on plastic encapsulated chips. Also, on such chips the heat dissipation per transistor junction is generally small enough to make the temperature drop from the junction to the chip, to be found from Eq. 3.14, extremely small. Hence, the chip can be modeled as an isothermal body, and the junction temperature becomes synonymous with the chip temperature. In many thermal design analyses for CMOS-based systems this equivalence of the junction temperature with the chip temperature is tacitly assumed.

TABLE 3.1 Thermal Properties of Semiconductors and Packaging Materials

Material	Thermal conductivity, λ [W/m K]	Thermal diffusivity, κ [10^{-6} m²/s]	Coefficient of thermal expansion, α [10^{-6} K^{-1}]
Semiconductors			
Silicon	150	89.2	2.5
Gallium arsenide	50	27	5.8
Inorganic substrates			
Alumina	30	12	6.0
Aluminum nitride	230		4.3
Diamond	2300	1300	1.2
Mullite	5		4.2
Silicon carbide	220	103	3.7
Silicon dioxide	1.38	0.83	0.5
Silicon nitride	16	9.65	
Organics			
Epoxy glass (xy)	0.3		15
Molding compound	0.7		20
Polyimide	0.4		50
Metallics			
Alloy 42 (Fe Ni)	16		6
Aluminum	230	97	23
Copper	400	117	17
Copper-Invar-copper			
xy	107		44
z	14		
Gold	317	127	14
Kovar	17.3		5.2
Molybdenum	140	54	5.0
Tungsten	170	68	4.6

TABLE 3.2 Thermal Conductivity of Solder and Interface Materials

Material	Thermal conductivity, λ [W/m K]
Solder	
63Sn/37Pb solder	50
5Sn/95Pb solder	36
Gap filler	
Air	0.026
Helium	0.152
Polymer adhesive	1 ~ 2
Silicone grease	0.2
Silicone rubber	0.2
Thermal paste	~1

3.2.4 Heat flow across interfaces

Let us consider in more detail the heat flow through an array of solder bumps bonded to the chip and the substrate (Fig. 3.5a). The heat flow from the circuits on the chip converges to the upper end of the bump, then expands in the substrate after exit from the bump. A similar situation exists at the interface between the Ball-Grid-Array (BGA) module and the wiring substrate. Due to the proximity of the circuit plane to the bumps any realistic modeling of heat conduction near and through the bumps requires detailed information about the circuit and the structure of the bump. However, a drastically simplified model gives us some insight about the effects of the parameters on heat conduction through the bump array.

Such a model, called *flux tube model,* is depicted in Fig. 3.5c. The heat flow descends from a hypothetical tube in medium 1 having the diameter d_p, passes through the model of a bump, which is a cylinder having the height δ_b and the diameter d_b, then expands into another hypothetical tube in medium 2 having the same diameter d_p. The diameter d_p is determined such that the cross-sectional area of the flux tube is equal to the chip (or module for BGA) area divided by the number of bumps.

A similar model can be conceived for the interface between components in mechanical contact, such as the one between the module and

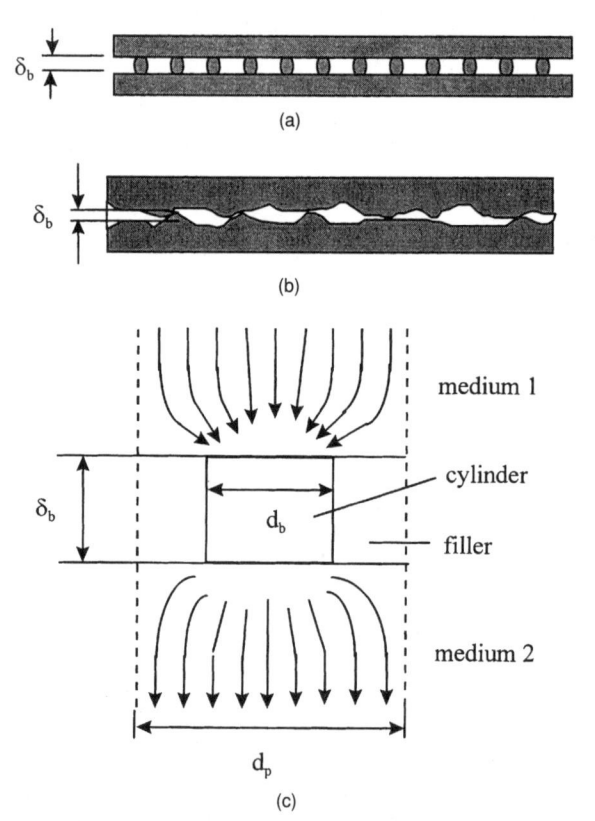

Figure 3.5 Thermal interfaces; (*a*) solder bump array; (*b*) contact interface; (*c*) flux tube model.

the spring-loaded heatsink (Fig. 3.5b). The number of contacts at such interface is determined by the roughness of the mating surfaces and the pressure applied to the interface. A flux tube model is defined in terms of statistically averaged area per contact spot. We consider a situation where the material of surface 1 is harder than that of surface 2; besides, surface 1 is smoother than surface 2. Then, once brought into contact, only roughness peaks on surface 2 yield under the contact pressure. In a flux tube model, the cylinder corresponds to the peak of surface 2 in contact with surface 1, and the gap height δ_b is the roughness height of surface 2 modified by the plastic deformation of roughness peak.

The thermal resistance through a flux tube is given by the following equation.

$$R_b = \frac{(1-\varepsilon)^{1.5}}{2d_b}\left(\frac{1}{\lambda_1}+\frac{1}{\lambda_2}\right) + \frac{4\delta_b}{\lambda_s \pi d_b^2} \qquad (3.18)$$

where $\varepsilon = d_b/d_p$, and λ_1, λ_2, and λ_s are the thermal conductivity of medium 1, 2, and the cylinder, respectively. The first term on the right-hand side of Eq. 3.18 represents the resistance for contraction and expansion of heat flow in media 1 and 2, and is valid for $0 < \varepsilon \le 0.3$ [21].

The heat flow through the filler medium encounters the resistance,

$$R_f = \frac{4\delta_b}{\lambda_f \pi (d_p^2 - d_b^2)} \qquad (3.19)$$

where λ_f is the thermal conductivity of the filler medium.

The total resistance across the flux tube interface is

$$R = \frac{1}{(1/R_b) + (1/R_f)} \qquad (3.20)$$

In the following examples, we introduce an assumption that $d_b = \delta_b$, and show the numerical values in $R_b\lambda_s d_b$ and $R_f\lambda_q d_b$, which are nondimensional. Figure 3.6a shows an example for the solder bump connecting silicon and alumina ceramic. Assume that the d_b is fixed, so that $\varepsilon \to 0$ means $d_p \to \infty$. The free space between the silicon and the alumina substrate is filled by air or polymer underfill. It is shown that the heat conduction through the air gap is negligible in a practically important range of ε, while the polymer underfill makes the heat conduction across the gap almost as important as that through the bumps. The contraction and expansion resistance makes 36 to 26 percent contributions to R_b.

Figure 3.6b shows another example for the case where the components made from aluminum and epoxy are brought into contact. The R_b

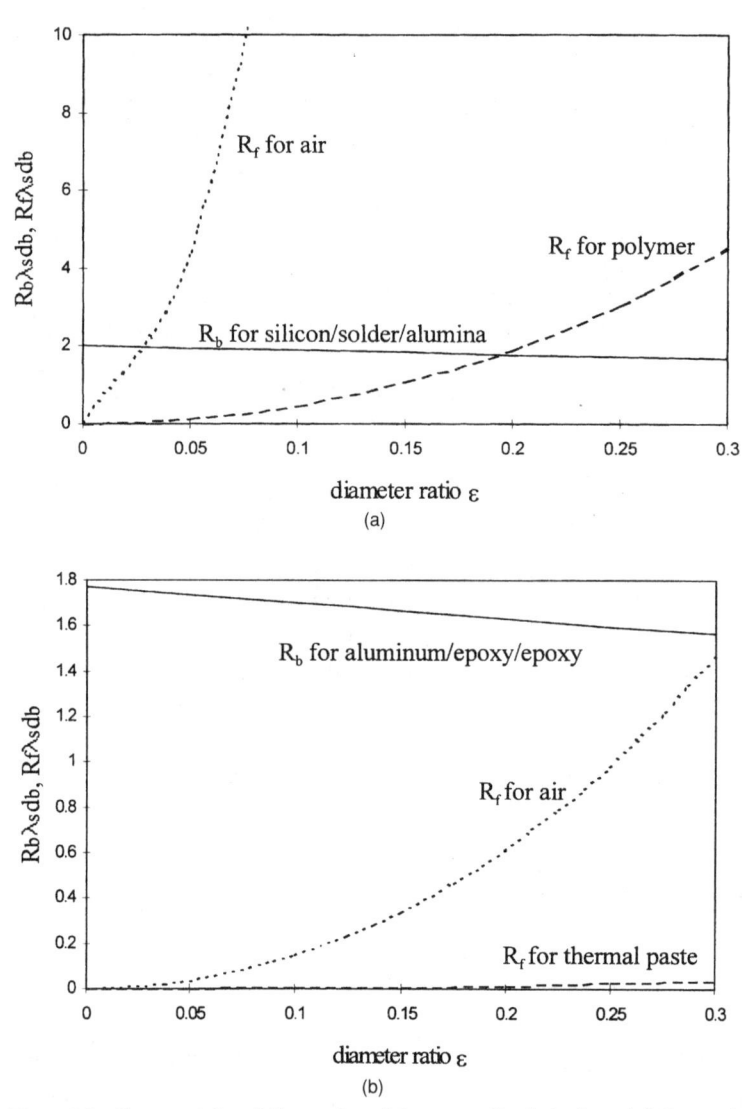

Figure 3.6 Components of thermal resistance on the interface; (a) thermal resistance components on silicon/solder bump/alumina ceramic (R_b) and interstitial medium (R_f); (b) thermal resistance components on aluminum/epoxy contact (R_b) and interstitial medium (R_f).

is dominated by the resistance component on the epoxy cylinder. A common presumption is that only 2 to 3 percent of the nominal area is in actual contact at a mechanically assembled interface. This range is translated to 0.14 to 0.17 in ε. In this range the total thermal resistance is dominated by R_f, even where the filler medium is air. In such a circumstance, the thermal conductivity of the filler medium and the gap height are dominant parameters. These parameters are generally difficult to be estimated accurately; hence, the experimental determination of total thermal resistance is necessary.

Expansion of heat flow also occurs in a heat spreader attached to the chip or the module. In modeling the function of the heat spreader, the parameters neglected in the preceding models need to be taken into account: they are the shape of the heat source, the shape and the thickness of the heat spreader, and the boundary condition on the spreader's surfaces. More discussion is developed later in Sec. 3.2.8.

3.2.5 Heat conduction in wiring substrates

The wiring substrate serves as a heat spreader for the chips and the modules on it. The heat-spreading function of the wiring substrate is important, in particular, in small systems, where relatively large thermal resistance on heat flow paths over the module causes a large share of heat flow to the substrate.

To estimate the effect of heat spreading in the wiring substrate we need to know its thermal conductivity. However, there have been few data or methods to predict thermal conductivity that are directly applicable to a case in hand. Indeed, the wiring substrate is an example of materials that pose sheer difficulty in modeling their thermal behavior.

A combination of analytical and experimental approaches is necessary to estimate the thermal conductivity of a particular wiring substrate. The analysis can be performed only on models where the complexity of wiring patterns is reduced to a minimum. Figure 3.7

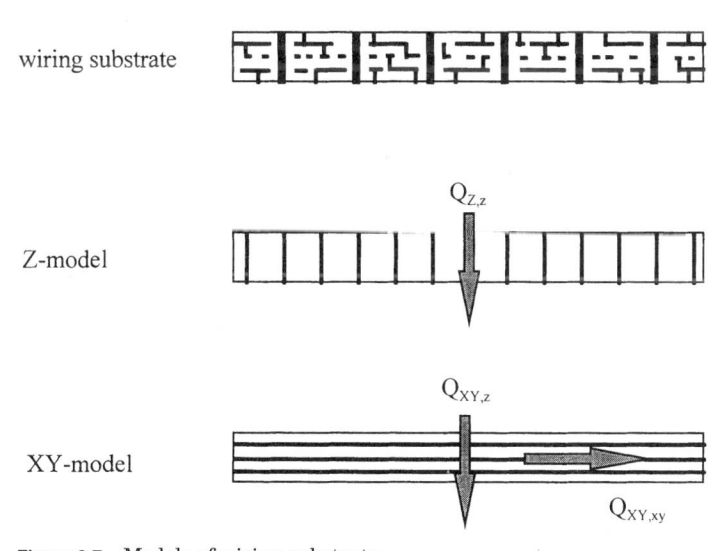

Figure 3.7 Models of wiring substrate.

shows such models. In the Z-model only through-hole vias are assumed to be present in the insulator matrix, and the equivalent thermal conductivity in the z-direction is written as

$$\lambda_{Z,z} = \lambda_M a_M + \lambda_I(1 - a_M) \qquad (3.21)$$

where λ_M and λ_I are the thermal conductivity of the metal and the insulator, respectively, and a_M is the sum of cross-sectional areas of the via metal per unit area of the substrate.

In the XY-model only power and ground layers are assumed to be present in the insulator matrix. The equivalent thermal conductivity in the z-direction is written as

$$\lambda_{XY,z} = \frac{1}{(\gamma_M/\lambda_M) + (1 - \gamma_M)/\lambda_I} \qquad (3.22)$$

where γ_M is the ratio of the summed thickness of metal to the substrate thickness. The equivalent thermal conductivity in the plane is written as

$$\lambda_{XY,xy} = \lambda_M \gamma_M + \lambda_I(1 - \gamma_M) \qquad (3.23)$$

Numerical examples are given as follows. A printed circuit board has through-hole vias with a number density 25 per 1 cm² board area, and four power and ground layers, each layer being a 50-μm-thick copper plane. The via-hole has a diameter 0.43 mm, and its inner surface is electroplated by 15-μm-thick copper. Referring to Table 3.1, $\lambda_M = 400$ W/m K and $\lambda_I = 0.3$ W/m K (epoxy glass). The equivalent thermal conductivity values given by Eqs. 3.21, 3.22, and 3.23 are, respectively, $\lambda_{Z,z} = 2.3$ W/m K, $\lambda_{XY,z} = 0.3$ W/m K, and $\lambda_{XY,xy} = 32$ W/m K. Absence of continuous metal connections for the heat flow Q_Z in the XY-model reduces $\lambda_{XY,z}$ below the level of $\lambda_{Z,z}$ and $\lambda_{XY,z}$ by factors of 10 and 100, respectively. This implies that signal lines, which are invariably thin lines and separated by the insulator on most of their length, have a secondary effect on heat conduction in the printed circuit board.

3.2.6 Forced convective heat transfer from flat surfaces

The thermal resistance for heat flow from the solid surface to the nearby coolant is determined by the heat transfer process occurring in the boundary layer developed on the surface. The boundary layer is a manifestation of the viscosity of coolant, which brings the coolant velocity (u) from the mainstream velocity at the outer edge of the boundary layer (U) to zero on the solid surface (Fig. 3.8).

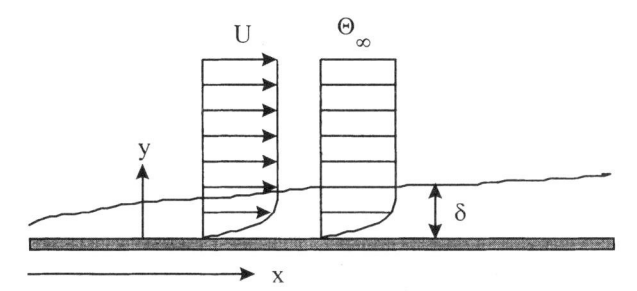

Figure 3.8 Boundary layer on flat plate.

Within the boundary layer the coolant temperature (T) also varies from the mainstream temperature (T_∞) to the surface temperature (T_s). Where the coolant is air, the profile of $\Theta = T_s - T$ closely matches the velocity profile (Fig. 3.8). The heat flux on the surface (q_s [W/m²]) is determined by Fourier's law

$$q_s = -\lambda_f \left(\frac{\partial T}{\partial y}\right)_{y=0} = \lambda_f \left(\frac{\partial \Theta}{\partial y}\right)_{y=0} \qquad (3.24)$$

where λ_f [W/m K] is the thermal conductivity of coolant, and y is the coordinate taken normal to the surface. The profile of Θ may be approximated by a parabolic function which is 0 at $y = 0$, and increases to $\Theta_\infty = T_s - T_\infty$ and has zero gradient ($\partial \Theta / \partial y = 0$) at $y = \delta$ (the boundary layer thickness). With this profile, Eq. 3.24 is now transformed into the heat-transfer coefficient as

$$h \equiv \frac{q_s}{\Theta_\infty} = \frac{2\lambda_f}{\delta} \ [\text{W/m}^2 \ \text{K}] \qquad (3.25)$$

Equation 3.25 dictates that, in order to increase the heat transfer coefficient, thereby reducing the thermal resistance on the component surface, we have to employ the means to decrease the boundary layer thickness. The boundary layer thickness is a function of the velocity of the main stream (U) and the streamwise coordinate along the surface (x). This function takes different forms for different flow configurations and regimes.

Figure 3.9 shows two principal flow configurations found on electronic modules: parallel flow and impingement flow. Figure 3.9 also shows three regimes of flow: laminar, transitional, and turbulent flows. There are other factors that affect the boundary layer development: separation and recirculation of flow on the sides of the module and the point of transition from laminar to turbulent flow.

There are some general rules for the boundary layer thickness.

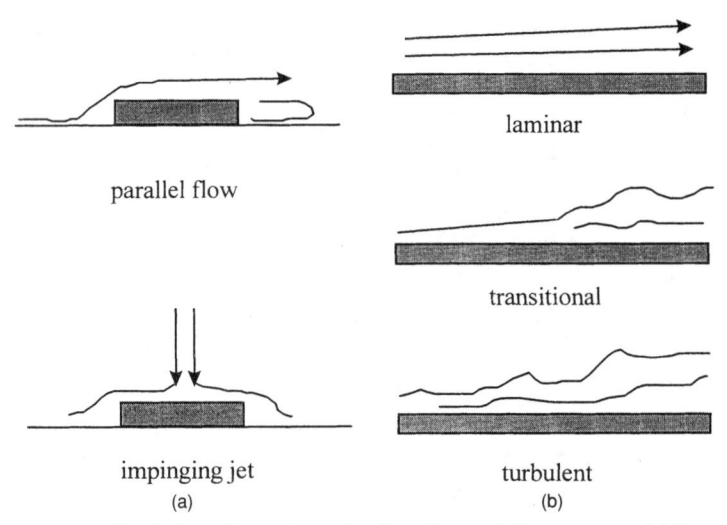

Figure 3.9 Typical configurations of coolant flow and flow regimes; (a) flow configurations; (b) flow regimes.

1. The boundary layer becomes thinner as the coolant velocity increases.
2. It is extremely thin where it starts to develop, and grows in the streamwise direction. Hence, the average thickness of the boundary layer is less, where the flow path length is made shorter. This means that the smaller the size of the component, the less the average thickness of the boundary layer. Also, an impinging jet of the coolant yields higher heat transfer coefficients than parallel flow, because the flow path length is reduced to half the component length.
3. The turbulent boundary layer has an internal structure that produces a very high temperature gradient next to the surface, while the total thickness becomes larger than the laminar boundary layer. Hence, for the turbulent boundary layer, δ in Eq. 3.25 has to be regarded as the sublayer thickness that is much smaller than the total boundary layer thickness.

Nondimensional representation of the heat transfer coefficient is the Nusselt number defined as

$$\mathrm{Nu} \equiv \frac{hL}{\lambda_f}$$

where L is the length scale representing the size of the component. Using Eq. 3.25, we write $\mathrm{Nu} = 2L/\delta$; namely, the Nusselt number represents the ratio of the component size to the boundary layer thickness. The effects of the flow configuration, regime, and velocity on the boundary layer thickness are summarized in the correlation of Nu versus the Reynolds number,

$$\mathrm{Re} \equiv \frac{UL}{\nu}$$

where ν [m²/s] is the kinematic viscosity of the coolant. Note that ν has the same dimension with the thermal diffusivity κ which appeared in Sec. 3.2.2. The ratio of ν to the thermal diffusivity of the coolant, κ_f, is called the Prandtl number,

$$\mathrm{Pr} \equiv \frac{\nu}{\kappa_f}$$

which is a measure of similarity between the velocity (u) and temperature (Θ) distributions in the boundary layer. Air has Pr = 0.7, so that there is close similarity between the profiles of u and Θ. In most liquids, the temperature variation is contained in a layer which is thinner than the boundary layer, hence, Pr > 1. Various heat transfer data are correlated in the form,

$$\mathrm{Nu} = C\,\mathrm{Re}^m\,\mathrm{Pr}^n \qquad (3.26)$$

where C is a numerical constant or given in terms of the geometric dimensions of the heat source and the coolant supply duct, and m and n are constants. Table 3.3 summarizes C, m, and n for two basic flow configurations. The correlation for impinging jet heat transfer was derived from a more elaborate correlation for a single round jet

TABLE 3.3 Constants in Heat Transfer Correlation, Eq. 3.26

Configuration	C	m	n	Applicable ranges	Source
Parallel flow on flat plate	0.664	1/2	1/3	$\mathrm{Re} \leq 5 \times 10^5$	Incropera & DeWitt [10]
	$0.037\,\mathrm{Re}^{4/5} - 871$		1/3	$5 \times 10^5 < \mathrm{Re} < 10^8$	
	0.037	4/5	1/3	$\mathrm{Re} \geq 10^8$	
Impinging jet	$5.18\,(D/L)^{1/2} \times (1 - 2.2\,D/L)$	1/2	0.42	$2 \leq H/D \leq 5$ $5 \leq L/D \leq 10$ $10^4 \leq \mathrm{Re} \leq 5 \times 10^4$	Reduced from Martin [13]

obtained by Martin [13]. In reducing the original correlation, the applicable parametric ranges were narrowed into those of practical interest in electronics cooling.

3.2.7 Forced convective heat transfer from modules on board

Once mounted on the printed circuit board, heat transfer from the module is affected by the geometric parameters depicted in Fig. 3.10. In addition to the size of the module [b (height) × l (length) × w (width)], there are such parameters as the separation distance between the neighboring modules (row spacing s, column spacing s'), the location of the module [in row number n ($1 \leq n \leq N$) counted from the entrance of the coolant channel], and the channel height (H). The real situation is far more complex than that portrayed in Fig. 3.10; particularly in small systems, the geometric regularity in placement of modules and other components exists only to a certain extent. Most of the pertinent data reported in the literature are useful for the design of big systems, where a number of equal-sized modules are mounted on the board with a regular placement pitch. A brief account of such knowledge base is given below, with an intention to provide some conceptual guides for designers of small systems as well.

Figure 3.11 presents a collection of the curves of heat transfer coefficient (h) versus the air velocity (U). The data "single module" was reported by Chang, Shyu, and Fang [2], with a single heated block measuring 6 × 6 × 2 cm ($b/l = 1/3$) in a channel having a relative height $H/b = 3$. The curve "sparse flat packs" correlates the data obtained by Wirtz and Dykshoorn [20]. Their model of flat pack (block) had a

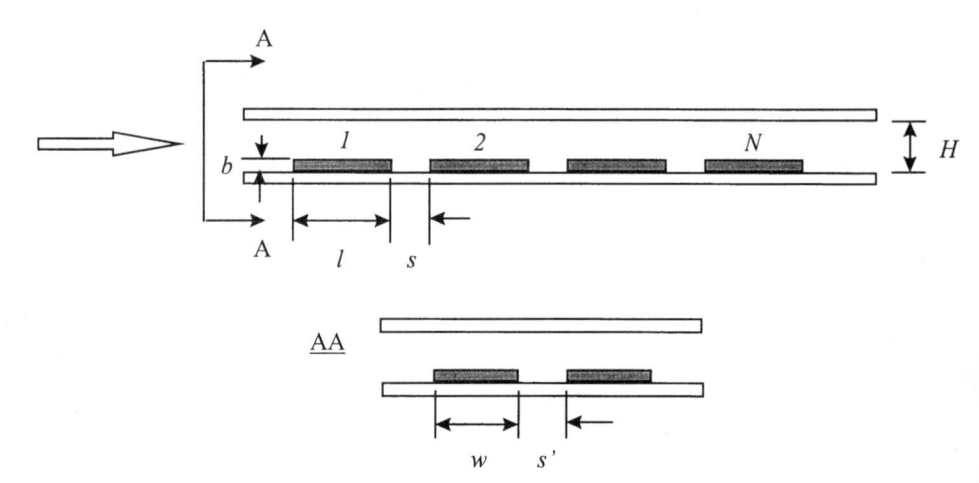

Figure 3.10 Module array on board.

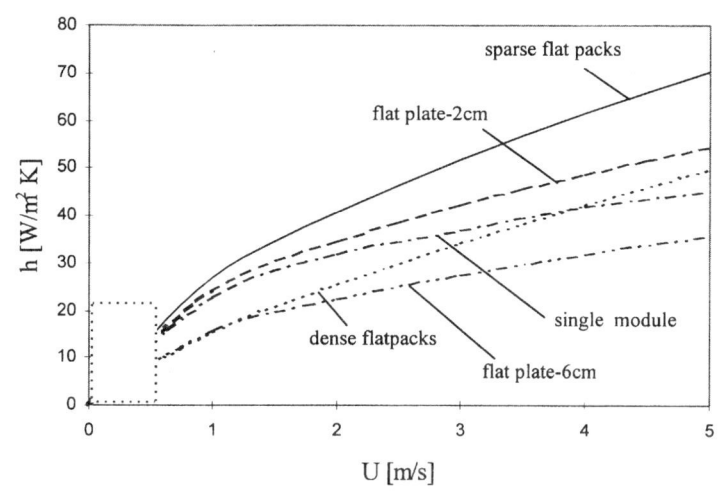

Figure 3.11 Typical heat transfer coefficients for forced convection air cooling of modules and flat plates.

dimension 2.54 × 2.54 × 0.64cm, hence, the ratio $b/l = 0.23$. Other parameters were $H/b = 3.6$ and $s/l = 1$. The curve "densely flatpacks" correlates the data obtained by Sparrow, Neithammer, and Chaboki [19]. The block was 2.667 × 2.667 × 1cm, and placed in a channel having $H/b = 2.67$. The spacing between the blocks was $s/l = 0.25$. Those data and correlations were discussed by Moffat and Ortega [15]. Also shown in Fig. 3.11 are the curves for heat transfer from a 2.54-cm-long flat plate ("flat plate-2cm") and a 6-cm-long flat plate ("flat plate-6cm"), drawn using the first formula of Table 3.3 and the physical properties of air listed in Table 3.4. What we see in Fig. 3.11 are interpreted as follows.

1. The smaller the size of the module, the higher the heat transfer coefficient. See the sparse flatpacks and the flat plate-2cm versus the single module and the flat plate-6cm.
2. The coolant stream in the channel is accelerated where the module partially blocks the channel. Acceleration of coolant flow has an

TABLE 3.4 Thermophysical Properties of Single-Phase Coolant at 300K

Coolant	Density, ρ [kg/m³]	Specific heat, c_p [kJ/kg K]	Kinematic viscosity, ν [10^{-6} m²/s]	Thermal conductivity, λ [W/m K]	Thermal diffusivity, κ [10^{-6} m²/s]	Prandtl number, Pr	Coeff. thermal expansion, β [K^{-1}]
Air	1.16	1.00	15.9	0.0263	22.5	0.7	0.0033 (300K)
FC77	1590	1.17	0.28	0.057	0.03	9.25	0.0014
Water	1000	4.18	0.86	0.613	0.15	5.7	0.00028

effect of reducing the boundary layer thickness, thereby increasing h. The flat plate formula gives h for the plate in open space; hence, its underprediction of h compared to the sparse flat packs and the single module performance partly reflects the effect of flow acceleration. A similar effect of flow acceleration is brought by the development of boundary layer on the upper channel wall, which displaces the coolant toward the center of the channel.

3. Where the modules are sparsely populated, the flow separated behind the module is developed into a turbulent wake before it reaches the next module. The increased level of turbulence is effective in producing high h. This partially explains the difference between the dense flatpacks and the sparse flatpacks.

4. Where the modules are densely packed on the board, coolant masses are trapped between the modules, and the front and back surfaces of the module do not serve as effective heat transfer surfaces. This may also be attributed to the difference between the sparse flatpacks and the dense flatpacks.

5. The preceding observations in (2), (3), and (4) lead us to consider how the location of the module affects h. Where the causes (2) and (3) prevail, the heat transfer coefficient (h_n) tends to increase on downstream rows (increasing n). On the other hand, where the cause (4) prevails, h_n decreases with increasing n. In any circumstances, however, it has been proved that the variation of h_n becomes practically negligible after the second or third row.

The heat transfer coefficient discussed earlier relates the heat flux to a temperature difference between the module surface and the coolant over the module; namely, T_∞ in Θ_∞ of Eq. 3.25 is the temperature of the coolant just outside the boundary layer developed on the module of interest. This temperature may be called *local free stream temperature*. The local free stream temperature increases as the coolant picks up heat from the modules in a row. Its increase from the inlet of the channel to the location of the Nth module is

$$(\Delta T)_{\text{mix}} = \frac{1}{c_p M} \sum_{n=1}^{N-1} Q_n \quad [\text{K}] \quad (3.27)$$

where c_p [J/kg K] is the specific heat at constant pressure of the coolant, M [kg/s] is the mass flow rate of the coolant per column of the modules, and Q_n [W] is the heat dissipation rate of the nth module.

The term $(1/c_p M)$ is the equivalent thermal resistance, as noted in regard to Fig. 3.2c and Eq. 3.8. This thermal resistance, however, has to be amplified by a certain factor to account for imperfect mixing of the coolant in the channel. Namely, Eq. 3.27 holds only where the coolant is thoroughly mixed. In actual circumstances the coolant is never thoroughly mixed, but has a temperature distribution in the channel. Figures 3.12a and b depict thoroughly mixed flow and actual flow, respectively. Figure 3.12b shows T_∞ for the Nth module ($T_{\infty,N}$), and the temperature distribution in the boundary layer on the Nth module. The coolant mass of relatively high temperature is called *thermal wake*. We write the difference between the surface temperature of the Nth module ($T_{s,N}$) and the coolant temperature at the channel inlet (T_0) as

$$T_{s,N} - T_0 = \Delta T + (T_{s,N} - T_{\infty,N}) \qquad (3.28)$$

$$\Delta T \equiv T_{\infty,N} - T_0 = \sigma \, (\Delta T)_{\text{mix}} \qquad (3.29)$$

$$T_{s,N} - T_{\infty,N} = \frac{Q_N}{h_N A_N} \qquad (3.30)$$

where σ is the multiplying factor to account for the effect of thermal wake on the local free stream temperature, h_N and A_N are the heat transfer coefficient and the heat transfer area on the Nth module, respectively, and Q_N is the heat dissipation rate from the Nth module. The value of σ depends on several factors such as the distance between the upstream heat sources and the Nth module, the coolant velocity, and the level of turbulence. When these factors are taken into account, σ has to be a function of n, and placed inside the summation on the right-hand side of Eq. 3.29. However, for the sake of brevity and with reasonable accuracy, we assume that σ is constant over the modules over a row of modules. Figure 3.13 shows a thermal resistance network viewed from the Nth module to the channel inlet with the assumption of constant σ. For a conservative estimate, σ may be assumed as high as 2 to 3.

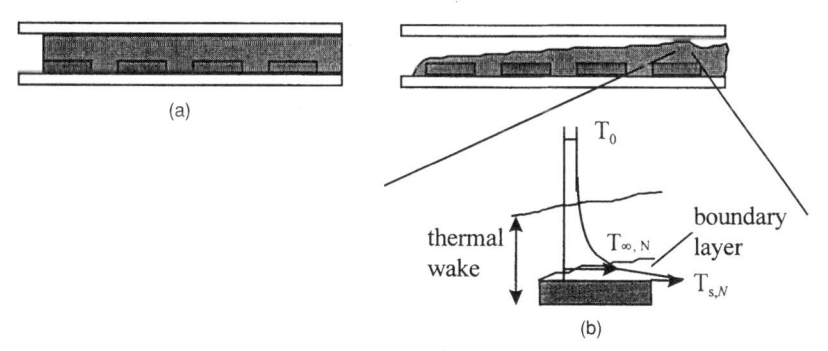

Figure 3.12 Mixing of warm air with free stream; (a) perfectly mixed flow; (b) thermal wake.

$$(\sigma/c_p\, M)\left(\sum_{n=1}^{N-1} Q_n \Big/ Q_N\right)$$

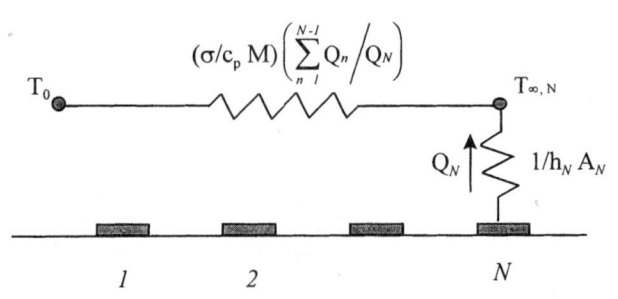

T_0 ———/\/\/\——— $T_{\infty, N}$

Q_N , $1/h_N A_N$

1   2   N

Figure 3.13 Thermal resistance from the module on the downstream row to the inlet air.

3.2.8 Conjugate heat transfer on heat spreader

Heat transfer at the board level is made even more complex if a significant portion of heat is conducted from the module to the board. The heat from the module spreads in the board and is eventually transferred to the coolant from the upper and lower surfaces of the board. The effect of the board as a heat spreader is particularly large, where the number of high-power modules is one or a few, and the back side of the board is exposed to the coolant. Heat conduction inside the board and convective heat transfer on the surface of the board are coupled in what we call *conjugate mode* of heat transfer. Similar conjugate heat transfer problems exist, where heat spreaders by design are attached to chips or modules.

Figure 3.14 shows a model of the heat spreader, where the heat source and the spreader are both replaced by circular disks having respectively equal surface areas; the heat source disk has a radius a,

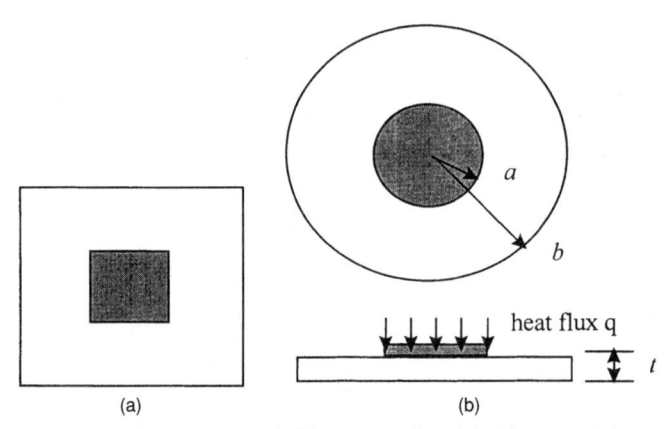

Figure 3.14 Convection-cooled heat spreader; (a) chip or module on heat spreader; (b) circular model.

and the spreader disk has a radius b. For the assembly of modules and a board, the spreader area may be a per-module area of the board.

The nondimensional parameter involved in the conduction/convection conjugate heat transfer problem is the Biot number,

$$\text{Bi} = \frac{hL}{\lambda}$$

where h [W/m² K] is the heat transfer coefficient on the spreader surface, L [m] is the length scale, and λ [W/m K] is the thermal conductivity of the spreader material. The physical meaning of Biot number is developed as follows. We consider a situation where a heat flux (q [W/m²]) is conducted in the spreader material by a distance L, then transferred to the coolant from the surface. The temperature difference between the spreader surface and the coolant ($\Theta_{s/c}$) is of the order of q/h, while the temperature drop in the spreader (Θ_{sp}) is of order qL/λ. Therefore, Bi represents $\Theta_{sp}/\Theta_{s/c}$. Small Bi means that the heat spreader is nearly isothermal and its performance is largely determined by the surface heat transfer. These two measures of temperature, Θ_{sp} and $\Theta_{s/c}$, are called *temperature scales*.

Fisher, Zell, Sikka, and Torrance [6] proposed a formula that allows estimation of the heat-spreading function. Their formula was used to produce curves in Fig. 3.15. The Biot number (Bi) was defined using the radius of the heat source (a) as the length scale and was assumed to be equal on the top and bottom surfaces of the spreader disk. The side surface of the spreader was assumed adiabatic. The temperature becomes maximum at the center of the heat source, and in terms of the temperature rise from the environment, it is denoted as Θ_{max}. The vertical axis of Fig. 3.15 is the nondimensional temperature $\theta_{max} \equiv \Theta_{max}/\Theta_{sp}$, where L in previously defined Θ_{sp} is replaced by a. The horizontal axis is the ratio of the spreader radius to the source radius, $b^* = b/a$. Another parameter is the ratio of the spreader thickness (t) to a, $t^* = t/a$.

Figure 3.15a shows a curve for $t^* = 0.4$ and Bi = 0.1. Actual examples producing these nondimensional values are: a 20 × 20-mm module bonded to a substrate having $\lambda = 3$ W/m K; and h = 30 W/m² K, typically produced on a substrate of 0.16-m length ($b^* = 4$) by a 10-m/s air flow. For a typical value of $\theta_{max} = 2$, the heat flux of $q = 1$ W/cm² (10,000 W/m²) produces $\Theta_{max} = 67$K. Expansion of the spreader area beyond $b^* = 2.5$ does not produce noticeable improvement in Θ_{max} due to poor thermal conductivity of the substrate.

Figure 3.15b shows another example, where $t^* = 0.1$ and Bi = 0.001. Typically, a 1-mm-thick aluminum spreader attached to a 20 × 20-mm module, and h = 30 W/m² K, produce this set of t^* and Bi. The values of θ_{max} are large, but actual temperatures are held small. For instance, set

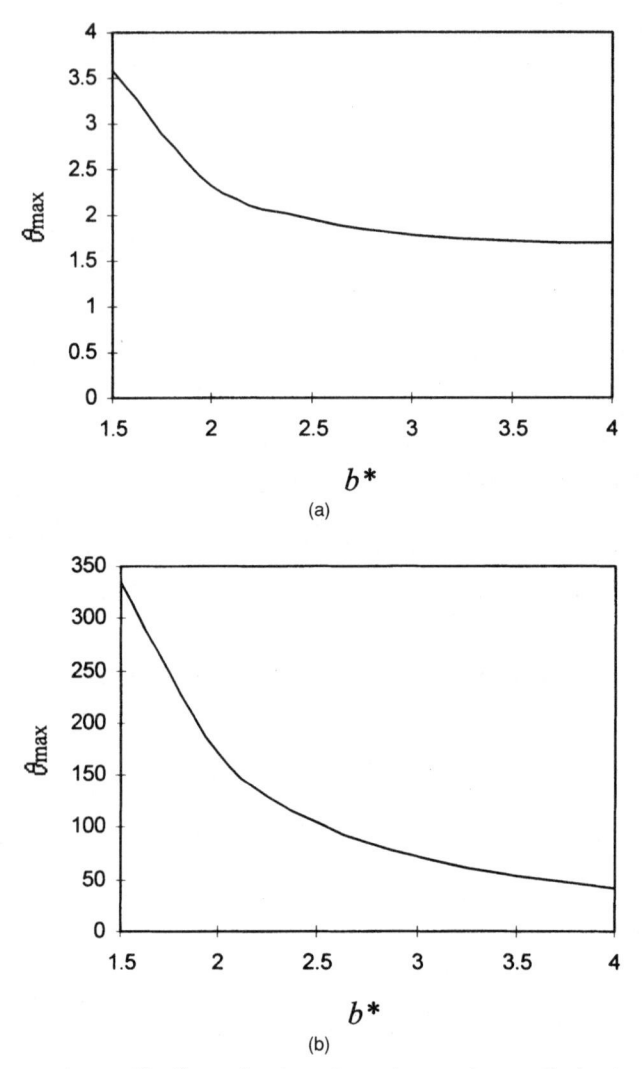

Figure 3.15 Nondimensional maximum temperature on the heat source (Θ_{max}) versus heat spreader radius ($b^* = b/a$); (a) Bi = 0.1 and nondimensional spreader thickness t^* ($= t/a$) = 0.4; (b) Bi = 0.001 and nondimensional spreader thickness t^* ($= t/a$) = 0.1.

$\theta_{max} = 100$. For $q = 1$ W/cm^2, $\Theta_{sp} = 0.435$K, so that $\Theta_{max} = 43.5$K. High thermal conductivity of aluminum makes expansion of the spreader area effective in reducing the heat source temperature.

3.2.9 Flow of coolant in system box

In air-cooled systems heat from the modules and other components is collectively carried away by the cooling air which is driven through the system box. The air temperature at the exit of the system box (T_e) is computed from the heat balance equation,

$$\sum Q = c_p M(T_e - T_0) \tag{3.31}$$

where $\sum Q$ [W] is the sum of heat generation rates in the box, c_p [J/kg K] is the specific heat at constant pressure of the coolant, M [kg/s] is the coolant flow rate, and T_0 [K] is the coolant temperature at the inlet to the system box.

Equation 3.31 is applicable regardless of geometric details of the flow path. It is also applicable to any coolant. To apply the equation, we first define the control volume. In the preceding example the control volume was a system box. To find more detailed information about the temperature in the box, we confine our view on divided zones in the box and apply the heat balance equation on those zones. Figure 3.16 shows such a control volume in the box (V_{sub}) and the equation applied to V_{sub}, where the subscript *sub* to $\sum Q$ and M means the association of these quantities with V_{sub}.

Equation 3.31 can be used to determine the required coolant flow rate (M). The coolant temperature at the exit has to be held below a critical level. The critical level of T_e is determined from the reliability requirement on modules in the system box; namely, T_e has to be below the critical junction temperature (T_j^*) by a certain degree. The margin between T_e and T_j^* depends on the thermal resistance from the junction to the coolant. In order to deliver the required coolant flow rate we

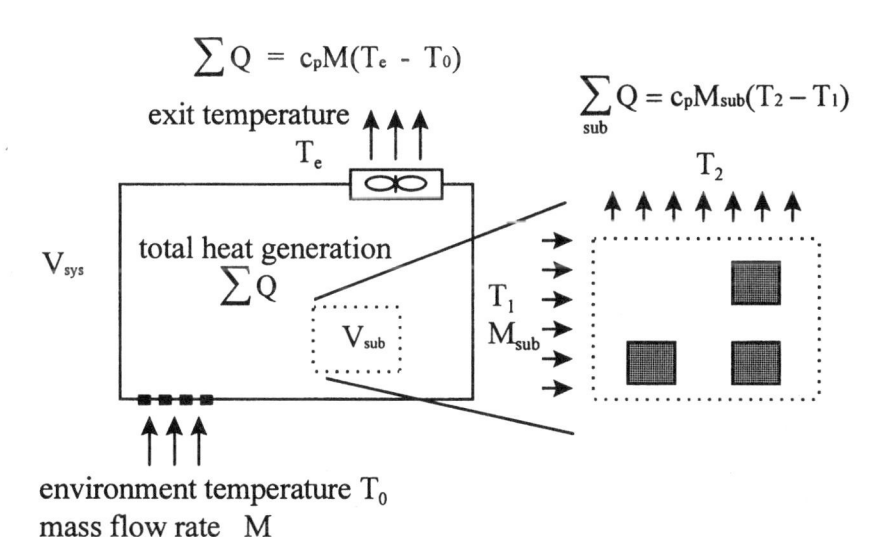

Figure 3.16 Application of Eq. 3.31 to control volumes.

have to estimate the flow resistance in the system box, then select a proper coolant-moving device such as a fan or a blower.

The flow resistance in the system box depends on the placement of components in the system box. The component placement, in turn, determines the distribution of coolant flow within the box, which has to be designed so as to ensure cooling of individual modules and other components. The design needs estimation of coolant flow rates through various control volumes in the box.

The coolant flow is driven by a pressure difference. More precisely, the pressure in this statement is the total pressure (P), defined as

$$P = p + \tfrac{1}{2}\rho u^2 \tag{3.32}$$

where p [Pa] is the static pressure, ρ [kg/m^3] is the density of coolant, and u [m/s] is the coolant velocity. The second term on the right-hand side of Eq. 3.32 is called the *dynamic pressure.*

The total pressure is preserved along the flow path, if there is no loss of kinetic energy in the coolant flow. However, the coolant has viscosity and thereby loses its kinetic energy in various modes. The loss of kinetic energy reduces the total pressure along the flow path; this loss has to be compensated by the work input from a mechanical device such as a fan, a blower, or a pump.

Consider two cross sections in the flow path, sections 1 and 2, which have equal cross-sectional areas, and assume that there is no exchange of coolant mass between the flow path and the surroundings. Then, the coolant velocity is the same at two cross sections. From Eq. 3.32, we see that the difference in static pressure between the cross sections is equal to the difference in total pressure. In this way, we can measure the loss of total pressure from the measurement of static pressure difference called *pressure drop* (Δp).

Figure 3.17 is the representation of such a situation, where p_1 and p_2 are the static pressures at sections 1 and 2, respectively. What causes

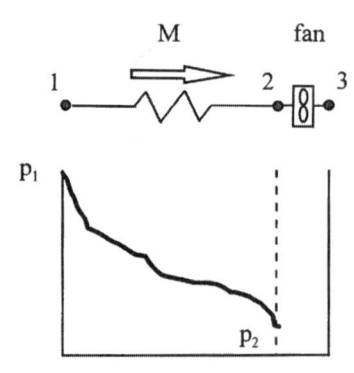

Figure 3.17 Resistance to coolant flow and pressure drop.

the loss of kinetic energy is captured as a resistance to the flow. However, the resistance is generally a function of the coolant velocity; hence, the flow-resistance network is nonlinear. In order to reflect its physical cause, the flow-resistance is defined using the dynamic pressure rather than the flow rate or the velocity. The resistance coefficient (ζ) is defined as

$$\zeta = \frac{\Delta p}{\frac{1}{2}\rho u^2} \tag{3.33}$$

where $\Delta p \equiv p_1 - p_2$

Figure 3.18 shows flow path configurations that present major resistance to the coolant flow: (a) filters installed at the coolant inlet; (b) sharp bends, for example, formed between the exit of the board array and the box wall; (c) contractions and expansions created between components of different sizes; and (d) boards with modules. Due to geometric diversity in the internal structure of the system box, there are few data of ζ which can be directly applicable to any particular situations. Table 3.5 gives a list of typical values of ζ for the path configurations of Fig. 3.18.

3.2.10 Natural convection from flat surfaces

The density of fluid (ρ [kg/m³]) decreases with increasing temperature (T [°C]). In the temperature range of electronics cooling, the relationship between ρ and T is given by

$$\rho = \rho_\infty [1 - \beta(T - T_\infty)] \tag{3.34}$$

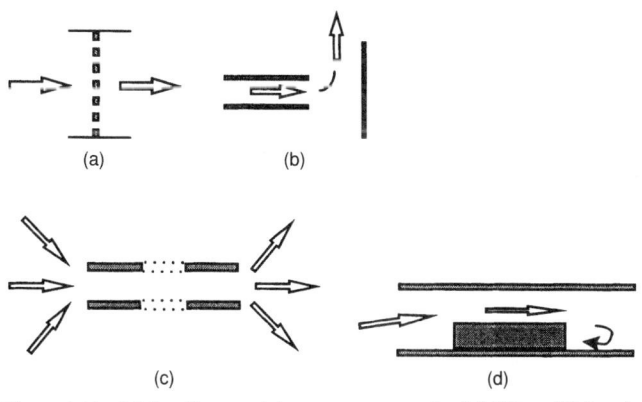

Figure 3.18 Major flow resistance components; (a) filter; (b) bend; (c) contraction and expansion; (d) block.

TABLE 3.5 Typical Values of Resistance Coefficient

Configuration	ζ
Filter	9.4*
Bend	2.5
Contraction, expansion	6.5
Board; narrow channel	2–3**
Board; wide channel	0.1–0.2**

* Ishizuka [11] for the opening to total area ratio = 0.4.
** Reduced from Moffat, Arvizu, and Ortega [14].

where ρ_∞ is the density at the environmental temperature T_∞ [°C], and β [K^{-1}] is the thermal expansion coefficient. Table 3.4 includes the values of β of typical coolants. Heat transfer from the heat source produces the density difference in the coolant, thereby the buoyancy force which drives natural convection. The maximum driving potential is the temperature difference between the surface of the heat source and the environment, $(T_s - T_\infty)$. The magnitude of buoyancy force per unit volume of coolant is written as

$$f_b = \rho_\infty g \beta (T_s - T_\infty) \tag{3.35}$$

where g is the gravitational acceleration (9.81 m/s²).

On the vertical heated surface the boundary layer develops (Fig. 3.19). Suppose that the air velocity in the boundary layer is at the level w. The time required for coolant to pass the height of the heat source (L) is $t = L/w$. During this time period the heat diffuses in the coolant having the thermal diffusivity κ, and the front of the temperature vari-

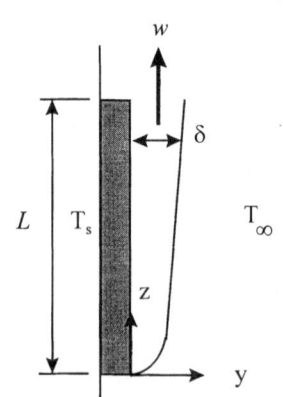

Figure 3.19 Natural convection boundary layer on vertical plate.

ation reaches the edge of the boundary layer δ. From the discussion on heat diffusion developed in Sec. 3.2.2, we have $\delta \approx \sqrt{\kappa L/w}$, or writing for w, $w \approx \kappa L/\delta^2$. Using this w, the viscous force per unit area of the heat source surface is written as

$$F_v = \mu \frac{w}{\delta} = \mu\kappa \frac{L}{\delta^3} \tag{3.36}$$

The buoyancy force acting on a volume of coolant which has unit base area on the surface and extends by δ is f_b of Eq. 3.35 times δ,

$$F_b = f_b \delta = \rho_\infty g\beta(T_s - T_\infty)\delta \tag{3.37}$$

The ratio F_b/F_v determines the state of natural convection under a given set of temperature difference and heat source size. Replacing δ by the given length scale L, we have the definition of the Rayleigh number,

$$\mathrm{Ra} \equiv \frac{g\beta(T_s - T_\infty)L^3}{\kappa\nu} \approx \frac{F_b}{F_v} \tag{3.38}$$

where $\nu = \mu/\rho_\infty$ is used.

A correlation between the Nusselt number ($\mathrm{Nu} \equiv hL/\lambda$) and Ra is given by Churchill and Chu [3], which covers a wide range of Ra including both laminar and turbulent regimes.

$$\mathrm{Nu} = 0.68 + \frac{0.670\,\mathrm{Ra}^{1/4}}{[1 + (0.492/\mathrm{Pr})^{9/16}]^{4/9}} \tag{3.39}$$

For air ($\mathrm{Pr} = 0.7$), Eq. 3.39 becomes

$$\mathrm{Nu} = 0.68 + 0.513\,\mathrm{Ra}^{1/4} \tag{3.40}$$

Figure 3.20 shows examples of heat transfer coefficients in air computed from Eq. 3.40 for a 2-cm-long and a 10-cm-long plate. The difference between the performance of the two plates reflects the effect of the boundary layer thickness which is larger on the latter.

For heat transfer from horizontal surfaces, an upward facing surface (Fig. 3.21a) and a downward facing surface (Fig. 3.21b), the correlations are given in the form

$$\mathrm{Nu} = C\,\mathrm{Ra}^m \tag{3.41}$$

The values of C and m are listed in Table 3.6. The length scale in these correlations is $L \equiv A_s/P$, where A_s and P are the area and the perimeter of the surface, respectively.

142   Chapter Three

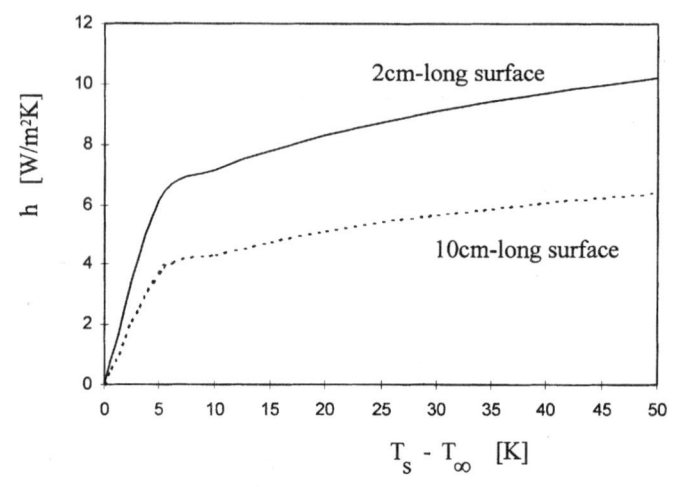

Figure 3.20 Typical heat transfer coefficients for natural convection air-cooling of vertical plates.

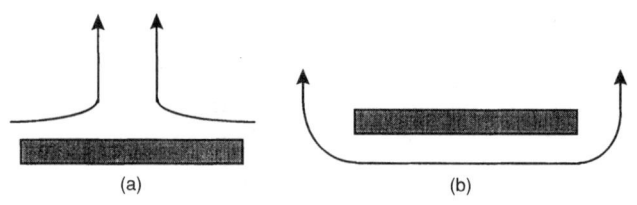

Figure 3.21 Natural convection on horizontal surfaces; (a) upward facing; (b) downward facing.

3.2.11 Natural convection heat transfer in constrained space

Equations 3.39 to 3.41 are applicable for the surface facing an ample free space. Where other components are in close distance from the heat source, the intensity of convection is reduced due to the increased flow resistance in constrained space. Also, where multiple heat-generating components are present, thermal plumes from lower components place upper components in a higher temperature environment.

TABLE 3.6 Constants in Equation (3.41) (Goldstein, Sparrow, and Jones [7]; Lloyd and Moran [12])

Surface orientation	C	m	Applicable range
Upward	0.54	¼	$10^4 \leq Ra \leq 10^7$
	0.15	⅓	$10^7 \leq Ra \leq 10^{11}$
Downward	0.27	¼	$10^5 \leq Ra \leq 10^{10}$

Figure 3.22a shows a pair of boards carrying module arrays. A simplest model of this situation is a pair of heated vertical plates separated by a distance s. The heat generation from modules is spread out over the plate surface having the height L. One model depicted in Fig. 3.22b has the heat flux divided evenly on both sides of the plate, $q/2$ and $q/2$, which will be referred to as the *symmetric model*. In the other model in Fig. 3.22c, the *asymmetric model,* the entire heat flux q leaves on one side of the board, and the other side is left adiabatic. The symmetric model assumes infinite thermal conductivity for the board and the asymmetric model assumes zero conductivity. Actual situations fall between these two models.

According to Bar-Cohen and Rohsenow [1], the local Nusselt number at the top of the plate is given in the form,

$$\mathrm{Nu}_s = \left[\frac{C_1}{\mathrm{Ra}_s\, s/L} + \frac{C_2}{(\mathrm{Ra}_s\, s/L)^{2/5}} \right]^{-1/2} \qquad (3.42)$$

where $\mathrm{Nu}_s \equiv hs/\lambda$, and $\mathrm{Ra}_s \equiv g\beta q^* s^4/\lambda\kappa\nu$. For the symmetric model, $q^* = q/2$, and for the asymmetric model, $q^* = q$. The constants are

$$C_1 = 48 \text{ (symmetric)}, 24 \text{ (asymmetric)}$$

$$C_2 = 2.51$$

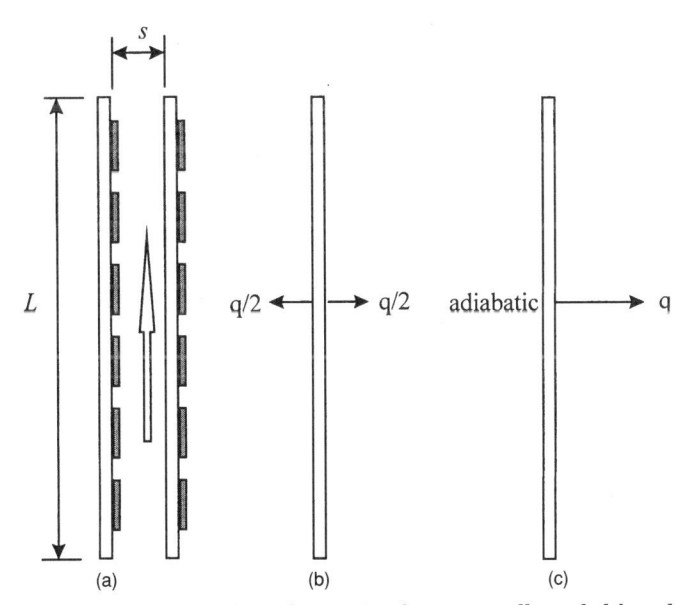

Figure 3.22 Model of heat dissipation from naturally cooled board array; (a) board pair; (b) symmetric model; (c) asymmetric model.

Numerical examples are shown in Fig. 3.23. The board height is set as $L = 20$ cm, and the module near the top end of the plate is assumed to have a size 2×2 cm. The examples are shown in terms of the thermal resistance $R = 1/hA$, where $A = 4$ cm^2, and the plate spacing s. The effect of space constraint becomes obvious when the spacing is reduced below around 10 mm, which corresponds to $s/L < 0.05$. Another noteworthy point is the effect of thermal conductivity of the board. Thermal vias (Fig. 3.24) imbedded in the board may help to bring the thermal resistance closer to that of the symmetric model.

The design option for whether the system box is entirely sealed (Fig. 3.25a) or partially open to the environment (Fig. 3.25b) has great influence on the module temperature. Heat flow from inside the sealed box to the environment encounters large thermal resistance of natural convection on the module surface, the internal surface, and the external surface of the box. Allowing the coolant from the exterior through the vent holes substantially reduces the role of thermal resistance on the box surfaces. Although the number of research reports in the literature on natural convection in the enclosure with and without vents is

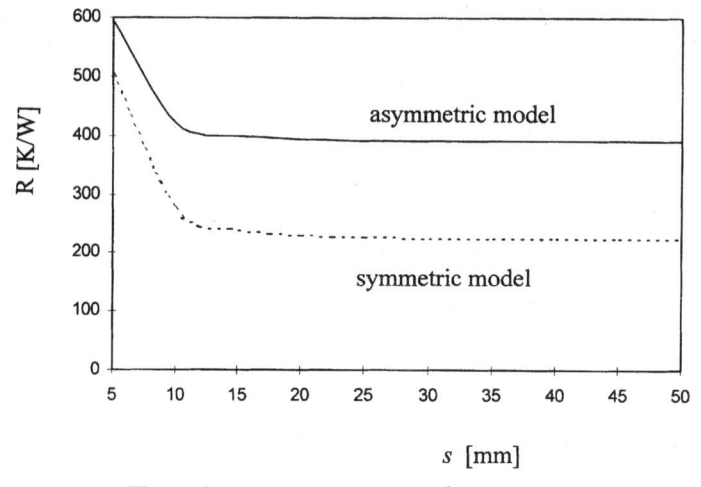

Figure 3.23 Thermal resistance on a 2×2 cm^2 module near the top end of natural convection air-cooled board array; board size 20×20 cm^2, and the total heat dissipation/board = 20W; s = board spacing.

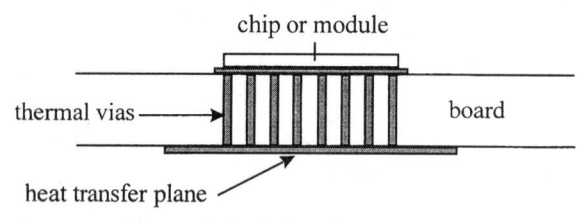

Figure 3.24 Thermal vias in board.

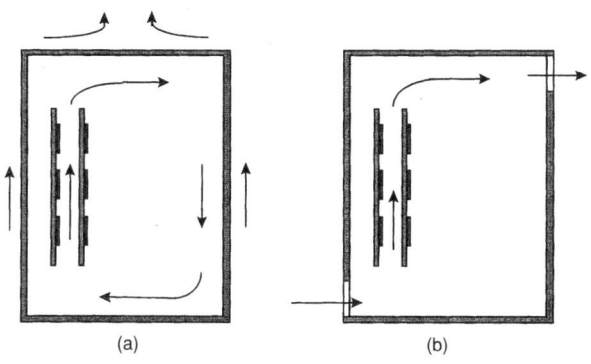

Figure 3.25 Natural convection in system box; (a) sealed box; (b) box with vents.

increasing, there are still few that can be immediately applicable to the design task on hand. One needs to conduct experiment and numerical analyses to see the feasibility of natural convection cooling for every particular case.

3.2.12 Radiation heat transfer

Heat transfer by radiation is a sole mode of cooling in a vacuum environment. It also plays important roles in natural convection cooling of electronic equipment. The radiative heat flux (q_R [W/m^2]) from the surface of temperature T_s [K] to the environment of temperature T_∞ [K] is given by

$$q_R = \varepsilon\sigma(T_s^4 - T_\infty^4) \tag{3.43}$$

where ε is the emissivity of the surface, and $\sigma = 5.670 \times 10^{-8}$ W/m^2 K^4 (Stefan-Boltzmann constant). The emissivity of 0.9–1 may be assumed for the module encapsulation, wiring boards, and most system box materials. Highly polished bare metallic surfaces have small ε, for instance, 0.04 for highly polished aluminum surface.

Usually T_s is not known; hence, where radiation heat transfer is coupled with convective heat transfer, we have to solve nonlinear problems. Fortunately, the temperature difference in electronics cooling is generally small compared to the absolute temperature level measured in Kelvin. Hence, we write

$$T_s^4 \approx T_\infty^4 + 4T_\infty^3(T_s - T_\infty)$$

and from Eq. 3.43 the radiation heat transfer coefficient (h_R) is written as

$$h_R = 4\varepsilon\sigma T_\infty^3 \tag{3.44}$$

Through the use of h_R and Eq. 3.4, radiation heat transfer can be incorporated in the linear thermal resistance network.

Equation 3.44 allows immediate evaluation of radiation heat transfer. Setting $\varepsilon = 1$ and $T_\infty = 300K$, we have $h_R = 6$ W/m² K, which is comparable to the heat transfer coefficients for natural convection cooling shown in Fig. 3.20.

Where the heat source sees the environment only through an opening, the radiation thermal resistance is given as

$$R = \frac{1}{h_R F A} \qquad (3.45)$$

where F is the view factor, which is a view angle from the source to the opening divided by the total angle (in most cases 2π radian). For detailed information about F, refer to the heat transfer textbook (for instance, Incropera and DeWitt [10]).

3.2.13 Liquid cooling

3.2.13.1 Single-phase liquid cooling. Although Secs. 3.2.6 to 3.2.9 cover both air and single-phase liquid cooling, this section highlights the benefit of liquid cooling. Figure 3.26 shows two schemes of liquid cooling that have been employed in commercial computers: (a) immersion cooling by Fluorinert coolant FC77 and (b) indirect water cooling. In the direct immersion cooling, the bare chips or the modules are in direct contact with the coolant. The coolant for direct immersion cooling must not only be chemically stable on its own but also compatible with other materials used in the electronic package. These requirements narrow down the selection of coolant to a family of perfluorinated coolants (FC) supplied by the 3M Company. FC77 is one such

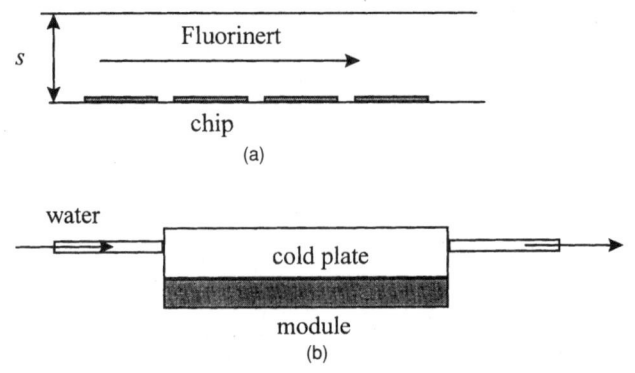

Figure 3.26 Schemes of liquid cooling; (*a*) direct immersion liquid cooling; (*b*) indirect water cooling.

coolant. In the indirect water cooling the water flows in the cold plate, and the cold plate is in contact with the module encapsulation containing chips.

The liquid coolant's large density—hence, large heat capacity—helps to suppress the temperature rise in the coolant path. Because of this characteristic, the path of liquid coolant can be made narrow, while allowing high heat dissipation from the chip array or the multichip module. To estimate heat transfer in channels, the following Nusselt number formula (with a modification on Dittus-Boelter correlation [5]) can be used,

$$\mathrm{Nu}_D = 0.023\, \mathrm{Re}_D^{4/5}\, \mathrm{Pr}^{0.4} \qquad (3.46)$$

where $\mathrm{Nu} \equiv hD/\lambda$, and $\mathrm{Re}_D \equiv VD/\nu$. If the coolant channel has a circular cross section, D is its diameter. For parallel-plate channels, $D \equiv 2s$, where s is the plate spacing. Equation 3.46 is valid for turbulent flows, namely, for $\mathrm{Re}_D > 3000$. In most cases of liquid cooling the flow is in the turbulent regime.

By way of example, we consider a heat transfer surface having an area 10×10 cm^2, which is a typical size of multichip module in mainframe computers. The coolant channel is a parallel-plate channel having the plate spacing s. The equivalent thermal resistance of Eq. 3.8, namely, the capacity of coolant flow to absorb heat, is plotted in Fig. 3.27a with s on the horizontal axis. For both FC77 and water, the velocity is set at 2 m/s, which is a practical upper bound for liquid flow in industrial equipment. It is shown that, even with a 1-mm-high channel, the liquid coolant can absorb heat dissipation of the order of 1 kW at the expense of only a few degrees of temperature rise. If the coolant is air flowing at 10 m/s in a 25-mm-high channel, the equivalent thermal resistance is 0.034 K/W. This example illustrates that the liquid coolant is a solution where the space around the modules has to be made as tight as possible. A typical example of such tight constraint on the space is found in three-dimensional packaging of modules. The Cray-2 supercomputer of Cray Research Inc. is built in three-dimensional packaging and is cooled by FC77.

The liquid coolant is also capable of producing high heat transfer coefficients in narrow channels. Figure 3.27b shows the heat transfer coefficient versus the plate spacing s. If air is used with a 10-m/s velocity in a 5-mm-high channel, the heat transfer coefficient is 57 W/m^2 K, two orders of magnitude lower than those obtained with FC77 and water.

3.2.13.2 Boiling cooling. Boiling is a liquid-to-vapor phase change at a heated surface and is broadly categorized as either pool boiling or flow

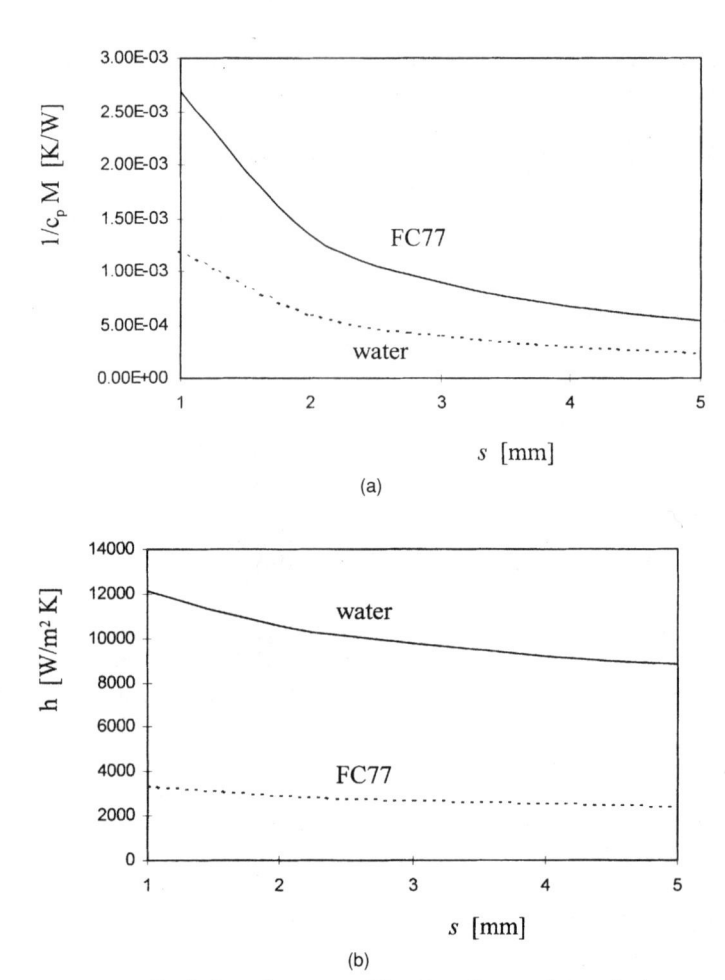

Figure 3.27 Typical performance of liquid cooling; coolant velocity = 2 m/s; channel cross section 10 cm × s; s = plate spacing (= channel height); (a) equivalent thermal resistance for temperature rise in coolant flow; (b) typical heat transfer coefficients.

boiling. Direct immersion of chips in the coolant best utilizes the high heat transfer coefficient brought by boiling of the coolant. For reasons of chemical compatibility of chip and substrate materials, FC coolants and cryogenic coolants such as nitrogen and helium are considered suitable for microelectronic applications. Among them, FC coolants can be used in a room-temperature environment. This is a great advantage of FCs over cryogenic coolants, as far as the requirement for refrigeration systems and the ease of cooling system operation are concerned. Table 3.7 shows the boiling point and the heat of vaporization for typical FC coolants and water. Also included in Table 3.7 is the dielectric

TABLE 3.7 Properties of Fluorocarbon Boiling Coolants and Water (Danielson et al. [4])

	FC87	FC72	FC77	Water
Boiling point (°C)	30	56	97	100
Heat of vaporization (J/kg)	87927	87927	83740	2257044
Dielectric constant	1.71	1.72	1.75	78.0

constant, which is relevant to the electrical performance of direct-immersion cooled modules.

Figure 3.28 shows three principal types of boiling/immersion cooling approaches. In pool boiling as in Fig. 3.28a, the coolant circulation is driven by the buoyancy of the vapor bubbles. When the coolant is driven by a pump, forced-convection flow affects the dynamics of bubble nucleation and growth. Forced-convection boiling is used to cool a densely stacked array of cards as shown in Fig. 3.28b. Cooling chips using an array of impinging liquid jets as in Fig. 3.28c have the benefit of producing a uniform temperature distribution over the chip array. In these approaches, the principal heat transfer surface is the back side of the chip, namely, opposite to the side where integrated circuits are provided. The back side is polished during wafer processing and, if left untreated, poses certain challenges to boiling cooling, as discussed later.

The temperature potential governing the boiling heat transfer is the difference between the surface temperature and the saturation temperature of the coolant, ΔT_{sat} (wall superheat). When the pressure in the vessel is kept as close as possible to one atmosphere, as is almost always the case for a system using FC coolant, ΔT_{sat} is the difference

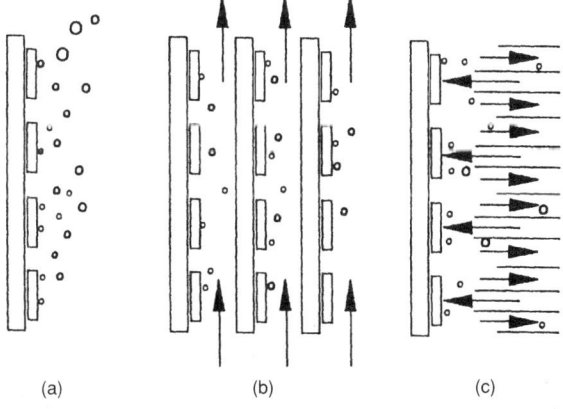

Figure 3.28 Types of boiling immersion cooling; (a) pool boiling; (b) forced convection boiling; (c) jet impingement boiling.

between the surface temperature and the boiling point of the coolant. Figure 3.29 presents a typical boiling curve showing the qualitative relationship between ΔT_{sat} and the heat flux, q. Such a curve is obtained in the laboratory by changing the power input, usually in small steps, to a heater attached to or imbedded in the heat transfer surface.

The line A-B delineates the single-phase convective-heat transfer regime. An incremental increase in the heat flux above the end point of the convective heat transfer line, B, results in the onset of fully established nucleate boiling, and ΔT_{sat} decreases by a certain amount, denoted here as ΔT_{ex} (and called *temperature overshoot* or *temperature excursion*). Further increase in heat flux causes only a small change in ΔT_{sat} up to a certain point D. An increase in the number of bubble nucleation sites with increasing heat flux, together with a rising intensity of bubble formation at active nucleation sites, produces a steep slope of the curve CD. Increasing the heat flux beyond D causes ΔT_{sat} to increase substantially, and at a certain point E, ΔT_{sat} jumps to a level at which the surface temperature goes beyond the tolerable range for microelectronic applications. D is at the point of departure from nucleate boiling and is often referred to as the DNB point. The heat flux at E is known as the critical heat flux (CHF). The phenomena occurring beyond point D are governed by intensified coalescence of bubbles on the heat transfer surface, which ultimately leads to the blanketing of the heat transfer surface by a vapor layer at point E.

The temperature overshoot (ΔT_{ex}) has to be reduced as much as possible. A large ΔT_{ex} means that the chip will reach a high temperature and then undergo a rapid cooling immediately following the onset of nucleate boiling. Also a large ΔT_{ex} implies that the probability of bub-

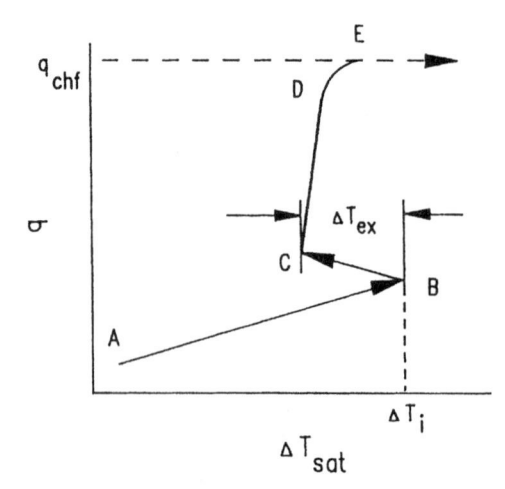

Figure 3.29 Typical q versus ΔT_{sat} curve showing convective heat transfer (A-B), established boiling (C-D), and near CHF (D-E) regimes.

ble nucleation is low for a given combination of surface material, surface finish, and coolant. On a multichip module this means a high probability of finding conditions where bubbles are formed on some chips but not on others. The combination of FC coolant and the polished silicon surface is likely to produce such erratic behavior in boiling, due to the high wetting capability of the FC coolant and the low number density of bubble nucleation sites on the polished surface.

The boiling heat transfer rate on the curve CD can be approximated by

$$q = C_{sf}(\Delta T_{sat})^n \qquad (3.47)$$

where C_{sf} is a constant, and $n = 3$ (usual value) or greater. The C_{sf}, in particular, depends on the combination of coolant and surface material, and the roughness of the heat transfer surface. Generally, rough surfaces yield large C_{sf}, hence, large heat transfer coefficients ($q/\Delta T_{sat}$).

In order to prevent large temperature overshoot and increase boiling heat transfer coefficient, studies were conducted to see the effects of porous structures bonded to the chip. Figure 3.30 shows the boiling curves obtained with a dendrite layer [18], a copper heat sink having vertical channels [18], and a porous copper stud [17]. The coolants FC86 and FC72 in this figure are almost identical in thermal properties and have the boiling point at 56°C. The data in Fig. 3.30 indicate the superior performance of FC boiling on the porous structure. For example, at $\Delta T_{sat} = 10K$, namely the surface temperature at 66°C, the heat flux based on the chip area can be in a range 20 to 70 W/cm². This is produced by the enhanced boiling activity on the porous structure plus the increased actual heat transfer area.

Equally large enhancement of heat transfer and virtual elimination of temperature overshoot can be obtained by jet impingement boiling. With FC72 jets directed to bare chips at the nozzle exit velocity of 7 m/s, a heat flux of the order of 100 W/cm² was reported [16].

3.2.13.3 Heat exchangers. The heat pumped out by the liquid coolant has to be discharged to the atmospheric air through the air-cooled heat exchanger. In boiling cooling, the generated vapor is led to the air-cooled heat exchanger, condensed to liquid phase, releasing latent heat to the air, then the condensate is returned to the liquid pool.

Figure 3.31 shows a schematic of an air-cooled heat exchanger. For the sake of brevity we consider the case of single-phase liquid cooling. Our task is to determine the size of the heat exchanger required to release a given heat dissipation from the electronic system (Q [W]) to the air flow through the heat exchanger. The size of the heat exchanger is represented here by the frontal area (A_F [m²]) and the depth (b [m]).

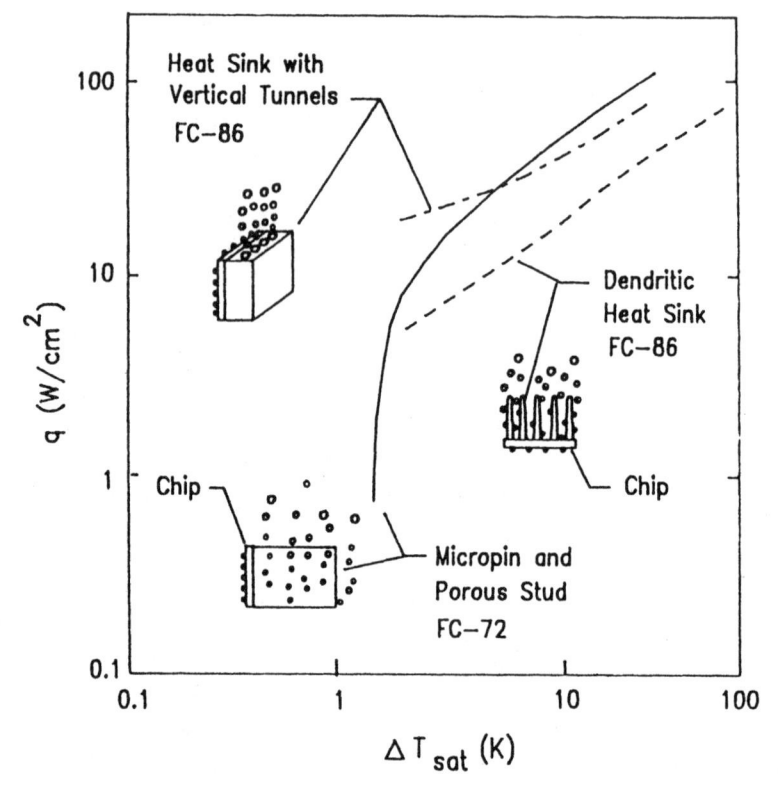

Figure 3.30 Pool boiling curve for various heat sink configurations (adapted from Incropera [9]).

Within the volume $A_F b$, the liquid flow paths have the heat transfer area A_l [m²], and the air-flow paths have the area A_a [m²]. The area-to-volume ratios ($A_l/A_F b$, $A_a/A_F b$) and the area ratio (A_a/A_l) depend largely on the type of heat exchanger described in Sec. 3.4.3.

The mass flow rate of liquid coolant (M_l [kg/s]) is given from the requirement for cooling the electronic equipment. If the tubes connecting the liquid-cooled electronic equipment and the air-cooled heat exchanger are thermally insulated, the difference between the coolant inlet and outlet temperatures ($T_{l,1}$ and $T_{l,2}$, respectively) is given by

$$T_{l,1} - T_{l,2} = \frac{Q}{c_{pl} M_l} \tag{3.48}$$

where c_{pl} is the specific heat of liquid coolant.

The air temperature at the inlet of the heat exchanger is the environmental temperature, denoted here as $T_{a,1}$. The air velocity (v_a [m/s]) through the heat exchanger is usually set at the level of several meters

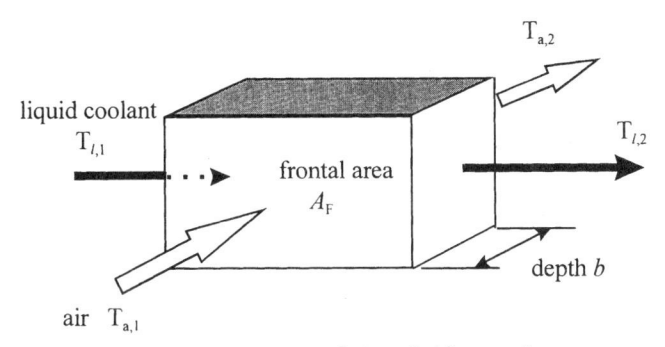

Figure 3.31 System parameters of air-cooled heat exchanger.

per second. The mass flow rate of air (M_a [kg/s]) through the heat exchanger is $M_a = \rho_a v_a A_F$. In the same form with Eq. 3.48 we write

$$T_{a,2} - T_{a,1} = \frac{Q}{c_{pa} M_a} \qquad (3.49)$$

where c_{pa} is the specific heat of air.

Heat transfer from the liquid to the air has to overcome the thermal resistance on the internal surface of the coolant path ($1/h_l A_l$) and that on the air-cooled surface ($1/h_a A_a$), where h_l and h_a are the heat transfer coefficient on the liquid coolant path and that on the air-cooled surface, respectively. As we have seen in the example in the first part of this section, $h_a \approx (1/10 - 1/100) h_l$. To reduce the air-side thermal resistance to a level comparable to that on the liquid side, A_a must be made many times larger than A_l. The performance of the air-cooled heat exchanger depends to a large extent on how large A_a can be accommodated in a compact space.

On the basis of heat transfer area on the liquid coolant path, we define the overall thermal resistance (U [W/m² K]) as

$$U = \frac{1}{1/h_l + (A_l/h_a A_a)} \qquad (3.50)$$

The temperature potential which drives the heat flow from the liquid coolant to the air is the logarithmic temperature difference

$$\Delta T_{lm} = \frac{(T_{l,1} - T_{a,1}) - (T_{l,2} - T_{a,2})}{\ln \left[(T_{l,1} - T_{a,1})/(T_{l,2} - T_{a,2}) \right]} \qquad (3.51)$$

which is derived considering the variation of coolant and air temperatures along the flow paths. Using U and ΔT_{lm}, we write

$$Q = A_l U \Delta T_{lm} \qquad (3.52)$$

Equations 3.48, 3.49, and 3.52 do not provide a single solution for the required size of the heat exchanger. Instead, the problem is to find a minimum heat exchanger volume from solutions that satisfy these equations. For a particular type of heat exchanger, the area ratio A_a/A_l is fixed. The velocity of the liquid coolant is in a range 1 to 2 m/s, and that of the air around 5 m/s. Hence, the frontal area A_F and the depth b of the heat exchanger are treated as virtual variables, and the product $A_F b$ is minimized.

3.3 Thermal Analysis and Design

3.3.1 Objectives of thermal design

The criteria for thermal design of electronic equipment are given in terms of the upper bound for the junction temperature (T_J^*) and that for the variation of junction temperature in the equipment (ΔT_J^*). There have been widely held notions about what these indexes mean to the reliability and performance of the equipment. They are enumerated as follows, together with cautionary comments.

1. As T_J^* is lowered, corrosive reactions of impurity elements that attack electronic devices and wires on the chip are deterred. Namely, the cooler the equipment is held, the longer the equipment survives. Based on this notion, T_J^* is defined at low levels for expensive large systems, for instance, in a 55 to 60°C range for mainframe computers. For consumer electronic equipment the junction temperature of around 100°C is often allowed.

We need a large volume of failure data to establish the design criteria. As a matter of fact, the introduction of new products has been outpacing the accumulation of temperature-related failure data. There are few concrete data bases that justify the setting of T_J^* at particular levels. The prevailing idea is "stick to the criteria which have worked well by now."

2. The characteristics of electronic devices are affected by the temperature. Hence, the variation of junction temperatures increases the probability of timing mismatch between devices, thereby slowing down the signal processing. In principle, this is true. However, due to complexity of signal flows, and also due to the effects of other factors such as the dimensional tolerance in fabrication of devices on device characteristics, there is little ground on which a concrete value is assigned to ΔT_J^*. Again, the "following-the-precedence" idea prevails, and ΔT_J^* is usually set in a range 10 to 15°C.

The comments made in (1) and (2) do not necessarily propose to abolish the presently prevailing design criteria. Instead, they are stated not only to raise caution about the pitfalls but also to point out the multiple aspects of thermal management.

1. The existing criteria alone may not be sufficient to guarantee the reliability of new generations of electronic equipment. As increasingly finer structural features are involved in the packaging, additional criteria may have to be defined to guarantee the structural integrity of components and systems. For instance, the temperature difference between the BGA module and the wiring board needs to be held below a certain threshold in order to reduce stress and strain on the solder joints between the module and the board.

2. Since the criterion for T_J^* itself is hazy to a certain extent, it is not justifiable to attempt to meet the criterion with unduly high accuracy. If more tolerance for T_J^*, say, 5K higher, means substantial reduction of manufacturing cost, it is worth pursuing alternative designs allowed under a modified T_J^*. This does not mean to recommend lax design practice, but comes out of the very nature of electronic packaging design. That is, the packaging design is an art of synthesizing the results of analysis, experiment, prototype testing, and cost evaluation. The thermal criteria need be defined from a viewpoint that extends over whole aspects of packaging design.

3. As electronic devices find their ways into diverse applications, some of them operate in a high-temperature environment. In high-temperature applications, the environment temperature already exceeds the traditional T_J^* of 50 to 100°C. If the conventional criteria are adopted, we need refrigeration systems, which are often not affordable due to required space and cost. The temperature acceleration test—normally conducted to find correlation between T_J^* and the failure rate—is not feasible, because it is likely to cause failure mechanisms of totally different nature, such as meltdown of solder joints. The only way left for the packaging design of high-temperature devices is the identification of possible failure mechanisms on the basis of laboratory experiment and analysis, and the development of countermeasures to failure through materials selection, heat-flow management, and structural design. This so-called "Physics-of-Failure Approach" is universally important for the packaging design of all electronic equipment; however, more time and resources are needed to develop it into a standard practice in the electronics industry.

3.3.2 Breadth and depth of thermal analysis

There are multiple ways to perform thermal analysis; a proper way has to be opted in response to the requirement for the breadth and depth of results. The breadth means the number of variable parameters and the ranges of their variations set in the analysis. The depth means the spatial and temporal resolutions of the solution.

In an industrial environment, the constraint on resources to perform analysis often sets the bound on the quality of analytical results. In planning the thermal analysis we have to first review the resources at our disposal. There are three types of resources to be considered.

1. *Human resources.* The human resources are the available man-hours and the level of expertise of engineers in thermal analysis.
2. *Time.* The time afforded for analytical work may be some months for the design of large systems, a month for the design of small systems, and only one day for emergency troubleshooting.
3. *Tools*
    a. *Utility of knowledge base.* For some design problems the data, the formulas, and the analytical solutions are readily available in the open literature. But, for many problems they are absent.
    b. *Numerical simulation tools.* The number of commercially available numerical simulation codes is rapidly growing. Many of them can be run on desktop and even on laptop computers. Interpretation of simulation results needs human expertise.
    c. *Experimental facilities.* Makeshift mock-ups with simple instruments serve the purpose to capture the global features of coolant flow. The laboratory facility and sophisticated instruments are required to find detailed temperature distributions on, for example, the prototype multichip module.

The available resources need to be mobilized in an effective manner. Understanding the work steps helps to streamline the resource mobilization. The following sections (from 3.3.3 to 3.3.6) describe the work steps that have been customarily followed by design analysts, but in most instances without definite perceptions about them. For the sake of brevity we consider only steady-state cases. For the analysis of transient heat transfer problems, most of what is described in the following can also be applied by substituting the word *temporal resolution* for *spatial resolution.*

3.3.3 Possible cooling modes

The modeling of a system by one solid lump is too crude for any design purposes, but helps to develop a perspective about the option of cooling mode. It is assumed that the heat generated by the components is completely diffused inside the model lump. Heat transfer from the surface area (A) determines the temperature of the lump. For the rule-of-thumb evaluation we do not need to evaluate the heat transfer coefficients using the formulas. Table 3.8 provides a guide for such a task,

Thermal Design, Analysis, and Measurement of Electronics Packaging   157

TABLE 3.8   Ranges of Heat Transfer Coefficients

Fluid	Mode	Heat transfer coefficient (W/m² K)
Air	Natural convection	3–12
Air	Forced convection	10–100
Fluorinert; liquid	Natural convection	100–300
Fluorinert; liquid	Forced convection	200–3000
Fluorinert	Boiling	2000–6000
Water	Forced convection	3000–7000

where the ranges of heat transfer coefficients are shown for different cooling media and modes.

For instance, we think of a laptop computer hermetically sealed in a system box having the surface area, typically, $A = 0.1$ m². We ask how much heat generation is allowed by passive cooling, that is, natural convection cooling by air. From Table 3.8, a typical value of heat transfer coefficient may be picked up as $h = 5$ W/m² K. From Eq. 3.3 the thermal resistance is $R = 1/hA = 2$ K/W. The criterion is set such that the surface temperature of the laptop has to be below 40°C even in the sultry environment of 30°C (= T_0). Then, the available temperature difference is $\Delta T = 10$K, and allowable heat dissipation rate is $Q = \Delta T/R = 5$ W.

To allow higher rates of heat generation we may have to consider forced convection cooling. To estimate the capability of forced convection cooling, we introduce an additional feature to the model; an air duct having the cross-sectional area $A_c = 10$cm² and the length $L = 30$ cm. The heat transfer area is $A = L \sqrt{4\pi A_c} = 0.03$ m². We pick up a value $h = 50$ W/m² K from Table 3.8. Setting $\Delta T = 10$K, we have $Q = 14$ W. We have to check the air temperature at the exit of the duct, which can be computed using Eq. 3.31 and the physical properties of air in Table 3.4. The air velocity is held at a moderate level $v = 2$ m/s, then the mass flow rate through the duct is $M = \rho v A_c = 0.00232$ kg/s. The air temperature at the exit is $T_e = T_0 + Q/c_p M = 36$°C.

As described by the preceding examples, the utility of Table 3.8 is in the order of magnitude evaluation of allowable heat dissipation. Table 3.4 and Eq. 3.31 also have to be used to identify possible modes of cooling. For compact systems where the constraint on space is paramount, the temperature rise in coolant becomes a matter of concern rather than the thermal resistance on the heat transfer surface. In such cases, large heat capacity provided by liquid makes liquid cooling the only viable option.

3.3.4   Scaling

The important step in the initial phase of the thermal analysis is to ask in what spatial resolution we need to know the temperature distribution to meet the objective of the analysis. The spatial resolution means the length scale with which we divide the physical space into subzones.

For instance, consider a case where the objective is to find a temperature distribution that will be used for the thermal stress analysis of solder joints connecting a BGA module to a substrate (Fig. 3.32). The stress analysis, performed either numerically or experimentally, is normally conducted on the assumption that the BGA module and the substrate are at different temperatures, but temperature variations within the BGA module and the substrate are rarely questioned. In this case, the length scale to be applied in the x-y plane is the side length of the BGA module (L). Finer spatial resolutions are not of primary importance. In the z-direction, however, a much smaller length scale equal to the height of the module (H) needs to be employed. Beyond H there are different materials, and large temperature variations are expected.

Where the objective is to determine the chip temperature, the shortest distance where large temperature variation is expected to occur is near the chip, for instance, the thickness of the module encapsulation. This is the case where the length scale is not determined by the size of the object of ultimate interest. Instead, it is determined from the size of the material which surrounds the object and poses appreciable thermal resistance.

As pointed out in the beginning of this chapter, thermal analysis must start from the system-level analysis. With rapid growth of computer performance and simulation codes, we are tempted to assume that any thermal problems will become tractable without resorting to stepwise analysis and modeling. Such days may not come so soon. Suppose that the system is a cube having the side length L_{sys}. When the entire three-dimensional space is divided by the scale L, the number of nodal points is $(L_{sys}/L)^3$. For instance, for small to medium-size systems, we may set $L_{sys} = 30$ cm. With $L = 1$ mm, we have to work with 2.7×10^7 nodes. This size of node matrix is on the border of being manageable by the most advanced massively parallel supercomputing. The matrix size goes far beyond our reach when we try to apply the once-for-all approach to the analysis of larger systems in finer resolutions.

Figures 3.33 and 3.34 show examples of the thermal analysis performed on the laptop computer [8]. Figure 3.33 shows two examples of

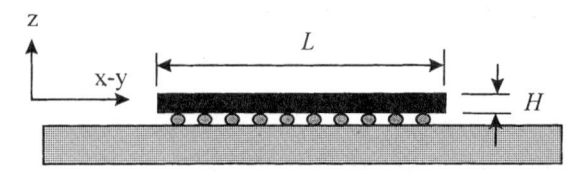

Figure 3.32 Ball Grid Array module mounted on substrate.

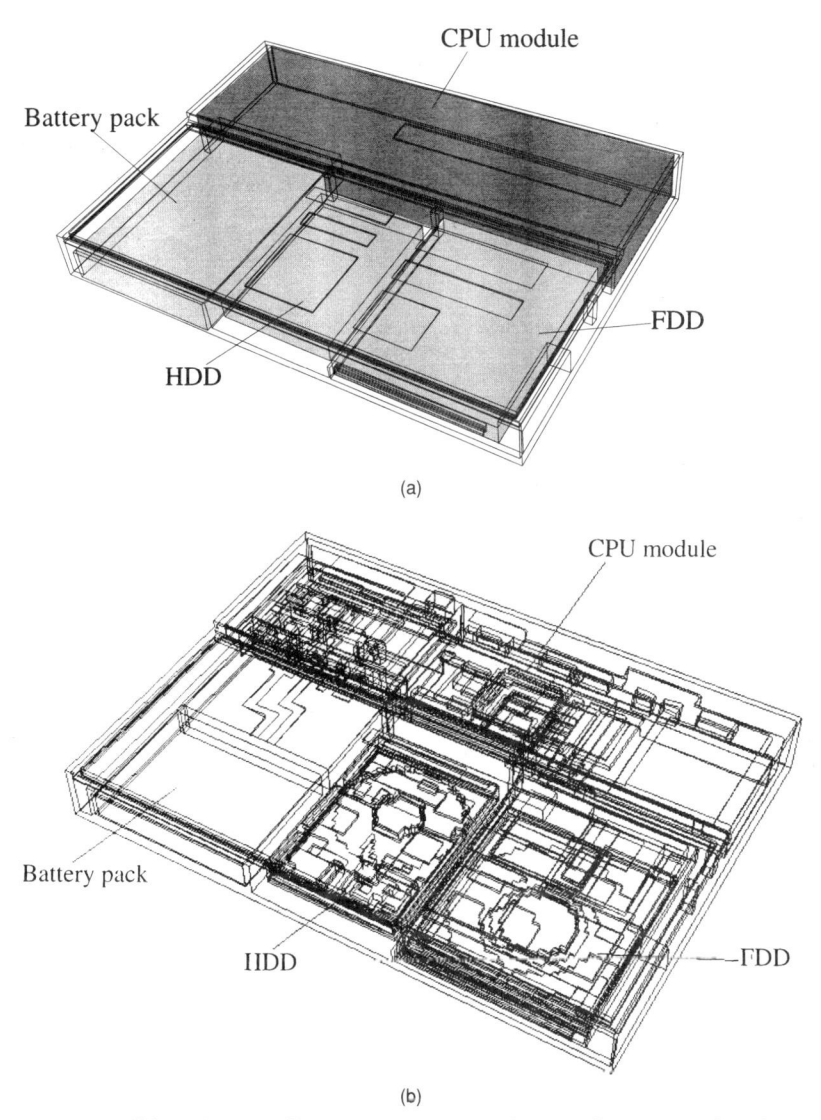

Thermal Design, Analysis, and Measurement of Electronics Packaging 159

Figure 3.33 Discretization of laptop computer into elements for numerical analysis (*Courtesy of K. Hisano, Toshiba Corporation*); (*a*) discretization with 111,132 grid points (line drawing); (*b*) discretization with 957,944 grid points (bit-map drawing).

space discretization. Figure 3.33*a* illustrates major hardware components in rectangular lumps; there are smaller features on the back side of the boards, but they are hidden in line drawing format. Figure 3.33*b* includes detailed features in the components; it is shown in bit-map format, so that the features on the back side of the boards are also included. Figures 3.34*a* and *b* show temperature distributions corre-

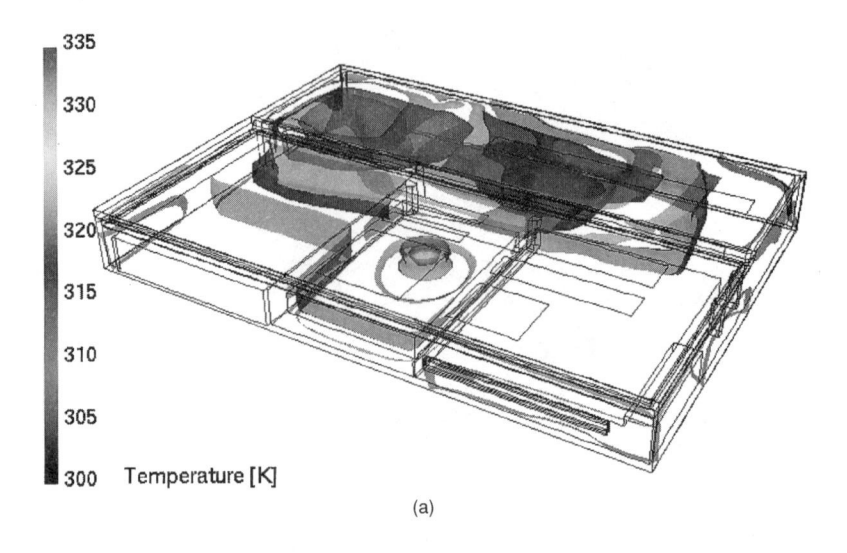

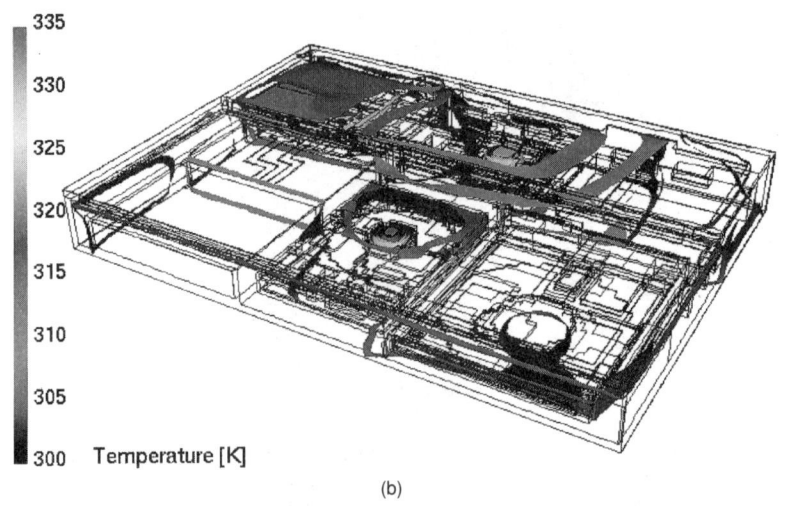

Figure 3.34 Temperature distributions in isotherm imaging (*Courtesy of K. Hisano, Toshiba Corporation*); (a) for the discretization of Fig. 3.33a; (b) for the discretization of Fig. 3.33b.

sponding to Figs. 3.33a and b, respectively. Figure 3.34b indicates the need to apply fine-length scales to the zones containing the CPU module and the hard disc drive. To increase the resolution from that of Fig. 3.34b, the high-temperature zones have to be carved out and zoomed up for the analysis due to the constraint on the computer capacity. Comparison of the results obtained with the divisions of Figs. 3.33a and b, although not entirely visible from the prints, indicates that the coarse mesh of Fig. 3.33a suffices to define the boundary temperatures

Thermal Design, Analysis, and Measurement of Electronics Packaging 161

for the zoomed-up zones. Figure 3.35 shows the temperature distribution on the CPU module found by the zoomed-up analysis.

3.3.5 Modeling

Modeling is an art to simplify the geometry, the thermal and fluid-mechanical boundary conditions, and to find the equivalent thermal properties of complex physical structures. These aspects of modeling are discussed next.

3.3.5.1 Model of geometry. Geometrical features less than the applied length scale (L) are ignored, and the actual structures are smeared even in the volume $L \times L \times L$. The volume will be called hereafter *lump*, or *control volume*. The latter naming is appropriate when we write the heat balance equation about the volume as we see in Sec. 3.3.6 ("Formulation"). In the following these two terms will be used interchangeably.

The number of lumps involved in the system is of the order of $(L_{sys}/L)^3$, where L_{sys} is the representative length of the system. The largest length scale is L_{sys}. The utility of modeling the system box by one lump was already discussed in Sec. 3.3.3 ("Possible Cooling Modes"). As we decrease the length scale for finer division of the physical structure, fewer structural features are involved in one lump. Ultimately, with application of very fine length scales, the control volume becomes a *mesh volume* or *grid volume*, where only one medium is contained and the variations of temperature and velocity (if the medium is fluid) are small.

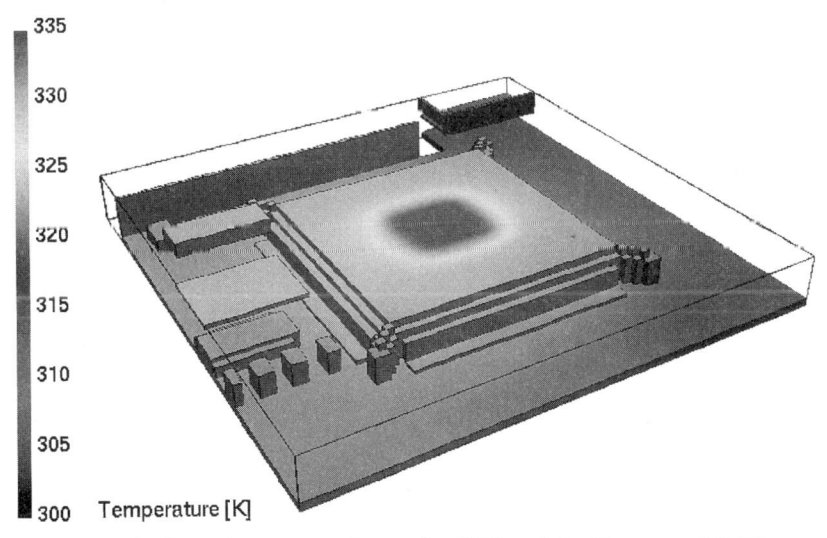

Figure 3.35 Isotherm image on and near the CPU module (*Courtesy of K. Hisano, Toshiba Corporation*).

There are cases where the negligence of geometric features in certain directions reduces the dimensions of the physical space in the model. Modeling an array of DIP packages by a row of two-dimensional ribs is a typical example (Fig. 3.36a and b). The two-dimensional model is permissible where the length scale in the x-direction L_x is much greater than the scales in other directions L_y and L_z, not only in the geometry of component arrangement but also in the coolant flow. The rib model is further simplified to a parallel-plate channel model, where the channel height is employed as L_z (Fig. 3.36c). Another technique to reduce the space dimension is to replace a square by a circular disc, as we have seen in Fig. 3.14. This technique is frequently employed in heat transfer and stress analysis of modules and other components.

3.3.5.2 Equivalent transport properties. Where the lump is surrounded by large thermal resistance on its surfaces, the temperature variation within the lump can be ignored. This is equivalent to assuming that the lump has infinitely large thermal conductivity; hence, it is treated

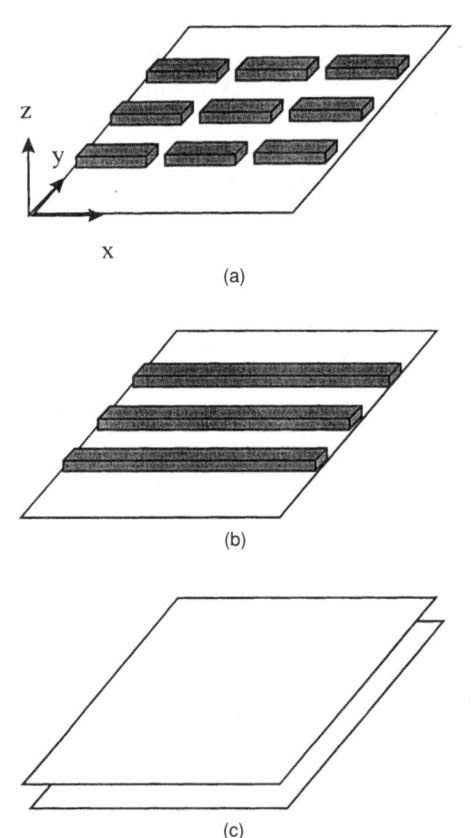

Figure 3.36 Modeling geometry; the case of a board with a DIP module array; (a) DIP array; (b) two-dimensional rib model; (c) parallel-plate channel model.

as an isothermal body. In another extreme case, a planar heat source has uniform heat generation but its thermal conductivity in the plane is very small. The heat-flux distribution on such heat source is uniform, uninfluenced by the thermal situation in the neighboring lumps.

In general, to estimate the equivalent transport properties of the lump we have to look into the detailed features that are ignored in the geometry modeling. In exchange for the need to work on detailed structures we assume a simple thermal boundary condition. A typical example was already discussed in Sec. 3.2.5 ("Heat Conduction in Wiring Substrate"). There, the vias and laminations in the substrate were considered in a one-dimensional heat-flow situation in order to estimate the equivalent thermal conductivity. As this example indicates, the estimation of the equivalent thermal properties is itself the thermal analysis performed on the subsystem under simplified thermal boundary condition.

The thermal conductivity relates the heat flux to the linear temperature gradient in Fourier's law. Hence, the concept of equivalent thermal conductivity holds well where the temperature distribution is supposed to be nearly linear in the direction of heat flow. Where the temperature distribution in the control volume is likely to have strong nonlinearity, the analysis based on the equivalent thermal conductivity may produce only crude results. This is the case in the control volume assumed in the coolant flow. The transport properties of the coolant are the thermal diffusivity (κ), the kinematic viscosity (ν), and the specific heat (c_p). The κ and ν are the properties that control heat and momentum transfer across the stream. It is in the cross-stream direction where large variations of temperature and velocity occur; hence, the concept of equivalent transport properties is in general not feasible there. The difficulty is usually bypassed by employing either one of the following two schemes of control volume setting. One is to have the volume bounded by the solid walls (Fig. 3.37a), where the use of heat transfer coefficient and resistance coefficient (Eq. 3.33) obviates the need to assume the equivalent transport properties for the heat and momentum balance equations. The other is to employ fine spatial divisions (Fig. 3.37b), in particular, near the solid surface, where the

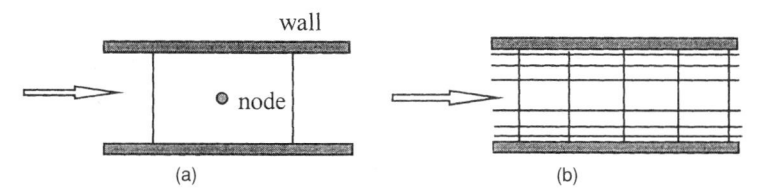

Figure 3.37 Discretization of coolant domain; (a) wall-to-wall lump; (b) fine grid for CFD simulation.

temperature and velocity variations become maximum, so that genuine values of κ and ν can be used without any approximations. With such fine spatial divisions, the analysis is based on the fundamental equations of heat and momentum balance for fluids. The equations are solved numerically by Computational Fluid Dynamics (CFD) codes. In recent years, the commercially available CFD codes are growing rapidly in number, sophistication, and user-friendliness. The power of CFD simulation is in its capability to provide solutions on the picture display with details of temperature and velocity distributions.

It should be noted, however, even in CFD simulations, that some modeling is involved. The reason for the modeling need stems from the constraint on the computational resource. Namely, there is a lower bound for the size of control volume, which allows the computational load to be in manageable range by the computers. In particular, in the simulation of turbulent flows, the presently feasible minimum control volume is still too large for a spectrum of small scale turbulent motions of fluid. The effects of such subscale turbulent motion on heat and momentum transfer in the control volume are modeled in the *turbulence model*. Sufficient amount of know-how has been developed about the turbulence model in fully developed turbulent flows in the paths of simple geometry. The coolant flow in electronic equipment, however, is often in the transition state between laminar and fully developed turbulent regimes. Besides, the flow paths are often tortuous, and the flow separation occurs behind modules and other obstacles. The application of presently available turbulence models to the analysis of coolant flow is an act of approximation. Therefore, the result of CFD simulation has to be benchmarked experimentally.

3.3.5.3 Boundary condition. The thermal boundary condition needs to be assumed on the exterior of the model. The model is an entire actual system as we have seen in Figs. 3.33 and 3.34, or a part of the equipment as illustrated by Fig. 3.35. For each control volume next to the exterior of the model, the thermal boundary condition is specified in one of the following forms: uniform temperature, uniform heat flux, and uniform heat transfer coefficient. The uniform temperature condition on the boundary of the control volume facing the coolant is equivalent to the assumption of infinite heat capacity of coolant and uniform heat transfer coefficient on the boundary. Whether the uniform temperature or the uniform heat flux condition is appropriate for a particular control volume depends on the equivalent thermal conductivity of the neighboring control volumes, which was discussed previously. The heat transfer coefficient in laminar flow regime is sensitive to whether the entire boundary of the model is close to the uniform temperature or the uniform heat flux condition. However, the difference in heat transfer coefficient is usually embedded in other uncertainties resulting from the geometry and property modeling. Most of the heat transfer correlations shown in this chapter are based on the uniform temperature boundary condition.

The uncertainty in solution for the control volumes facing the boundary is inevitable. On the boundary, there is a discontinuity of spatial resolution (Fig. 3.38), and the same boundary condition has to be assumed for all the control volumes next to a boundary. We assume that the uncertainty from the boundary condition is diffused out in the distance between the boundary and the deep inside of the zoomed-up zone. Note that the distance in numerical simulation is measured in the number of grids between two points. Hence, with finer grids the diffusion takes place in a shorter physical distance. Numerical simulation conducted with different grid scales may determine which distance between the boundary and the point of ultimate interest is appropriate.

3.3.6 Formulation

The formulation is the representation of heat transfer process in the system by a set of simultaneous algebraic equations. Each equation in the equation set describes the balance of heat and momentum on a control volume. The balance equation for the volume shown in Fig. 3.39 is written as

$$\sum_{i=1}^{N} \Phi_i + Q_s = 0 \tag{3.53}$$

where Φ_i is the rate of mass flow, heat flow, or momentum flow through the boundary segment i (positive when directing inward), and Q_s is the source term.

In the heat and momentum balance equations the Φ_i is related to the potential difference $(P_{I'} - P_I)$ and the resistance on the link between the

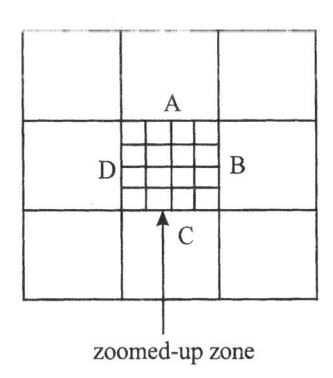

Figure 3.38 Finer discretization of zoomed-up local zone.

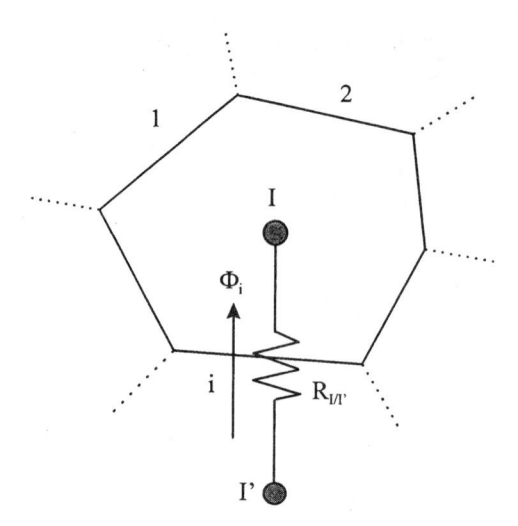

Figure 3.39 Control volume for writing balance equations.

nodal points ($R_{II'}$; I in the volume and I' in one of the neighboring volumes),

$$\Phi_i = \frac{P_{I'} - P_I}{R_{II'}} \qquad (3.54)$$

Substituting Eq. 3.54 into 3.53, and writing balance equations for all nodal points, we have simultaneous algebraic equations, which can be written in the form

$$[A]\{P\} = \{B\} \qquad (3.55)$$

where **A** is the conductance matrix, the element of which is $\pm 1/R_{II'}$, **P** is the column matrix for the potential at nodal points, and **B** is the matrix containing the source and boundary terms.

The analysis of heat transfer in the coolant flow requires a set of three matrix equations to determine temperature, static pressure, and velocity: the heat balance equation where **P** is the temperature matrix; the momentum balance equation where **P** is the static pressure matrix; and the mass conservation or vorticity equation where **P** is the velocity matrix. The matrix **A** in the heat balance equation contains the velocity terms, and that in the momentum balance equation contains the squared velocity terms (Eq. 3.33). Therefore, we have to solve nonlinear simultaneous equations.

3.3.7 Solution

The heat balance equation for solid components is linear; hence, the solution of Eq. 3.55 is straightforward. The matrix inversion, the

Gauss-Seidel iteration, and the successive overrelaxation are the typical methods of solution.

The equations for the coolant flow have to be reduced to linear equations. One way of linearization is to write the square velocity term as $u^{(k-1)}u^k$, where $u^{(k-1)}$ is the value obtained by the previous iteration, and u^k is the value to be determined by the current iteration. The Newton-Raphson iteration is effective in accelerating the convergence to the solution. The left-hand side of Eq. 3.55 for momentum balance is set as $\mathbf{F} \equiv [\mathbf{A}]\{\mathbf{P}\} - \{\mathbf{B}\}$, and the derivative matrix $\mathbf{F}'$ is formed of the elements $F'_{I,J} \equiv \partial F_{I,J}/\partial u_{I/J}$, where $u_{I/J}$ is the coolant velocity directing from the node I to J. The new value at the kth iteration is determined from the following equation substituting the values at the previous $(k-1)$th iteration into $\mathbf{F}$ and $\mathbf{F}'$,

$$[\mathbf{F}'^{(k-1)}]\{\mathbf{P}^{(k)}\} = -[\mathbf{F}^{(k-1)}] + [\mathbf{F}'^{(k-1)}]\{\mathbf{P}^{(k-1)}\} \tag{3.56}$$

For the convergence criterion it is recommended to use the balance equation, Eq. 3.53, rather than the percentage variation of the variables in iteration; namely, the convergence is reached when

$$\left| \sum_{i=1}^{N} \Phi_i + Q_s \right|_{\max} \leq \varepsilon \tag{3.57}$$

where the suffix max means the control volume where the value is maximum, and ε is a small number specified referring to the confidence level of the solution.

The confidence level of the solution can be checked by performing the sensitivity analysis. In the sensitivity analysis the parameters that are considered influential to the final result are varied in possible ranges, and the upper and lower bounds of the solution are identified. Typical examples of such parameters are: the resistance coefficient for coolant flow, the Nusselt number–Reynolds number correlation, the thermal conductivity of materials, the length scale applied to divide the physical space into a set of control volumes, and the thermal boundary condition.

Finally, where the method of superposition (Sec. 3.2.2) is applicable, we can construct a solution from a set of building-block solutions. For instance, in Fig. 3.38, the zoomed-up zone has four boundaries, A, B, C, and D. We solve a heat conduction problem with the condition that the temperature on the boundary A is unity, and zero on the others. The solution is one of the building-block solutions, which is denoted here as $\{\Theta_A\}$. Likewise, we have building-block solutions $\{\Theta_B\}$, $\{\Theta_C\}$, and $\{\Theta_D\}$. For the general situation where the boundary temperatures are T_A, T_B, T_C, and T_D on the boundaries A, B, C, and D, respectively, the temperature distribution in the zone is given by

$$\{\mathbf{T}\} = T_A \{\Theta_A\} + T_B \{\Theta_B\} + T_C \{\Theta_C\} + T_D \{\Theta_D\} \tag{3.58}$$

Once the building-block solutions are obtained and stored, they can be used repeatedly to find solutions for different thermal boundary conditions. We have such convenience where the heat transfer is a linear process. Heat conduction in solid components is a linear process. Heat transfer in forced convection coolant flow is also a linear process. However, the momentum transfer in the coolant flow is in general nonlinear; hence, the velocity solutions cannot be superposed. Natural convection heat transfer is a nonlinear process, because the temperature distribution is coupled with the velocity distribution.

3.4 Hardware Devices for Cooling

The thermal design has to be implemented by selecting proper devices for cooling. In this section, the air-cooled heat sinks, the fans and blowers, the liquid-cooling hardware, the heat pipes, and the thermoelectric coolers are explained.

3.4.1 Air-cooled heat sinks

As we have seen in Table 3.8, the heat transfer coefficient for forced convection heat transfer to air flow is utmost of the order of 100 W/m² K. This is translated to a thermal resistance of about 25 K/W on the surface of a 4-cm² module. Using the formula for fully turbulent flow in Table 3.3, we find that, to achieve 100 W/m² K, we need to have an air velocity of 12.7 m/s over the 2-cm-long surface. This example illustrates a practical limit to the air-cooling without heat sinks. The heat sink is to provide extended surfaces for heat transfer from the module to the air flow. The increase of heat transfer area offsets low heat-transfer coefficients of air cooling, thereby producing low thermal resistance (see Eq. 3.3). An important geometric parameter of the heat sink is the ratio of actual heat transfer area (A) to the base area (A_b), $r_A \equiv A/A_b$.

Four typical types of heat sinks are shown in Fig. 3.40. Among them the plate-fin heat sink (Fig. 3.40a) has been most popular because of the ease of manufacturing. Extrusion forming of aluminum is the common method of manufacturing plate-fin heat sinks. The extrusion-

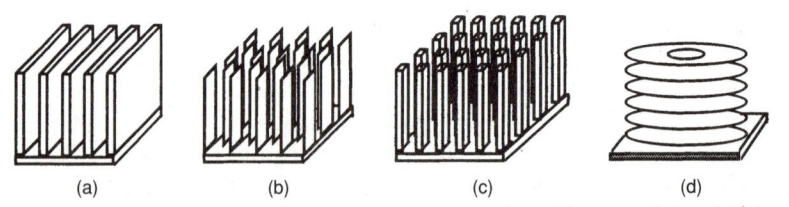

Figure 3.40 Typical air-cooled heat sinks; (a) plate fins; (b) serrated fins; (c) pin fins; (d) disc fins.

formed aluminum heat sink for a 4-cm² module has a dimension, typically, 1 mm fin thickness, 1–2 mm spacing between the fins, 1–1.4 cm fin height, hence, r_A = 8–16. Recently, finer fins have been adopted for cooling of high-power modules; the fin is typically 0.15 mm thick and 12 mm high, and the placement pitch of fins is 1 mm. An area ratio of 25 is obtained by such fine fins. The manufacturing of fine-fin heat sinks needs blazing or bonding process to implant fins on the base metal.

Serrated fins (Fig. 3.40b) are designed to increase the heat transfer coefficient on the fins. We have seen in Sec. 3.2.6 that the shorter the streamwise length of the heat transfer surface, the thinner the average boundary layer thickness; hence, the higher heat transfer coefficient results.

The pin-fin heat sink (Fig. 3.40c) and the disc-fin heat sink (Fig. 3.40d) have least sensitivity of the heat-transfer performance to the orientation of coolant flow. This is a great advantage in small systems where the constraint on the space does not necessarily guarantee a rectified unidirectional flow in front of the heat sink. The pin-fin heat sink is often coupled with a small axial fan, with the fan positioned directly above the pin array to enhance heat transfer.

The geometric parameters involved in the heat sinks are numerous and the coolant flow in and around the fin array is complex. Due to the involvement of many parameters in heat transfer from the heat sink, there are few concise formulas that allow the prediction of heat transfer performance. The prediction has to be based on elaborate analytical models, CFD simulations, experiments, or the data supplied by the vendors. The most difficult part of the modeling is the prediction of air flow through the fin array. Almost without exception, there is open space around the heat sink, which allows some coolant mass to bypass the heat sink. Also, some of the coolant mass that once enters the fin array escapes to the free stream in the middle of the inter-fin flow paths. Due primarily to the increasing rate of this bypass flow, the relationship between the thermal resistance (R) and the air velocity in front of the heat sink (V_F) takes the form depicted in Fig. 3.41. For most of the commercially available heat sinks, the flattening of R begins when V_F is increased above 3 to 5 m/s, depending on the number density of the fins.

Finer thickness of the plate fins or finer diameter of the pin fins allows the plantation of a greater number of fins on a given base area, providing a larger actual heat transfer area. However, the heat transfer enhancement levels off with reducing the fin thickness. Figure 3.42 depicts a situation where a part of the fin area near the tip does not serve as an effective heat transfer area. The local temperature of the fin decreases from the base temperature T_b toward the tip, as the

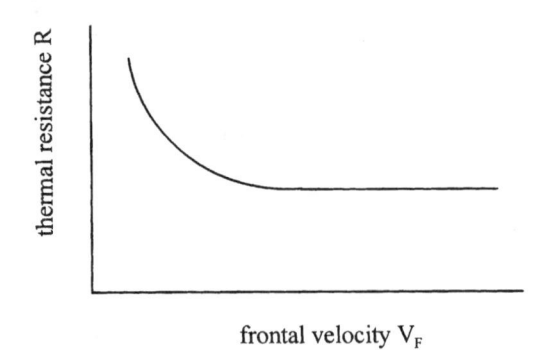

Figure 3.41 Level-off of air-cooled heat sink performance with increasing air velocity.

coolant takes away the heat from the fin. Since this means decreasing temperature potential to drive the heat from the fin to the coolant, the local heat transfer also decreases toward the fin tip. Where the fin temperature is close to the coolant temperature, the rate of heat transfer drops to a negligible level. The temperature drop in the fin depends on, among others, the thermal conductivity of the fin material (λ_F). If we use an imaginary material having infinitely large thermal conductivity, the fin temperature becomes T_b everywhere, and the heat transfer rate from such fin is an upper bound for the performance of real fins of the same dimensions. The fin efficiency (η_F) is defined as $\eta_F \equiv Q/Q_b$, where Q is the actual heat transfer rate, and Q_b is the heat transfer rate for the fin having infinite λ_F. For the plate fin having the height H_F and the thickness t_F, the fin efficiency is given by

$$\eta_F = \frac{\tanh(mH_F)}{mH_F} \tag{3.59}$$

where $m \equiv \sqrt{2h/\lambda_F t_F}$, h is the heat transfer coefficient, t_F the fin thickness, and H_F the fin height. For a 0.15-mm-thick aluminum fin ($\lambda_F =$ 230 W/m K, from Table 3.1) in the air flow giving $h = 100$ W/m² K (almost a practical upper bound), the fin efficiency becomes 0.5 at $H =$

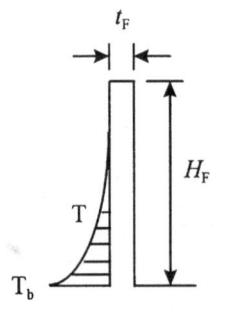

Figure 3.42 Temperature distribution on the fin; the case of low fin efficiency.

17 mm. This numerical example indicates that the efficiency of aluminum fins remains at a reasonably high level in practical situations.

The benefits of the heat sink are impaired when the bonding of the heat sink to the heat source is not appropriate. Today, besides thermal paste whose λ is listed in Table 3.2, various bonding materials are commercially available. Easy-to-use thermal pads and tapes have the specific thermal resistance in a range 0.7 to 1.5K cm^2/W. High-performance polymer adhesive provides 0.3K cm^2/W.

3.4.2 Fans and blowers

The characteristic of the air-moving device is determined using the test apparatus of Fig. 3.43. The apparatus is composed of a large chamber, a valve that controls the air flow into the chamber, a pressure gauge to monitor the static pressure in the chamber (against the room pressure, Δp), and a flow meter to measure the volumetric flow of air through the chamber (V). The test device is installed on a wall of the chamber, draws air from the chamber, and discharges it to the room (suction mode). When the valve is totally open, the resistance to the flow is minimal, so that Δp is close to zero, and V reaches maximum. Squeezing the valve reduces both V and the internal pressure of the chamber, thereby increasing Δp. Changing the valve opening successively, we obtain a curve in the Δp-V graph, called *characteristic curve*.

The most popular type of air-moving device for electronics cooling is the tube-axial fan (Fig. 3.44a). The tube-axial fan occupies little space, is low in cost, and is durable. Its characteristic is depicted by a solid curve ("axial fan characteristic") in Fig. 3.45. The actual characteristic curve depends on the diameter and the rotation speed of the fan. Fans of various diameters are available from vendors, from as small as around 20 mm. Small fans are often coupled with finned heat sinks to provide locally enhanced cooling for high-powered modules.

For large systems, squirrel cage blowers are often used (Fig. 3.44b). The blower's characteristic curve is schematically shown in Fig. 3.45 by

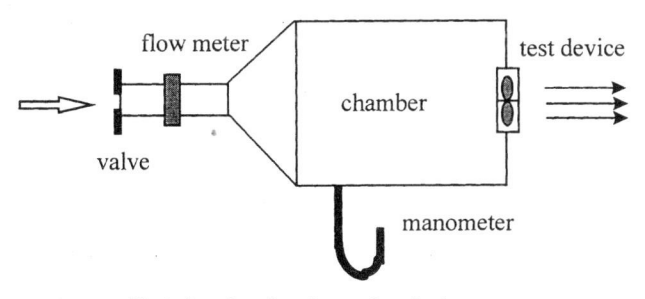

Figure 3.43 Test chamber for air-moving device.

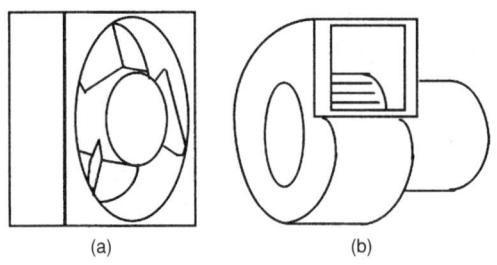

Figure 3.44 Air-moving devices; (a) tube-axial fan; (b) squirrel cage blower.

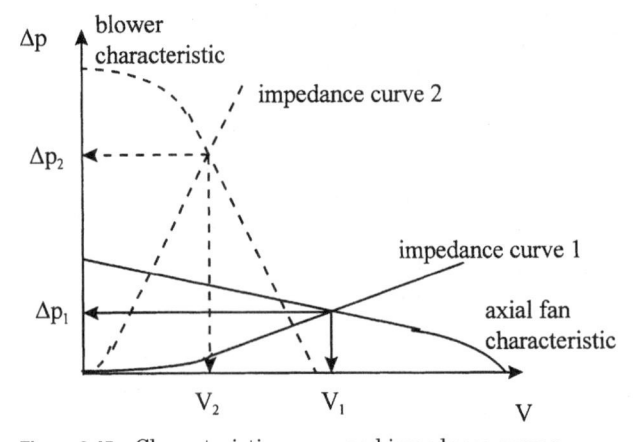

Figure 3.45 Characteristic curves and impedance curves.

a broken curve ("blower characteristic"). The curve illustrates that the blower is capable of overcoming relatively large flow resistance in the coolant path. The blower's actual characteristic depends on the size and the rotation speed.

When installed in electronic equipment, the air-moving device operates at a particular point on the characteristic curve. The operation point is at the intersection between the device characteristic curve and the flow resistance curve in the Δp-V graph. The flow resistance is represented by a near-parabolic curve called the *impedance curve*.

The importance of choosing the proper device is illustrated by two impedance curves in Fig. 3.45. The volumetric flow rate V_1 can be obtained by the fan having the characteristic shown by a solid curve, but not by the blower having the characteristic curve shown by a broken curve. On the other hand, the pressure drop Δp_2 can be overcome by the blower, but not by the fan. Of course, this is a simplified account of the device selection. In actual designs there are many options open to the designer. Even within the same type of devices there are such optional items as the blade design, the diameter of the device, and the use of multiple devices in parallel or serial installation.

The correctness in device selection has to be confirmed by experiment, using either a mock-up of equipment or real equipment. Experimental confirmation is necessary, because the device characteristics are likely to depart from the vendor-supplied data. The deviation is possibly large in small systems, where fans are installed in constrained space. Proximity of solid objects to the fan distorts the approaching flow of coolant to the fan. Uneven velocity distribution prior to the fan inlet generally lowers the fan performance.

One of the important design items for air-cooled equipment is the management of acoustic noise. There are three causes of noise generation from the air-moving device: (1) the aerodynamic cause from the rotation of vanes; (2) the mechanical vibration of the device; and (3) the resonance in cavities in the coolant path. The analysis of frequency spectrum of the noise helps to identify major causes of noise generation, thereby to find the measures to lower the noise level.

3.4.3 Liquid-cooling hardware

The basic components of the single-phase liquid-circulation system are the pump, the piping, the reservoir, and the air-cooled heat exchanger (Fig. 3.46). Proper matching of the flow resistance in the piping and the pump delivery capability is the first step of design. There are more steps that are necessary to ensure reliable operation of the cooling system. The reservoir must have enough capacity to calm pressure surges

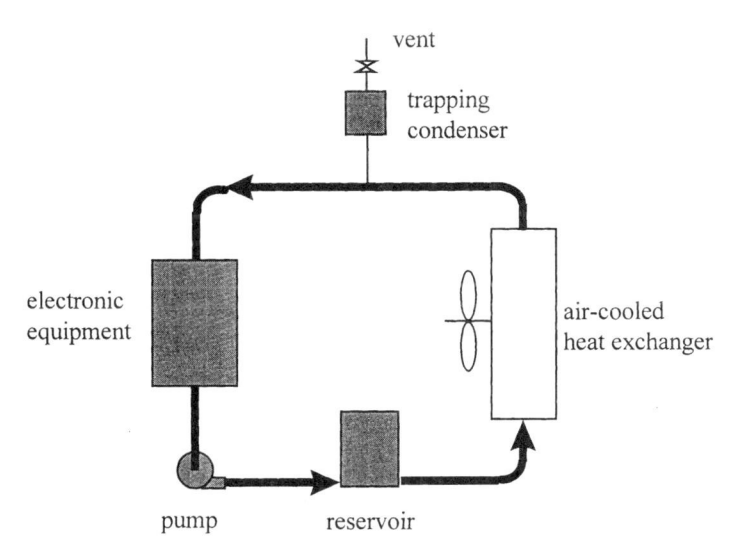

Figure 3.46 Liquid coolant circulation loop.

in the coolant flow. The presence of air and other foreign gases in the piping virtually lowers the pump performance. They have to be trapped and kept from the main piping. Dirt has to be removed by the filter. The piping needs to be disconnected with ease from the electronic module or the system box in time when such disconnection is needed to replace defunct components.

The design of the air-cooled heat exchanger is likely to produce an appreciable impact on the overall system size and the cost. Oversized heat exchangers mean not only penalties on the system's physical size, but also extra cost incurred by redundant liquid coolant in the heat exchanger. As we have seen in Sec. 3.2.13.3, the air-side heat transfer area has to be at least a few tens times the area on the liquid side. Hence, the important index to represent the compactness of the air-cooled heat exchanger is the air-side heat transfer area per unit volume of the heat exchanger (A_{air}/V_{hx} [m^2/m^3]). The A_{air}/V_{hx} is largely determined by the manufacturing method of the heat exchanger.

Two types of heat exchanger are depicted in Fig. 3.47: the tube-in-fin heat exchanger (Fig. 3.47a) and the compact radiator (corrugated-fin heat exchanger, Fig. 3.47b). The manufacturing of tube-in-fins is a relatively simple process: holes are punched through aluminum plates, copper tubes are inserted in those holes, then expanded from inside by pushing balls through the tubes. U-bends are then blaze-bonded to complete paths for liquid coolant. Aluminum plates serve as the extended surfaces to dissipate heat from the tubes to the air flow. Due to its simple manufacturing process the tube-in-fin heat exchanger is low cost; however, it is difficult to increase the (A_{air}/V_{hx}) ratio above the level of 600 m^2/m^3.

The compact radiator is manufactured by sandwiching corrugated metal strips between flat tubes, and blaze-bonding the fins and the tubes in the furnace. The ratio (A_{air}/V_{hx}) can be higher than 700 m^2/m^3 by employing finer corrugation pitch and flat tubes. The cost of the unit is higher than that of the tube-in-fin; however, it depends on the volume of purchase order from the vendor.

The phase change of coolant produces significant benefits to thermal management of electronic equipment, as we have seen in Sec. 3.2.13. However, it requires extra caution in the design of coolant circulation

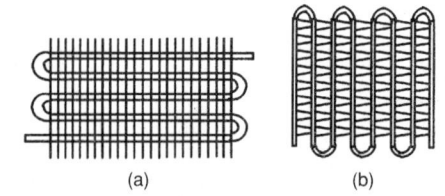

Figure 3.47 Typical air-cooled heat exchangers; (a) tube-in-fin heat exchanger; (b) corrugated-fin heat exchanger.

system. The internal pressure of the cooling system has to be maintained close to the atmospheric pressure by installing a properly designed reservoir. Excessive pressure poses undue mechanical stress on the system structure and causes leakage of coolant. Low internal pressure, on the other hand, causes leakage of air into the system. The air is noncondensable; hence, it tends to accumulate inside the tubes of the air-cooled heat exchanger and decreases the condensation heat-transfer coefficient. The possibility of allowing air and other noncondensable gases in the circulation loop has to be lowered as much as possible. In particular, the Fluorinert coolant has a high solubility of air and water, so that degassing is needed after filling the coolant in the circulation system. Degassing requires a trapping condenser which traps the coolant vapor, while allowing noncondensable gases to escape from the system (Fig. 3.46).

3.4.4 Heat pipes

The principle of heat pipe operation is depicted in Fig. 3.48. The working fluid inside the tube undergoes phase change, thereby transporting thermal energy in the form of latent heat. The working fluid turns into vapor in the evaporator section, the vapor flows through the adiabatic section, then turns into liquid in the condenser section releasing latent heat to the cooling air through the heat sink. The condensate returns to the evaporator section by capillary action of the wick or the grooves provided on the internal wall of the pipe. A high rate of heat transport by phase change makes the equivalent thermal conductivity of the heat pipe many times larger than those of metallic rods. Water is the most effective working fluid due to its large latent heat, and also its large surface tension force that helps the capillary action of the wick or the grooves. The saturation temperature of water inside the pipe has to be lowered to a level that is compatible to electronics cooling applications. This is achieved by charging water in the pipe in vacuumed condition. Copper tubes are normally used for reasons of materials compatibility with water.

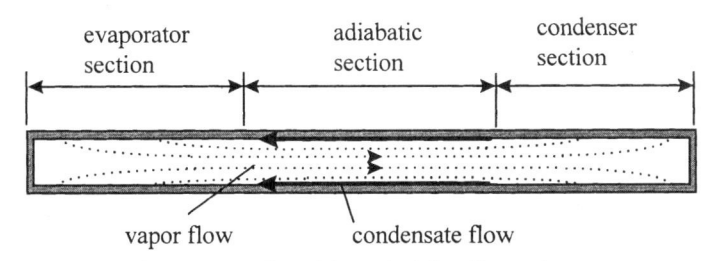

Figure 3.48 Structure and working principle of heat pipe.

The function of the heat pipe is to transport heat from the heat source to the air-cooled heat sink, the latter requiring a large space that is often not available near the heat source. The overall thermal resistance from the heat source to the air over the condenser section is almost dominated by the thermal resistance on the air-cooled heat sink. The heat pipe used in recent generations of laptop computers thermally connects the microprocessor module to the wide-area aluminum heat spreader from where the heat is dissipated to the air (Fig. 3.49a). The overall thermal resistance is typically 4 to 6 K/W. For high-duty applications, a few heat pipes are arranged in parallel and connected to the array of plate fins (Fig. 3.49b). Because of large heat transfer area provided by the fin array the thermal resistance on the air-cooled end approaches that on the tube-in-fin heat exchanger. The overall thermal resistance in this case may be lowered to the order of 0.2 to 0.4 K/W.

3.4.5 Thermoelectric coolers

The thermoelectric cooler has the structure shown in Fig. 3.50. The P and N type semiconductor pellets are sandwiched between two ceramic plates—one of the plates in contact with the heat source and the other with the air-cooled heat sink. The P and N pellets are electrically connected in series. They also provide parallel heat-flow paths from the heat source to the heat sink. The DC source supplies the flow of electrons through the pellets. At the P-N junction on the heat-source side, the electrons jump to higher energy states, absorbing thermal energy.

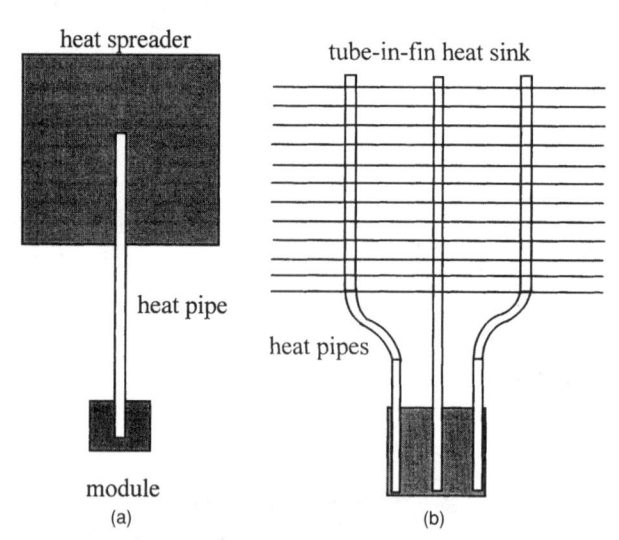

Figure 3.49 Typical installation forms of heat pipes; (a) heat pipe with heat spreader; (b) heat pipes with air-cooled heat exchanger.

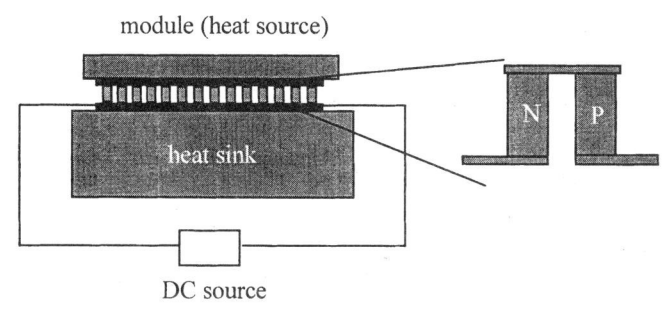

Figure 3.50 Thermoelectric cooler; basic structure.

On the heat-sink side the electrons settle back to lower states, releasing thermal energy to the heat sink. This heat pump mechanism is called the *Peltier effect*.

The heat pump performance of the thermoelectric cooler depends on the number of pellets and the electric power supplied from the DC source. The heat pump capacity can be increased by using multiple numbers of pellets. On the other hand, the pellet has to be made thin to reduce the electrical resistance; hence, the total thickness of the cooler is in a few millimeter range. This means that the heat sink has to be located in close proximity to the heat source.

The ratio of the heat pump capacity to the power input—that is, the heat pump efficiency—is largely determined by the pellet material and the temperature range of operation (i.e., the temperature difference between the heat source and the heat sink, ΔT) for a given heat pump head. Presently, bismuth-telluride is used in commercial coolers, which affords a heat pump efficiency of around several tens percentage points for $\Delta T = 30$ to ~ 60°C. The heat sink has to handle the heat dissipation, which grows many times higher than that from the heat source to be cooled with increasing ΔT.

In certain applications, thermoelectric coolers have definite advantages: their installation and operation are relatively simple, so that the coolers are fit for compact equipment such as CCD cameras, and also for electronic and optical devices in robots which operate in a hostile environment. The heat source temperature can be controlled in a small range and in short time; hence, thermoelectric coolers are coupled with temperature-sensitive devices such as laser diodes.

3.5 Measurements

Measurement is an integral part of thermal management; measurements of thermal and flow fields in electronic equipment are conducted for various purposes:

1. Confirmation of thermal design
   a. In the product development phase
   b. In routine product screening (or rating)
2. Diagnosis of overheating (troubleshooting)
   a. For prototype products
   b. In the field
3. Monitoring of thermal state during operation

The requirements for accuracy, and spatial and temporal resolutions of data depend on the purpose of measurement. Detailed measurements of temperature, coolant velocity, and acoustic noise are required during the product development phase, (1.a) and (2.a). The product screening (1.b) requires temperature measurement at one or a few locations. In this category we may include the measurement of case-to-air thermal resistance of modules conducted by semiconductor vendors. For diagnosis in the field (2.b), the global temperature map provides useful information. For systems that normally rely on natural convection cooling, in situ monitoring of the chip temperature is used to turn on a cooling fan in the event of impending overheating. This section summarizes the sensors and the methods of measurement which are employed in various phases from product development to field service.

3.5.1 Temperature measurements

3.5.1.1 Thermocouples. The thermocouple utilizes the electric voltage produced at a junction of two unlike metals, which varies in proportion to the temperature (the *Seebeck effect*). Figure 3.51 shows a schematic of the temperature sensor, composed of two wires of metal A and metal B, and a millivolt meter. The wires are joined at two ends—one is placed at a measurement spot and another kept at a fixed reference temperature (T_0). The difference in electric potential between the two ends is measured by the millivolt meter and converted to the temperature (T_1) of the measurement spot.

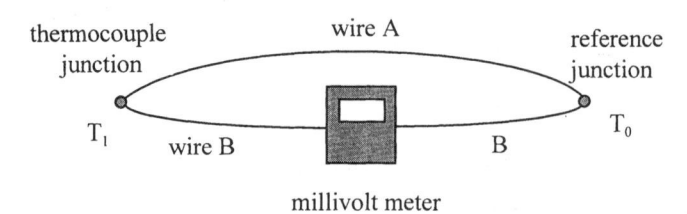

Figure 3.51 Temperature measurement by thermocouple.

The thermocouple pairs listed in Table 3.9 are commonly used in electronics cooling applications. A commercial thermometer set has a thermocouple encased in a sheath for the convenience of handling. To measure the temperature of a tiny spot, or measure temperatures simultaneously at many locations, the handwork is necessary to form thermocouple junctions from wires and bond them to the object surface.

3.5.1.2 Thermistors. The electric resistance of material generally changes with the temperature. In particular, the electric resistances of some metal oxides show high sensitivity to the temperature. The *thermistor* is the resistor element made from such materials. Thermistors have higher temperature sensitivity than thermocouples and, unlike thermocouples, do not require any reference temperature source. On the other hand, the electric resistance of the lead wires that connect the thermistor element and the temperature indicator has to be tailored to ensure correct measurement of temperature.

3.5.1.3 Sensors embedded in the chip. The forward voltage drop at the emitter/base (e/b) junction is a function of temperature. Hence, by feeding a small current through the e/b junction, the temperature can be monitored. Another type of temperature sensor is the resistor deposited on the chip. The change of electric resistance is used to measure the temperature. These embedded sensors can be used to monitor the chip temperature during actual service and generate the warning when the temperature exceeds a threshold. A part of the chip area has to be spared, and an attendant circuitry has to be fabricated, for such in situ monitoring.

On-chip sensors are also incorporated in the thermal chip. The thermal chip is specifically designed to determine the junction-to-case or the junction-to-air thermal resistance. The whole chip area is occupied by heater elements and sensors. The thermal chip is a convenient tool to evaluate the packaging design in the product development phase.

3.5.1.4 IR thermal imaging. The radiative energy emitted from the object has a spectrum that varies with the temperature. In the temperature range where electronic equipment operates, the radiative energy spectrum in the infrared (IR) band has a high sensitivity to the temper-

TABLE 3.9 Thermocouples

Type	Materials	Useful temperature range
T	Copper-constantan	–260 to 350°C
J	Iron-constantan	–150 to 750°C
K	Chromel-alumel	–250 to 1250°C

ature change. The IR rays in the wave length bands 3 to 5 µm or 8 to 12 µm are collected by the optical lenses and the intensity is converted to the electrical signal by the detector. High sensitivity of the IR detector is gained at cryogenic temperatures; hence, liquid nitrogen has been used to create a cryogenic environment around the detector. In some recent designs thermoelectric coolers and other refrigeration mechanisms are employed to eliminate the need to install liquid nitrogen bins.

The IR thermal imaging produces temperature maps of the exterior of the object (Fig. 3.52). It offers convenience to grasp the global feature of the thermal field, thereby devising means of improving the thermal design. The information gained from the IR image is qualitative, if no provisions are worked on the surface of the object. To obtain quantitative information, the object surface has to be coated by the substance having a known emissivity close to unity; besides, at least one thermocouple needs to be attached to the surface to define a reference temperature of the IR image.

The drawback of the IR thermal imaging is the difficulty in accessing the interior of the electronic equipment and components. Removal of the system box to measure the temperature distribution on the printed circuit board, or removal of the encapsulation from the module to find the temperature distribution on the chip, modifies the genuine thermal field. The interpretation of IR image in such cases involves analytical and numerical work.

3.5.1.5 Quick diagnostic measures. Besides IR thermal imaging there are other measures that allow quick assessment of thermal field on the surface of the object. Temperature-sensitive paints, labels, and microcapsule-encapsulated liquid crystals provide convenience in applying them to the surface of the modules and other components.

3.5.2 Air-flow measurements

The measurements concerning air flow are required mainly in the product-development phase. Using mock-ups or real components and systems the flow resistance and the coolant distribution are measured.

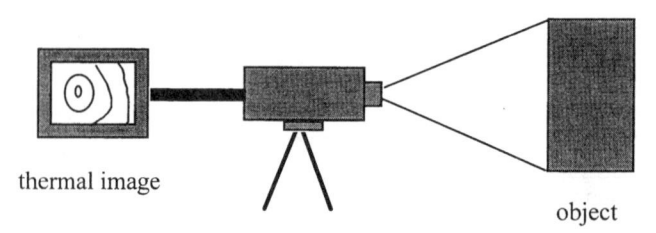

Figure 3.52 Infrared (IR) thermal imaging system.

3.5.2.1 Flow resistance. Experimentation using the duct system depicted in Fig. 3.53 provides flow resistance data for module arrays and other components. The room air is drawn to the duct by the blower. To ensure smooth entry of air flow, the duct inlet is provided with the bell mouth. The manometer measures the static pressure difference between a location downstream of the object and the room air (Δp). The orifice flow meter measures the flow rate through the duct. The anemometer at the duct inlet measures the approach velocity of air (U). The data of Δp and U are reduced to the resistance coefficient ($\zeta \equiv \Delta p / \frac{1}{2} \rho U^2$), which will be used later in the design analysis.

3.5.2.2 Velocity. The approach velocity can also be found from the volumetric flow rate divided by the cross-sectional area of the duct. The anemometer, however, provides more specific information about flow near the object.

There are two types of anemometer most commonly employed in the velocity measurement: the hot-wire anemometer (Fig. 3.54a) and the Pitot-static tube (Fig. 3.54b).

The sensor of the hot-wire anemometer is a platinum wire stretched between the tips of a pair of needles. The wire is heated by passing electric current through it, and its temperature is determined by the heat transfer to the air flow. The electrical resistance of the wire changes with the wire temperature, which in turn depends on the air velocity. The hot-wire is placed as one of the resistors of the bridge circuit, by which the voltage drop on the wire is measured under constant current condition, or the change of electric current that is prompted to keep the wire temperature constant is detected. Because of its small probe size, the hot-wire anemometer enables the measurement of detailed features of velocity distribution.

The Pitot-static tube measures the total pressure and the static pressure simultaneously. Based on Eq. 3.32 the velocity is determined. This is a robust classic instrument.

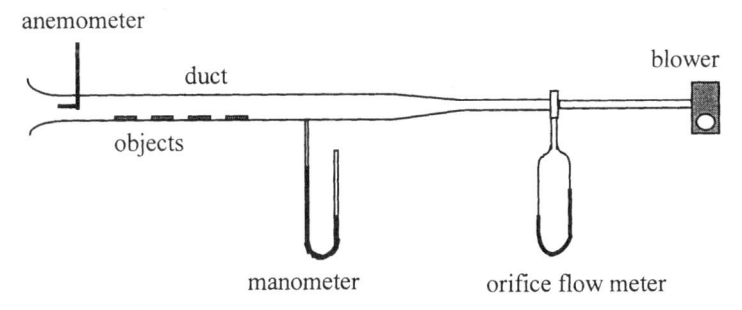

Figure 3.53 Duct system for measurement of pressure drop.

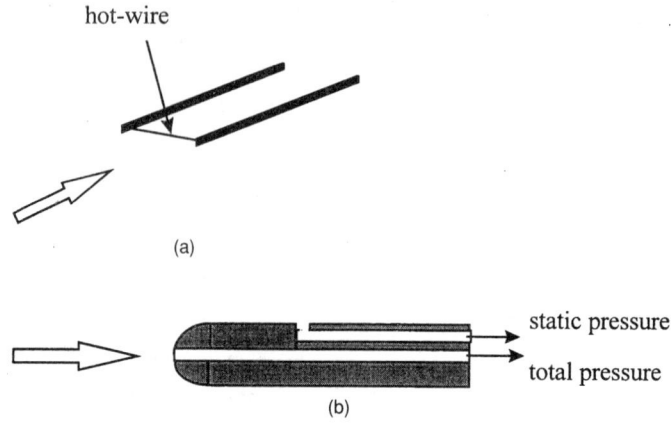

Figure 3.54 Anemometer probes; (a) hot-wire probe; (b) Pitot-static tube.

3.5.2.3 Flow visualization. The flow visualization provides information about the global feature of the coolant flow. Figure 3.55 shows a basic set required for flow visualization: a transparent box, a sheet light, a recording device, and flow tracers. The flow tracers can be provided by incense smoke, dry-ice mist, and particles of various kinds.

The flow visualization serves the purpose of improving the coolant flow path design. Components' placement, location of coolant inlets and outlets, installation of flow deflectors, and other structural factors can be changed in a mock-up box to see their effects on the coolant flow pat-

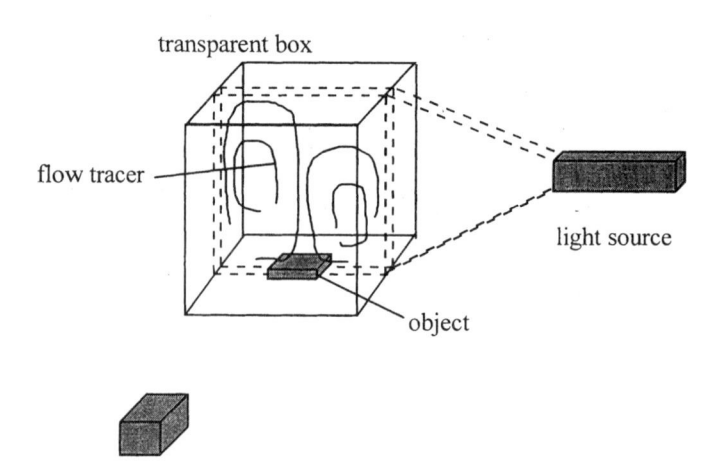

Figure 3.55 Flow visualization system.

tern. The flow visualization also serves the purpose of benchmarking the results of CFD simulations.

3.5.2.4 Acoustic noise. The acoustic noise is collected using the microphone and processed into the weighted sum of sound pressure. The weighting is based on the frequency spectrum that reflects the sensitivity of human ears to the frequency of sound. The sound pressure level is expressed in decibel,

$$L_p = 20 \log (p/p_0) \quad [\text{dB}]$$

where p is the sound pressure and p_0 is the reference pressure set at 2×10^{-5} N/m².

In order to identify sources of noise in the electronic equipment, background noise has to be eliminated. This is done by placing the equipment in a soundproof room where external noise and reflection of sound waves on the walls are eliminated.

3.5.3 Thermal resistance measurements

The junction-to-case thermal resistance ($R_{J/c}$) on the module can be measured only by the use of thermal chips or sensor-carrying chips. The thermal resistance on the module surface has to be reduced to a negligible level compared to $R_{J/c}$. In an attempt to realize this situation the test module is mounted on a water-cooled heat sink. This arrangement of the module and the heat sink directs the heat flow in a particular direction, so that the thermal resistance obtained in this manner serves only as a guide. The real thermal resistance is likely to be altered by the change of heat-flow pattern in the module that is caused by the heat transfer on the module surface.

The junction-to-coolant thermal resistance ($R_{J/a}$) and the case-to-air thermal resistance ($R_{c/a}$) can be measured using the duct system of Fig. 3.53. For naturally cooled modules, the test module is placed in a chamber. The $R_{c/a}$ is determined by measuring the temperature of the module surface by a thermocouple or from an IR thermal image. Such measurement is based on the assumption that the module surface is nearly isothermal.

The measurement of thermal resistance on the module level, as described previously, and also on the board and system levels, almost without exception requires modifications of the structure of the test object and the thermal boundary condition on the object surface. Such modifications may lead to thermal designs with either a too narrow or excessive margin against the target temperature specifications. Their effects have to be evaluated using the analytical tools and the modeling techniques described in this chapter.

3.6 Summary

The objectives of thermal design are customarily specified in terms of the upper bounds for the junction temperature and the junction temperature variation in the equipment. The upper bounds are typically set at 50–100°C for the junction temperature, depending on the expected service life of products, and 10K for the junction temperature variation. These criteria have been adopted in the industry over many years without any serious incidents that required fundamental review of them. However, recent acceleration of packaging technology and diversification of applications of electronics products are creating new challenges to the thermal design. The foremost challenge is the need to broaden our scope in defining the thermal design criteria. Not only the junction temperature, but the temperature distributions at the module and board levels as well, need to be incorporated in the design criteria, in order to guarantee the structural integrity of such components as ball-grid-array modules and chip-on-board devices. As the application of electronics technology diversifies, the thermal environment for electronic devices also diversifies, some exceeding the traditional upper bounds for the junction temperature. If we try to follow the classical methodology of reliability engineering, we need to amass the failure data every time new environment and applications develop for the product. This is no longer a feasible approach in the face of shrinking product development cycles as well as accelerating introduction of new materials and packaging schemes. The time of field operation and the body of failure data are often too short and small to rely entirely on the statistical methods in our attempt to establish the thermal criteria.

The only feasible alternative to the classical approach is to identify possible failure mechanisms and formulate the criteria to prevent them. The importance of this so-called Physics-of-Failure approach has just started receiving appreciation. The formulation of criteria is itself an involving work, but this is a natural reflection of the fact that the design of electronic equipment now requires an integrated approach, not a once-through serial execution of independent tasks. In the past, the thermal design has been left near the downstream end of the design work, and executed under a popular notion that "a 10K increase of junction temperature doubles the failure rate." Such an approach is now being phased out (Sec. 3.3.1).

In the era of integrated design we have to devise certain effective means to learn the art of thermal design. At the first level of learning we capture the global image of the thermal design and understand what knowledge segments serve as the building blocks of the design. This chapter is written with an intention to provide such first-level learning. In order to serve this intention, the coverage of relevant data, formulas, and literature is cut to a minimum. Also, many of the cautionary comments about the accuracy and applicable range of formulas, which are the norm in the heat transfer text books, are spared. Learning about the details is left to subsequent levels of learning, for which the references are provided at the end of the chapter. At several places in this chapter it is emphasized that, in our attempt to understand the heat transfer process in electronic equipment, we first place our view on the whole equipment, then narrow our attention to local details. The same approach may be adopted in the learning process, placing our attention first on the global feature of the subject, then, to the details of scientific bases of knowledge building blocks.

Section 3.2 explained the elements of heat transfer while tracing the heat flow in electronic equipment from the transistor junctions to the atmospheric air. Heat generated at the transistor junctions undergoes spreading and contraction as it flows in the chip, and from the chip to the wiring substrate. The solution for heat spreading from a point source in the semi-infinite medium is used to consider the meaning of the junction temperature (Sec. 3.2.2). The junction temperature is the temperature at a point removed from the actual junction by a distance of the order of $\sqrt{\kappa \Delta t}$, where κ is the thermal diffusivity of the chip material, and Δt is the switching cycle time. Where the chip is surrounded by the media posing large thermal resistance, the chip temperature is synonymous with the junction temperature (Sec. 3.3.3).

The heat crosses the interfaces between different media as it flows from the chip to the module encapsulation, the wiring substrate, and to the heat sink. The thermal resistance at the interfaces is often dominated by the medium filling the interstitial gap (Sec. 3.2.4).

Heat conduction in the wiring substrate is becoming an important part of the heat transfer process as the system becomes more compact, leaving less room for direct heat removal from the module surface. The thermal conductivity of the substrate may be evaluated by taking into account the contributions from vias, power, and ground layers (Sec. 3.2.5). Thermal vias are effective means of reducing the board-level thermal resistance, where the surface heat transfer coefficient is small such as on the naturally cooled board (Sec. 3.2.11).

The thermal resistance for convective heat transfer from the component surface to the coolant is posed by the boundary layer developed on the surface. The thinner the boundary layer, the less the thermal resistance. Hence, it is useful to understand how the boundary layer thickness is related to the operative and geometric parameters in both forced convection and natural convection cooling (Secs. 3.2.6 and 3.2.10). The boundary layer on the component of interest is affected by the presence of other components in its proximity that impose spatial constraints on the coolant flow. Also, the heat dissipation from nearby components creates thermal wakes that envelop the components in downstream rows. There are some qualitative guides to estimate the effects of spatial constraint and thermal wake on heat transfer (Secs. 3.2.7 and 3.2.11). For quantitative estimation we have to resort to numerical simulations and experiments.

Heat spreading at the module and board levels, either intentionally designed or resulting naturally, in certain circumstances plays substantial roles in the formation of temperature field in the equipment. The Biot number (hL/λ, λ = thermal conductivity of solid) is the parameter that indicates the effect of heat spreading on the module temperature. The smaller the Biot number, the more effective the substrate is in heat spreading (Sec. 3.2.8).

Heat is disposed from the system to the atmospheric air by any one of the following means: through surface heat transfer from the sealed system box (Sec. 3.2.11); by driving the air flow through the system box by the fan, the blower, or the buoyancy force (Secs. 3.2.9 and 3.2.11); or by use of liquid coolant as an intermediate heat carrier between the module and the air-cooled heat exchanger (Sec. 3.2.13). The option of heat disposal scheme depends largely on the volumetric heat density which has to be transported from the internal space of the system to the atmosphere. Liquid cooling, in particular, which accompanies phase change, is a sole solution where the internal space is squeezed tight in three-dimensional packaging of modules.

Before embarking on the thermal design analysis, we need to review the available resources at our disposal to achieve the objective of the analysis. The resources to be considered are: the human resource, the time, the knowledge base, the numerical simulation tools, and the experimental facilities (Sec. 3.3.2). The planning about the depth and breadth of analysis is important for efficient execution of the design work.

Identification of possible cooling modes is the first step of the design (Sec. 3.3.3), which requires only the ranges of heat transfer coefficients for various heat transfer modes and the heat balance equation for coolant flow (Eq. 3.31). The design analysis for a particular cooling mode consists of four steps: scaling, modeling, formulation, and solution.

The scaling is the division of the physical domain into lumps or control volumes (Sec. 3.3.5). The constraint on the computational and experimental resources often necessitates the stepwise execution of the analysis. We successively narrow our scope from the system to local zones of interest, while increasing the spatial resolution of the analysis.

The modeling replaces the actual geometric configurations of lumps by those having lesser dimensions, substitutes the thermal properties of the materials by the equivalent properties, and approximates the boundary conditions by simpler ones (Sec. 3.3.5). In developing the model it is necessary to set a goal regarding whether we need to identify an upper and/or lower bound of the solution, or a most likely solution.

The formulation (Sec. 3.3.6) is the writing of balance equations for heat, mass, and momentum transfer on the control volume. For the solution of balance equations (Sec. 3.3.7), the well-established methods are available. For linear problems such as heat conduction in the solid, there is a way to reduce the computational load by using the method of superposition. That is, once a body of building-block solutions is created, those elementary solutions can be superposed to find a temperature distribution for any specified thermal boundary condition. The resistance network for coolant flows in the system box is nonlinear. Also, the natural convection cooling is a nonlinear process. For the solution of those nonlinear problems, the Newton-Raphson iteration method is applied. Where radiation heat transfer is involved, Eq. 3.44 is the approximation to incorporate a radiation thermal resistance into a linear thermal resistance network (Sec. 3.2.13).

The quality of thermal design depends much on the proper choice of various cooling devices (Sec. 3.4). The working principle, the basic structure, and the related topics are described for air-cooled heat sinks (Sec. 3.4.1), fans and blowers (Sec. 3.4.2), liquid-cooling hardware (Sec. 3.4.3), heat pipes (Sec. 3.4.4), and thermoelectric coolers (Sec. 3.4.5).

Measurements of thermal and flow fields are an integral part of thermal management (Sec. 3.5). The objectives of measurement are the benchmarking of the thermal analysis in the product development phase, the product screening, the diagnosis of overheating, and the in situ monitoring. The objective and the nature of required information determine the measurement method to be employed. The interpretation of the measured data requires some analysis. In particular, the thermal resistance data supplied by the component vendors often need modifications, when the components are to be used in environment that departs significantly from the vendor's test apparatus.

Today, the number of technical publications on thermal management is rapidly increasing. Unfortunately, most of the information published in the literature is not directly applicable to a specific case on hand. However, it helps one's job to see in the published literature what models were used, what methods were employed, and how much resource was involved. This chapter is intended to provide certain basic viewpoints that may help to explore further the art of thermal management.

3.7 Nomenclature

The units are all metric, but different scaled units are used depending on the context; for example, m^2 and cm^2 for area, s (second) and ns (nanosecond) for time. Some symbols below are accompanied by the units to give the idea about their dimensions. For the sake of brevity, the explanations about suffixes are omitted except for those that appear in multiple places in the text.

A	constant in Eq. 3.9, Sec. 3.2.2 [$s^{3/2}$ K]
A'	constant in Eq. 3.11, Sec. 3.2.2 [m]
A	area
a	heat source radius, Sec. 3.2.8
a_b	cross-sectional area of solder bump, Sec. 3.2.3
a_M	summed cross-sectional area of vias per unit area of substrate, Sec. 3.2.5
B	constant in Eqs. 3.9 and 3.11 [m^2s]
Bi	Biot number, $= hL/\lambda$
b	module height, Sec. 3.2.7; heat spreader radius, Sec. 3.2.8
C	heat capacitance [J/K]
c	specific heat [J/kg K]
c_p	specific heat at constant pressure [J/kg K]
D	pipe diameter, or twice the plate spacing, Sec. 3.2.13
d	diameter
F	force, Sec. 3.2.10; view factor, Sec. 3.2.12
f_b	force per unit volume, Sec. 3.2.10
g	gravitational acceleration
H	channel height, Sec. 3.2.7; module height, Sec. 3.3.4
h	heat transfer coefficient [W/m^2 K]
L	length
M	mass [kg], Sec. 3.2.1; mass flow rate [kg/s], Secs. 3.2.7, 3.2.9, and 3.2.13
N	total number
n	module count from coolant inlet
Nu	Nusselt number, $= hL/\lambda_f$
P	heat dissipation per pulse [J], Sec. 3.2.2; total pressure [Pa], Sec. 3.2.9; potential, Sec. 3.3.6
P	perimeter of heat transfer surface, Sec. 3.2.10
p	static pressure; sound pressure, Sec. 3.5.2 [Pa]
Δp	pressure drop [Pa]
Pr	Prandtl number
Q	heat generation rate, heat transfer rate [W]
q	heat generation rate of point source [W], Sec. 3.2.2; heat flux [W/m^2], in other sections
R	thermal resistance [K/W]
Ra	Rayleigh number, Eq. 3.38
Re	Reynolds number, $= UL/\nu$
r	radial coordinate; area ratio, Sec. 3.4.1

s	distance, Sec. 3.2.2; spacing between modules, Sec. 3.2.7; spacing between boards, Sec. 3.2.11
T	temperature
T_J	junction temperature
T_J^*	upper bound of T_J
T_s	surface temperature
T_∞	mainstream temperature
T_0	environment temperature
ΔT	temperature difference
t	time; heat spreader thickness, Sec. 3.2.8
U	free stream velocity of coolant, Sec. 3.2.6; overall heat transfer coefficient, Sec. 3.2.13
u	local coolant velocity
u_j	source solution, Eq. 3.9, Sec. 3.2.2
V	volumetric flow rate, Sec. 3.4.2
V_F	frontal air velocity
V_{hx}	heat exchanger volume
v_a	air velocity, Sec. 3.2.13
x	streamwise distance, Sec. 3.2.6

Greek symbols

β	thermal expansion coefficient [K^{-1}]
δ	thickness
ε	error bound, Sec. 3.2.2; bump-to-flux tube diameter ratio, Sec. 3.2.4; emissivity, Sec. 3.2.12
Φ	transfer rate across control volume boundary
γ_M	ratio of summed metal layer thickness to substrate thickness, Sec. 3.2.5
η_F	fin efficiency
κ	thermal diffusivity [m^2/s]
λ	thermal conductivity [W/m K]
ν	kinematic viscosity [m^2/s]
Θ	temperature difference = $T_s - T$; elementary solution, Sec. 3.3.7
θ	ratio of temperature scales, Sec. 3.2.8
ρ	density [kg/m^3]
σ	thermal wake factor, Eq. 3.29, Sec. 3.2.7; Stefan-Boltzmann constant, Sec. 3.2.12
ζ	resistance coefficient

3.8 References

3.8.1 References cited in the text

1. Bar-Cohen, A., and W. M. Rohsenow. "Thermally Optimum Spacing of Vertical Natural Convection Cooled Parallel Plates," *ASME Journal of Heat Transfer,* vol. 106, 1984, pp. 116–123.
2. Chang, M. J., R. J. Shyu, and L. J. Fang. "An Experimental Study of Heat Transfer from Surface Mounted Components to a Channel Air Flow," ASME Paper 87-HT-75, 1987.
3. Churchill, S. W., and H. H. S. Chu. "Correlating Equations for Laminar and Turbulent Free Convection from a Vertical Plate," *International Journal of Heat and Mass Transfer,* vol. 18, 1975, pp. 1323–1329.
4. Danielson, R. D., L. Tousignant, and A. Bar-Cohen. "Saturated Pool Boiling Characteristics of Commercially Available Perfluorinated Liquids," *Proc. ASME/JSME Thermal Engineering Joint Conference,* vol. 3, ASME, 1987, pp. 419–430.
5. Dittus, F. W., and L. M. K. Boelter. University of California, Berkeley, *Publications on Engineering,* vol. 2, 1930, p. 443.
6. Fisher, T. S., F. A. Zell, K. K. Sikka, and K. E. Torrance. "Efficient Heat Transfer Approximation for the Chip-on-Substrate Problem," *ASME Journal of Electronic Packaging,* vol. 118, 1996, pp. 271–279.
7. Goldstein, R. J., E. M. Sparrow, and D. C. Jones. "Natural Convection Mass Transfer Adjacent to Horizontal Plates," *International Journal of Heat and Mass Transfer,* vol. 16, 1973, pp. 1025–1035.
8. Hisano, K., H. Iwasaki, and M. Ishizuka. "Thermal Analysis of Portable Electronic Equipment," *Thermal Science and Engineering (Journal of Heat Transfer Society of Japan),* 5(2), 1997, pp. 7–12. Also, through private communications.
9. Incropera, F. P. "Liquid Immersion Cooling of Electronic Components," *Heat Transfer in Electronic and Microelectronic Equipment,* A. Bergles, ed., pp. 407–444, New York: Hemisphere Publishing Corporation, 1990.
10. Incropera, F. P., and D. P. DeWitt. *Fundamentals of Heat and Mass Transfer.* New York: John Wiley & Sons, Inc., 1996, p. 356 for Table 3.3, pp. 718–728 for more about view factors. This textbook on heat transfer includes more in-depth explanations about the formulas and the experimental data used in this chapter, and some examples about electronics cooling.
11. Ishizuka, M. "Thermal Design Approaches for Electronic Equipment by Use of Personal Computers," *Computers and Computing in Heat Transfer Science and Engineering,* W. Nakayama and K. T. Yang, eds., pp. 391–407, New York: CRC Press/Begell House, 1993.
12. Lloyd, J. R., and W. R. Moran. "Natural Convection Adjacent to Horizontal Surfaces of Various Forms," ASME Paper 74-WA/HT-66, 1974.
13. Martin, H. "Heat and Mass Transfer between Impinging Gas Jets and Solid Surfaces," *Advances in Heat Transfer,* J. P. Hartnett and T. F. Irvine, Jr., eds., vol. 13, pp. 1–60, New York: Academic Press, 1977.
14. Moffat, R. J., D. E. Arvizu, and A. Ortega. "Cooling Electronic Components: Forced Convection Experiments with an Air-Cooled Array," *23rd AIChE/ASME National Heat Transfer Conference,* Denver, Col.: 1985.
15. Moffat, R. J., and A. Ortega. "Direct Air-Cooling of Electronic Components," chapter 3, in *Advances in Thermal Modeling of Electronic Components and Systems,* vol. 1, A. Bar-Cohen and A. D. Kraus, eds., New York: Hemisphere Publishing Corporation, 1988.
16. Mudawar, I., and D. C. Wadsworth. "Critical Heat Flux from a Simulated Chip to a Confined Rectangular Impinging Jet of Dielectric Liquid," *International Journal of Heat and Mass Transfer,* vol. 34, 1991, pp. 1465–1479.
17. Nakayama, W., T. Nakajima, and S. Hirasawa. "Heat Sink Studs Having Enhanced Boiling Surfaces for Cooling of Microelectronic Components," ASME Paper 84-WA/HT-89, 1984.
18. Oktay, S. "Departure from Natural Convection (DNC) in Low Temperature Boiling Heat Transfer Encountered in Cooling Micro-Electronic LSI Devices," *Heat Transfer—1982,* vol. 4, pp. 113–118, New York: Hemisphere, 1982.

19. Sparrow, E. M., J. E. Nithammer, and A. Chaboki. "Heat Transfer and Pressure Drop Characteristics of Arrays of Rectangular Modules Encountered in Electronic Equipment," *International Journal of Heat and Mass Transfer,* vol. 25, 1982, pp. 961–973.
20. Wirtz, R. A., and P. Dykshoorn. "Heat Transfer from Arrays of Flat Packs in a Channel Flow," *Proc. 4th Electronic Packaging Conference,* 1984, pp. 318–326.
21. Yovanovich, M. M., and V. W. Antonetti. "Application of Thermal Contact Resistance Theory to Electronic Packages," chapter 2 in *Advances in Thermal Modeling of Electronic Components and Systems,* vol. 1, A. Bar-Cohen and A. D. Kraus, eds., New York: Hemisphere Publishing Corporation, 1988.

3.8.2 General sources of technical information

3.8.2.1 Annual volumes

22. Bar-Cohen, A., and A. Kraus, eds., *Advances in Thermal Modeling of Electronic Components and Systems,* ASME Press.

3.8.2.2 About recent trends of thermal management published by the present author

23. Nakayama, W. "Information Processing and Heat Transfer Engineering; Some Generic Views on Future Research Needs," *Cooling of Electronic Systems,* S. Kakac, H. Yuncu, and K. Hijikata, eds., pp. 911–943, Netherlands: Kluwer Academic Publishers, 1994.
24. Nakayama, W. "Heat Transfer Engineering in Systems Integration: Outlook for Closer Coupling of Thermal and Electrical Designs of Computers," *IEEE Trans. Components, Packaging, and Manufacturing Technology,* vol. 18, 1995, pp. 818–826.
25. Nakayama, W. "Thermal Management of Electronic Equipment; Research Needs in the Mid-1990s and Beyond," *Applied Mechanics Reviews,* **49**(10), part 2, 1996, pp. S167–S174.
26. Nakayama, W. "Liquid-Cooling of Electronic Equipment: Where Does It Offer Viable Solutions?," *Advances in Electronic Packaging,* E. Suhir, Y. C. Lee, M. Shiratori, and G. Subbarayan, eds., vol. 2, pp. 2045–2052, ASME, 1997.

3.8.2.3 Books on thermal management (in chronological order)

27. Seely, J., and R. Chu. *Heat Transfer in Microelectronic Equipment,* New York: Marcel Dekker, 1972.
28. Scott, A. W. *Cooling of Electronic Equipment,* New York: John Wiley & Sons, 1974.
29. Steinberg, D. S. *Cooling Techniques for Electronic Equipment,* New York: John Wiley & Sons, 1980.
30. Kraus, A. D., and A. Bar-Cohen. *Thermal Analysis and Control of Electronic Equipment,* Washington, D.C.: Hemisphere Publishing Corporation, 1983.
31. Ellison, G. N. *Thermal Computations for Electronic Equipment,* New York: Van Nostrand Reinhold Co., 1984.

3.8.2.4 Journals

32. *ASME Journal of Electronic Packaging.*
33. *ASME Journal of Heat Transfer.*
34. *Electronics Cooling, Flomerics Limited.*
35. *IEEE Transactions on Components, Packaging, and Manufacturing Technology,* Parts A, B, and C.
36. *International Journal of Heat and Mass Transfer.*

3.8.2.5 Conferences where thermal management is one of primary agenda

37. ITHERM (*Intersociety Conference on Thermal and Thermomechanical Phenomena in Electronic Systems;* held in even years).
38. SEMI-THERM (*IEEE annual symposium; Semiconductor Thermal Measurement and Management Symposium*).
39. InterPack (*International and Intersociety Electronic and Photonic Packaging Conference;* held in odd years).

Chapter
4

Mechanical Design, Analysis, Measurement, and Reliability of Electronic Packaging

4.1 Introduction

With the advance of very large scale integration technologies, millions of devices can be fabricated on a silicon chip (see Chap. 1). At the same time, demands to further reduce packaging signal delay and increase packaging density between communicating circuits have led to the use of very high power dissipation single-chip modules and multichip modules. The result of these developments has been a rapid growth in module level heat flux within the personal, workstation, midrange, mainframe, and super computers and telecommunication products. Thus, thermal (temperature, stress, and strain) management is vital for microelectronic packaging designs and analyses. How to determine the temperature distribution in the electronics components and systems (already shown in Chap. 3) is out of the scope of this chapter, which instead focuses on the determination of thermal stress and strain distributions in the electronic packaging.

With very few exceptions, substances expand when their temperature is raised and contract when cooled. The deformation (expansion or contraction) due to temperature change in the absence of mechanical loads is called *thermal strain*. The thermal strain is not exactly linear with temperature change; however, for first-order approximation and small temperature changes, this strain can be described as proportional to the temperature change. This proportionality is expressed by the *coefficient of linear thermal expansion*, which is defined as the change in length that a bar of unit length undergoes when its temperature is changed by one degree.

Thermal stresses occur when any portion of the thermal expansion or contraction in a structure is constrained. Basically, there are two different sets of constraints under which thermal stresses occur: external constraints and internal constraints. Thermal stresses due to external constraints are readily apparent. The most familiar example is a bar fixed at both ends and subjected to a temperature rise. In this case, the bar is at a state of compression, except at the fixed ends that are at a very complex state of stress.

However, thermal stresses due to internal constraints are not so obvious. A structure made of one material may be free to expand and yet have thermal stresses due to a nonuniform temperature distribution. On the other hand, a structure made of more than one material (i.e., a composite structure) may be uniformly heated, have no external constraints, and still have thermal stresses due to different coefficients of expansion and mechanical properties. In both of these cases the constraints occur within the structure.

An electronic package assembly is a typical example of a composite structure that undergoes thermal loadings. It consists of at least two different materials, and is subjected to nonuniform temperature distributions. Due to the geometry, material construction, and thermal expansion mismatch of different parts of the package, thermal stresses can occur inside the package while it is being manufactured and while it is being used.

The determination of thermal stresses in electronic packaging is not an easy task. Closed-form and semi-closed-form solutions for very simple geometry and temperature loading such as those given by [1–10] and Secs. 4.2.1 [11], 4.2.2 [11], and 4.2.3 [12] are very useful, but are very limited in applications and difficult to obtain. The finite element method [13–22] is one of the best candidates for obtaining approximate results for the thermal stresses and strains in electronic packages. However, as with any popular method, many of the finite element analyses performed are not properly executed due to a limited understanding of the physics and equations of thermomechanics for electronic packaging.

In this chapter, the governing equations of thermomechanics for electronic packaging are briefly mentioned and their assumptions and limitations are also highlighted. Various electronic packaging example problems involving design, analysis, measurement, and reliability are discussed.

4.2 Thermal Mechanical Boundary Value Problems

4.2.1 Governing equations for electronic packaging

The governing equations of the theory of thermal stress and strain for electronic packaging in Cartesian coordinates are given by [23–36]

$$\nabla^2 T = \frac{\rho C_v}{k}\frac{\partial T}{\partial t} - \frac{W}{k} \quad (4.1)$$

$$\frac{\partial^2 u}{\partial x^2} + \frac{\partial^2 v}{\partial x\, \partial y} + \frac{\partial^2 w}{\partial x\, \partial z} + (1-2\nu)\nabla^2 u + \frac{X_x}{(\lambda+G)} = 2(1+\nu)\alpha\frac{\partial T}{\partial x} \quad (4.2)$$

$$\frac{\partial^2 v}{\partial y^2} + \frac{\partial^2 u}{\partial x\, \partial y} + \frac{\partial^2 w}{\partial y\, \partial z} + (1-2\nu)\nabla^2 v + \frac{X_y}{(\lambda+G)} = 2(1+\nu)\alpha\frac{\partial T}{\partial y} \quad (4.3)$$

$$\frac{\partial^2 w}{\partial z^2} + \frac{\partial^2 u}{\partial x\, \partial z} + \frac{\partial^2 v}{\partial y\, \partial z} + (1-2\nu)\nabla^2 w + \frac{X_z}{(\lambda+G)} = 2(1+\nu)\alpha\frac{\partial T}{\partial z} \quad (4.4)$$

where

$$\nabla^2 = \frac{\partial^2}{\partial x^2} + \frac{\partial^2}{\partial y^2} + \frac{\partial^2}{\partial z^2} \quad (4.5)$$

$$\beta = \frac{\alpha E}{1-2\nu} \quad (4.6)$$

$$\lambda = \frac{E\nu}{(1+\nu)(1-2\nu)} \quad (4.7)$$

$$G = \frac{E}{2(1+\nu)} \quad (4.8)$$

Heat transfer and thermal stress in electronic packaging are usually applied in two stages, transient (power on/off) and steady state (during operation). In both cases, for the theory of thermal stresses and strains, the temperature distribution $T(x_i,t)$ in the package is calculated by solving the heat conduction equation, Eq. 4.1, with the prescribed initial and boundary conditions. The displacement components (u, v, and w) everywhere inside the package are then determined by solving Eq. 4.2, Eq. 4.3, and Eq. 4.4 with the prescribed stress/displacement boundary conditions and with the calculated temperature distribution as an imposed boundary condition. This temperature distribution is mathematically shown at the right-hand side of Eqs. 4.2 through 4.4 and is a known function. The thermal stresses (σ_x, σ_y, σ_z, τ_{xy}, τ_{yz}, τ_{zx}) can be obtained by

$$\sigma_x = \frac{\lambda}{\nu}\left[(1-\nu)\frac{\partial u}{\partial x} + \nu\left(\frac{\partial v}{\partial y} + \frac{\partial w}{\partial z}\right)\right] - \beta(T - T_0) \quad (4.9)$$

$$\sigma_y = \frac{\lambda}{\nu}\left[(1-\nu)\frac{\partial v}{\partial y} + \nu\left(\frac{\partial w}{\partial z} + \frac{\partial u}{\partial x}\right)\right] - \beta(T - T_0) \quad (4.10)$$

$$\sigma_z = \frac{\lambda}{\nu}\left[(1-\nu)\frac{\partial w}{\partial z} + \nu\left(\frac{\partial u}{\partial x} + \frac{\partial v}{\partial y}\right)\right] - \beta(T - T_0) \quad (4.11)$$

$$\tau_{xy} = G\left(\frac{\partial u}{\partial y} + \frac{\partial v}{\partial x}\right) \quad (4.12)$$

$$\tau_{yz} = G\left(\frac{\partial v}{\partial z} + \frac{\partial w}{\partial y}\right) \quad (4.13)$$

$$\tau_{zx} = G\left(\frac{\partial w}{\partial x} + \frac{\partial u}{\partial z}\right) \quad (4.14)$$

and are shown in Fig. 4.18 of Sec. 4.3.1.

The strains (ε_x, ε_y, ε_z, γ_{xy}, γ_{yz}, γ_{zx}) can be obtained by

$$\epsilon_x = \frac{\partial u}{\partial x} \quad (4.15)$$

$$\epsilon_y = \frac{\partial v}{\partial y} \quad (4.16)$$

$$\epsilon_z = \frac{\partial w}{\partial z} \quad (4.17)$$

$$\gamma_{xy} = \frac{\partial u}{\partial y} + \frac{\partial v}{\partial x} \quad (4.18)$$

$$\gamma_{yz} = \frac{\partial v}{\partial z} + \frac{\partial w}{\partial y} \quad (4.19)$$

$$\gamma_{zx} = \frac{\partial w}{\partial x} + \frac{\partial u}{\partial z} \quad (4.20)$$

and their physical interpretation and limitations are discussed in Sec. 4.4.

In Eqs. 4.1–4.20, u, v, and w are the displacement components in the x-, y-, and z-directions, respectively; X_x, X_y, and X_z are the body force components in the x-, y-, and z-directions, respectively; σ_x is the normal stress acting in the x-direction, σ_y is the normal stress acting in the y-direction, σ_z is the normal stress acting in the z-direction; τ_{xy} is the shear stress acting in the y-direction of the plane normal to the x-axis, τ_{yz} is the shear stress acting in the z-direction of the plane normal to the y-axis, τ_{zx} is the shear stress acting in the x-direction of the plane normal to the z-axis; ε_x is the normal strain acting in the x-direction, ε_y is the normal strain acting in the y-direction, ε_z is the normal strain acting in the z-direction; γ_{xy} is the shear strain acting in the y-direction of the plane normal to the x-axis, γ_{yz} is the shear strain acting in the z-direction of the plane normal to the y-axis, γ_{zx} is the shear strain acting in the x-direction of the plane normal to the z-axis; α is the thermal coefficient of linear expansion; κ is the heat conductivity; ν is the Poisson's ratio; λ is the Lame's constant; E is the Young's modulus; G is the shear modulus; T is the instantaneous absolute temperature; T_0 is the reference temperature; ρ is the mass density; C_v is the heat capacity per unit mass; t is the time-independent variable; and W is the heat generation per unit time per unit volume.

Closed-form solutions of Eqs. 4.1–4.4 for most of the electronic packaging boundary-value problems are very difficult to obtain and they are usually approximated by the finite element method, which is discussed in Sec. 4.6. In this section, the closed-form solutions of a few very simple electronic packaging problems are presented.

4.2.2 Example I—deflection of a printed circuit board

Figure 4.1 shows a ($a \times b \times h$) 18″ × 12″ × 0.062″ (46 cm × 30 cm × 0.57 cm) PCB subjected to a uniform distributed load of w = 0.013 psi (89.57 Pa). It is simply supported on two opposite edges ($x = 0$ and $x = a$) and is free to move along the other two edges ($y = \pm b/2$). The Young's modulus of the PCB is E = 2000 ksi (13.78 MPa) and the Poisson's ratio is $\nu = 0.28$. It is interesting to determine the deflection at locations 1 through 6, and the bending and twisting moments and transverse shear forces at location 1, Fig. 4.1.

In this case, T and Xs are zero in Eqs. 4.1–4.4. The bending moments (M_x and M_y, shown in Fig. 4.2) are defined as

$$M_x = \int_{-h/2}^{h/2} \sigma_x z \, dz \qquad M_y = \int_{-h/2}^{h/2} \sigma_y z \, dz \qquad (4.21)$$

the twisting moment (M_{xy}) as

$$M_{xy} = \int_{-h/2}^{h/2} \tau_{xy} z \, dz \qquad (4.22)$$

198   Chapter Four

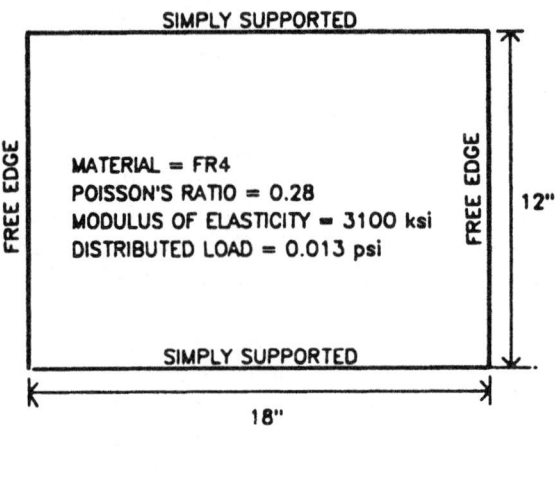

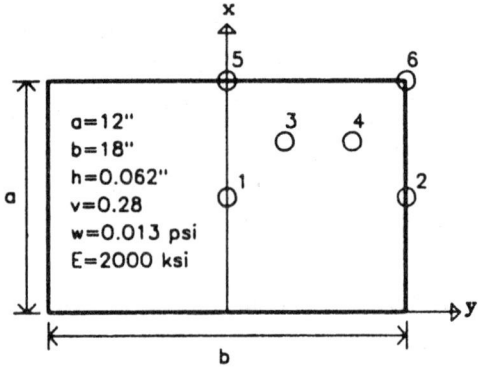

Figure 4.1 Boundary value problem of a PCB.

and the shearing forces (Q_x and Q_y) as

$$Q_x = \int_{-h/2}^{h/2} \tau_{xz}\, dz \qquad Q_y = \int_{-h/2}^{h/2} \tau_{yz}\, dz \qquad (4.23)$$

Then, Eq. 4.4 leads to the following equation governing the deflection (w) of the PCB

$$\frac{\partial^4 w}{\partial x^4} + 2\frac{\partial^4 w}{\partial x^2 \partial y^2} + \frac{\partial^4 w}{\partial y^4} = \frac{w}{D}. \qquad (4.24)$$

The boundary conditions for the present problem are

$$(w)_{x=0 \text{ and } x=a} = 0 \qquad (4.25)$$

$$\left(\frac{\partial^2 w}{\partial x^2} + \nu \frac{\partial^2 w}{\partial y^2}\right)_{x=0 \text{ and } x=a} = 0 \qquad (4.26)$$

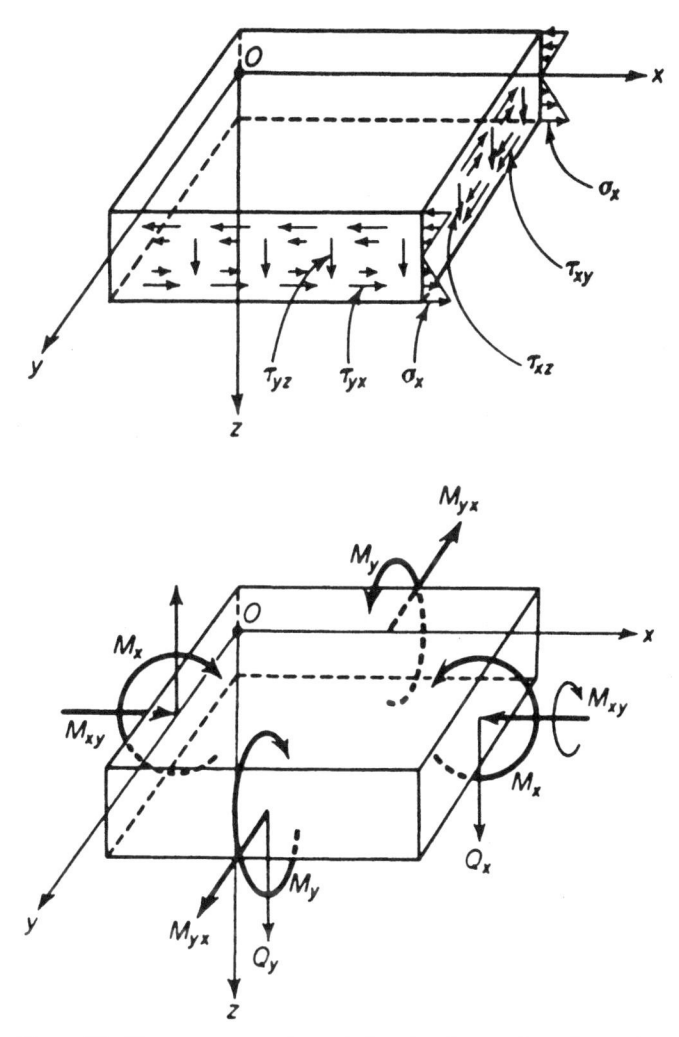

Mechanical Design/Analysis/Measurement/Reliability of Electronic Packaging 199

Figure 4.2 Stresses, moments, and shearing forces at a volume element.

$$\left[\frac{\partial^3 w}{\partial y^3} + (2-\nu)\frac{\partial^3 w}{\partial x^2 \partial y}\right]_{y=\pm b/2} = 0 \qquad (4.27)$$

$$\left(\frac{\partial^2 w}{\partial y^2} + \nu \frac{\partial^2 w}{\partial x^2}\right)_{y=\pm b/2} = 0 \qquad (4.28)$$

It can be shown that the closed-form solutions for the deflection (w), the bending moments (M_x and M_y), the twisting moment (M_{xy}), and the transverse shear forces (Q_x and Q_y) are given as follows:

Chapter Four

$$w = \frac{wa^4}{D} \sum_{m=1,3,5,\ldots}^{\infty} \left(\frac{4}{\pi^5 m^5} + A_m \cosh \frac{m\pi y}{a} + B_m \frac{m\pi y}{a} \sinh \frac{m\pi y}{a} \right) \sin \frac{m\pi x}{a}$$

(4.29)

$$M_x = -w(\pi a)^2 \sum_{m=1,3,5,\ldots}^{\infty} m^2 \left\{ -\frac{4}{m^5 \pi^5} + (\nu - 1) A_m \cosh \frac{m\pi y}{a} \right.$$
$$\left. + B_m \left[(\nu m^2 - 1) \frac{m\pi y}{a} \sinh \frac{m\pi y}{a} + 2\nu m^2 \cosh \frac{m\pi y}{a} \right] \right\} \sin \frac{m\pi x}{a} \quad (4.30)$$

$$M_y = -w(\pi a)^2 \sum_{m=1,3,5,\ldots}^{\infty} m^2 \left\{ -\frac{4\nu}{m^5 \pi^5} + (1 - \nu) A_m \cosh \frac{m\pi y}{a} \right.$$
$$\left. + B_m \left[2m^2 \cosh \frac{m\pi y}{a} + (m^2 - \nu) \frac{m\pi y}{a} \sinh \frac{m\pi y}{a} \right] \right\} \sin \frac{m\pi x}{a} \quad (4.31)$$

$$M_{xy} = w(\pi a)^2 (1 - \nu) \sum_{m=1,3,5,\ldots}^{\infty} m^2 \left[A_m \sinh \frac{m\pi y}{a} \right.$$
$$\left. + B_m \left(\sinh \frac{m\pi y}{a} + \frac{m\pi y}{a} \cosh \frac{m\pi y}{a} \right) \right] \cos \frac{m\pi x}{a} \quad (4.32)$$

$$Q_x = -w\pi^3 a \sum_{m=1,3,5,\ldots}^{\infty} m^3 \left\{ -\frac{4}{\pi^5 m^5} \right.$$
$$\left. + B_m \left[2m^2 \cosh \frac{m\pi y}{a} + (m^2 - 1) \frac{m\pi y}{a} \sinh \frac{m\pi y}{a} \right] \right\} \cos \frac{m\pi x}{a} \quad (4.33)$$

$$Q_y = -w\pi^3 a \sum_{m=1,3,5,\ldots}^{\infty} m^3 B_m$$
$$\cdot \left[(3m^3 - 1) \sinh \frac{m\pi y}{a} + (m^2 - 1) \frac{m\pi y}{a} \cosh \frac{m\pi y}{a} \right] \sin \frac{m\pi x}{a} \quad (4.34)$$

where

$$A_m = \frac{4}{m^5 \pi^5} \frac{\nu(1+\nu) \sinh \alpha_m - \nu(1-\nu) \alpha_m \cosh \alpha_m}{\Delta_m} \quad (4.35)$$

$$B_m = \frac{4}{m^5 \pi^5} \frac{\nu(1-\nu) \sinh \alpha_m}{\Delta_m} \quad (4.36)$$

$$\Delta_m = (3+\nu)(1-\nu) \sinh \alpha_m \cosh \alpha_m - (1-\nu)^2 \alpha_m \quad (4.37)$$

$$D = \frac{Eh^3}{12(1-v^2)} \tag{4.38}$$

$$\alpha_m = \frac{m\pi b}{2a} \tag{4.39}$$

The numerical results of the deflection at locations 1 to 6 of the PCB are shown in Table 4.1. The bending and twisting moments and transverse shear forces at location 1 are given in Table 4.2. These results were obtained by summing the preceding series up to $m = 15$ [11], and will be used to verify the finite element results shown in Sec. 4.6.2.

4.2.3 Example II—thermal stress in a package subjected to an instantaneous heat source

In [4], some closed-form solutions and useful curves and charts which relate the variables, temperature, stress, displacement and material property for a semi-infinite elastic substrate subjected to a uniform temperature distribution over a rectangular region on its surface have been given. Also, useful charts of the thermoelasticity responses for a substrate with finite thickness subjected to axially symmetrical tem-

TABLE 4.1 Deflections of the PCB

Locations (Fig. 4.1)	Coordinates (x, y)	Closed-form solution	Finite-element solution
1	(6, 0)	0.10076	0.10080
2	(6, 9)	0.11679	0.11690
3	(9, 3)	0.07236	0.07237
4	(9, 6)	0.07496	0.07498
5	(12, 0)	0	0
6	(12, 9)	0	0

Deflections (w) (in)

TABLE 4.2 Moments and Forces in the PCB

	Bending moments (in-lb/in) M_x	M_y	Transverse shear forces (lb/in) Q_x	Q_y	Twisting moment (in-lb/in) M_{xy}
Closed-form solution	0.05932	0.22974	0.00011	0.00254	0.00001
Finite-element solution	0.05926	0.23000	0.00031	0.00246	0.00004

202　Chapter Four

perature distributions have been given in [5]. In this book, the time history responses of the temperature, displacement, and stress are given for a package subjected to an instantaneous source of heat, and a heat source varying harmonically as a function of time. These heat sources could be coming from the chip or chip resistor/capacitor inside a package or a substrate, for example, chip-first packaging, buried chip resistor/capacitor substrate, fine pitch quad flat pack, and ball grid array.

Figure 4.3 schematically shows a package that is subjected to an instantaneous heat source, $W/(\rho C_v)$, at its center. Because of the nature of the problem, spherical coordinates are used and the heat source is at its origin. For a long time after the heat source applied ($t \to \infty$), and a long distance away from the origin ($R \to \infty$), the temperature should be vanished, i.e., $T \to 0$. In this case, it can be shown that the temperature distribution, $T(R, t)$, which satisfied Eq. 4.1 is given by [12]

$$\frac{\rho C_v \pi^{3/2}}{W} T = \tau^{-3/2} e^{-(R^2/\tau)} \qquad (4.40)$$

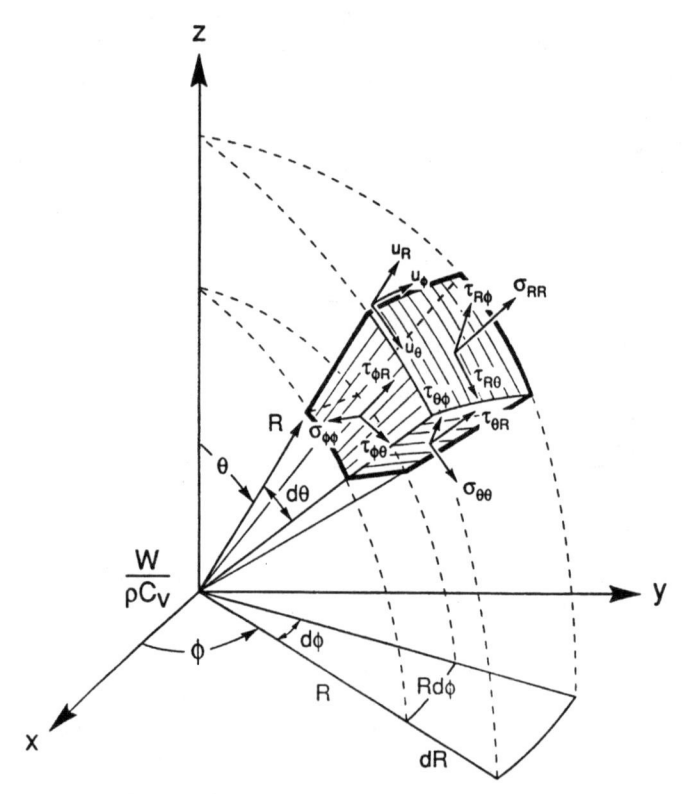

Figure 4.3 Package with an instantaneous heat source acting at the origin.

where

$$\tau = \frac{4k}{\rho C_v} t \qquad (4.41)$$

and

$$R = \sqrt{x^2 + y^2 + z^2} \qquad (4.42)$$

Equation 4.40 is plotted in Figs. 4.4 and 4.5 for various times (right after the instantaneous heat source), and distances (away from the origin). It can be seen from Fig. 4.4 that the temperature (T) is decreasing as the time (t or τ) is increasing. However, it is noted that for small values of τ, the temperature distributions raise very fast to the peak and then decrease to smaller values. Figure 4.5 shows the temperature versus R for various values of τ. It can be seen that the temperature is decreasing as the distance (R) from the origin is increasing.

Due to symmetry (Fig. 4.3), all the shear stress components, $\tau_{\theta\phi} = \tau_{\theta R} = \tau_{R\phi} = 0$, and the normal displacement components, $u_\theta = u_\phi = 0$. The nonzero displacement component, u_R, and the nonzero stress components, σ_{RR}, $\sigma_{\theta\theta}$, and $\sigma_{\phi\phi}$ can be determined by solving Eqs. 4.2–4.4 with the last term on the right-hand side (temperature distribution, T) given by Eq. 4.40. It should be noted that u_R, u_θ, and u_ϕ are the displacement components in the R-, θ-, and ϕ-directions, respectively; σ_{RR}

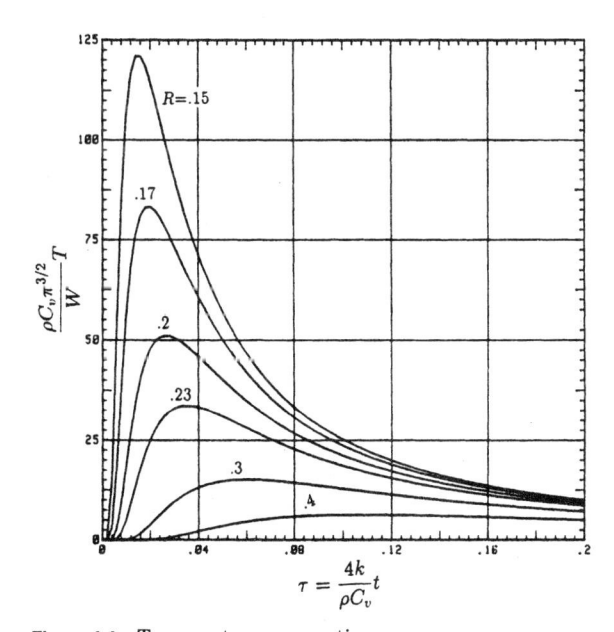

Figure 4.4 Temperature versus time.

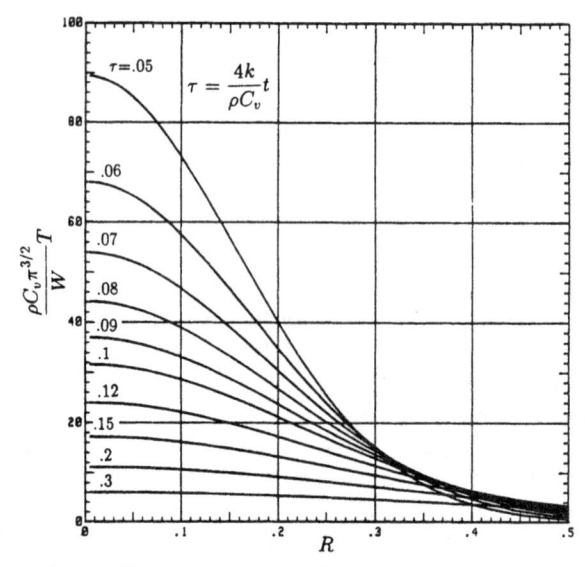

Figure 4.5 Temperature versus R.

is the normal stress acting in the R-direction; $\sigma_{\theta\theta}$ is the normal stress acting in the θ-direction; $\sigma_{\phi\phi}$ is the normal stress acting in the ϕ-direction; $\tau_{R\theta}$ is the shear stress acting in the θ-direction of the plane normal to the R-axis; $\tau_{\theta\phi}$ is the shear stress acting in the ϕ-direction of the plane normal to the θ-axis; and $\tau_{\phi R}$ is the shear stress acting in the R-direction of the plane normal to the ϕ-axis.

For the present problem, at $t = 0$ and $R = 0$, $u_R = 0$. Also, at $R \to \infty$ and any instant of t, $\sigma_{RR} = \sigma_{\theta\theta} = \sigma_{\phi\phi} = 0$. Finally, at $t \to \infty$ and any distance of R from the origin, $\sigma_{RR} = \sigma_{\theta\theta} = \sigma_{\phi\phi} = 0$. In this case, it can be shown that the displacement component, $u_R(R, t)$, and stress components, $\sigma_{RR}(R, t)$, $\sigma_{\phi\phi}(R, t) = \sigma_{\theta\theta}(R, t)$, are given by, respectively,

$$\frac{4\pi(1-\nu)\rho C_v}{\alpha(1+\nu)W} u_R = \frac{1}{R^2}\left[\text{erf}\left(\frac{R}{\sqrt{\tau}}\right) - \frac{2R}{\sqrt{\pi\tau}}\exp\left(-\frac{R^2}{\tau}\right)\right] \qquad (4.43)$$

$$\frac{2\pi(1-\nu)\rho C_v}{\alpha EW} \sigma_{RR} = -\frac{1}{R^3}\left[\text{erf}\left(\frac{R}{\sqrt{\tau}}\right) - \frac{2R}{\sqrt{\pi\tau}}\exp\left(-\frac{R^2}{\tau}\right)\right] \qquad (4.44)$$

$$\frac{4\pi(1-\nu)\rho C_v}{\alpha EW} \sigma_{\theta\theta} = \frac{1}{R^3}\left[\text{erf}\left(\frac{R}{\sqrt{\tau}}\right) - \left(1 + \frac{2R^2}{\tau}\right)\frac{2R}{\sqrt{\pi\tau}} \times \exp\left(-\frac{R^2}{\tau}\right)\right]$$

$$(4.45)$$

where

$$\text{erf}\left(\frac{R}{\sqrt{\tau}}\right) = \frac{2}{\sqrt{2\pi}}\int_0^{R/\sqrt{\tau}} e^{-\zeta^2}\, d\zeta \qquad (4.46)$$

Mechanical Design/Analysis/Measurement/Reliability of Electronic Packaging 205

Equation 4.43 is plotted in Figs. 4.6 and 4.7. It can be seen from Fig. 4.6 that for all the distances from the origin, the radial displacement (u_R) is decreasing as the time is increasing. Figure 4.7 shows the radial displacement versus the distance (R). It can be seen that for all the values of time (τ), the displacements raise to their maximum at a small distance from the origin and then decrease.

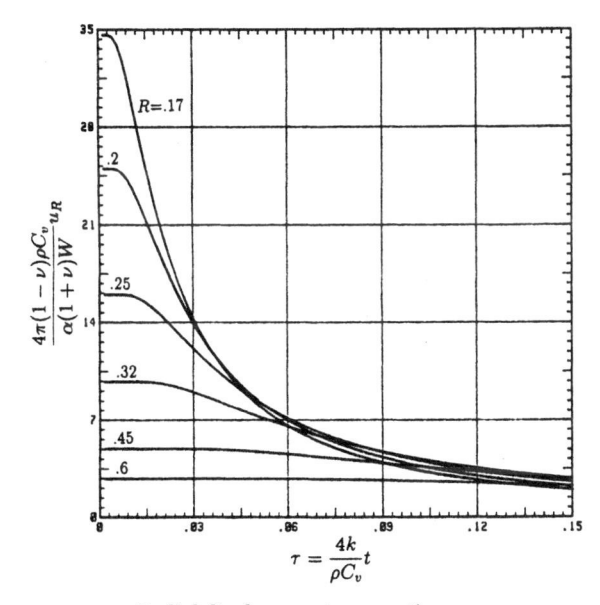

Figure 4.6 Radial displacement versus time.

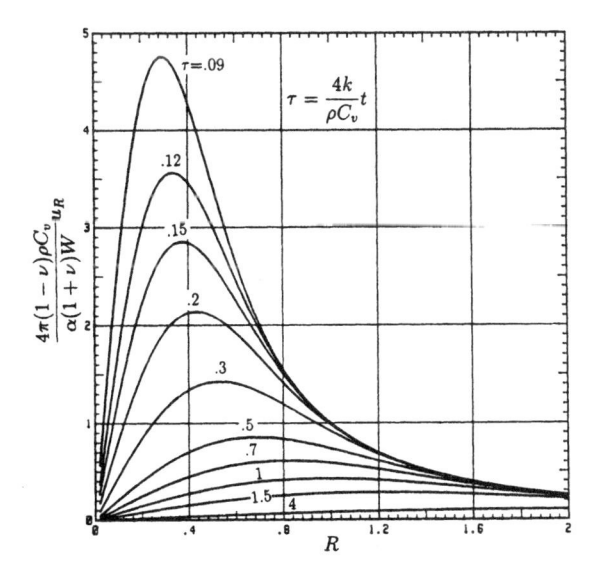

Figure 4.7 Radial displacement versus R.

The radial stress (σ_{RR}) in Eq. 4.44 is plotted versus time in Fig. 4.8 for various values of R, and is plotted versus R in Fig. 4.9 for various values of τ. It can be seen that in all the cases, the radial stress is in compression, and is decreasing when the time after the instantaneous heat source applied and the distance away from the heat source are increasing.

The circumferential stress, $\sigma_{\phi\phi}(R, t) = \sigma_{\theta\theta}(R, t)$, Eq. 4.45, versus the time (τ) is plotted in Fig. 4.10 for various values of distance (R) away from the heat source. It is interesting to note that the circumferential stress is in tension while the values of τ are very small (right after the heat source applied), and then changes into compression. The circumferential stress versus the distance (R) is plotted in Fig. 4.11 for various values of τ. It can be seen that the circumferential stress is in compression while the values of R are small (close to the heat source), then changes into tension, and finally vanishes at large R.

Example 1: Let us consider an Al_2O_3 package (10.16 cm × 10.16 cm), 1.27 cm thick. The physical and mechanical properties of the ceramic are: $k = 25$ Watts/mK, $\rho = 3970$ kg/m^3, $\alpha = 6 \times 10^{-6}$/K, $\nu = 0.21$, $E = 310$ GPa, and $C_v = 765$ joule/kgK. There is a chip at the center of the package and the chip dissipates an instantaneous heat source, $W = 1000$ joule. It is interesting to determine the temperature (T), displacement (u_R), and stresses (σ_{RR}, $\sigma_{\theta\theta}$, $\sigma_{\phi\phi}$) at $R = 0.0025$ m and $t = 0.1$ s. From the figures or from equations, we have $T = 1480$ K, $u_R = 2.8 \times 10^{-5}$ m, $\sigma_{RR} = 5.65$ GPa, $\sigma_{\theta\theta} = \sigma_{\phi\phi} = 2.66$ GPa, and the effective stress = 2.99 GPa.

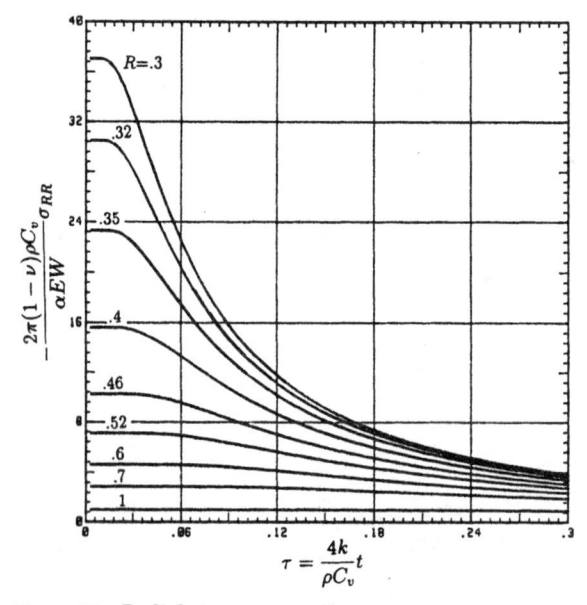

Figure 4.8 Radial stress versus time.

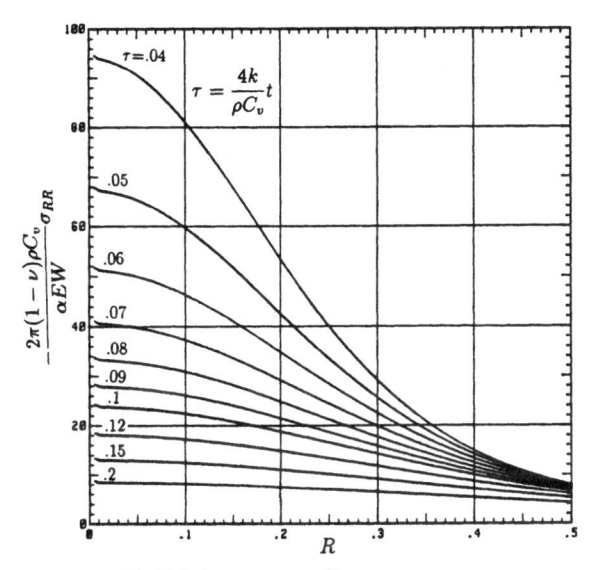

Mechanical Design/Analysis/Measurement/Reliability of Electronic Packaging 207

Figure 4.9 Radial stress versus R.

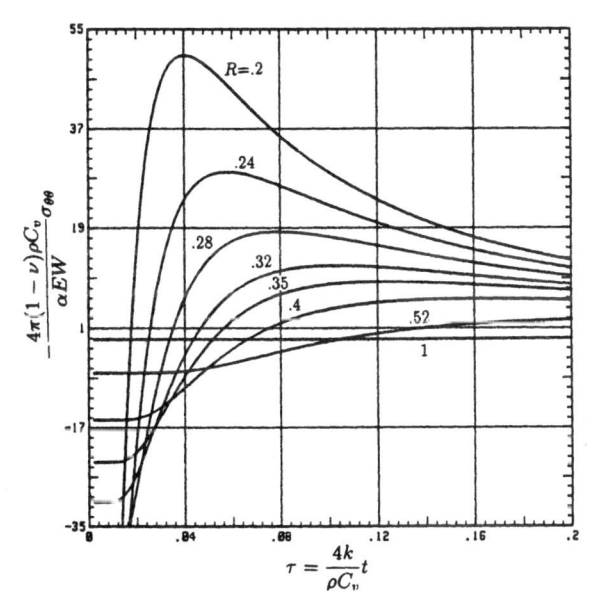

Figure 4.10 Circumferential stress versus time.

Example 2: Let us change the package material in Example 1 from Al_2O_3 to aluminum nitride and keep all other conditions the same. In this case, the physical and mechanical properties of the AlN are: $k = 200$ Watts/mK, $\rho = 3260$ kg/m^3, $\alpha = 4 \times 10^{-6}$/K, $\nu = 0.28$, $E = 325$ GPa, and $C_v = 745$ joule/kgK. Again, from the figures or equations, we have $T = 323$ K, $u_R = 2.1 \times 10^{-6}$ m, $\sigma_{RR} = 0.42$ GPa, $\sigma_{\theta\theta} = \sigma_{\phi\phi} = 1.5$ GPa, and the effective stress = 1.1 GPa.

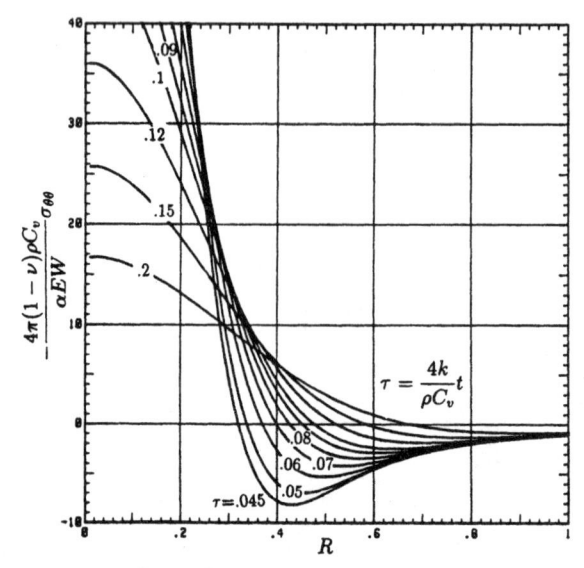

Figure 4.11 Circumferential stress versus R.

By comparing the results between these two packages, it can be seen that the temperature, displacement, and stresses in the AlN package are lower than those in the Al_2O_3. This is because the heat conductivity of AlN (200 Watts/mK) is higher than that (25 Watts/mK) of the Al_2O_3. Also, the thermal coefficient of linear expansion of AlN (4×10^{-6}/K) is lower than that (6×10^{-6}/K) of the Al_2O_3.

4.2.4 Example III—thermal stress in a package subjected to a harmonic heat source

In this case, the heat source acting at the origin is varying harmonically with time, that is, $(W/\rho C_v) \cos \omega t$, where ω is the circular frequency of the heat source, Fig. 4.12. The heat source repeats itself after a period (p) given by $\omega p = 2\pi$. The frequency (f) is the reciprocal of the period, $f = 1/p = \omega/2\pi$, and is measured in cycles per second. It can be shown that the temperature distribution, $T(R, t)$, displacement component, $u_R(R, t)$, and stress components, $\sigma_{RR}(R, t)$, $\sigma_{\phi\phi}(R, t) = \sigma_{\theta\theta}(R, t)$, are given by

$$\frac{4\pi k}{W} T = R^{-1} e^{-\xi} \cos(\omega t - \xi) \quad (4.47)$$

$$\frac{4\pi\omega(1-\nu)\rho C_v}{\alpha(1+\nu)W} u_R = R^{-2} \sin \omega t$$

$$- [(1+\xi)\sin(\omega t - \xi) + \xi \cos(\omega t - \xi)]R^{-2}e^{-\xi} \quad (4.48)$$

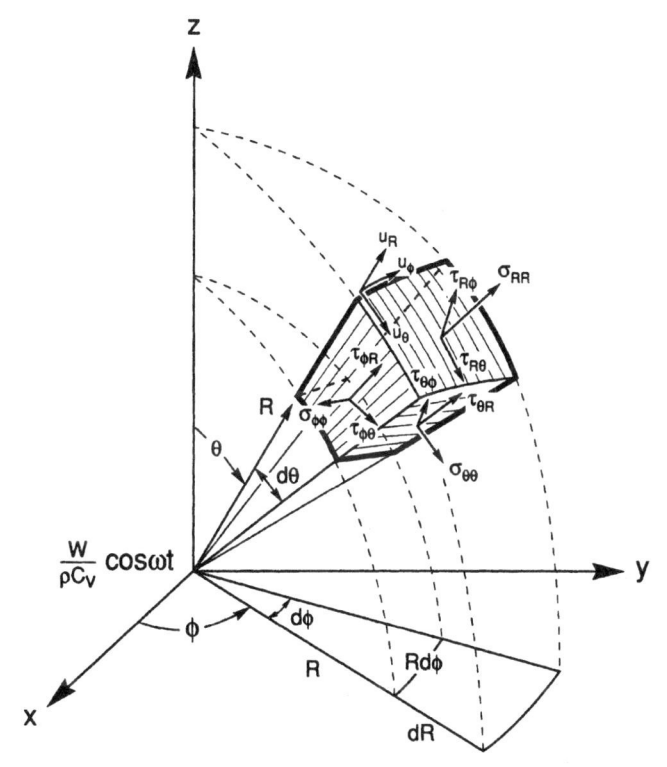

Figure 4.12 Package with a harmonic heat source acting at the origin.

$$\frac{2\pi\omega(1-\nu)\rho C_v}{\alpha EW}\sigma_{RR} = -R^{-3}\sin\omega t + R^{-3}e^{-\xi}$$
$$\times [(1+\xi)\sin(\omega t - \xi) + \xi\cos(\omega t - \xi)] \quad (4.49)$$

$$\frac{4\pi\omega(1-\nu)\rho C_v}{\alpha EW}\sigma_{\theta\theta} = R^{-3}\sin\omega t - 2\xi^2 R^{-3}e^{-\xi} \times \cos(\omega t - \xi) - R^{-3}e^{-\xi}$$
$$\times [(1+\xi)\sin(\omega t - \xi) + \xi\cos(\omega t - \xi)] \quad (4.50)$$

where

$$\xi = R\sqrt{\frac{\omega\rho C_v}{2k}} \quad (4.51)$$

In Eqs. 4.47–4.51, ξ is called the phase angle. With these equations, the time history thermoelasticity responses can be readily determined. For example, let $\omega = 0.0727$ and $\sqrt{\rho C_v/2k} = 10$, then the T, u_R, σ_{RR}, and $\sigma_{\phi\phi} = \sigma_{\theta\theta}$ versus ωt are plotted in Figs. 4.13–4.16 for various values of distance (R) away from the heat source. It can be seen that T, u_R, σ_{RR},

210 Chapter Four

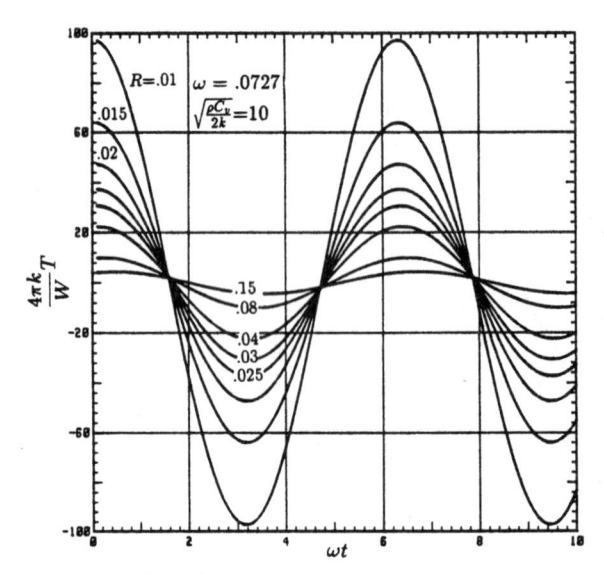

Figure 4.13 Temperature versus ωt.

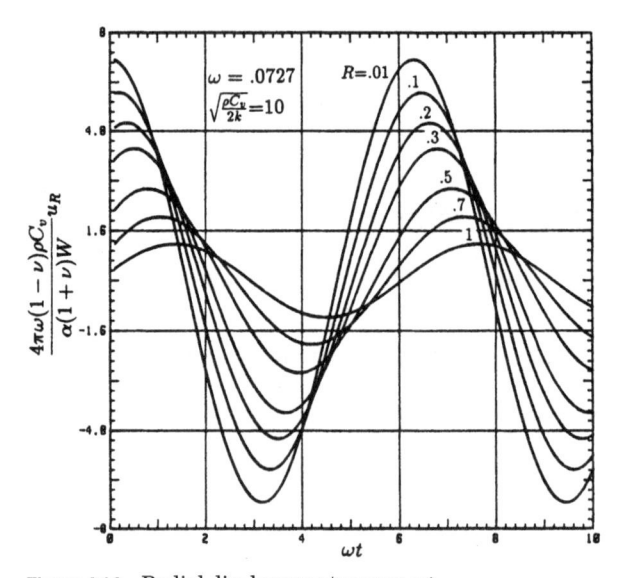

Figure 4.14 Radial displacement versus ωt.

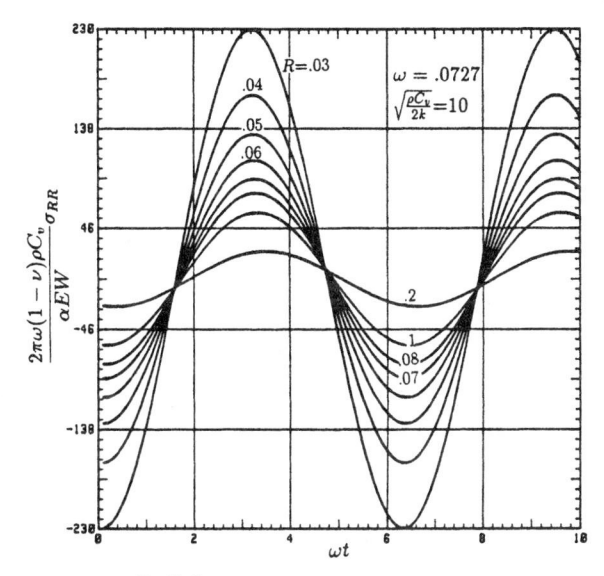

Figure 4.15 Radial stress versus ωt.

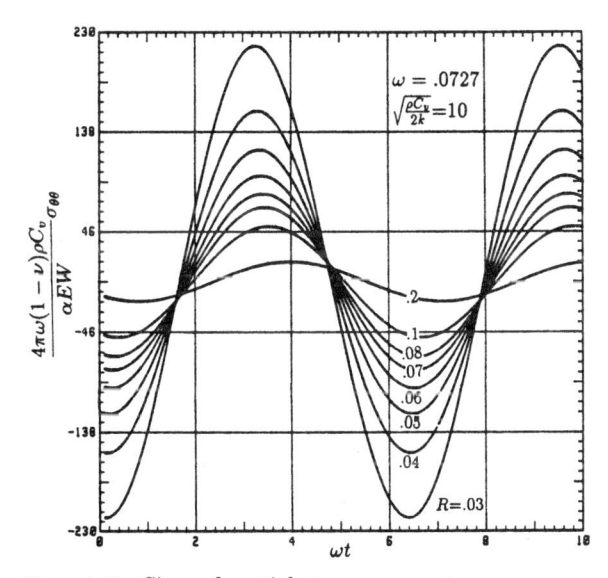

Figure 4.16 Circumferential stress versus ωt.

and $\sigma_{\theta\theta}$ vary harmonically with time. Also, they have higher magnitudes when they are near the heat source (smaller R) and vanish when they are far away from the heat source (very large R).

4.3 Analysis of Stress

4.3.1 Three-dimensional stress state

A *stress* is defined as the force, T (per unit area, A), which the part lying on the positive side of a surface element (the side on the positive side of the outer normal) exerts on the part lying on the negative side (Fig. 4.17). There are nine stress components acting at any point of the electronic package. For a system of rectangular Cartesian coordinates (x, y, z), there are three normal stress components (σ_x, σ_y, σ_z) and six shearing stress components (τ_{xy}, τ_{yz}, τ_{zx}, τ_{yx}, τ_{zy}, τ_{xz}). The notations for the stress components on the sides of this element and the directions taken as positive are as indicated in Figure 4.18. The positive normal stress produce tension and the negative normal stress produces compression. The stress components are usually arranged in a matrix format.

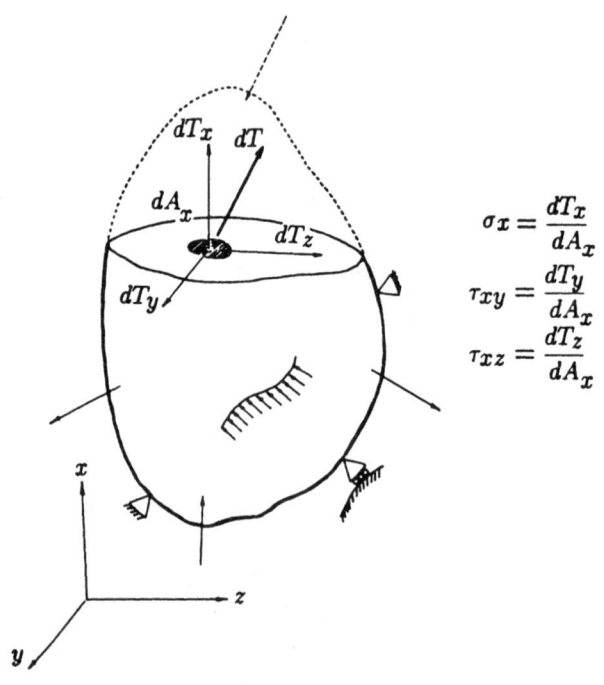

$$\sigma x = \frac{dT_x}{dA_x}$$

$$\tau_{xy} = \frac{dT_y}{dA_x}$$

$$\tau_{xz} = \frac{dT_z}{dA_x}$$

Figure 4.17 Definition of stress at a point.

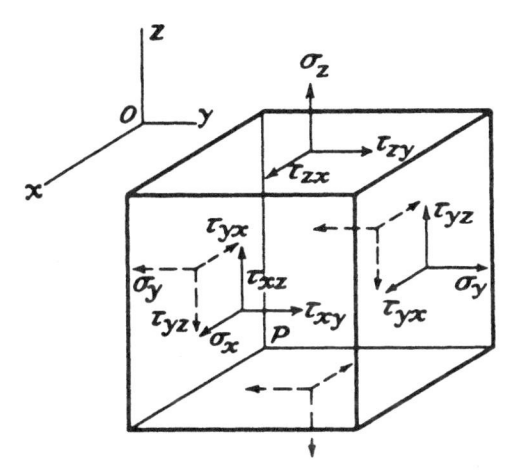

Mechanical Design/Analysis/Measurement/Reliability of Electronic Packaging 213

Figure 4.18 Notation of stress at a volume element.

$$\sigma = \begin{bmatrix} \sigma_x & \tau_{xy} & \tau_{xz} \\ \tau_{yx} & \sigma_y & \tau_{yz} \\ \tau_{zx} & \tau_{zy} & \sigma_z \end{bmatrix} \quad (4.52)$$

4.3.2 Transformation of coordinates

Let us consider another set of rectangular Cartesian coordinates (x', y', z') with the same origin but oriented differently from the previous coordinates (x, y, z), as shown in Fig. 4.19. Also, let the direction cosines of the x'-axis with respect to the x-axis, y-axis, and z-axis, be, respectively, l_1, m_1, and n_1; those of the y'-axis be l_2, m_2, and n_2; and those of the z'-axis be l_3, m_3, and n_3. Then the stress components with respect to the new coordinates (x', y', z') are given by

$$\begin{bmatrix} \sigma_{x'} & \tau_{x'y'} & \tau_{x'z'} \\ \tau_{y'x'} & \sigma_{y'} & \tau_{y'z'} \\ \tau_{z'x'} & \tau_{z'y'} & \sigma_{z'} \end{bmatrix} = \begin{bmatrix} l_1 & m_1 & n_1 \\ l_2 & m_2 & n_2 \\ l_3 & m_3 & n_3 \end{bmatrix} \begin{bmatrix} \sigma_x & \tau_{xy} & \tau_{xz} \\ \tau_{yx} & \sigma_y & \tau_{yz} \\ \tau_{zx} & \tau_{zy} & \sigma_z \end{bmatrix} \begin{bmatrix} l_1 & l_2 & l_3 \\ m_1 & m_2 & m_3 \\ n_1 & n_2 & n_3 \end{bmatrix} \quad (4.53)$$

By means of equilibrium, it can be shown that $\tau_{xy} = \tau_{yx}$, $\tau_{zy} = \tau_{yz}$, and $\tau_{yz} = \tau_{zy}$, and the matrix in Eqs. 4.52 and 4.53 is symmetrical.

4.3.3 Principal stress and direction in three dimensions

For a general state of stress at any point in a structure there exist three mutually perpendicular planes at that point on which the shearing stresses vanish. The remaining normal stress components on these

214 Chapter Four

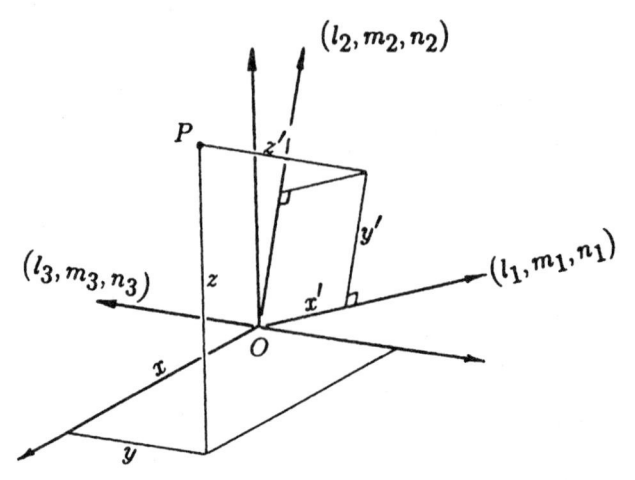

Figure 4.19 Coordinate transformation in 3-D.

three planes are called *principal stresses*. These three planes are called *principal planes* and the three mutually perpendicular axes that are normal to the three planes are called *principal axes,* that is, the principal axes coincide with the three principal stress directions. Consequently, the principal stresses are, by definition, the stresses that act perpendicularly to the principal planes. These principal stresses can be determined by solving

$$\sigma^3 - I_1\sigma^2 + I_2\sigma - I_3 = 0 \tag{4.54}$$

where

$$\begin{aligned}
I_1 &= \sigma_x + \sigma_y + \sigma_z \\
I_2 &= \sigma_x\sigma_y + \sigma_y\sigma_z + \sigma_z\sigma_x - \tau_{xy}^2 - \tau_{yz}^2 - \tau_{zx}^2 \\
I_3 &= \sigma_x\sigma_y\sigma_z + 2\tau_{xy}\tau_{yz}\tau_{zx} - \sigma_x\tau_{yz}^2 - \sigma_y\tau_{zx}^2 - \sigma_z\tau_{xy}^2
\end{aligned} \tag{4.55}$$

The three roots of Eq. 4.54 will give the three principal stresses (σ_1, σ_2, σ_3) of the given stress field (σ_x, σ_y, σ_z, $\tau_{xy} = \tau_{yx}$, $\tau_{yz} = \tau_{zy}$, and $\tau_{zx} = \tau_{xz}$). It can be shown that σ_1, σ_2, and σ_3 are not only orthogonal but also real.

The principal direction of the principal stress σ_i ($i = 1, 2, 3$) can be determined by the following simultaneous equations:

$$\begin{aligned}
(\sigma_i - \sigma_x)l_i - \tau_{xy}m_i - \tau_{zx}n_i &= 0 \\
-\tau_{xy}l_i + (\sigma_i - \sigma_y)m_i - \tau_{yz}n_i &= 0 \\
-\tau_{zx}l_i - \tau_{yz}m_i + (\sigma_i - \sigma_z)n_i &= 0 \\
l_i^2 + m_i^2 + n_i^2 &= 1
\end{aligned} \tag{4.56}$$

The solutions (l_i, m_i, n_i) of these equations will give the direction of the principal stress σ_i. In Eq. 4.56, l_i, m_i, and n_i are the direction cosines of a unit normal vector coinciding with the principal axis of σ_i.

4.3.4 Example I—transformation of stress in three dimensions

Figure 4.20 shows the state of stresses at a point of an electronic package with respect to Cartesian coordinates (x, y, z). It is interesting to determine the stress components with respect to the other Cartesian coordinates (x', y', z'), which is defined by the following direction cosines: $l_1 = 2/3$, $m_1 = 2/3$, $n_1 = -1/3$; $l_2 = -2/3$, $m_2 = 1/3$, $n_2 = -2/3$; $l_3 = -1/3$, $m_3 = 2/3$, $n_3 = 2/3$. From Eq. 4.53, we have

$$\begin{bmatrix} \sigma_{x'} & \tau_{x'y'} & \tau_{x'z'} \\ \tau_{y'x'} & \sigma_{y'} & \tau_{y'z'} \\ \tau_{z'x'} & \tau_{z'y'} & \sigma_{z'} \end{bmatrix} = \frac{1}{9} \begin{bmatrix} 2 & 2 & -1 \\ -2 & 1 & -2 \\ -1 & 2 & 2 \end{bmatrix} \begin{bmatrix} 75 & 25 & 0 \\ 25 & 60 & -25 \\ 0 & -25 & 50 \end{bmatrix} \begin{bmatrix} 2 & -2 & -1 \\ 2 & 1 & 2 \\ -1 & -2 & 2 \end{bmatrix}$$

$$= \begin{bmatrix} 98.89 & -0.56 & -1.11 \\ -0.56 & 62.22 & -0.56 \\ -1.11 & -0.56 & 23.89 \end{bmatrix} \text{ MPa} \qquad (4.57)$$

4.3.5 Example II—principal stresses in three dimensions

Let us consider the stress state in Example I and in view of Eqs. 4.54 and 4.55, we have

$$\sigma^3 - 185\sigma^2 + 10{,}000\sigma - 146{,}875 = 0 \qquad (4.58)$$

$\sigma_{xx} = 75$ MPa; $\sigma_{yy} = 60$ MPa; $\sigma_{zz} = 50$ MPa;

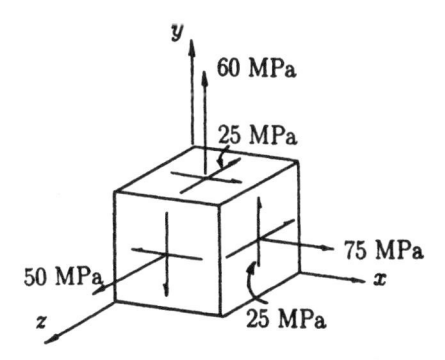

$\tau_{xy} = 25$ MPa; $\tau_{yz} = -25$ MPa; $\tau_{zx} = 0$ MPa;

Figure 4.20 3-D stress state at a volume element.

The three roots of this equation are the three principal stresses, $\sigma_1 = 98.914$ MPa, $\sigma_2 = 62.222$ MPa, $\sigma_3 = 23.864$ MPa.

The direction cosines for $\sigma_1 = 98.914$ MPa can be determined by Eq. 4.56 as follows:

$$(75 - 98.914)l_1 - 25m_1 = 0$$
$$-25l_1 + (98.914 - 60)m_1 + 25n_1 = 0 \tag{4.59}$$
$$25m_1 + (98.914 - 50)n_1 = 0$$
$$l_1^2 + m_1^2 + n_1^2 = 1$$

It can be shown that the direction cosines for the principal stress σ_1 are $l_1 = 0.6813$, $m_1 = 0.6517$, $n_1 = -0.3331$. The same procedure can be used to determine the principal directions of principal stresses σ_2 ($l_2 = 0.6517$, $m_2 = -0.3331$, $n_2 = 0.6814$) and σ_3 ($l_3 = 0.3331$, $m_3 = -0.6814$, $n_3 = -0.6518$).

4.3.6 Two-dimensional stress state

For a two-dimensional plane stress state (σ_x, σ_y, and τ_{xy}) in the xy-axes, the stress components $\sigma_{x'}$, $\sigma_{y'}$, $\tau_{x'y'}$ with respect to new $x'y'$-axes rotated counter-clockwise through an angle θ with respect to the xy-axes are given by (Figure 4.21),

$$\sigma_{x'} = \frac{\sigma_x + \sigma_y}{2} + \frac{\sigma_x - \sigma_y}{2} \cos 2\theta + \tau_{xy} \sin 2\theta \tag{4.60}$$

$$\sigma_{y'} = \frac{\sigma_x + \sigma_y}{2} - \frac{\sigma_x - \sigma_y}{2} \cos 2\theta - \tau_{xy} \sin 2\theta \tag{4.61}$$

$$\tau_{x'y'} = -\frac{\sigma_x - \sigma_y}{2} \sin 2\theta + \tau_{xy} \cos 2\theta \tag{4.62}$$

The principal stresses are given by

$$\sigma_1 = \frac{\sigma_x + \sigma_y}{2} + \sqrt{\tfrac{1}{4}(\sigma_x - \sigma_y)^2 + \tau_{xy}^2} \tag{4.63}$$

$$\sigma_2 = \frac{\sigma_x + \sigma_y}{2} - \sqrt{\tfrac{1}{4}(\sigma_x - \sigma_y)^2 + \tau_{xy}^2} \tag{4.64}$$

and the principal directions of the principal stresses are given by

$$\theta_p = \tfrac{1}{2} \tan^{-1} \frac{2\tau_{xy}}{\sigma_x - \sigma_y} \tag{4.65}$$

where the angle θ_p is measured counterclockwise from the positive x-axis. There are two values of θ_p that are 90° apart, one between 0° and

Mechanical Design/Analysis/Measurement/Reliability of Electronic Packaging 217

90° and the other between 90° and 180°. The maximum shear stress τ_{max} and the algebraically minimum shear stress τ_{min} are given by

$$\tau_{max} = \sqrt{\tfrac{1}{4}(\sigma_x - \sigma_y)^2 + \tau_{xy}^2} \qquad (4.66)$$

$$\tau_{min} = -\tau_{max} \qquad (4.67)$$

and the directions of the planes on which they act are given by

$$\theta_s = \tfrac{1}{2} \tan^{-1} \frac{\sigma_y - \sigma_x}{2\tau_{xy}} \qquad (4.68)$$

where the angle θ_s is measured counterclockwise from the positive x-axis. There are two values of θ_s that are 90° apart, one between 0° and 90° and the other between 90° and 180°. It can be shown that the planes of maximum shear stress occur at 45° to the principal planes.

4.3.7 Example III—principal stress in two dimensions

The given state of stresses in Fig. 4.21 are: $\sigma_x = 12$ MPa; $\sigma_y = -18$ MPa; $\tau_{xy} = -4.5$ MPa. The stress components, $\sigma_{x'}$, $\sigma_{y'}$, $\tau_{x'y'}$, with respect to new $x'y'$-axes rotated 20° counterclockwise with respect to the xy-axes are given by Eqs. 4.60–4.62: $\sigma_{x'} = 5.598$ MPa, $\sigma_{y'} = -11.60$ MPa, $\tau_{x'y'} = -13.08$ MPa. The principal stresses can be obtained from Eqs. 4.63 and 4.64: $\sigma_1 = 12.66$ MPa and $\sigma_2 = -18.66$ MPa. The principal directions (Eq. 4.65) are: 8.35° and 98.35° counterclockwise from the x-axis. The maximum shear stress (Eq. 4.66) is 15.66 MPa.

4.4 Analysis of Strains

4.4.1 Infinitesimal strains in Cartesian coordinates

The geometric interpretation of infinitesimal strain components (ε_x, ε_y, ε_z, γ_{xy}, γ_{yz}, γ_{zx}) in Eqs. 4.15–4.20 can be found in Fig. 4.22. The normal strain ε_x is defined by

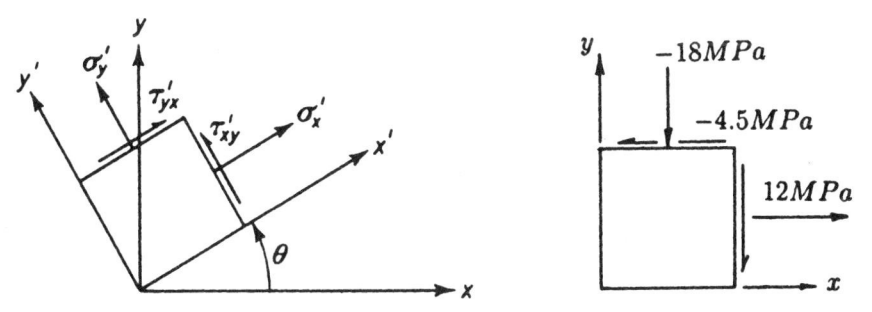

Figure 4.21 Coordinate transformation in 2-D stress state.

UNDEFORMED → **DEFORMED**

$$\epsilon_{xx} = \frac{\partial u}{\partial x}$$

$$\epsilon_{yy} = \frac{\partial v}{\partial y}$$

$$\gamma_{xy} = \frac{\partial v}{\partial x} + \frac{\partial u}{\partial y}$$

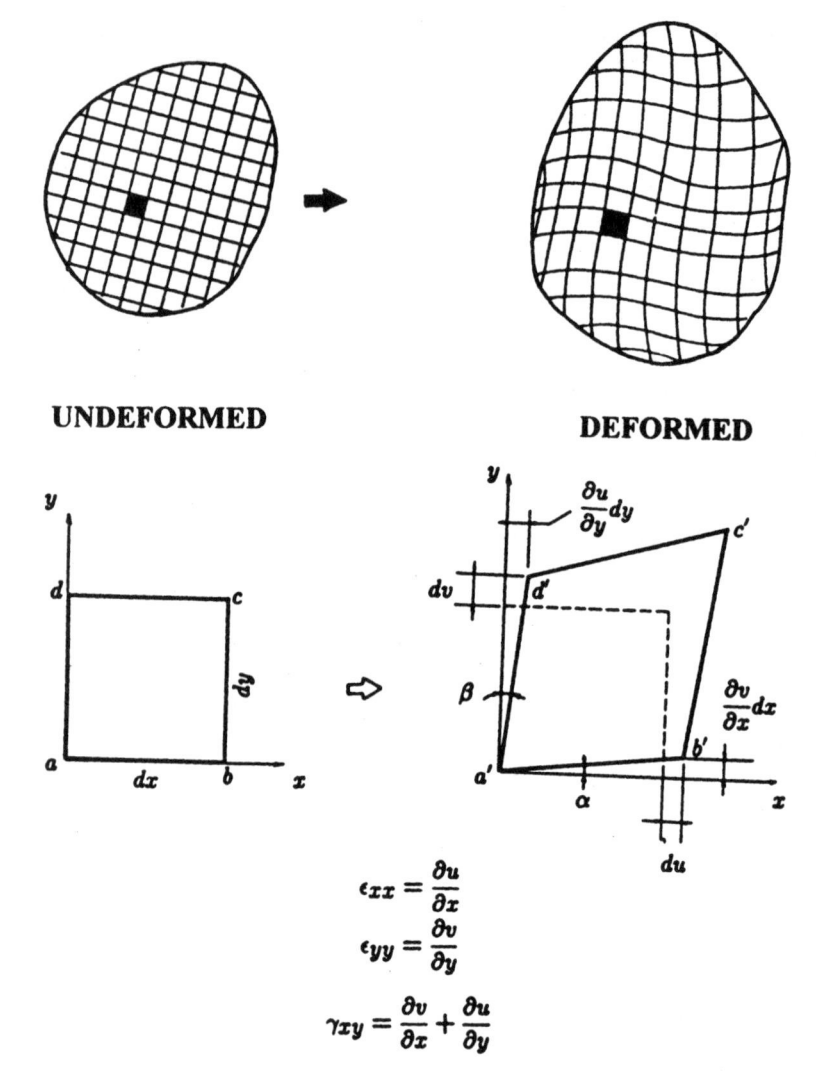

Figure 4.22 Infinitesimal deformation of an area element in the xy-plane.

$$\varepsilon_x = (a'b' - ab)/ab \tag{4.69}$$

where

$$(a'b')^2 = (dx + du)^2 + \left(\frac{\partial v}{\partial x} dx\right)^2$$

$$= (dx)^2 \left[1 + 2\left(\frac{\partial u}{\partial x}\right) + \left(\frac{\partial u}{\partial x}\right)^2 + \left(\frac{\partial v}{\partial x}\right)^2\right]$$

or

$$a'b' = dx\left[1 + 2\left(\frac{\partial u}{\partial x}\right) + \left(\frac{\partial u}{\partial x}\right)^2 + \left(\frac{\partial v}{\partial x}\right)^2\right]^{0.5} \cong dx\left[1 + \left(\frac{\partial u}{\partial x}\right)\right] \qquad (4.70)$$

and

$$ab = dx \qquad (4.71)$$

Thus,

$$\epsilon_x = \left(\frac{\partial u}{\partial x}\right) \qquad (4.72)$$

The *shear strain* is defined as the lateral displacement du divided by the original length dy (simply shear), shown in Fig. 4.23,

$$\begin{aligned}\text{Shear strain} &= \frac{du}{dy}\\ &= \tan \gamma\\ &\approx \gamma \text{ for small displacements}\\ &= \text{reduction in } \angle dab\end{aligned} \qquad (4.73)$$

At a general state of strain; the engineering shear strain is defined as the reduction in the right angle (Figs. 2.22 and 2.23):

$$\begin{aligned}\text{Shear strain } \gamma_{xy} &= \text{Reduction in right angle}\\ &= \alpha + \beta\\ &\approx \tan \alpha + \tan \beta\\ &= \left(\frac{\partial v}{\partial x} dx\right)\frac{1}{dx} + \left(\frac{\partial u}{\partial y} dy\right)\frac{1}{dy}\end{aligned} \qquad (4.74)$$

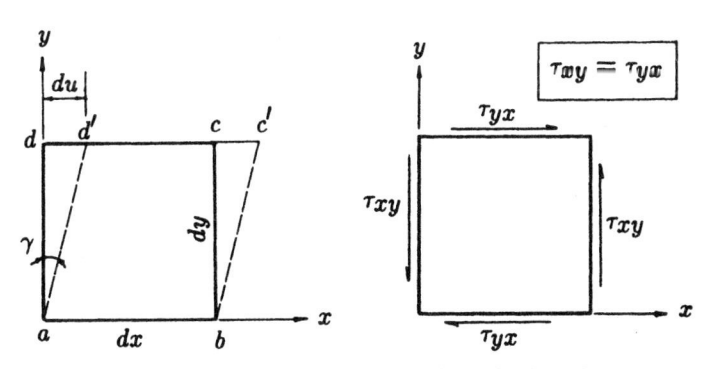

Figure 4.23 Shear strain defined in terms of *simply shear* deformation.

Hence,

$$\gamma_{xy} = \left(\frac{\partial v}{\partial x} + \frac{\partial u}{\partial y}\right) \qquad (4.75)$$

Just as for the general state of stress, for a general state of strain at any point in a structure there exist three mutually perpendicular planes at that point on which the shearing strains vanish. The remaining normal strain components on these three planes are called *principal strains*. These three planes are called *principal planes* and the three mutually perpendicular axes that are normal to the three planes are called *principal axes*, that is, the principal axes coincide with the three principal strain directions. Consequently, the principal strains are, by definition, the strains that act perpendicularly to the principal planes. These principal strains can be determined by solving

$$\epsilon^3 - J_1\epsilon^2 + J_2\epsilon - J_3 = 0 \qquad (4.76)$$

where

$$\begin{aligned}
J_1 &= \epsilon_x + \epsilon_y + \epsilon_z \\
J_2 &= \epsilon_x\epsilon_y + \epsilon_y\epsilon_z + \epsilon_z\epsilon_x - \tfrac{1}{4}(\gamma_{xy}^2 + \gamma_{yz}^2 + \gamma_{zx}^2) \\
J_3 &= \epsilon_x\epsilon_y\epsilon_z + \tfrac{1}{4}(\gamma_{xy}\gamma_{yz}\gamma_{zx} - \epsilon_x\gamma_{yz}^2 - \epsilon_y\gamma_{zx}^2 - \epsilon_z\gamma_{xy}^2)
\end{aligned} \qquad (4.77)$$

The three roots of Eq. 4.76 will give the three principal strains (ϵ_1, ϵ_2, ϵ_3) of the given strain field (ϵ_x, ϵ_y, ϵ_z, $\gamma_{xy} = \gamma_{yx}$, $\gamma_{yz} = \gamma_{zy}$, and $\gamma_{zx} = \gamma_{xz}$). It can be shown that ϵ_1, ϵ_2, and ϵ_3 are not only orthogonal but also real.

The principal direction of the principal strain ϵ_i ($i = 1, 2, 3$) can be determined by the following simultaneous equations:

$$\begin{aligned}
2(\epsilon_i - \epsilon_x)l_i - \gamma_{xy}m_i - \gamma_{zx}n_i &= 0 \\
-\gamma_{xy}l_i + 2(\epsilon_i - \epsilon_y)m_i - \gamma_{yz}n_i &= 0 \\
-\gamma_{zx}l_i - \gamma_{yz}m_i + 2(\epsilon_i - \epsilon_z)n_i &= 0 \\
l_i^2 + m_i^2 + n_i^2 &= 1
\end{aligned} \qquad (4.78)$$

The solutions (l_i, m_i, n_i) of these equations will give the direction of the principal strain ϵ_i. In Eq. 4.78, l_i, m_i, and n_i are the direction cosines of a unit normal vector coinciding with the principal axis of ϵ_i.

For a two-dimensional plane strain state (ϵ_x, ϵ_y, γ_{xy}) in the xy-plane, the strain components $\epsilon_{x'}$, $\epsilon_{y'}$, $\gamma_{x'y'}$ with respect to new $x'y'$-axes rotated counterclockwise through an angle θ with respect to the xy-axes are given by

$$\epsilon_{x'} = \frac{\epsilon_x + \epsilon_y}{2} + \frac{\epsilon_x - \epsilon_y}{2} \cos 2\theta + \tfrac{1}{2}\gamma_{xy} \sin 2\theta \qquad (4.79)$$

$$\epsilon_{y'} = \frac{\epsilon_x + \epsilon_y}{2} - \frac{\epsilon_x - \epsilon_y}{2} \cos 2\theta - \tfrac{1}{2}\gamma_{xy} \sin 2\theta \qquad (4.80)$$

$$\gamma_{x'y'} = -\frac{\epsilon_x - \epsilon_y}{2} \sin 2\theta + \tfrac{1}{2}\gamma_{xy} \cos 2\theta \qquad (4.81)$$

The principal strains are given by

$$\epsilon_1 = \frac{\epsilon_x + \epsilon_y}{2} + \tfrac{1}{2}\sqrt{(\epsilon_x - \epsilon_y)^2 + \gamma_{xy}^2} \qquad (4.82)$$

$$\epsilon_2 = \frac{\epsilon_x + \epsilon_y}{2} - \tfrac{1}{2}\sqrt{(\epsilon_x - \epsilon_y)^2 + \gamma_{xy}^2} \qquad (4.83)$$

and the principal directions of the principal strains are given by

$$\theta_p = \tfrac{1}{2} \tan^{-1} \frac{\gamma_{xy}}{\epsilon_x - \epsilon_y} \qquad (4.84)$$

where the angle θ_p is measured counterclockwise from the positive x-axis and has two values differing by 90°. For isotropic materials, the principal directions of strain coincide with those of stress. The maximum shear strain and the algebraically minimum shear strain are given by

$$\gamma_{max} = \sqrt{(\epsilon_x - \epsilon_y)^2 + \gamma_{xy}^2} \qquad (4.85)$$

$$\gamma_{min} = -\gamma_{max} \qquad (4.86)$$

and the directions of the planes on which they act occur at $\theta_p \pm \pi/4$.

For example, an element is at a state of plane strain with $\epsilon_x = 340 \times 10^{-6}$, $\epsilon_y = 110 \times 10^{-6}$, $\gamma_{xy} = 180 \times 10^{-6}$. Determine the strains at an element rotated through an angle $\theta = 30°$ from the positive x-axis. Also determine the principal strains and maximum shear strains and the directions associated with these strains.

By means of Eqs. 4.79–4.81, we have $\epsilon_{x'} = 360 \times 10^{-6}$, $\epsilon_{y'} = 90 \times 10^{-6}$, and $\gamma_{x'y'} = -110 \times 10^{-6}$. By means of Eqs. 4.82–4.84, we have $\epsilon_1 = 370 \times 10^{-6}$, $\theta_{p1} = 19°$, and $\epsilon_2 = 80 \times 10^{-6}$, $\theta_{p2} = 109°$. Also, by means of Eqs. 4.85 and 4.86, we have that the angle to the plane having the positive maximum shear strain ($\gamma_{max} = 290 \times 10^{-6}$) is 154.2° and the angle to the plane having the algebraically minimum shear strain ($\gamma_{min} = -290 \times 10^{-6}$) is 64°.

4.4.2 Infinitesimal strains in cylindrical coordinates

The normal strains and the engineering shearing strains with respect to cylindrical coordinates (r, θ, z), shown in Fig. 4.24 are given by,

$$\epsilon_{rr} = \frac{\partial u_r}{\partial r} \tag{4.87}$$

$$\epsilon_{\theta\theta} = \frac{u_r}{r} + \frac{1}{r}\frac{\partial u_\theta}{\partial \theta} \tag{4.88}$$

$$\epsilon_{zz} = \frac{\partial u_z}{\partial z} \tag{4.89}$$

$$\gamma_{r\theta} = \left(\frac{1}{r}\frac{\partial u_r}{\partial \theta} + \frac{\partial u_\theta}{\partial r} - \frac{u_\theta}{r}\right) \tag{4.90}$$

$$\gamma_{zr} = \left(\frac{\partial u_r}{\partial z} + \frac{\partial u_z}{\partial r}\right) \tag{4.91}$$

$$\gamma_{z\theta} = \left(\frac{1}{r}\frac{\partial u_z}{\partial \theta} + \frac{\partial u_\theta}{\partial z}\right) \tag{4.92}$$

where u_r, u_θ, and u_z are the displacement components in the cylindrical coordinates (r, θ, z), ε_{rr}, $\varepsilon_{\theta\theta}$, and ε_{zz} are the normal strain components acting in the r-direction, θ-direction, and z-direction, respectively. $\gamma_{r\theta}$ is the shear strain acting in the θ-direction of the plane normal to the r-axis; γ_{zr} is the shear strain acting in the r-direction of the plane normal to the z-axis; and $\gamma_{\theta z}$ is the shear strain acting in the z-direction of the plane normal to the θ-axis.

4.5 Mechanical Behavior of Packaging Materials

4.5.1 Stress-strain relation—elastic

By definition, an elastic material may be deformed and will return to its original configuration upon release of the deforming loads or temperatures (Fig. 4.25). The qualifying adjective of *linear* means that the load-deflection law may be represented by a linear relationship. The concept of linear elasticity was first announced by Robert Hooke in 1676 and then generalized by Cauchy into the statement that the components of stress are linearly related to the components of strain. For linear isotropic elastic materials in Cartesian coordinates, the stress-strain relations are:

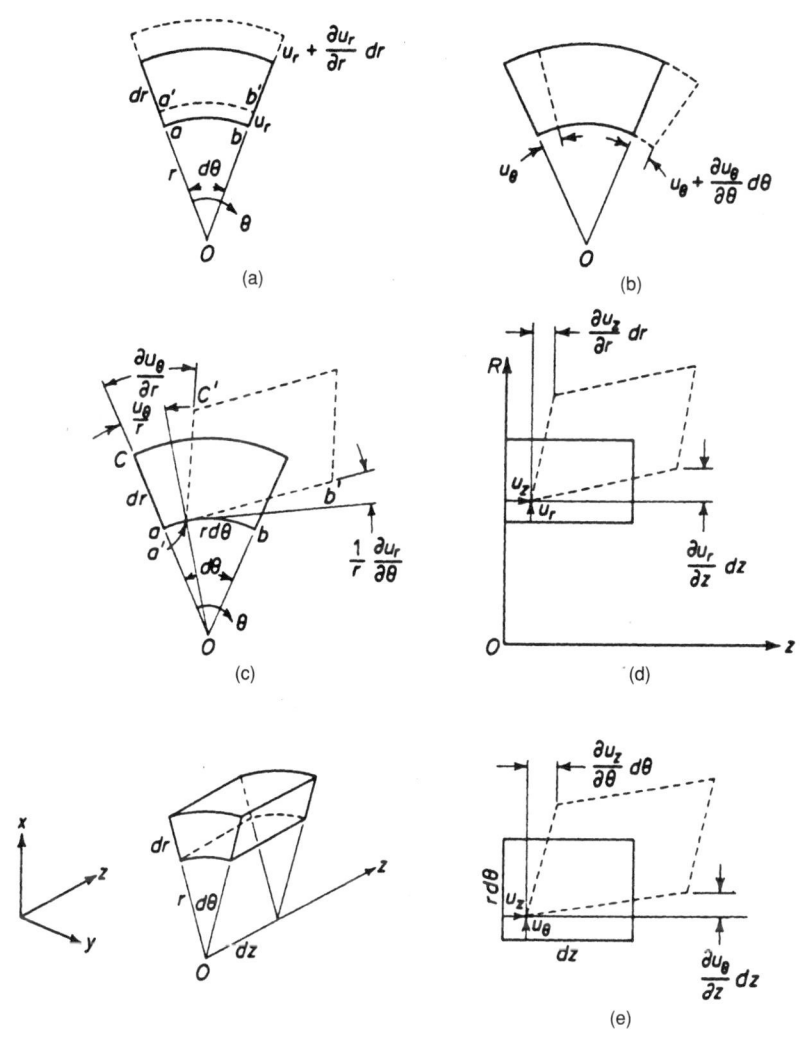

Figure 4.24 Displacement in cylindrical coordinates (after Secher, *Elasticity in Engineering*, Dover Publications, New York).

$$\varepsilon_x = \frac{1}{E}\left[\sigma_x - \nu(\sigma_y + \sigma_z)\right] + \alpha(T - T_0) \qquad (4.93)$$

$$\varepsilon_y = \frac{1}{E}\left[\sigma_y - \nu(\sigma_x + \sigma_z)\right] + \alpha(T - T_0) \qquad (4.94)$$

$$\varepsilon_z = \frac{1}{E}\left[\sigma_z - \nu(\sigma_x + \sigma_y)\right] + \alpha(T - T_0) \qquad (4.95)$$

$$\gamma_{xy} = \frac{1}{G}\tau_{xy} \qquad (4.96)$$

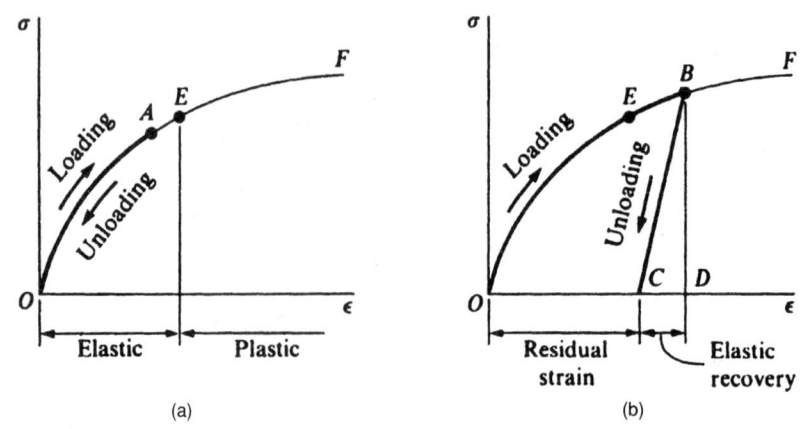

Figure 4.25 Stress-strain curves: (a) small (elastic) loading, (b) loading beyond elastic limit.

$$\gamma_{yz} = \frac{1}{G} \tau_{yz} \qquad (4.97)$$

$$\gamma_{zx} = \frac{1}{G} \tau_{zx} \qquad (4.98)$$

or

$$\sigma_x = \lambda e + 2G\epsilon_x - \beta(T - T_0) \qquad (4.99)$$

$$\sigma_y = \lambda e + 2G\epsilon_y - \beta(T - T_0) \qquad (4.100)$$

$$\sigma_z = \lambda e + 2G\epsilon_z - \beta(T - T_0) \qquad (4.101)$$

$$\tau_{xy} = G\gamma_{xy} \qquad (4.102)$$

$$\tau_{yz} = G\gamma_{yz} \qquad (4.103)$$

$$\tau_{zx} = G\gamma_{zx} \qquad (4.104)$$

where β, λ, and G are given in Eqs. 4.6, 4.7, and 4.8, respectively; E is the Young's modulus; ν is the Poisson's ratio; α is the thermal coefficient of linear expansion; T_0 is the reference temperature; T is the instantaneous absolute temperature; and $e = \varepsilon_x + \varepsilon_y + \varepsilon_z$ represents the volumetric strain (expansion) at a point. The mean stress (σ) at a point can be defined by $\sigma = \sigma_x + \sigma_y + \sigma_z$ and is related to the volumetric strain through the bulk modulus (K), which is defined as $K = e/3\sigma$. Table 4.3 shows the relationships among these modules.

TABLE 4.3 Relationship Between Elastic Modules

Given	E	G	ν	K
E, G	—	—	$\left(\dfrac{E}{2G} - 1\right)$	$\dfrac{EG}{3(3G - E)}$
E, ν	—	$\dfrac{E}{2(1 + \nu)}$	—	$\dfrac{E}{3(1 - 2\nu)}$
E, K	—	$\dfrac{3EK}{(9K - E)}$	$\dfrac{1}{2}\left(1 - \dfrac{E}{3K}\right)$	—
G, ν	$2G(1 + \nu)$	—	—	$\dfrac{2G(1 + \nu)}{3(1 - 2\nu)}$
G, K	$\left(\dfrac{9KG}{3K + G}\right)$	—	$\dfrac{3K - 2G}{2(G + 3K)}$	—
ν, K	$3K(1 - 2\nu)$	$\dfrac{3K(1 - 2\nu)}{2(1 + \nu)}$	—	—

4.5.2 Stress-strain relation—plastic

Most electronic packaging materials obey Hooke's law only at a certain range of small deformation and stress. Once a material is stressed beyond an elastic limit (Point E in Fig. 4.25), Hooke's law no longer applies. Some other material laws such as plasticity and creep need to be considered. Unlike elastic deformation, plastic deformation (residual strain) is not a reversible process, and depends not only upon the initial and final states of loading (e.g., temperature) but also upon the loading path by which the final state is achieved. Several aspects of real material behavior, such as the Bauschinger effect [a specimen initially stressed in tension often yields at a much reduced stress when restressed in compression, that is, the yield stresses in tension and compression are not the same (see Fig. 4.26)], cyclic hardening, plastic anisotropy, elastic hysteresis, and so forth, can be modeled (at different levels of sophistication) by the theories of plasticity [37–48]. In this section, some well-established and most often used theories are briefly discussed.

4.5.2.1 Yield surface.
The *yield surface* is defined as the surface in stress space with stress components as coordinates. Within the yield surface, the stress vector may change without any plastic strain increment; stress increments beginning from points in the surface, if directed toward the exterior, imply plastic strain increments.

4.5.2.2 Initial yield surface.
For isotropic plasticity, the initial yield surface must be independent of the orientation of the reference axes. By choosing the axes of the principal stresses as the reference axes, the

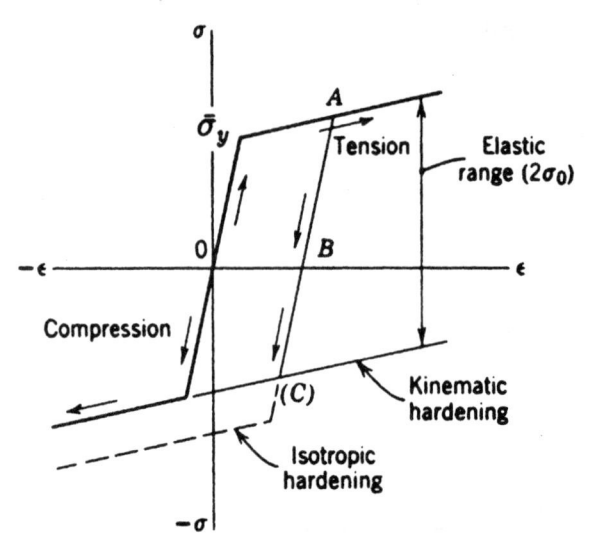

Figure 4.26 A stress-strain curve with Bauschinger effect ($\sigma_0 = \bar{\sigma}_y$).

initial yield surface may be expressed in terms of the principal stresses and represented by a surface in a stress space with σ_1, σ_2, σ_3 as coordinate axes. Thus the initial yield function may appear as

$$f(\sigma_1, \sigma_2, \sigma_3) = 0 \qquad (4.105)$$

Furthermore, experiment indicates that the hydrostatic pressure has very little effect on the plastic deformation. Hence, the initial yield condition may be expressed in terms of the deviatoric stress invariants in the form

$$f(\bar{I}_2, \bar{I}_3) = 0 \qquad (4.106)$$

where

$$\bar{I}_2 = \tfrac{1}{2} S_{ij} S_{ij} \qquad (4.107)$$

$$\bar{I}_3 = \tfrac{1}{3} S_{ij} S_{jk} S_{ki} \qquad (4.108)$$

$$S_{ij} = \sigma_{ij} - \tfrac{1}{3} \sigma_{\beta\beta} \delta_{ij} \qquad (4.109)$$

$\bar{I}_2$ and $\bar{I}_3$ are the deviatoric stress invariants and S_{ij} is the deviatoric stress tensor. Two simple yield surfaces (conditions) for the initial yield of isotropic materials with isotropic hardening (i.e., the tensile and compressive yield stresses are equal at all times) that have provided highly useful descriptions of many real materials are briefly discussed in the following.

Von Mises yield condition (distortion energy theory) is based on the assumption that yielding occurs when the second deviatoric stress invariant attains a prescribed value k.

$$f = \bar{I}_2 - k^2 = \tfrac{1}{2} S_{ij} S_{ij} - k^2 = 0 \tag{4.110}$$

or

$$f = \tfrac{1}{6}[(\sigma_x - \sigma_y)^2 + (\sigma_y - \sigma_z)^2 + (\sigma_z - \sigma_x)^2 + 6(\tau_{xy}^2 + \tau_{yz}^2 + \tau_{zx}^2)] - k^2 = 0 \tag{4.111}$$

where k may be a function of plastic strain for strain-hardening materials and the relation of k to test data follows $k = \tau_y$ (τ_y is the yield stress in pure shear) or $k = \bar{\sigma}_y/\sqrt{3}$ ($\bar{\sigma}_y$ is the yield stress in uniaxial tension).

For example, the stress state at a point in a packaging component is $\sigma_x = 200$ MPa, $\sigma_y = 100$ MPa, $\sigma_z = -50$ MPa, $\tau_{xy} = 30$ MPa, $\tau_{yz} = \tau_{zx} = 0$. It is interesting to know if the component at that point exhibits yielding. The material of the component has a uniaxial yield stress $\bar{\sigma}_y = 500$ MPa. From Eq. 4.111 we have

$$f = \tfrac{1}{6}[100^2 + 150^2 + 250^2 + 6(30)^2] - \tfrac{1}{3}(500)^2 = 50{,}200 - 250{,}000 < 0$$

Thus, the material at that point of the component will not yield.

Tresca yield condition (maximum shear theory) is based on the assumption that yield occurs when the maximum shear stress reaches a limiting value k.

$$f = 4\bar{I}_2^3 - 27\bar{I}_3^2 - 36k^2\bar{I}_2^2 + 96k^4\bar{I}_2 - 64k^6 = 0 \tag{4.112}$$

or

$$f = [(\sigma_1 - \sigma_2)^2 - 4k^2][(\sigma_2 - \sigma_3)^2 - 4k^2][(\sigma_3 - \sigma_1)^2 - 4k^2] = 0 \tag{4.113}$$

where $k = \tau_y$ or $k = \bar{\sigma}_y/2$. Experimental data appear to favor the use of the von Mises yield condition for most of the materials.

4.5.2.3 Subsequent yield surface.
Continued loading beyond the initial yield surface leads to plastic deformation, which may be accompanied by changes in both size and shape of the yield surface. For perfect plasticity, the yield surface does not change during plastic deformation and the initial yield surface remains valid. For isotropic hardening, however, the size of the yield surface increases but the shape remains the same during loading (Fig. 4.27). To take such changes into account it is necessary to modify the initial yield surface and to define the subsequent yield surface, also known as the loading surface. A general form for the loading surface is given by

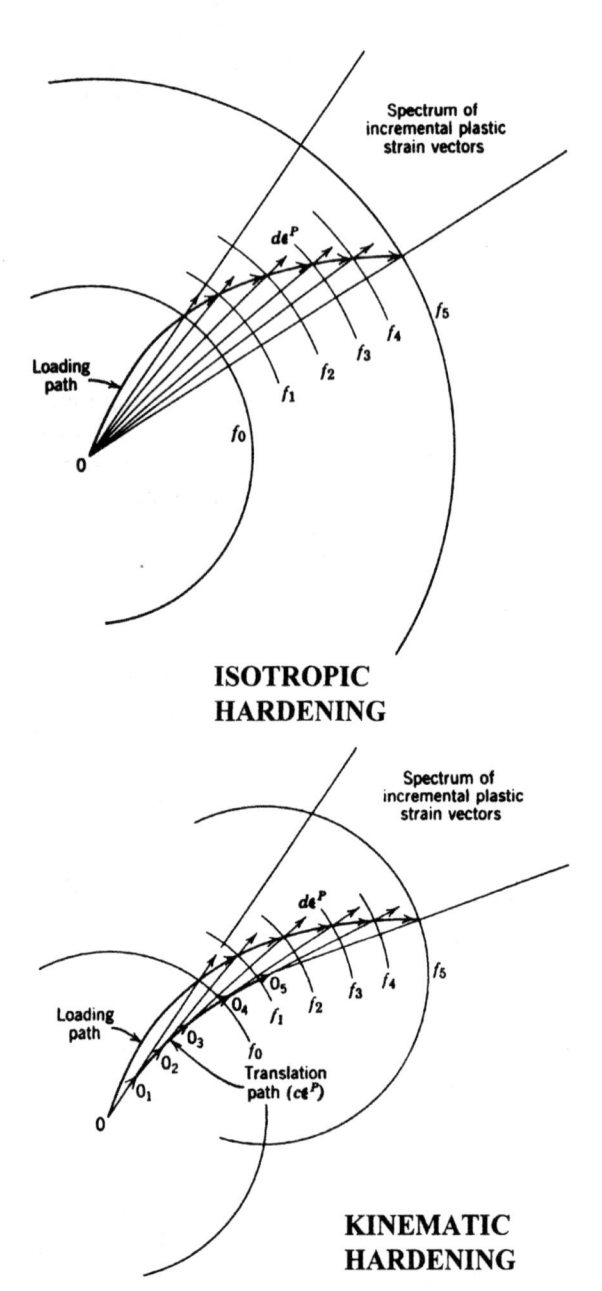

Figure 4.27 Isotropic hardening versus kinematic hardening in plasticity.

$$f(\sigma_{ij}, \epsilon^p_{ij}, k) = 0 \tag{4.114}$$

which depends not only upon the stresses σ_{ij} but also upon the plastic strain ϵ^p_{ij} and the work-hardening characteristics represented by the parameter k. Differentiating $f = 0$ by the chain rule of calculus, we have

$$df = \frac{\partial f}{\partial \sigma_{ij}} d\sigma_{ij} + \frac{\partial f}{\partial \epsilon^p_{ij}} d\epsilon^p_{ij} + \frac{\partial f}{\partial k} dk \tag{4.115}$$

where df, $d\sigma_{ij}$, $d\epsilon^p_{ij}$, and dk represent time differentials. If $f = 0$ and $df < 0$, a condition leading to an elastic state is implied, and it must follow that $d\epsilon^p_{ij} = dk = 0$. Thus

$$f = 0 \qquad \frac{\partial f}{\partial \sigma_{ij}} d\sigma_{ij} < 0 \tag{4.116}$$

is defined as *unloading*;

$$f = 0 \qquad \frac{\partial f}{\partial \sigma_{ij}} d\sigma_{ij} = 0 \tag{4.117}$$

is defined as *neutral loading*, since it implies that the stress-point remains on the initial yield surface; and

$$f = 0 \qquad \frac{\partial f}{\partial \sigma_{ij}} d\sigma_{ij} > 0 \tag{4.118}$$

is defined as *loading*, since it implies that the stress-point is moving outward from the current yield surface. For perfectly plastic materials, plastic flow occurs for $f = 0$, $(\partial f/\partial \sigma_{ij})d\sigma_{ij} = 0$, and the case $f = 0$, $(\partial f/\partial \sigma_{ij})d\sigma_{ij} > 0$ does not exist.

4.5.2.4 Constitutive equation: incremental theory. A general equation for determining the plastic stress-strain relation for any yield condition was proposed by Drucker. Based on his definition of work hardening materials ($d\sigma_{ij}d\epsilon_{ij} > 0$ upon loading; $d\sigma_{ij}d\epsilon^p_{ij} \geq 0$ on completing a cycle), he stated that the plastic strain increment vector must be normal to the yield or loading surface at a smooth point on that surface (Fig. 4.27), and must lie between adjacent normals at a corner point, that is,

$$d\epsilon^p_{ij} = d\lambda \frac{\partial f}{\partial \sigma_{ij}} \tag{4.119}$$

Equation 4.119 is called the normality principle of plasticity and can be applied to plastic anisotropic hardening; $d\lambda$ is a function that may depend on stress, strain, and strain history.

For isotropic hardening, the von Mises yield condition is given by

$$f = \tfrac{1}{2}S_{ij}S_{ij} - \tfrac{1}{3}\overline{\sigma}_y^2 \qquad (4.120)$$

Substituting Eq. 4.120 into Drucker's Eq. 4.119, we have

$$d\epsilon_{ij}^p = d\lambda S_{ij} \qquad (4.121)$$

or

$$d\epsilon_{ij}^p d\epsilon_{ij}^p = (d\lambda)^2 S_{ij}S_{ij} \qquad (4.122)$$

Defining the effective (equivalent) stress

$$\overline{\sigma} = \sqrt{(3/2)S_{ij}S_{ij}} \qquad (4.123)$$

$$\overline{\sigma} = (\sqrt{2}/2)\sqrt{(\sigma_1 - \sigma_2)^2 + (\sigma_2 - \sigma_3)^2 + (\sigma_3 - \sigma_1)^2} \qquad (4.124)$$

$$\overline{\sigma} = (\sqrt{2}/2)\sqrt{(\sigma_x - \sigma_y)^2 + (\sigma_y - \sigma_z)^2 + (\sigma_z - \sigma_x)^2 + 6(\tau_{xy}^2 + \tau_{yz}^2 + \tau_{zx}^2)}$$

$$(4.125)$$

and the effective (equivalent) plastic incremental strain

$$d\overline{\epsilon}_p = \sqrt{(2/3)d\epsilon_{ij}^p d\epsilon_{ij}^p} \qquad (4.126)$$

$$d\overline{\epsilon}_p = (\sqrt{2/3})\sqrt{(d\epsilon_1^p - d\epsilon_2^p)^2 + (d\epsilon_2^p - d\epsilon_3^p)^2 + (d\epsilon_3^p - d\epsilon_1^p)^2} \qquad (4.127)$$

$$d\overline{\epsilon}_p = (\sqrt{2/3})\sqrt{(d\epsilon_x^p - d\epsilon_y^p)^2 + (d\epsilon_y^p - d\epsilon_z^p)^2 + (d\epsilon_z^p - d\epsilon_x^p)^2 + d\gamma}$$

$$(4.128)$$

where

$$d\gamma = \tfrac{3}{2}[(d\gamma_{xy}^p)^2 + (d\gamma_{yz}^p)^2 + (d\gamma_{zx}^p)^2] \qquad (4.129)$$

then Eq. 4.122 becomes

$$d\lambda = \frac{3}{2}\frac{d\overline{\epsilon}_p}{\overline{\sigma}} \qquad (4.130)$$

Suppose there exists a universal stress-strain curve (which coincides with a uniaxial true stress versus true plastic strain curve for the material)

$$\overline{\sigma} = H\left(\int d\overline{\epsilon}_p\right) \qquad (4.131)$$

expressing an equivalent stress $\overline{\sigma}$ as a function H of an equivalent plastic strain increment $d\overline{\epsilon}_p$ integrated over the strain history. Then the slope (H') of the equivalent stress $\bar{\sigma}$ versus equivalent plastic strain $\int d\bar{\epsilon}_p$ curve is given by

$$H' = \frac{d\bar{\sigma}}{d\bar{\epsilon}_p} \tag{4.132}$$

Substituting Eq. 4.132 into Eq. 4.130, then Eq. 4.121, we have

$$d\lambda = \frac{3}{2} \frac{d\bar{\sigma}}{\bar{\sigma} H'} \tag{4.133}$$

and

$$d\epsilon_{ij}^p = \frac{3 d\bar{\sigma}}{2\bar{\sigma} H'} S_{ij} \tag{4.134}$$

Equation 4.134 is referred to as the Levy-Mises equation and is applied to problems of mostly plastic deformation (the elastic deformation is very small and is neglected). However, in most of electronic packaging problems the elastic strains cannot be neglected. In that case the incremental total strain is given by

$$d\epsilon_{ij} = d\epsilon_{ij}^e + d\epsilon_{ij}^p \tag{4.135}$$

where $d\epsilon_{ij}^e$ is the incremental elastic strain tensor and can be derived from Eqs. 4.99–4.104 with $\beta = 0$, and $d\epsilon_{ij}^p$ is given by Eq. 4.134. Thus Eq. 4.135 becomes

$$d\epsilon_{ij} = \frac{dS_{ij}}{2G} + (1 - 2\nu)\delta_{ij} \frac{d\sigma_{ij}}{3E} + \frac{3 d\bar{\sigma}}{2\bar{\sigma} H'} S_{ij} \tag{4.136}$$

Equation 4.136 is called the Prandtl-Reuss equation.

Equation 4.136 can only be applied for isotropic hardening materials. For kinematic hardening materials that obey the von Mises yield condition for the initial yield surface, Prager's loading function corresponding to a translation of this surface is expressed as

$$f = \tfrac{1}{2}(S_{ij} - \alpha_{ij})(S_{ij} - \alpha_{ij}) - \frac{\bar{\sigma}_y^2}{3} = 0 \tag{4.137a}$$

where α_{ij} represents the translation of the center of the initial yield surface in the stress space (see Fig. 4.27). For linear hardening, we have

$$\alpha_{ij} = c\epsilon_{ij}^p \tag{4.138}$$

and Eq. 4.137a becomes

$$f = \tfrac{1}{2}(S_{ij} - c\epsilon_{ij}^p)(S_{ij} - c\epsilon_{ij}^p) - \frac{\bar{\sigma}_y^2}{3} = 0 \tag{4.137b}$$

where c is a constant and ϵ_{ij}^p is a plastic strain tensor. From Drucker's normality principle of plasticity, Eq. 4.119, we have

$$d\epsilon_{ij}^p = \frac{3}{2} \frac{(S_{mn} - c\epsilon_{mn}^p)dS_{mn}}{c\bar{\sigma}_y^2} (S_{ij} - c\epsilon_{ij}^p) \tag{4.139}$$

The material constant c can be determined from the uniaxial test similarly to the isotropic hardening case and is equal to two-thirds of the slope (H') of the uniaxial stress-plastic strain diagram. Then we have Prager's constitutive equation for kinematic hardening materials:

$$d\epsilon_{ij}^p = \frac{9}{4} \frac{[S_{mn} - (2/3)H'\epsilon_{mn}^p]dS_{mn}}{H'\bar{\sigma}_y^2} (S_{ij} - \tfrac{2}{3}H'\epsilon_{ij}^p) \tag{4.140}$$

For elastoplastic solids with kinematic hardening, we have

$$d\epsilon_{ij} = \frac{dS_{ij}}{2G} + (1 - 2\nu)\delta_{ij}\frac{d\sigma_{ij}}{3E} + \frac{9}{4} \frac{[S_{mn} - (2/3)H'\epsilon_{mn}^p]dS_{mn}}{H'\bar{\sigma}_y^2} (S_{ij} - \tfrac{2}{3}H'\epsilon_{ij}^p) \tag{4.141}$$

Equation 4.141 is sometimes called the Reuss-Prager equation. The qualitative difference between isotropic and kinematic hardening can be illustrated for the same loading path in the π-plane ($\sigma_1 + \sigma_2 + \sigma_3 = 0$) in the stress space (Fig. 4.27). It can be seen that, for kinematic hardening, the initial yield surface translates in the π-plane without rotation and without change in size (the center of the yield surfaces f_0 moves from 0 to 0_1 for the subsequent yield surface f_1, etc.) On the other hand, isotropic hardening assumes a uniform expansion of the initial yield surface in the π-plane. Also, the resultant of the incremental plastic strain vectors predicted by isotropic hardening lags farther behind the resultant stress vector than the strain resultant predicted by Prager's kinematic hardening. For many other yielding and loading functions including the Bauschinger effect, see [37–48].

4.5.2.5 Constitutive equation: total strain theory. A very simple nonlinear stress-strain relation was provided by Hencky. He proposed a one-to-one correspondence between the stress and strain. Thus, the total plastic strain components are taken to be proportional to the corresponding deviatoric stress components, that is

$$\epsilon_{ij}^p = \Lambda S_{ij} \tag{4.142a}$$

where ϵ_{ij}^p is the total plastic strain and Λ is positive during loading and zero during unloading and may depend on stress and strain. For von Mises yield condition, Eq. (4.110), we use the same definition of equivalent stress, Eq. (4.123) and define the following equivalent total plastic strain:

$$\bar{\epsilon}_p = \sqrt{\tfrac{2}{3}\epsilon_{ij}^p \epsilon_{ij}^p} \qquad (4.142\text{b})$$

Then, we have

$$\Lambda = \frac{3}{2} \frac{\bar{\epsilon}_p}{\bar{\sigma}} \qquad (4.142\text{c})$$

and

$$\epsilon_{ij}^p = \frac{3}{2} \frac{\bar{\epsilon}_p}{\bar{\sigma}} S_{ij} = \frac{3}{2}\left(\frac{1}{S} - \frac{1}{E}\right) S_{ij} \qquad (4.142\text{d})$$

where E and S are, respectively, the Young's modulus and secant modulus of the uniaxial true stress versus true strain curve. Once again, the existence of a universal stress-strain curve is assumed and coincides with the uniaxial tension curve. Finally, the total strain-stress relation is given by

$$\epsilon_{ij} = \frac{S_{ij}}{2G} + (1 + 2\nu)\delta_{ij} \frac{\sigma_{ij}}{3E} + \frac{3}{2}\left(\frac{1}{S} - \frac{1}{E}\right) S_{ij} \qquad (4.142\text{e})$$

It should be pointed out that the total strain theory (deformation plasticity) is not acceptable physically, except that for the case of proportional loading (a sufficient condition) the total strain theory coincides with the incremental theory.

4.5.3 Stress-strain relation—creep

The stress-strain relations for the elastic and plastic materials previously described are obtained from tension tests involving only static loading of the material specimens and were time independent. However, some electronic packaging materials (e.g., solders) develop additional strains over "long" periods of time and are said to *creep*. Creep is a mathematical model for rate-sensitive elastoplastic materials operating at elevated temperature. *Creep strain* may be broadly defined as elastoplastic time-dependent deformation under constant load at "high" temperature [49–75]. Since different materials have different melting temperatures, it is convenient to define a homologous temperature (the ratio of the test or use temperature to the melting temperature on an absolute temperature scale). In general, creep becomes of engineering significance at a homologous temperature greater than 0.5. For some electronic packaging materials such as solder, creep deformation becomes important even if it is at room temperature.

Figure 4.28 shows a typical family of creep curves which can be obtained from "long-time" uniaxial tests or shear tests at constant temperature and various stress conditions ($\sigma_4 > \sigma_3 > \sigma_2 > \sigma_1$). The slope of these curves ($d\varepsilon_i/dt = \tan \Psi_i$) is referred to as creep rate. At the first stage, the creep rate is undefined and the initial creep strain consists of either entirely elastic strain or partially elastic strain and partially plastic strain. During the second stage, the creep rate decreases with time because the effect of strain hardening is greater than that of annealing (recovery). These two effects are in equilibrium (balance) during the third stage, and the creep rate reaches essentially a steady state and changes very little with time. However, during the fourth stage, the creep rate increases rapidly with time until fracture occurs. This is because the reduced cross-sectional area (either due to necking or internal void formation) causes an increase in stresses.

Andrade's pioneering work on pure metals has had considerable influence on the thinking on the mechanisms of the second and third stages of creep. He considered the second and third stages of creep as being the superposition of transient creep (with a creep rate decreasing with time) and quasi-viscous creep (with a constant creep rate) processes which occur right after the sudden strain (first stage of creep), which results from applying the load (Fig. 4.28).

The second stage of creep is called *primary creep*. During this period, the primary creep is predominated by transient creep. For low temperatures and stresses, as in the creep of lead at room temperature, pri-

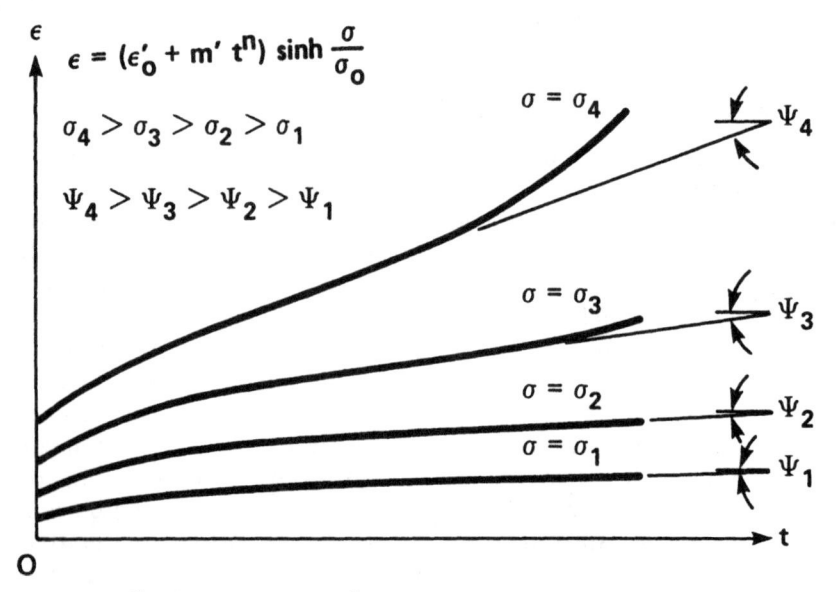

Figure 4.28 Constant stress creep-time curve.

mary creep is the predominant creep process. The third stage of creep is called *secondary creep* or *steady-state creep*. The average value of the creep rate during secondary creep is called the *minimum creep rate*. For example, Fig. 4.28 shows the angles $\Psi_4 > \Psi_3 > \Psi_2 > \Psi_1$ of the minimum creep rate at various creep curves. The fourth stage of creep is called *tertiary creep*, which is often associated with microstructural changes such as coarsening of precipitate particles, recrystallization, or diffusional changes in the phases that are present.

The family of creep curves shown in Fig. 4.28 can be plotted in various ways with certain objectives in mind. For mechanics analysts, the family of creep curves are usually plotted in the following two ways. Figure 4.29 shows the corresponding stresses versus minimum creep rates curve at a constant temperature. This curve can be used as the constitutive relation for steady-state creep analysis of structures at "high" temperature.

Another useful way to plot the family of creep curves (Fig. 4.28) generated at constant temperature and various stress conditions ($\sigma_4 > \sigma_3 > \sigma_2 > \sigma_1$) is shown in Fig. 4.30. These curves, which relate the stresses and creep strains at selected values of time for a constant temperature, are called *isochronous stress-strain curves* and can be used as the constitutive relations for creep strain analysis of structures.

$$\dot{\epsilon} = \epsilon_* \sinh \frac{\sigma}{\sigma_*}$$

$$\dot{\epsilon} = \frac{d\epsilon}{dt}$$

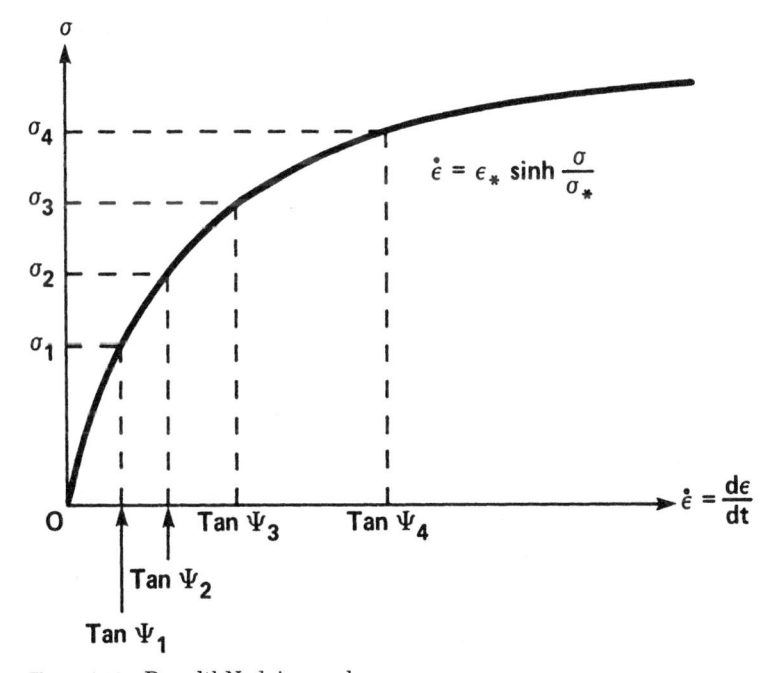

Figure 4.29 Prandtl-Nadai creep law.

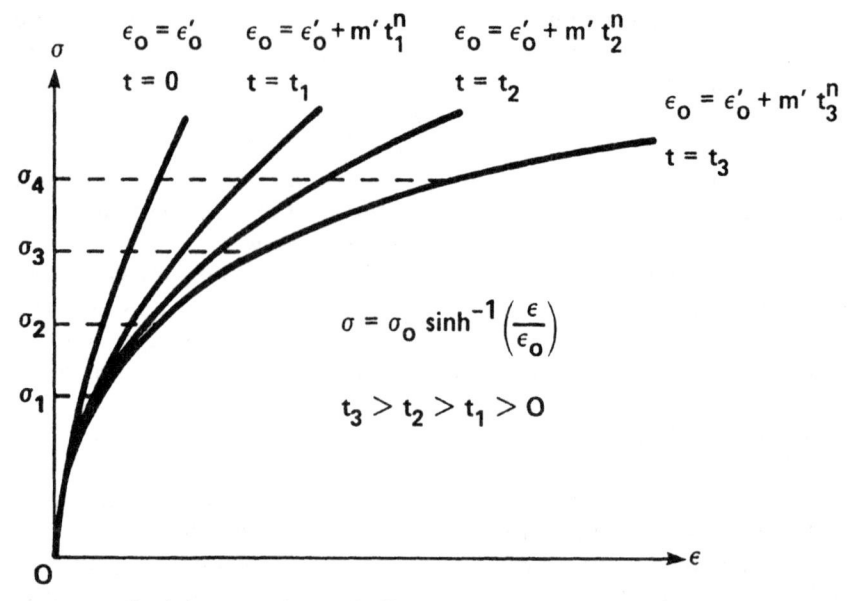

Figure 4.30 Isochronous stress-strain diagrams.

Some of the commonest empirical uniaxial constitutive equations for describing the second (primary) and third (secondary) stages of creep (Figs. 4.28–4.30) are presented herein. In the broadest sense, the creep strain (ϵ_c) is a function of applied load (or stress, σ), time (t), and temperature (T), that is,

$$\epsilon_c = p(\sigma, t, T) \tag{4.143}$$

which is usually assumed to be separable into

$$\epsilon_c = f(\sigma)g(t)h(T) \tag{4.144}$$

In Eq. 4.144, the stress-dependence term has been proposed by Norton, Prandtl, Dorn, and Garofalo as

$$f(\sigma) = \begin{cases} A\sigma^n & \text{Norton} & (4.145) \\ B(\sigma - F)^n & \text{Friction stress} & (4.146) \\ C \sinh(\alpha\sigma) & \text{Prandtl} & (4.147) \\ D \exp(\beta\sigma) & \text{Dorn} & (4.148) \\ E[\sinh(\gamma\sigma)]^n & \text{Garofalo} & (4.149) \end{cases}$$

The time-dependence term has been proposed by Bailey, Andrade, Graham, and Walles as

$$g(t) = \begin{cases} t & \text{Secondary creep} & (4.150) \\ \eta t^m & \text{Bailey} & (4.151) \\ (1 + \eta t^{1/3})e^{kt} & \text{Andrade} & (4.152) \\ \sum \eta_i t^{m_i} & \text{Graham and Walles} & (4.153) \end{cases}$$

The temperature-dependence term in Eq. 4.144 is usually associated with the Arrhenius law and has the form:

$$h(T) = G \exp(-\Delta H/RT) \qquad (4.154)$$

In Eqs. 4.145–4.154, ΔH is the activation energy, R is Boltzmann's constant, T is the absolute temperature, t is time, and all the remaining symbols other than σ are material constants that can be determined by fitting the creep curves from experiment. It can be seen from Eqs. 4.144 to 4.154 that there are many ways to write the creep equation.

In this section, the solder materials are assumed to obey the following constitutive equation (Fig. 4.28):

$$\epsilon = (\epsilon_o' + m't^n)\sinh\frac{\sigma}{\sigma_o} \qquad (4.155)$$

where ϵ_o', m', n, and σ_o are material constants of solders. The corresponding steady-state creep equation (Fig. 4.29)

$$\dot{\epsilon} = \frac{d\epsilon}{dt} = \epsilon_* \sinh\frac{\sigma}{\sigma_*} \qquad (4.156)$$

is called the Prandtl creep law and the ϵ_* and σ_* can be determined from Figs. 4.28 and 4.29. (More steady-state creep of solders is discussed in the next section.)

The isochronous stress-strain curves (Fig. 4.30)

$$\sigma = \sigma_o \sinh^{-1}\left(\frac{\epsilon}{\epsilon_o}\right) \qquad (4.157)$$

where $\epsilon_o = \epsilon_o' + m't^n$ with the variable, time (t) held constant, are used to obtain the tangent-modulus creep buckling load and the critical time to buckling of centrally loaded solder columns in the present section.

Guided by Shanley [34], the theoretical buckling load of centrally loaded columns is given by

$$\sigma = \frac{P_T}{A} = \frac{\pi^2 E_T}{(l/r)^2} \qquad (4.158)$$

where P_T is the tangent modulus buckling load, A is the cross-sectional area, l is the length of the column, r is the minimum radius of gyration, l/r is the slenderness ratio, and E_T is the tangent modulus. Shanley observed (from Fig. 4.30) that for a given constant stress, the slopes of the curves decrease with increasing values of time. Thus, he proposed to consider the column at any particular time as behaving according to the isochronous stress-strain curve for that value of time. According to this assumption, the slopes of the isochronous curves represent the tangent modulus, that is,

$$E_T = \frac{d\sigma}{d\epsilon} = \frac{\sigma_o}{\epsilon\sqrt{1+(\epsilon/\epsilon_o)^2}} \qquad (4.159)$$

Since $P_T = A\sigma$, then Eq. 4.157 becomes

$$\frac{\epsilon}{\epsilon_o} = \sinh\frac{P_T}{A\sigma_o} \qquad (4.160)$$

Substituting Eqs. 4.159 and 4.160 into Eq. 4.158, we have

$$\sqrt{\epsilon_o}\left(\frac{l}{r}\right) = \frac{\pi}{\sqrt{(P_T/\sigma_o A)}\sqrt{1+\sinh^2(P_T/\sigma_o A)}} \qquad (4.161)$$

Equation 4.161 is plotted in Fig. 4.31, where a correction factor (1.09) is included to offset the error introduced into the theory by the stress distribution in the column changing with time. With this curve and the values of l, r, A, the tangent modulus buckling load of a solder column is readily obtained. Also, this curve can be used to design the length, diameter, and materials of the centrally loaded columns.

Substituting Eq. 4.159 into Eq. 4.158 with $\epsilon_o = \epsilon'_o + m't^n$, we have

$$\sigma = \frac{\pi^2}{(l/r)^2}\frac{\sigma_o}{\sqrt{(\epsilon'_o + m't^n)^2 + \epsilon^2}} \qquad (4.162)$$

Eliminating creep strain ϵ by substituting Eq. 4.155 into Eq. 4.162, we have [73]

$$t_{cr} = \left[\frac{1}{m'}\left(\frac{\pi^2}{(l/r)^2}\frac{\sigma/\sigma_o}{\sqrt{1+\sinh^2(\sigma/\sigma_o)}} - \epsilon'_o\right)\right]^{1/n} \qquad (4.163)$$

as the equation for critical time t_{cr} to buckling of a solder column at a constant stress σ. Thus, for a given ϵ'_o, m', n, σ_o, l, and r the critical time to buckling can be readily determined [73].

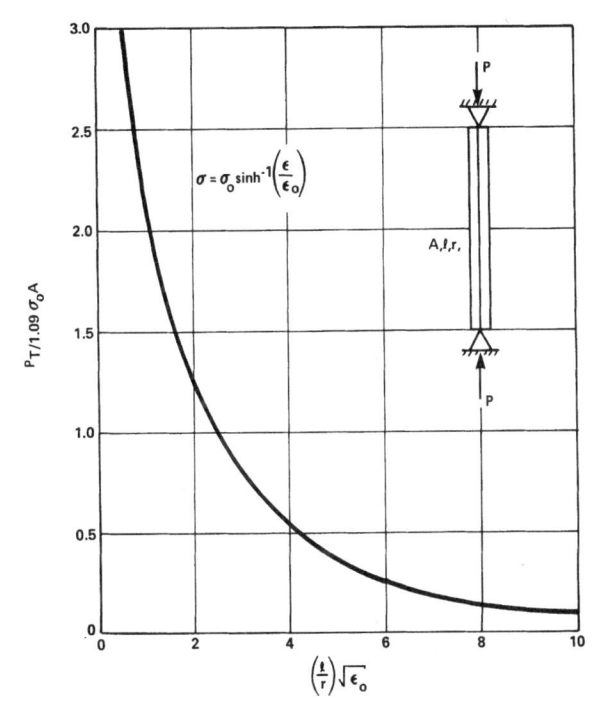

Figure 4.31 Tangent modulus buckling curve of solder column with creep.

4.5.4 Creep of solders under combined loads

For 96.5wt%Sn3.5wt%Ag, 97.5wt%Pb2.5wt%Sn, and 60wt%Sn40wt%Pb solders, it has been shown by Darveaux and Banerji [76] that the Garofalo-Arrhenius creep equation fits very well with the data, as seen in Figs. 4.32, 4.33, and 4.34. These curves can be used as the constitutive relations for steady-state creep analysis of solder structures at "high" temperature.

In this section, the load-deformation behavior of bending and twisting of thin-walled circular solder cylinder is studied. This kind of specimen is used for the steady-state creep test of solder materials under combined load [76–78]. The material is assumed to be incompressible and to follow the Garofalo-Arrhenius creep equation (see previous section). Hencky's total-strain theory [39], Section 4.5.2.5 is assumed to be valid. Due to the geometry of the structure and the loading conditions, $\sigma_r = \sigma_\theta = \tau_{rz} = \tau_{r\theta} = d(\gamma_{r\theta})/dt = d(\gamma_{rz})/dt = 0$. It is also assumed that the thickness of the solder cylindrical wall is so thin, compared with the inner radius, that all the nonzero stresses are uniformly distributed across the wall thickness.

240   Chapter Four

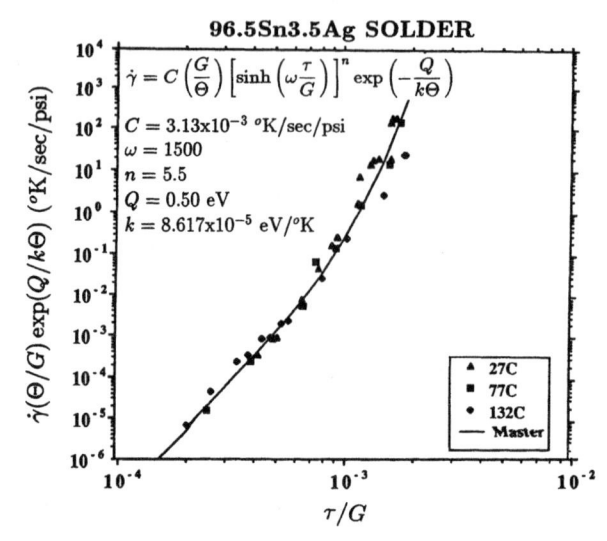

Figure 4.32 Steady-state creep curve for 96.5Sn3.5Ag (after Darveaux).

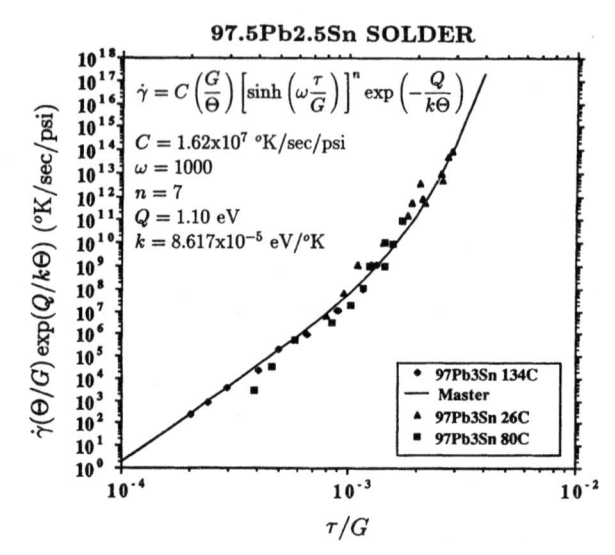

Figure 4.33 Steady-state creep curve for 97.5Pb2.5Sn (after Darveaux).

The Garofalo-Arrhenius steady-state creep (see Sec. 4.5.3) is generally expressed as:

$$\dot{\gamma} = \frac{d\gamma}{dt} = C\left(\frac{G}{\Theta}\right)\left[\sinh\left(\omega\frac{\tau}{G}\right)\right]^n \exp\left(\frac{-Q}{k\Theta}\right) \quad (4.164)$$

60Sn40Pb SOLDER

$$\dot{\gamma} = C\left(\frac{G}{\Theta}\right)\left[\sinh\left(\omega\frac{\tau}{G}\right)\right]^n \exp\left(-\frac{Q}{k\Theta}\right)$$

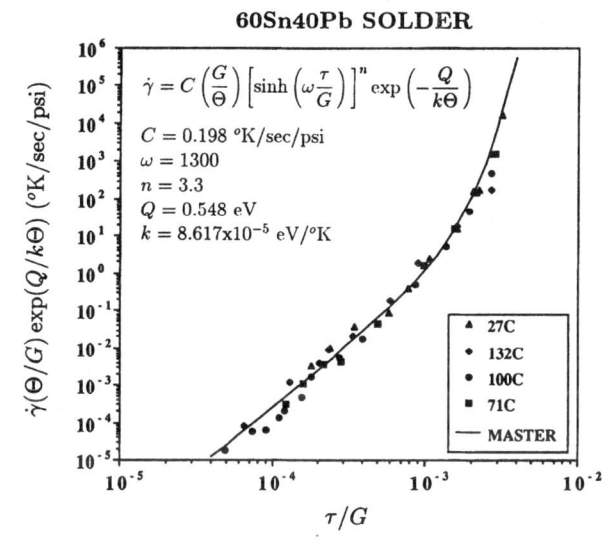

$C = 0.198$ °K/sec/psi
$\omega = 1300$
$n = 3.3$
$Q = 0.548$ eV
$k = 8.617 \times 10^{-5}$ eV/°K

- ▲ 27C
- • 132C
- • 100C
- ■ 71C
- — MASTER

Figure 4.34 Steady-state creep curve for 60Sn40Pb (after Darveaux).

where γ is the steady-state creep shear strain, $\dot{\gamma}$ is the steady-state creep shear strain rate, t is the time, C is a material constant, G is the temperature-dependent shear modulus, Θ is the absolute temperature (°K), ω defines the stress level at which the power law stress dependence breaks down, τ is the shear stress, n is the stress exponent, Q is the activation energy for a specific diffusion mechanism (for example, dislocation diffusion, solute diffusion, lattice self-diffusion, and grain boundary diffusion), and k is the Boltzmann's constant (8.617×10^{-5} eV/°K). For 96.5wt%Sn-3.5wt%Ag, 97.5wt%Pb-2.5wt%Sn, and 60wt%Sn-40wt%Pb solders, the material constants of Eq. 4.164 have been experimentally determined by Darveaux and Banerji [76] with a single hyperbolic sine function.

By using test data from [76] on the 96.5wt%Sn-3.5wt%Ag solder, Eq. 4.164 can be written as

$$\dot{\gamma} = \gamma_o\left[\sinh\left(\frac{\tau}{\tau_o}\right)\right]^{5.5} \qquad (4.165a)$$

where

$$\gamma_o = \frac{31(553 - \Theta)}{\Theta}\exp\left(\frac{-5802}{\Theta}\right) \qquad (4.166a)$$

and

$$\tau_o = 3687 - 6.67\Theta \qquad (4.167a)$$

By using test data from [76] on the 97.5wt%Pb-2.5wt%Sn solder, Eq. 4.164 can be written as

$$\dot{\gamma} = \gamma_o \left[\sinh\left(\frac{\tau}{\tau_o}\right) \right]^7 \quad (4.165b)$$

where

$$\gamma_o = \frac{1.62 \times 10^7 (1140 - \Theta)}{\Theta} \exp\left(\frac{-12{,}765}{\Theta}\right) \quad (4.166b)$$

and

$$\tau_o = 1710 - 1.5\Theta \quad (4.167b)$$

Again, by using test data from [76] on the 60Sn40Pb solder, Eq. 4.164 can be written as

$$\dot{\gamma} = \gamma_o \left[\sinh\left(\frac{\tau}{\tau_o}\right) \right]^{3.3} \quad (4.165c)$$

where

$$\gamma_o = \frac{1604(508 - \Theta)}{\Theta} \exp\left(\frac{-6360}{\Theta}\right) \quad (4.166c)$$

and

$$\tau_o = 3163 - 6.23\Theta \quad (4.167c)$$

If the 96.5Sn3.5Ag, 97.5Pb2.5Sn, and 60Sn40Pb solders obey the von Mises criterion, then Eq. 4.165a can be written as (for the 96.5Sn3.5Ag solder)

$$\dot{\epsilon} = \epsilon_o \left[\sinh\left(\frac{\sigma}{\sigma_o}\right) \right]^{5.5} \quad (4.168a)$$

where

$$\epsilon_o = \frac{18(553 - \Theta)}{\Theta} \exp\left(\frac{-5802}{\Theta}\right) \quad (4.169a)$$

and

$$\sigma_o = 6386 - 11.55\Theta \quad (4.170a)$$

Similarly, Eq. 4.165b can be written as (for the 97.5Pb2.5Sn solder)

$$\dot{\epsilon} = \epsilon_o \left[\sinh\left(\frac{\sigma}{\sigma_o}\right) \right]^7 \quad (4.168b)$$

Mechanical Design/Analysis/Measurement/Reliability of Electronic Packaging   243

where

$$\epsilon_o = \frac{1.4 \times 10^7 (1140 - \Theta)}{\Theta} \exp\left(\frac{-12{,}765}{\Theta}\right) \quad (4.169\text{b})$$

and

$$\sigma_o = 2962 - 2.6\Theta \quad (4.170\text{b})$$

Also, Eq. 4.165c can be written as (for the 60Sn40Pb solder)

$$\dot{\epsilon} = \epsilon_o \left[\sinh\left(\frac{\sigma}{\sigma_o}\right)\right]^{3.3} \quad (4.168\text{c})$$

where

$$\epsilon_o = \frac{926(508 - \Theta)}{\Theta} \exp\left(\frac{-6360}{\Theta}\right) \quad (4.169\text{c})$$

and

$$\sigma_o = 5478 - 10.79\Theta \quad (4.170\text{c})$$

In Eq. 4.168, σ is the uniaxial stress, $\dot{\epsilon}$ is the uniaxial steady-state creep strain rate. The unit for σ, τ, σ_o, and τ_o is in lb/in^2 (psi), the unit for γ_o and ϵ_o is in 1/sec, and the unit for the temperature (Θ) is in degrees Kelvin (°K) which is obtained by adding 273.16 to temperature in degrees Celsius (°C).

Equations 4.165 and 4.168 can only be applied to, respectively, pure shear and uniaxial tension conditions. For combined stresses state, it is necessary to define an effective steady-state creep strain rate ($\dot{\epsilon}_e$) and an effective stress (σ_e) as follows

$$\dot{\epsilon}_e = \sqrt{\tfrac{2}{3}\dot{\epsilon}_{ij}\dot{\epsilon}_{ij}} \quad (4.171)$$

$$\sigma_e = \sqrt{\tfrac{3}{2}S_{ij}S_{ij}} \quad (4.172)$$

where

$$S_{ij} = \sigma_{ij} - \tfrac{1}{3}\sigma_{pp}\delta_{ij} \quad (4.173)$$

In Eq. 4.171, $\dot{\epsilon}_{ij}$ is the steady-state creep strain rate tensor. In Eqs. 4.172 and 4.173, S_{ij} is the deviatoric stress tensor, σ_{ij} is the stress tensor, and δ_{ij} is the Kronecker delta. Assuming that there exists a universal stress-strain rate curve and it coincides with the uniaxial curve, Eq. 4.168, then, we have

$$\dot{\epsilon}_e = \epsilon_o \left[\sinh\left(\frac{\sigma_e}{\sigma_o}\right)\right]^n \quad (4.174)$$

In view of Fig. 4.32 (96.5Sn3.5Ag solder), it can be seen that the power law breaks down at approximately $\tau/G = 10^{-3}$. The stress exponent, $n = 5.5$, indicates a dislocation climb deformation mechanism. It occurs by dislocation glide (dislocations moving along slip planes and overcoming barriers by thermal activation) aided by vacancy diffusion. The glide step produces almost all the strain but the climb step controls the velocity. Since dislocation climb requires diffusion of vacancies or interstitials, the rate-controlling step is atomic diffusion. Due to its stress dependence, the activation energy ($Q = 0.5$ eV) is somewhat below the expected value of lattice (1.1 eV) or dislocation pipe diffusion (0.66 eV).

In view of Fig. 4.33 (97.5Pb2.5Sn solder), it can be seen that the power law breaks down at approximately $\tau/G = 10^{-3}$. The stress exponent, $n = 7$, suggests a dislocation pipe-assisted diffusion deformation mechanism. It occurs by dislocation glide (dislocations moving along slip planes and overcoming barriers by thermal activation) aided by vacancy diffusion. The glide step produces almost all the strain but the climb step controls the velocity. Since dislocation climb requires diffusion of vacancies or interstitials, the rate controlling step is atomic diffusion. The activation energy ($Q = 1.1$ eV) is very close to that of the lattice diffusion in lead (1.1 eV), however, which is somewhat inconsistent to $n = 7$ for pipe-assisted diffusion ($Q = 0.66$ eV).

In view of Fig. 4.34 (60Sn40Pb solder), it can be seen that the power law breaks down at approximately $\tau/G = 10^{-3}$. The stress exponent, $n = 3.3$, indicates a dislocation viscous-glide deformation mechanism (which involves dislocations moving along slip planes and overcoming barriers by thermal activation). Due to its stress dependence, the activation energy ($Q = 0.548$ eV) is somewhat below the expected value of solute interdiffusion.

Figure 4.35 shows a thin-walled circular solder cylinder subjected to a bending moment (M) and a twisting moment (T). Due to the geometry of the structure and the loading condition, the nonzero stress and strain rate components are σ_z (normal stress in the z-direction), $\tau_{\theta z}$ (shear stress in the z-direction of the plane normal to the θ-axis), $\dot{\epsilon}_r$ (steady-state creep normal strain rate in the r-direction), $\dot{\epsilon}_\theta$ (steady-state creep normal strain rate in the θ-direction), $\dot{\epsilon}_z$ (steady-state creep normal strain rate in the z-direction), and $\dot{\gamma}_{\theta z}$ (steady-state creep shear strain rate in the z-direction of the plane normal to the θ-axis). (See Fig. 4.24 in Sec. 4.4.2 for geometry interpretation of these strain components.) Consequently, Eqs. 4.171 and 4.172 become [70, 74, 75]

$$\dot{\epsilon}_e = \frac{\sqrt{2}}{3}\sqrt{(\dot{\epsilon}_r - \dot{\epsilon}_\theta)^2 + (\dot{\epsilon}_\theta - \dot{\epsilon}_z)^2 + (\dot{\epsilon}_z - \dot{\epsilon}_r)^2 + \tfrac{3}{2}\dot{\gamma}_{\theta z}^2} \qquad (4.175)$$

and

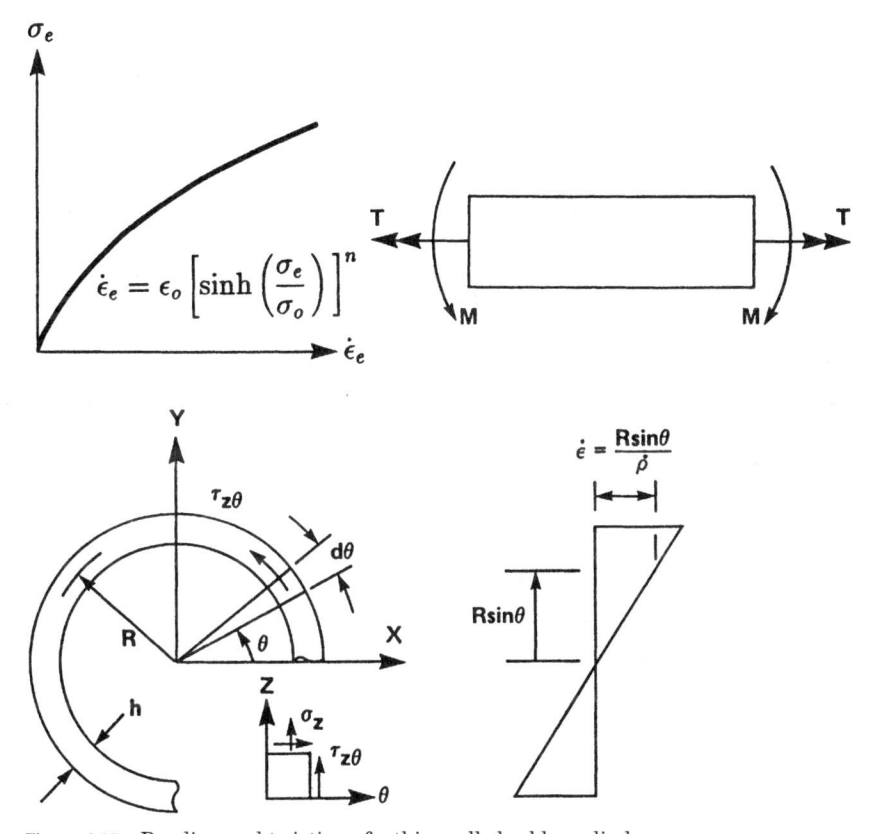

Mechanical Design/Analysis/Measurement/Reliability of Electronic Packaging 245

Figure 4.35 Bending and twisting of a thin-walled solder cylinder.

$$\sigma_e = \sqrt{\sigma_z^2 + 3\tau_{z\theta}^2} \tag{4.176}$$

Since $\sigma_r = \sigma_\theta = 0$, then $\dot{\epsilon}_r = \dot{\epsilon}_\theta$; and since the solder material is assumed to be incompressible, that is, $\dot{\epsilon}_r + \dot{\epsilon}_\theta + \dot{\epsilon}_z = 0$, then we have

$$\dot{\epsilon}_r = \dot{\epsilon}_\theta = -\tfrac{1}{2}\dot{\epsilon}_z \tag{4.177}$$

Equation 4.175 becomes

$$\dot{\epsilon}_e = \sqrt{\dot{\epsilon}_z^2 + \tfrac{1}{3}\dot{\gamma}_{z\theta}^2} \tag{4.178}$$

Substituting Eqs. 4.176 and 4.178 into Eq. 4.174, yields

$$\sqrt{\dot{\epsilon}_z^2 + \tfrac{1}{3}\dot{\gamma}_{z\theta}^2} = \epsilon_o \left[\sinh\left(\frac{\sqrt{\sigma_z^2 + 3\tau_{z\theta}^2}}{\sigma_o} \right) \right]^n \tag{4.179}$$

For the problem under consideration, Hencky's theory [39] simplifies to (see also section 4.5.2.5)

$$\frac{3\dot{\epsilon}_z}{\sigma_z} = \frac{\dot{\gamma}_{z\theta}}{\tau_{z\theta}} \tag{4.180}$$

By solving Eqs. 4.179 and 4.180, we have the following stress distributions

$$\sigma_z = \sigma_o \sqrt{1-\beta^2}\,\frac{\sin\theta\,\sinh^{-1}[K\sqrt{\beta^2+(1-\beta^2)\sin^2\theta}]^{1/n}}{\sqrt{\beta^2+(1-\beta^2)\sin^2\theta}} \tag{4.181}$$

and

$$\tau_{z\theta} = \frac{\sigma_o\beta}{\sqrt{3}}\,\frac{\sinh^{-1}[K\sqrt{\beta^2+(1-\beta^2)\sin^2\theta}]^{1/n}}{\sqrt{\beta^2+(1-\beta^2)\sin^2\theta}} \tag{4.182}$$

in which

$$K = \frac{\dot{\epsilon}_{em}}{\epsilon_o} > 0 \tag{4.183}$$

$$\beta = \frac{\dot{\gamma}_{z\theta}}{\sqrt{3}\dot{\epsilon}_{em}} \leq 1 \tag{4.184}$$

have been substituted. $\dot{\epsilon}_{em}$ is the maximum effective strain rate which is defined as

$$\dot{\epsilon}_{em} = \sqrt{\dot{\epsilon}_z^2(\theta = \pi/2) + \tfrac{1}{3}\dot{\gamma}_{z\theta}^2} \tag{4.185}$$

Thus, the steady-state strain rate components can be written as

$$\dot{\epsilon}_z = K\epsilon_o\sqrt{1-\beta^2}\,\sin\theta \tag{4.186}$$

and

$$\dot{\gamma}_{z\theta} = \sqrt{3}K\beta\epsilon_o \tag{4.187}$$

The equilibrium equations for M and T are defined by the relations (Fig. 4.35):

$$M = \int \sigma_z y\,dA \tag{4.188}$$

and

$$T = \int \tau_{z\theta} R\,dA \tag{4.189}$$

Substituting Eqs. 4.181 and 4.182 into Eqs. 4.188 and 4.189 yields, respectively,

Mechanical Design/Analysis/Measurement/Reliability of Electronic Packaging 247

$$\frac{M}{2\pi R^2 h\sigma_o} = \frac{2\sqrt{1-\beta^2}}{\pi}\int_0^{\pi/2} \frac{\sin^2\theta\, \sinh^{-1}[K\sqrt{\beta^2+(1-\beta^2)\sin^2\theta}]^{1/n}}{\sqrt{\beta^2+(1-\beta^2)\sin^2\theta}}\, d\theta$$

(4.190)

and

$$\frac{T}{2\pi R^2 h\sigma_o} = \frac{2\beta}{\sqrt{3}\pi}\int_0^{\pi/2} \frac{\sinh^{-1}[K\sqrt{\beta^2+(1-\beta^2)\sin^2\theta}]^{1/n}}{\sqrt{\beta^2+(1-\beta^2)\sin^2\theta}}\, d\theta \qquad (4.191)$$

Equations 4.190 and 4.191 are plotted in Fig. 4.36 (for the 96.5Sn3.5Ag solder, $n = 5.5$), Fig. 4.37 (for the 97.5Pb2.5Sn solder, $n = 7$) and Fig. 4.38 (for the 60Sn40Pb solder, $n = 3.3$) for a wide range of values of β and K that were introduced for the sake of convenience. Thus, for a given temperature Θ and a set of values of bending moment (M) and twisting moment T, the values of β and K can be read from either Fig. 4.36, 4.37, or 4.38 (depending on which solder material) and the stresses (σ_z, $\tau_{z\theta}$) and steady-state creep strain rate ($\dot{\epsilon}_z$, $\dot{\gamma}_{z\theta}$) can be obtained from Eqs. 4.181, 4.182, 4.186, and 4.187. The values of ϵ_o and σ_o for the 96.5Sn3.5Ag, 97.5Pb2.5Sn and 60Sn40Pb solders can be obtained from, respectively, Eqs. 4.169a, 4.170a; 4.169b, 4.170b; and 4.169c, 4.170c.

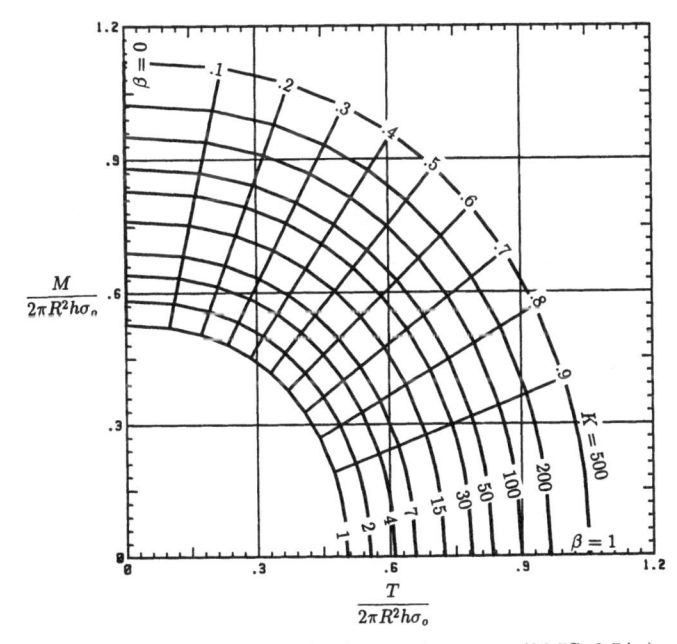

Figure 4.36 Bending and twisting interaction curves (96.5Sn3.5Ag).

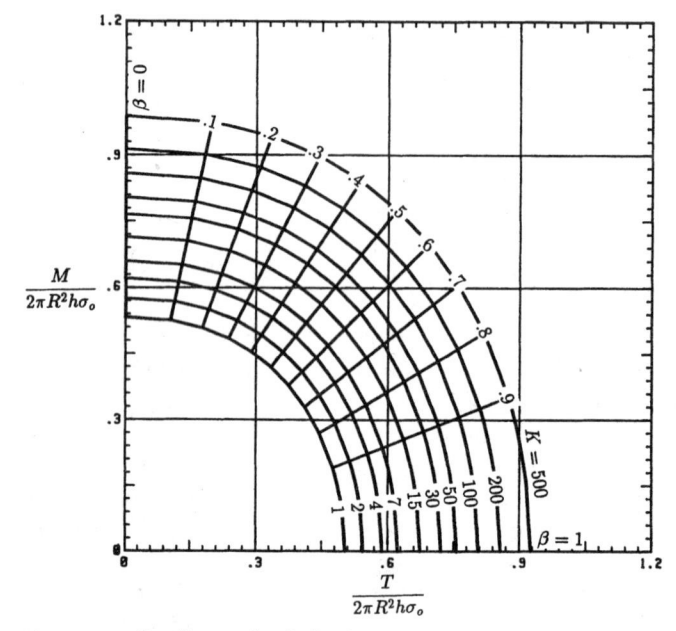

Figure 4.37 Bending and twisting interaction curves (97.5Pb2.5Sn).

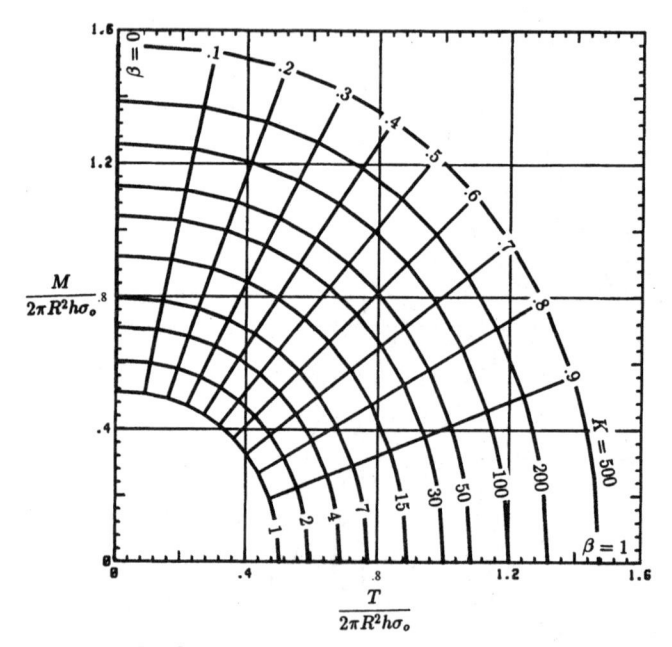

Figure 4.38 Bending and twisting interaction curves (60Sn40Pb).

It is noteworthy that curves of constant β are nearly radial lines so that, as long as the ratio of M and T remains constant, each volume element of the solder interconnect can be assumed with negligible error to be subjected to proportional loading (a sufficient condition for the total-strain theory coincides with the incremental theory). Similar behavior has been observed by Smith and Sidebottom [79] for the case of a solid circular cylinder subjected to tension and twisting moment. They have also shown that the error in using the incompressible solution to predict the behavior of tension-torsion members made of compressible materials is very small.

By considering the geometrically compatible deformation of the cylinder (Fig. 4.35), we have

$$\dot{\epsilon}_z = \frac{R \sin \theta}{\dot{\rho}} \qquad (4.192)$$

where

$$\frac{1}{\dot{\rho}} = \frac{d}{dt}\left(\frac{1}{\rho_r}\right) \qquad (4.193)$$

and

$$\dot{\gamma}_{z\theta} = R\dot{\varphi} = R\frac{d\dot{\varphi}}{dz} \qquad (4.194)$$

in which $1/\dot{\rho}$ is the curvature rate, $1/\rho_r$ is the curvature, ρ_r is the radius of curvature, $\dot{\varphi}$ is the twist rate per unit length, and $\dot{\varphi}$ is the total angle of twist rate.

Thus, Eqs. 4.181, 4.186, 4.188, and 4.192 lead to the following bending moment-curvature rate relation

$$\frac{M}{2\pi R^2 h \sigma_o} = \frac{2\sqrt{1-\beta^2}}{\pi}$$

$$\times \int_0^{\pi/2} \frac{\sin^2 \theta \, \sinh^{-1}\left[(R/\epsilon_o\dot{\rho})\sqrt{\beta^2/(1-\beta^2) + \sin^2\theta}\right]^{1/n}}{\sqrt{\beta^2 + (1-\beta^2)\sin^2\theta}} \, d\theta \qquad (4.195)$$

and is plotted in Fig. 4.39 (for the 96.5Sn3.5Ag solder, $n = 5.5$), Fig. 4.40 (for the 97.5Pb2.5Sn solder, $n = 7$), and Fig. 4.41 (for the 60Sn40Pb solder, $n = 3.3$). Thus, for a given Θ, M, and T, we can read the value of β and K from either Fig. 4.36, 4.37, or 4.38 (depending on which solder material) and then read the value of $R/\epsilon_o\dot{\rho}$ from either Fig. 4.39, 4.40, or 4.41 for the curvature rate. Also, Eqs. 4.182, 4.187, 4.189, and 4.194 lead to the twisting moment-twist rate per unit length relation

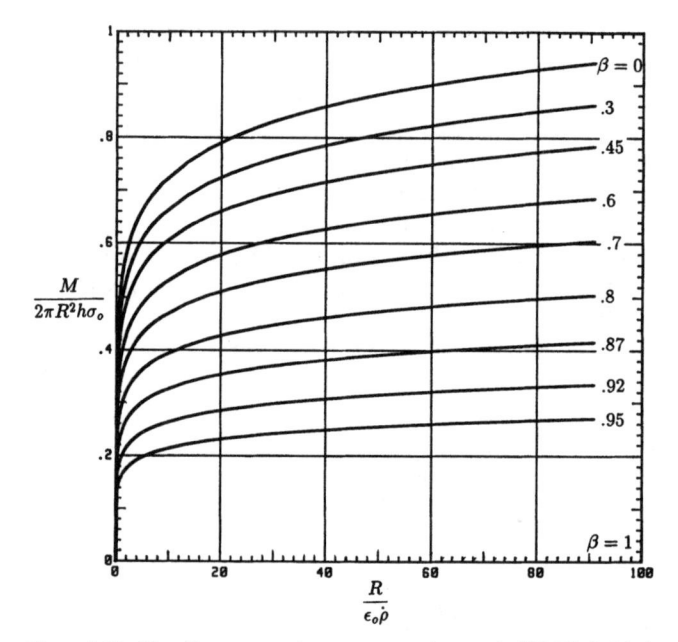

Figure 4.39 Bending moment versus curvature rate (96.5Sn3.5Ag).

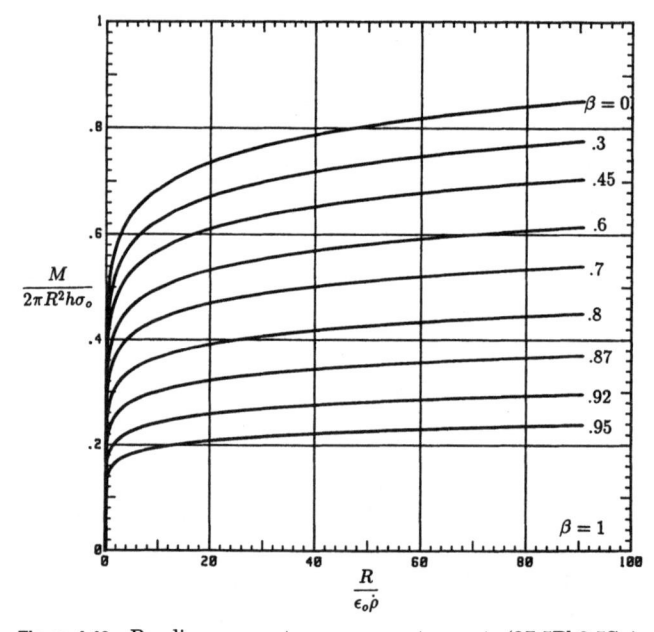

Figure 4.40 Bending moment versus curvature rate (97.5Pb2.5Sn).

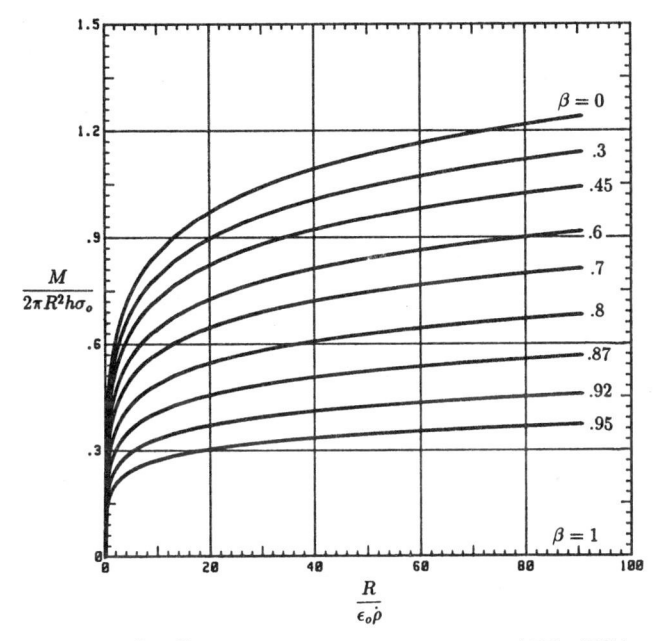

Mechanical Design/Analysis/Measurement/Reliability of Electronic Packaging 251

Figure 4.41 Bending moment versus curvature rate (60Sn40Pb).

$$\frac{T}{2\pi R^2 h \sigma_o} = \frac{2}{3\pi} \frac{R\dot{\phi}}{\epsilon_o}$$

$$\times \int_0^{\pi/2} \frac{\sinh^{-1}[(R\dot{\phi}/\sqrt{3}\epsilon_o)\sqrt{1-[1-(3K^2\epsilon_o^2/R^2\dot{\phi}^2)]\sin^2\theta}]^{1/n}}{\sqrt{(R\dot{\phi}/\sqrt{3}\epsilon_o)^2 + [K^2 - (R^2\dot{\phi}^2/3\epsilon_o^2)]\sin^2\theta}} d\theta \quad (4.196)$$

and is plotted in Fig. 4.42 (for the 96.5Sn3.5Ag solder, $n = 5.5$) and Fig. 4.43 (for the 97.5Pb2.5Sn solder, $n = 7$) and Fig. 4.44 (for the 60Sn40Pb solder, $n = 3.3$). Thus, for a given Θ, M, and T, we can read the value of β and K from either Fig. 4.36, 4.37, or 4.38 (depending on which solder material) and then read the value of $R\dot{\phi}/\sqrt{3}\epsilon_o$ from either Fig. 4.42, 4.43, or 4.44 for the twist rate per unit length.

4.5.5 TMA, DMA, DSC, and TGA of 63Sn37Pb solder

Most of the electronic packaging materials (e.g., solders) are temperature dependent. Their temperature-dependent material properties are usually obtained by TMA (thermal mechanical analysis), DMA (dynamic mechanical analysis), DSC (differential scanning calorimeter), and TGA (thermal gravimetric analysis). In this section, the TCE, modulus, T_g, moisture uptake, and melting point of a 63wt%Sn37wt%Pb (63Sn37Pb) solder are measured.

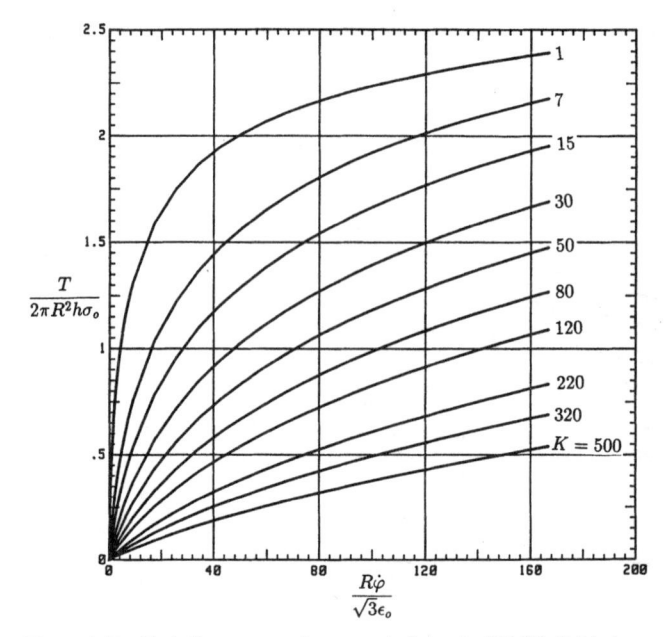

Figure 4.42 Twisting moment versus twist rate (96.5Sn3.5Ag).

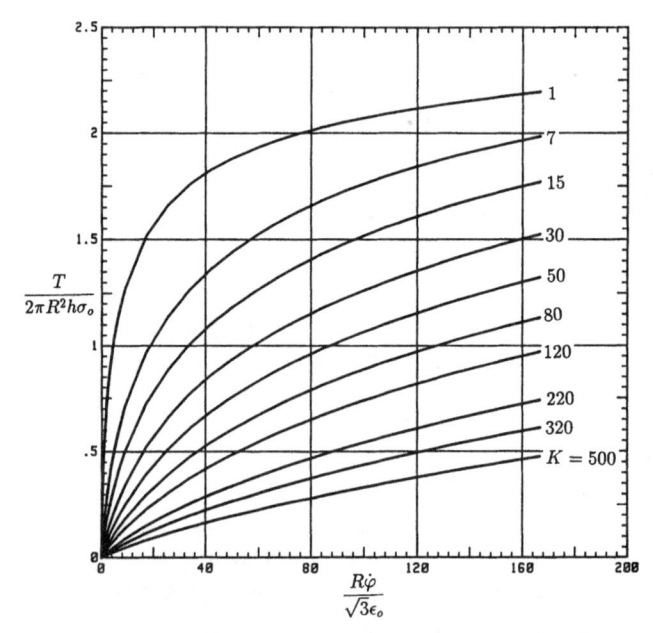

Figure 4.43 Twisting moment versus twist rate (97.5Pb2.5Sn).

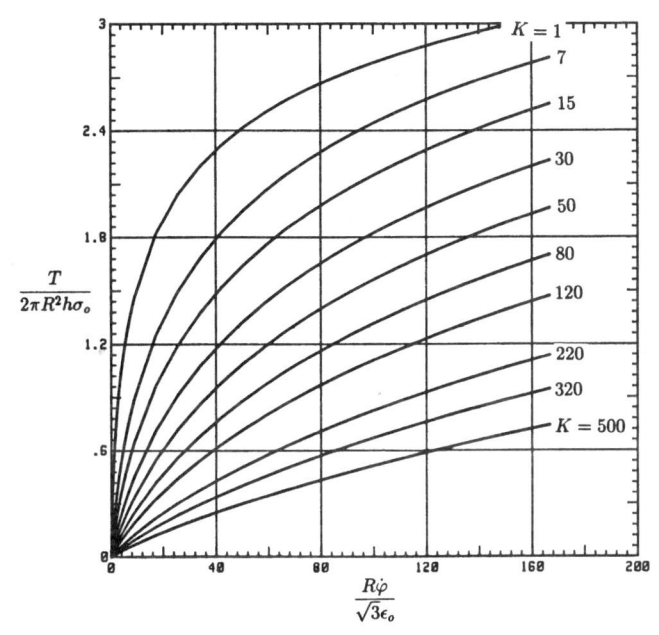

Figure 4.44 Twisting moment versus twist rate (60Sn40Pb).

4.5.5.1 TMA (thermal mechanical analysis). The objective of TMA is to measure the change in dimension of a sample (such as expansion or contraction) as the sample is heated, cooled, or held at a constant (isothermal) temperature. The instrument design consists of the platinum furnace system, which can be operated in the range from −170 to 1000°C (Fig. 4.45). The scan rate is from 0.1°C to 100°C/min. The specimen is mounted between a quartz platform and probe, then a static load is applied to the specimen while the dimension of the specimen is monitored by the linear variable differential transducer (LVDT) throughout the analysis (Fig. 4.46). The TCE of the material is obtained by the slope of the dimension-change vs temperature curve, and T_g of the material is obtained by the onset of the two different slopes of the curve. For 63Sn37Pb solder, the TCE is about 22.7×10^{-6}/°C (Fig. 4.47), and the T_g does not exist.

4.5.5.2 DMA (dynamic mechanical analysis). The objective of DMA is to measure mechanical properties, such as modulus as a function of time, temperature, frequency, stress, or combinations of these parameters. The instrument design consists of the force motor which can be programmed to apply constant stress, dynamic stress, or combinations of both (Figs. 4.45 and 4.48). The core rod applies stress to the sample and is held in place using an electromagnetic suspension. The ceramic fur-

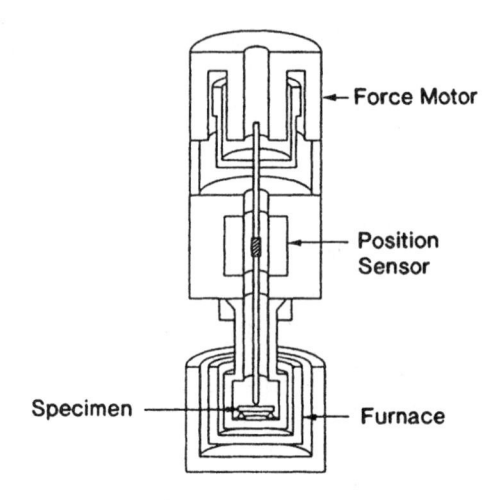

Figure 4.45 Schematic diagram of a commercial TMA (Thermal Mechanical Analyzer) and DMA (Dynamic Mechanical Analysis). TMA force motor applies static load while DMA force motor applies dynamic load on to specimen.

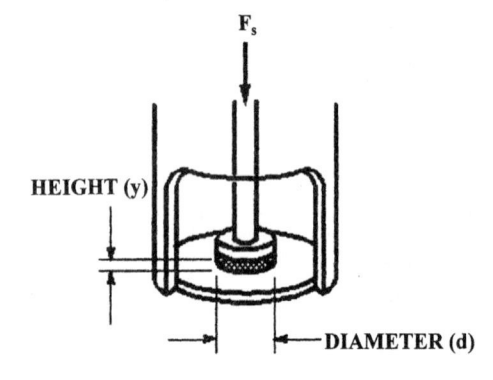

Figure 4.46 Outlook of quartz tube, probe and specimen of TMA for TCE measurement.

nace with platinum furnace element is capable of heating and cooling at a very high rate and also can be heated up to 1000°C. For electronic packaging materials, the flexural properties such as flexural modulus and dynamic mechanical properties such as storage modulus, loss modulus, and tangent delta (tanδ) can be obtained with DMA.

In a static-load condition, *flexural modulus E* is defined as:

$$E = \frac{F_s x^3}{4y^3 z \Delta} \tag{4.197}$$

where F_s is static load, x is the width (span) between two supports, Δ is the maximum static deflection of specimen, y is the thickness of specimen, and z is the width of specimen, Fig. 4.48.

Figure 4.49 shows the static flexural modulus E of the 63Sn37Pb solder as a function of temperatures. In dynamic stress analysis, *flexural storage modulus E_s* is a measure of the energy stored per cycle of deformation and can be expressed as:

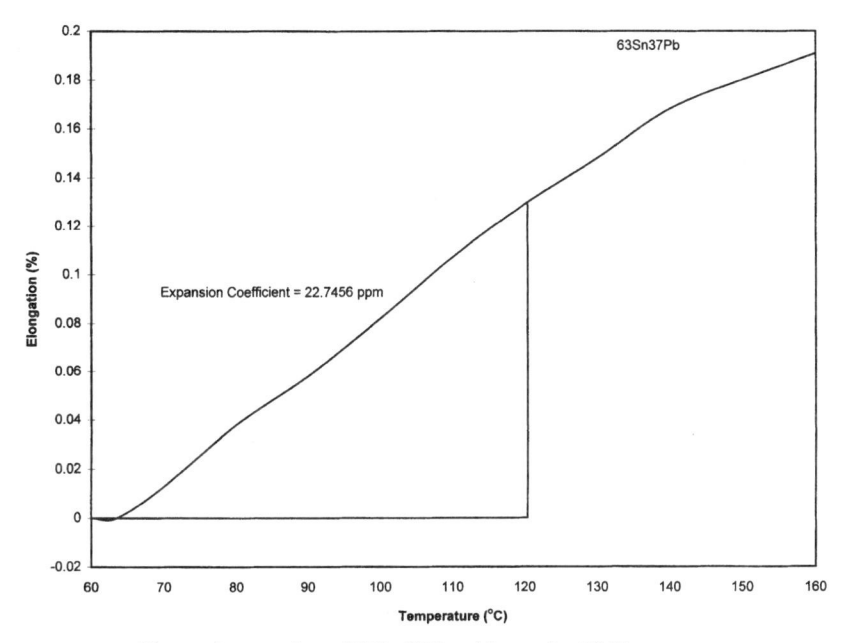

Mechanical Design/Analysis/Measurement/Reliability of Electronic Packaging 255

Figure 4.47 Thermal expansion of 63Sn37Pb solder under TMA.

$$E_s = \frac{F_d x^3 \cos\delta}{4y^3 z \nabla} \qquad (4.198)$$

where F_d is the dynamic load, δ is the phase angle, and ∇ is the maximum dynamic deflection of specimen.

Figures 4.50a and 4.50b show the dynamic flexural storage modulus, E_s, as a function of temperatures and static-dynamic load combinations. It can be seen that the dynamic modulus is different with different static loads—the higher the static load, the higher the modulus. This is physically impossible. The drawbacks of Eqs. 4.197 and 4.198 are stated at the end of this section. *Flexural loss modulus* E_1 is a measure of the energy lost per cycle of deformation and can be expressed as:

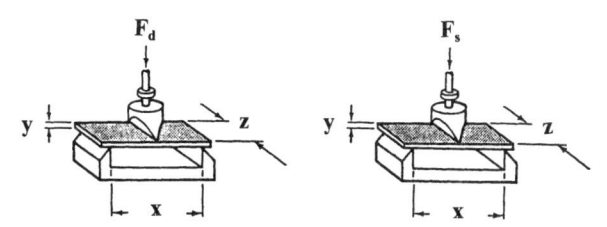

Figure 4.48 DMA three-point bending specimen setup for both static (F_s) and dynamic (F_d) loads.

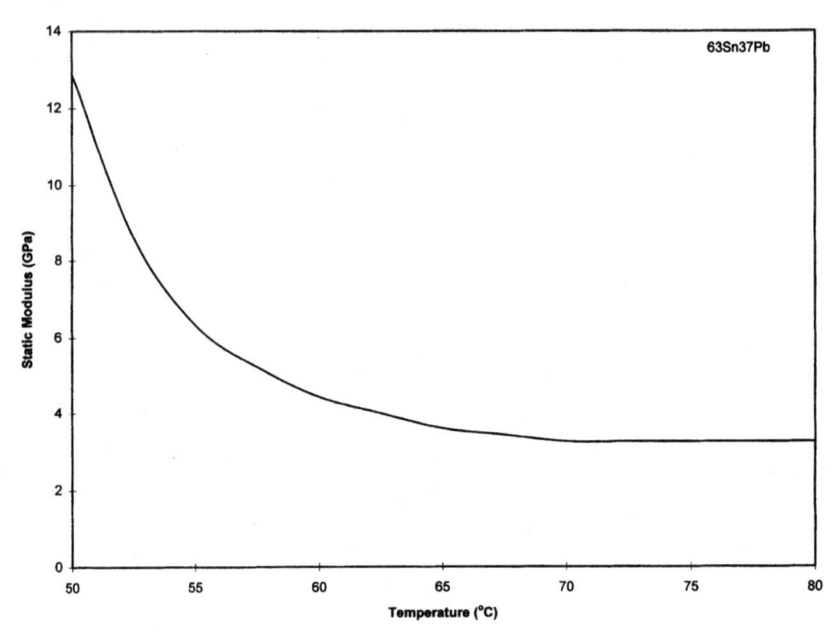

Figure 4.49 Static modulus of 63Sn37Pb solder under DMA three-point static load (8000 mN).

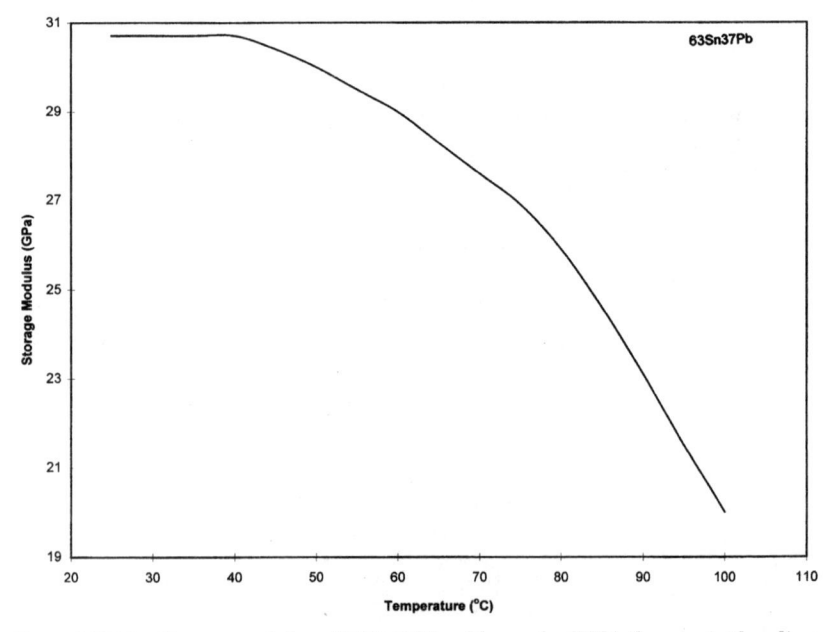

Figure 4.50 (a) Storage modulus of 63Sn37Pb solder under DMA three-point bending (static load = 7900 mN, dynamic load = 100 mN).

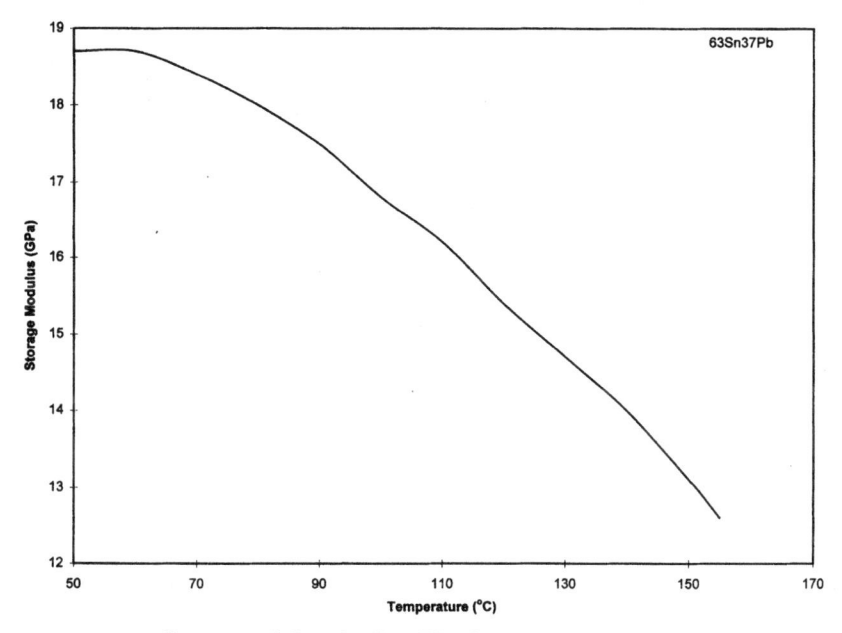

Mechanical Design/Analysis/Measurement/Reliability of Electronic Packaging 257

Figure 4.50 (b) Storage modulus of 63Sn37Pb solder under DMA three-point bending (static load = 1000 mN, dynamic load = 500 mN).

$$E_1 = \frac{F_d \sin\delta \, x^3}{4y^3 z \nabla} \tag{4.199}$$

Tangent delta (tan δ), which is a measure of material related damping property, can then be obtained by dividing E_1 by E_s.

$$\frac{E_1}{E_s} = \tan\delta \tag{4.200}$$

The temperature at the peak of a tan δ curve is often reported in literature as glass transition temperature (T_g). However, for solders, $E_1 = 0$, and T_g does not exit. It should be pointed out that Eqs. 4.197–4.200 are for linear elastic response of the materials. Plastic and creep behaviors are not included. Since 63Sn37Pb solder's yield stress and melting point are very low, the curves shown in Figs. 4.49, 4.50a, and 4.50b may not be accurate at high temperatures (i.e., there are already large amounts of plastic and creep strains in addition to the elastic strain).

4.5.5.3 DSC (differential scanning calorimeter). The objective of DSC is to measure the amount of energy (heat) absorbed or released by a sample as it is heated, cooled, or held at a constant (isothermal) temperature. The instrument design consists of two independent furnaces, one for the sample and one for the reference (Fig. 4.51). When an exothermic or endothermic change occurs in the sample material, energy is applied or removed to one or both furnaces to compensate for the energy change occurring in the sample. Since the system is always directly measuring energy flow to or from the sample, DSC can directly measure melting temperature, T_g, temperature onset of crystallization, and temperature onset of curing. The kinetic software enables one to analyze a DSC peak to obtain specific kinetic parameters that characterize a reaction process.

Any material reaction can be represented by the following equation:

$$A \xrightarrow{k} B + \Delta H \tag{4.201}$$

where A is the material before reaction, B is the material after reaction, ΔH is the heat absorbed or released, and k is the Arrhenius rate constant.

The Arrhenius equation is given by:

$$k = Z \exp(-E_a/RT) \tag{4.202}$$

where Z is the pre-exponential constant, E_a is the activation energy of the reaction, R is the universal gas constant (8.314 j/°C/mole), and T is the absolute temperature in degrees Kelvin.

The rate of reaction (dx/dt) can be directly measured by DSC and can be expressed as:

$$dx/dt = k(1-x)^n \tag{4.203}$$

Figure 4.51 Specimen set-up for DSC (Differential Scanning Calorimeter).

where dx/dt is the rate of reaction, x is the fraction reacted, t is time, k is Arrhenius rate constant, and n is the order of reaction.

Combining Eqs. 4.202 and 4.203 and assuming an nth-order reaction kinetics and constant program rate, activation energy, and pre-exponential constant, yields:

$$dx/dt = Z \exp(-E_a/RT)(1-x)^n \qquad (4.204)$$

The fraction reacted x is directly related to the fractional area of the DSC reaction peak. The kinetic parameters Z, E_a, and n are determined by using an advanced multilinear regression method (MLR). Figure 4.52 shows the heat flow versus temperature curve of the 63Sn37Pb solder. It can be seen that the melting point (peak) is about 182.83°C (very close to 183°C).

4.5.5.4 TGA (thermal gravimetric analysis). The objective of TGA is to measure the change in mass of a sample as the sample is heated, cooled, or held at a constant (isothermal) temperature. The instrument design consists of the microbalance which allows the sensitive measurement of weight changes as small as a few micrograms, furnace, and sample holder area (Fig. 4.53). The vertical analytical design also serves to isolate the balance mechanism from the furnace, eliminating temperature fluctuations that cause drifts and nonlinearity. This sys-

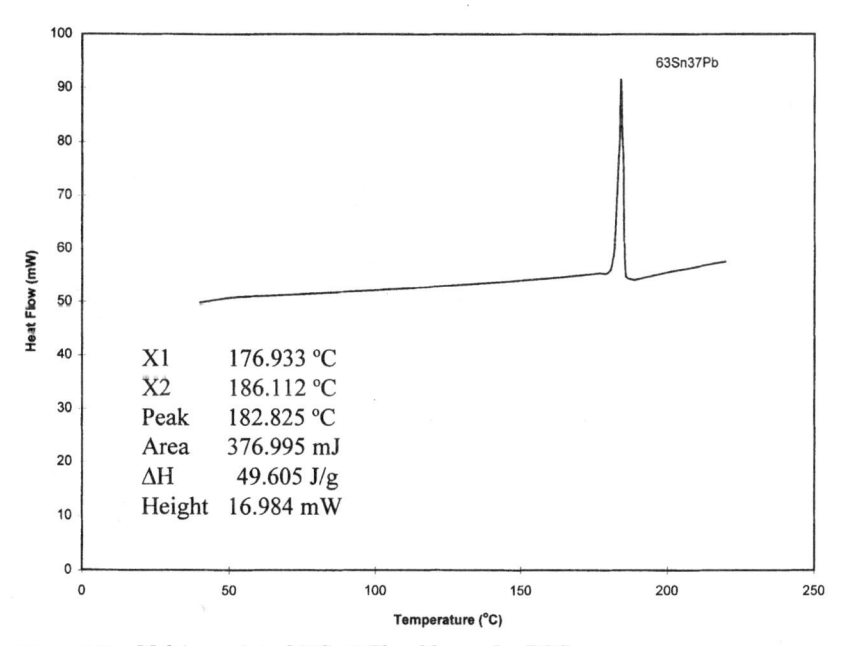

Figure 4.52 Melting point of 63Sn37Pb solder under DSC.

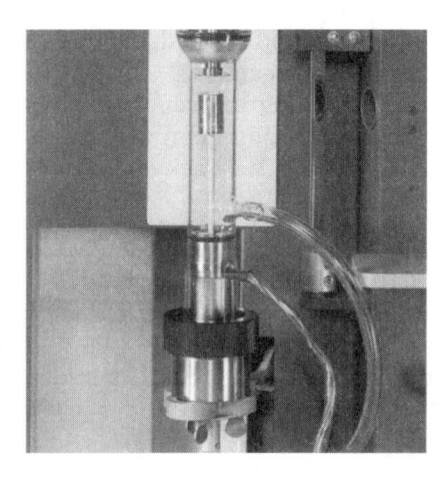

Figure 4.53 High sensitivity weight measurements with TGA (Thermal Gravimetric Analyzer).

tem can measure curie point, decomposition temperature, moisture uptake, and component separation.

The change in mass during thermal scan can be expressed as:

$$\frac{W_f - W_i}{W_i} \times 100\% \qquad (4.205)$$

where W_f is the final weight after thermal scan and W_i is the initial weight before thermal scan.

Figure 4.54 shows the moisture content of the 63Sn37Pb solder at dry and after 20 hours of steam-aging conditions. The test conditions are from 50 to 102°C at 40°C/min and then stay at 102°C for the remainder of the test. It can be seen from Fig. 4.54 that the moisture content of the 63Sn37Pb solder after 20 hours of steam aging is 21 percent more than before (the dry condition).

4.5.6 Effect of underfill materials on the mechanical and electrical performance of a functional solder-bumped flip-chip device

Solder-bumped flip chips on expensive substrates have been used since the 1960s [80–84]. The past few years have witnessed an explosive growth in the research efforts devoted to solder-bumped flip chips on low-cost substrates [85–103]. There are at least two major reasons that it works [97]. One is the printed circuit board (PCB) with sequential or built-up circuits, such as the DYCOstrate, plasma-etched redistribution layers (PERL), surface laminar circuits (SLC), film redistribution layer (FRL), interpenetrating polymer build-up structure system (IBSS), high-density interconnect (HDI), conductive adhesive bonded

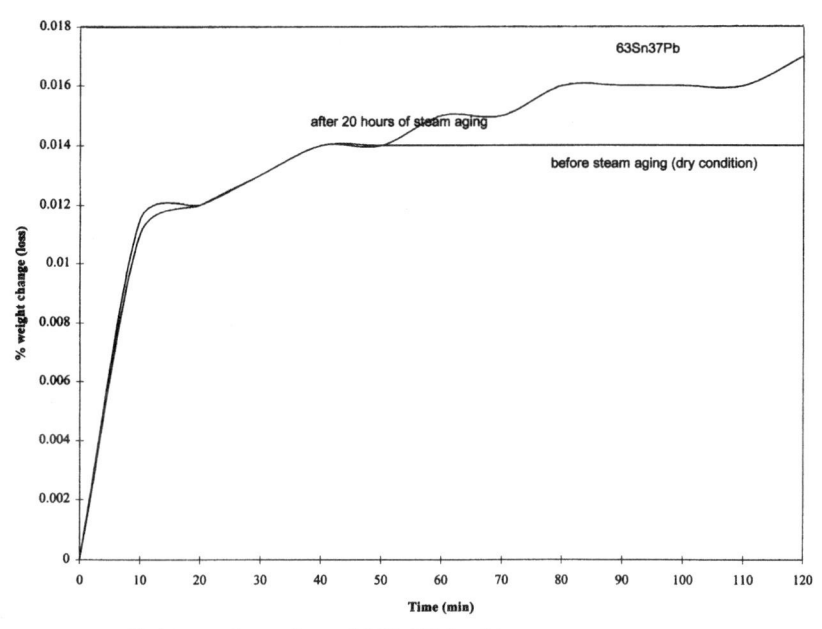

Figure 4.54 Moisture absorptions of 63Sn37Pb solder.

flex, sequential bonded films, sequential bonded sheets, and microfilled via technology.

The other reason is the underfill epoxy encapsulant used to reduce the effect of the global thermal expansion mismatch between the silicon chip and the organic substrate, that is, to reduce the stresses and strains in the flip-chip solder joints (since the chip and the substrate are tightly adhered by the underfill), and to redistribute the stresses and strains over the entire chip area that would otherwise be increasingly concentrated near the corner solder joints of the chip. Other advantages of underfill encapsulant are to protect the chip from moisture, ionic con taminants, radiation, and hostile operating environments such as thermal [93–96, 100–101], mechanical pull [100], mechanical shear [93], mechanical twist [94], and shock/vibration [102].

The important disadvantages of underfill encapsulant are to reduce manufacturing throughput and make rework very difficult. Even though the researches of reworkable underfill are very active, however, most of them are using solvent chemical (e.g., CFC issue), and most of the chips (e.g., passivation) and substrates (e.g., solder mask, via, and copper pads) are degraded or even damaged after being reworked. Also, fast-flow and fast-cure underfill encapsulants are on their way. However, the material properties of these underfills could be degraded (e.g., due to excessive/large voids, too high a TCE, and too low a Young's modulus) and, thus, affect the mechanical and electrical performance of solder-bumped flip-chip assemblies.

In this section, four different underfill encapsulants with different size and content of filler and epoxy are studied. Their curing conditions such as time and temperature are measured by a DSC unit. Their material properties such as the TCE, T_g, modulus, and moisture content are carried out using the TMA, DMA, and TGA. Their flow rate and mechanical (shear) strength in a solder-bumped flip chip on organic substrate are determined experimentally. Their effects on the electrical performance (voltage) of a functional flip-chip device are measured. (For more information about underfill materials, please refer to Chap. 5 of this book.)

4.5.6.1 Underfill materials. There are four different encapsulant materials under consideration: Underfill A, B, C, and D (Table 4.4) [86]. It can be seen that the filler contents are 40, 60, 60, and 70 percent for Underfill A, B, C, and D, respectively. For all these materials their filler is silica. The resin of Underfill A and B is a bis-phenol type epoxy; of Underfill C is a high-T_g bis-phenol type epoxy; and of Underfill D is cycloaliphatic epoxy. The filler sizes of Underfill A, B, C, and D are, respectively, 20, 20, 20, and 5 µm. All the underfill materials are defreezing at room temperature for one hour before measurements. For each test configuration the sample size is three.

4.5.6.2 Curing conditions. In order to determine the curing conditions of underfill materials, they are put into an aluminum pan (which will form a disc sample with dimensions 6.4 ± 0.2 mm in diameter and 1.6 ± 0.1 mm in height), weighed, and then put in Perkin Elmer DSC 7 equipment. Thermal scan is carried out at a 5°C/min heating rate ranging from 40 to 250°C. Based on kinetic calculations, the heat flow versus temperature curves can be converted to percent reaction versus time curves.

TABLE 4.4 Curing Conditions of Underfill A, B, C, and D

Type of underfill	Filler content and size	Curing condition temp. (°C)/time (min)		
A	40% <20µm	130/50	140/25	150/12
B	60% <20µm	130/60	140/35	150/12
C	60% <20µm	130/90	140/55	150/20
D	70% <5µm	140/90	150/50	160/35

Remark 1: Resin of A and B is bis-phenol type epoxy; Resin of C is high T_g bis-phenol type epoxy; while Resin of D is cycloaliphatic epoxy.
Remark 2: Filler of Underfill A, B, C, and D is silica.

Figure 4.55a shows the heat flow versus temperature curve of Underfill A. It can be seen that the curing temperature (peak) is about 138°C. Figure 4.55b shows a typical set of curves of Underfill A with different curing temperatures. It can be seen that at 130°C, Underfill A never reached a 100 percent cured, since its curing temperature is 138°C. Table 4.4 summarizes the average curing conditions of Underfill A, B, C, and D. It can be seen that, for all the underfill materials, they cured faster at higher temperatures. It should be pointed out that it takes only 20 minutes or less to cure Underfill A, B, and C at 150°C.

4.5.6.3 Material properties

TCE. The TCEs of Underfill A, B, C, and D (with sample dimensions 6.4 ± 0.2 mm in diameter and 1.6 ± 0.1 mm in height) are determined by the TMA in an expansion quartz system (60 to 200°C) at a 5°C/min heating rate. Figure 4.56 shows the typical elongation curves (up to 110°C) of Underfill A, B, C, and D; Table 4.5 summarizes the average results. It can be seen that the TCE is the largest for Underfill A (40% filler content) and smallest for Underfill D (70% filler content). In general, the higher the filler content, the lower the TCE. This is because the filler expands less than the epoxy resin. Concerning solder joint thermal-fatigue reliability, it is preferred to have lower-TCE underfill materials [93–96, 101–102].

Modulus. The modulus of Underfill A, B, C, and D is measured with a three-point bending specimen (3.0 ± .3 × 2.9 ± 0.3 × 19 ± 3 mm) in a

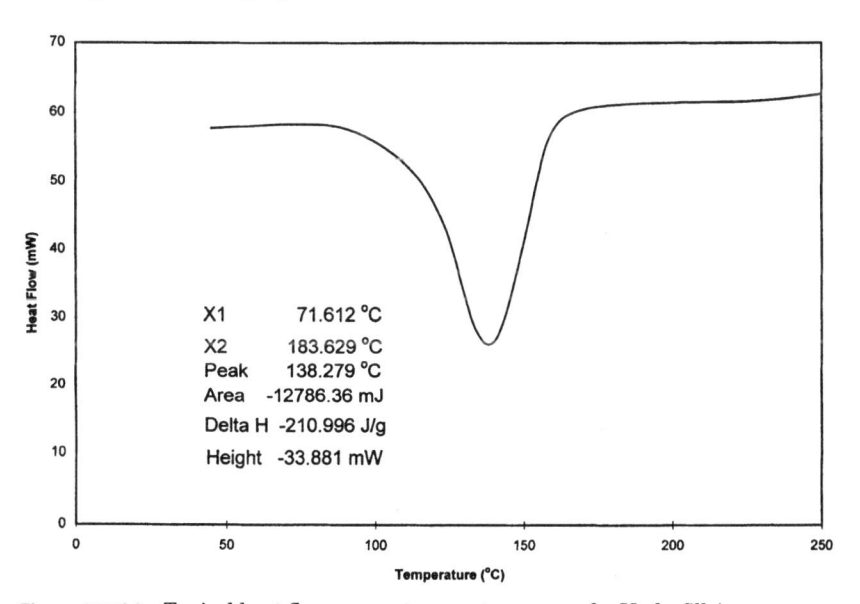

Figure 4.55 (a) Typical heat flow versus temperature curve for Underfill A.

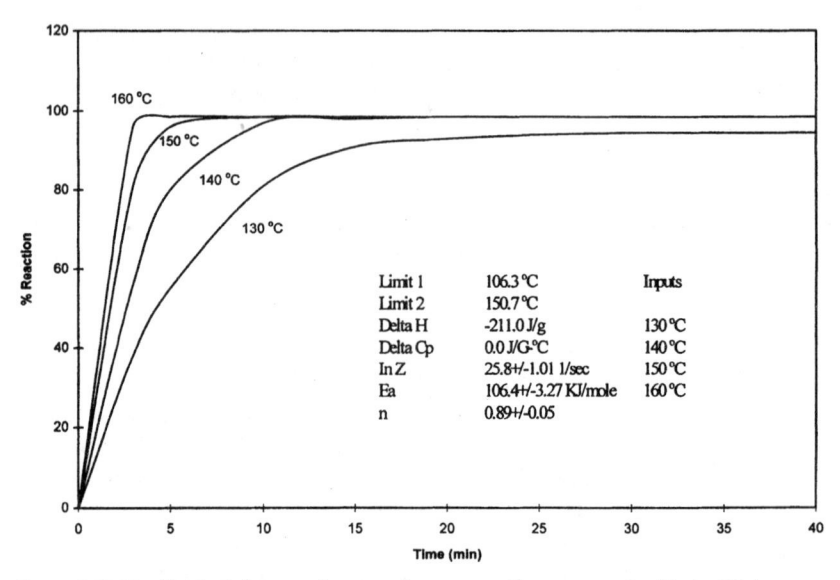

Figure 4.55 (b) Typical degree of conversion versus time curves for Underfill A.

DMA unit (55 to 200°C) at a heating rate of 5°C/min. The typical test results of static modulus (Eq. 4.197), storage modulus (Eq. 4.198), and loss modulus (Eq. 4.199) of Underfill A, B, C, and D as a function of temperature are shown in Fig. 4.57a, b, and c, respectively. The average results of the storage modulus are reported in Table 4.5. It can be seen

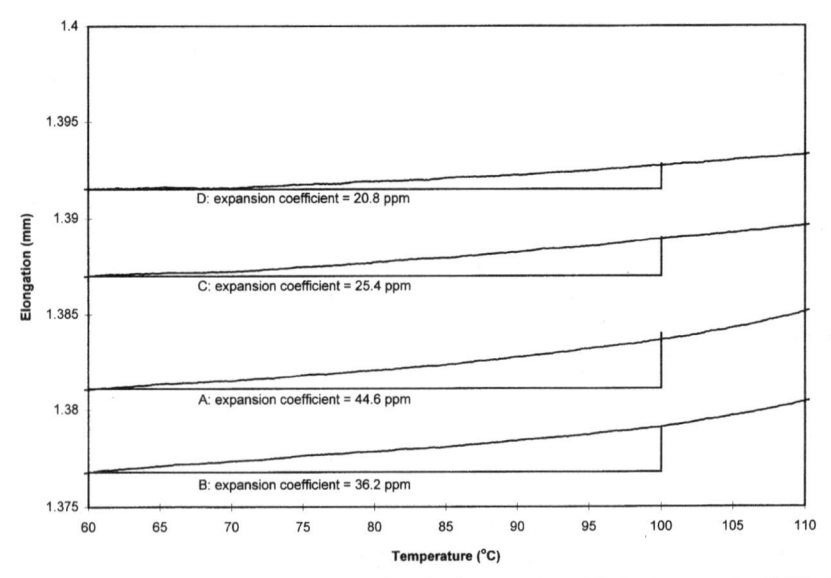

Figure 4.56 Typical elongation curves of Underfill A, B, C, and D for determining TCE.

TABLE 4.5 Material Properties of Underfill A, B, C, and D and Their Effects on Moisture Content

Type of underfill	Filler content and size	TCE (ppm)	T_g (°C)	Storage modulus (Gpa) 25°C	55°C	110°C	Moisture content (%) Dry	20 hours of steam aging
A	40% <20 μm	46.5	142	3.0	2.76	2.01	.09	.33
B	60% <20 μm	36.2	142	3.3	3.3	2.73	.05	.51
C	60% <20 μm	26.1	158	4.4	3.6	3.42	.04	.28
D	70% <5 μm	21	137	4.8	3.3	2.86	.044	.52

that the static modulus, storage modulus, and loss modulus of all the underfill materials is temperature dependent: the higher the temperatures, the lower the modules. Also, the higher the filler content (which acts like reinforcement in filler reinforced composites), the larger the modulus. At room temperature, the largest modulus is for Underfill D, since it has the most percentage of filler content and the fillers will increase stiffness. However, at higher temperatures, the largest modulus is for Underfill C, since it has the largest T_g and the filler content is not that much less than that of Underfill D.

T_g. The tangent delta (tanδ) of Underfill A, B, C, and D can be determined by dividing the loss modulus by the storage modulus (Eq. 4.200). Figure 4.58 shows the typical tangent delta curves of Underfill A, B, C,

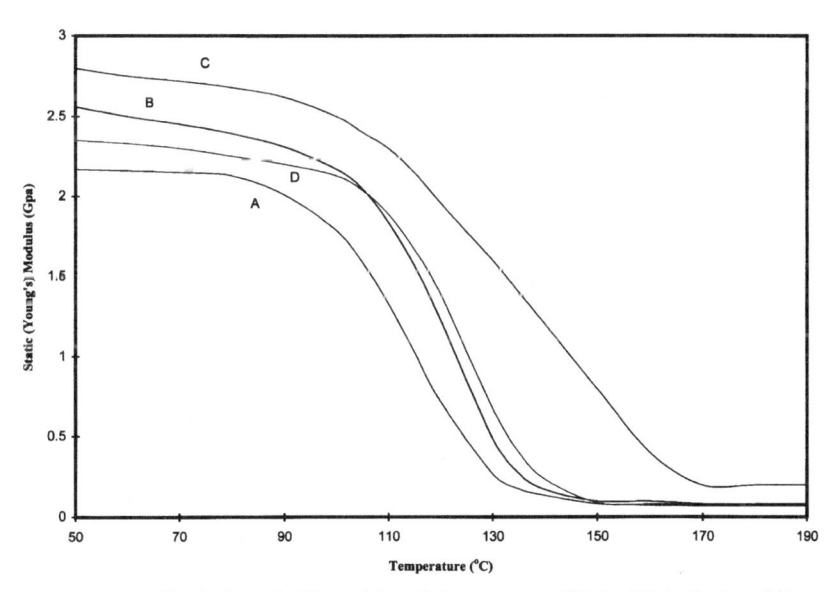

Figure 4.57 (a) Typical static (Young's) modulus curves of Underfill A, B, C, and D.

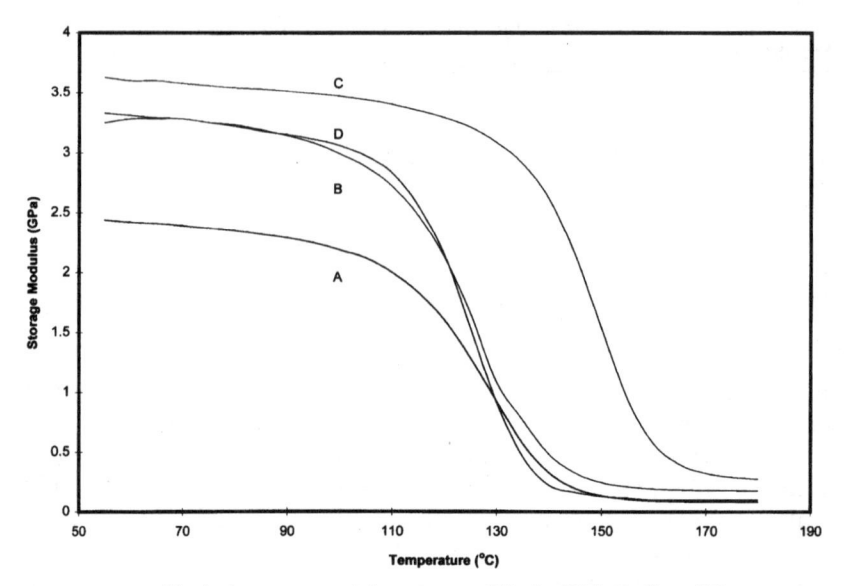

Figure 4.57 (b) Typical storage modulus curves of Underfill A, B, C, and D.

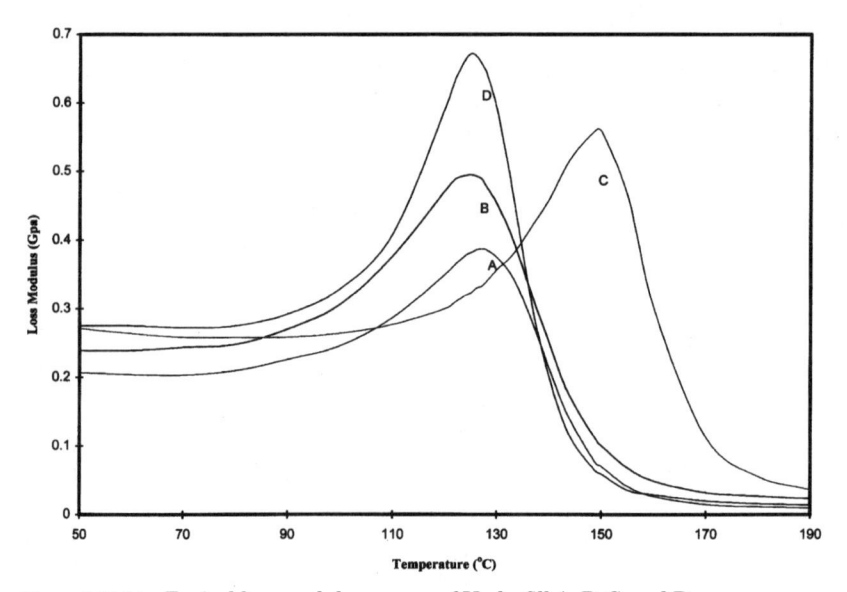

Figure 4.57 (c) Typical loss modulus curves of Underfill A, B, C, and D.

and D (T_g can be obtained from the temperature corresponding to the peak of the curve), and the average values of T_g are shown in Table 4.5. It can be seen that the largest T_g is for Underfill C, since its resin is a high-T_g bis-phenol type epoxy. The smallest T_g is for Underfill D, since its resin is cycloaliphatic epoxy.

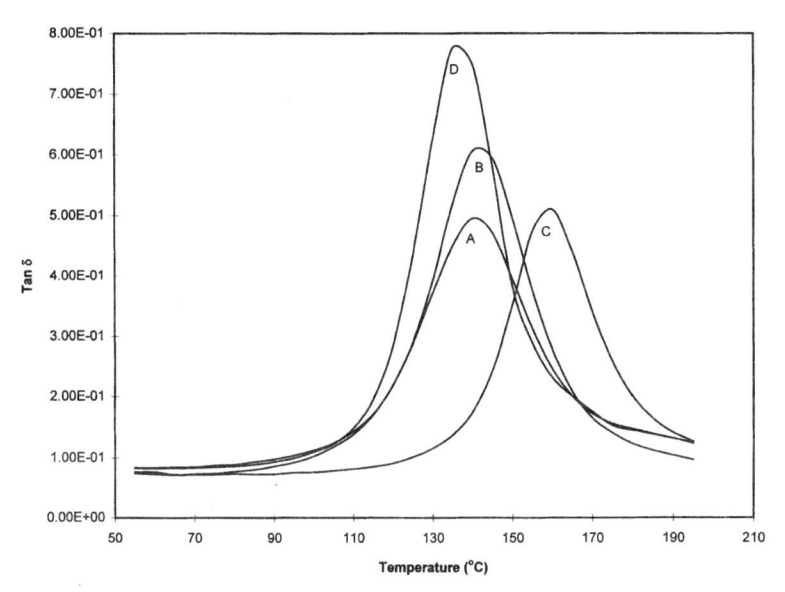

Mechanical Design/Analysis/Measurement/Reliability of Electronic Packaging 267

Figure 4.58 Typical tangent delta (tanδ) curves of Underfill A, B, C, and D for determining T_g of Underfill A, B, C, and D.

Moisture content. Two sets of tests are carried out to determine the moisture content: one is for a dry specimen and the other is for a steam-aging specimen. The steam-aging specimen is prepared under steam evaporation for 20 hours in a closed hot water bath. All the specimen dimensions are 6.4 ± 0.2 mm in diameter and 1.6 ± 0.1 mm in height. Weight losses of Underfill A, B, C, and D are measured with the TGA equipment made by Perkin Elmer under 110°C for 4 hours.

Figures 4.59 and 4.60, respectively, show the typical percent weight loss (moisture content) of Underfill A, B, C, and D before and after 20 hours of steam aging. Their average moisture contents are shown in Table 4.5. It can be seen that the moisture content of Underfill A, B, C, and D after 20 hours of steam aging is at least three times more than before (the dry condition). This is especially true for higher filler content Underfill B and D, since porosity increases when the amount of filler increases. It is noted that the amount of moisture content of Underfill C is the lowest. This could be due to its high $T_g = 158°C$.

4.5.6.4 Underfill flow rate. The flow rates of Underfill A, B, C, and D are measured underneath a functional solder-bumped flip chip on a BT (bismaleimide triazine) substrate. The dimensions of the chip are 6.3 × 3.6 mm and have 32 bumps. The BT substrate has two metal layers and 32 vias (0.254 mm in diameter). The standoff height of the solder bumps is 0.06 ± 0.02 mm.

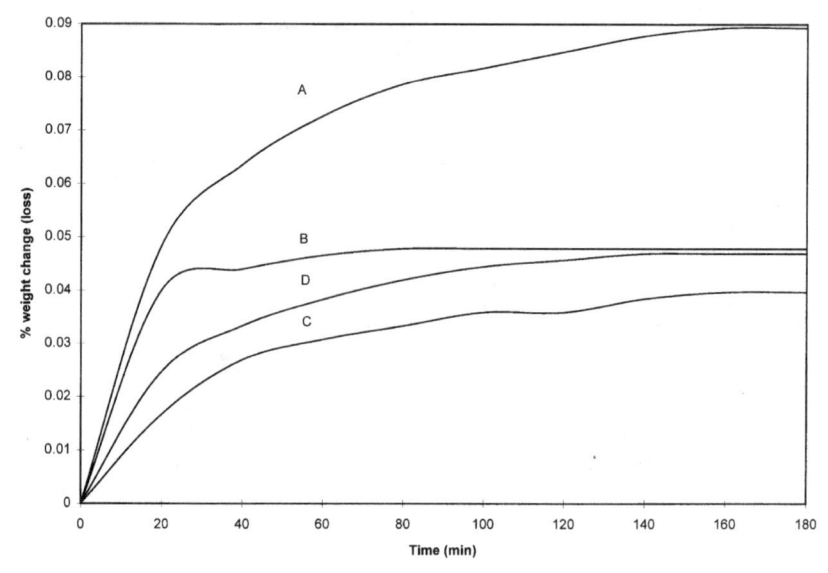

Figure 4.59 Moisture content of Underfill A, B, C, and D at dry condition.

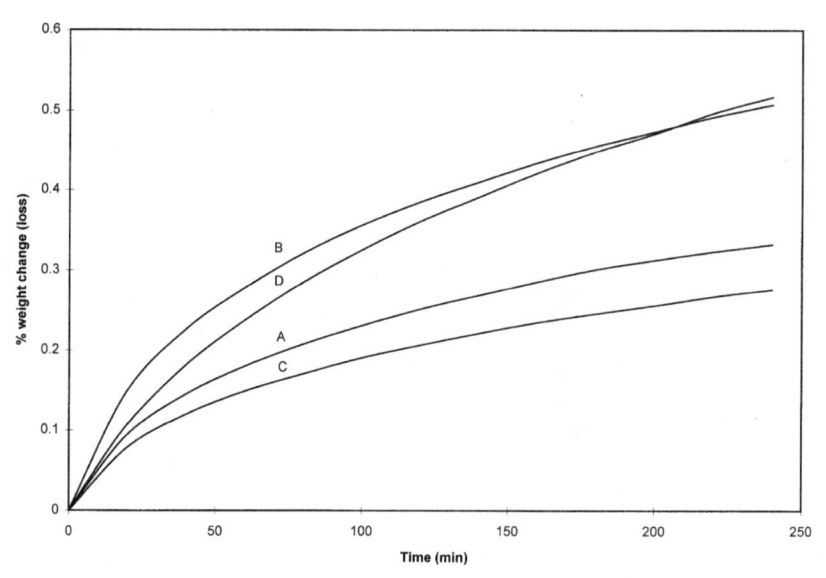

Figure 4.60 Moisture content of Underfill A, B, C, and D after 20 hours of steam aging.

The assembled flip chip is placed on a hot plate at 80°C. Approximately 0.025 cm^3 of room-temperature underfill is beaded around two chip sides using a syringe in an "L" pattern. The time for the underfill to completely fill the gap is recorded. There is no underfill-materials flow to the bottom side of the substrate.

The flow rates for Underfill A, B, C, and D are shown in Table 4.6. It can be seen that the flow rate is strongly affected by the filler content—the more the filler content (i.e., the higher viscosity), the smaller the flow rate. In this study, it is noted that Underfill C has the lowest flow rate, since it has the largest T_g.

4.5.6.5 Mechanical performance.
The Royce Instruments Model 550 is used to perform the mechanical shear tests. The shear wedge is placed against one edge of the solder-bumped flip chip with underfill on the BT substrate, which is clamped on the stage. A push of the wedge is applied to shear the chip/bumps/underfill away (a destructive test).

The average test results of Underfill A, B, C, and D are shown in Table 4.7, and a typical load-deflection curve of solder-bumped flip chip with Underfill C is shown in Fig. 4.61. It can be seen from Table 4.7 that the shear force is about the same for all the underfills under consideration. It can also be seen that, for all the cases, the shear force of the solder-bumped flip chip with underfill drops about 25 percent after 20 hours of steam aging. Thus, the moisture content has significant influence on mechanical performance of underfilled flip-chip assemblies.

4.5.6.6 Electrical performance.
The electrical performance of Underfill A, B, C, and D is carried out under a 5-volt DC by using a Kepco power supply. The average voltage readout of the solder-bumped flip chip with and without underfills is summarized in Table 4.8. It can be seen that, for all the underfills under consideration, there is almost no difference in voltage readout between the ones with underfill at dry condition and

TABLE 4.6 Flow Rate of Underfill A, B, C, and D

Type of underfill	Filler content and size	Flow rate (mm/s)
A	40% <20 μm	.167
B	60% <20 μm	.133
C	60% <20 μm	.003
D	70% <5 μm	.118

Remark: Standoff is 0.07 ± 0.02 mm for flow rate measurement.

TABLE 4.7 Effects of Moisture on the Mechanical Performance of Underfill A, B, C, and D

	Shear force (kgf)	
Type of underfill	Dry condition	20 hours of steam aging
A	43.15	33.77
B	40.45	31.25
C	43.45	34.2
D	41.55	33.15

Remark: Mechanical performance was based on the shear force in shear test.

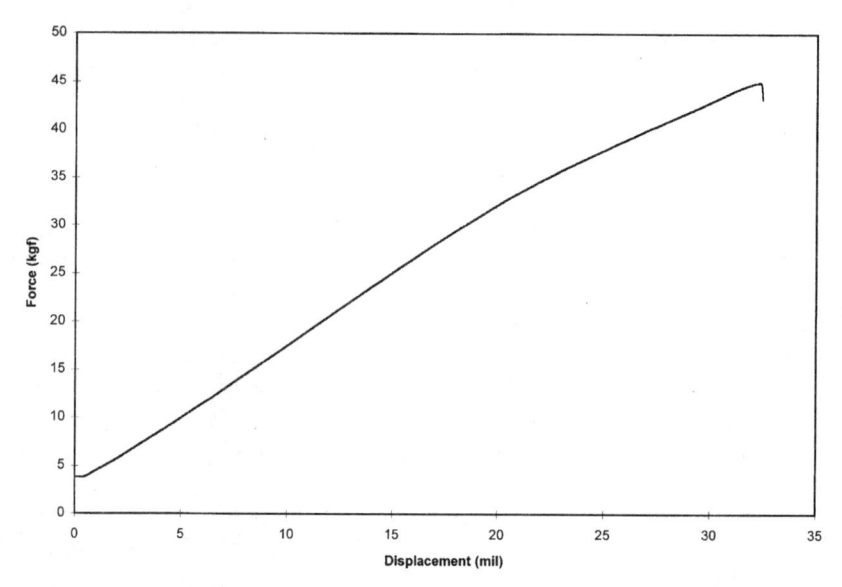

Figure 4.61 Shear force versus displacement curve for Underfill C.

the ones without underfill. Also, for solder-bumped flip chip with underfills, the difference in voltage readout between the ones before and the ones after 20 hours of steam aging is insignificant.

4.5.6.7 Summary. Four different underfills with different resin epoxy and size and filler content have been studied. Their curing condition and material properties have also been measured using DSC, TMA, DMA, and TGA equipment. Furthermore, their flow rate and effects on the mechanical (shear) and electrical (voltage) performance in a solder-bumped functional flip chip on an organic substrate have been experimentally determined. Some important results are summarized as follows.

TABLE 4.8 Electrical Performance of Underfill A, B, C, and D

	\multicolumn{6}{c}{Voltage readout (volt)}			
	With underfill at dry condition		With underfill after 20 hours of steam aging		Without underfill	
Type of underfill	High	Low	High	Low	High	Low
A	4.05	3.88	4.03	3.87	4.05	3.88
B	4.07	3.94	4.05	3.93	4.06	3.92
C	4.05	3.88	4.04	3.85	4.05	3.91
D	4.1	3.94	4.1	3.92	4.08	3.93

Remark: Input voltage is 5 V.

1. For all the underfills considered, curing time is temperature dependent—they cured faster at higher temperatures.
2. For all the underfills considered, the higher the filler content, the lower the TCE.
3. In this study, the cycloaliphatic epoxy leads to lower T_g underfill materials than the bis-phenol type epoxy.
4. For all the underfills considered, the modulus is temperature dependent—the higher the temperature, the lower the modulus.
5. For all the underfills considered, the moisture content after 20 hours of steam aging is at least three times more than the dry condition. In this study, the higher the T_g, the lower the moisture content. Also, the higher the filler content, the higher the moisture content.
6. For all the underfills considered, the flow rate is strongly affected by the filler content—the more the filler content, the smaller the flow rate. In this study, the higher the T_g, the lower the flow rate.
7. For all the underfills considered, the moisture content significantly affects the mechanical performance of solder-bumped flip-chip assemblies. The shear force of the underfilled functional chip drops about 25 percent after 20 hours of steam aging.
8. For all the underfills considered, the effects of underfill on the electrical (voltage readout) performance of the functional chip are insignificant.

4.6 Finite Element Analysis

4.6.1 Nonlinear finite element formulation

The thermal stresses and strains in electronic packaging components and systems with materials governed by Eq. 4.136 and Eq. 4.144 are very difficult to obtain. The finite element method can be one of the best candidates for obtaining approximate results for the thermal stresses and deformations in electronic packages and interconnects.

The basic concept of the finite element method is that a boundary-value problem can be decomposed into a finite number of regions (elements). For each element, trial function approximations of displacement components are used in conjunction with variational principles or Galerkin's approximation and matrix methods to transform the boundary-value problem into a system of simultaneous algebraic equations. Since the method may be applied to individual discrete elements of the continuum, each element may be given distinct physical and material properties, thus achieving very general descriptions of a continuum as a whole. This feature of the finite element method is very attractive to practicing analysts who deal with composite structures such as electronic components and systems.

In formulating the nonlinear finite element method by the principle of incremental minimum potential energy, the incremental displacement field is represented by interpolation functions together with incremental generalized displacements at a finite number of nodal points in each element. (It should be noted that although plasticity is formulated in this section, creep will also be formulated by the same procedures.) In matrix form, the assumed incremental displacement may be written as (Fig. 4.62)

$$[\Delta u] = [A][\Delta s] \qquad (4.206)$$

where $[\Delta u]$ is an incremental displacement column matrix, $[\Delta s]$ is a nodal incremental displacement column matrix, and $[A]$ is a shape function matrix. In Fig. 4.62, for the sake of simplicity, all the Δs in front of u, s, and the like, have been omitted. The corresponding incremental strain column matrix is

$$[\Delta \epsilon] = [L][\Delta u] \qquad (4.207)$$

where $[L]$ is the differential operator matrix.

The corresponding incremental stress column matrix is

$$[\Delta \sigma] = [D][\Delta \epsilon] - [D][\Delta \epsilon_0] + [\sigma_0] \qquad (4.208)$$

where $[\Delta \epsilon_0]$ and $[\sigma_0]$ are the incremental column matrix of initial strains and the column matrix of initial stresses, respectively, and $[D]$

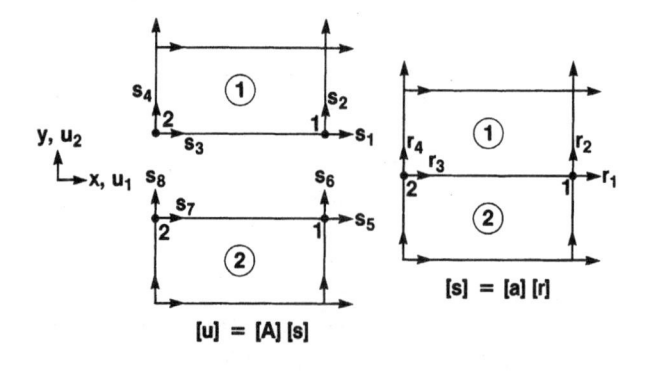

Where [u] = Displacement Column Matrix
[s] = Nodal Displacement Column Matrix
[A] = Shape Function Matrix
[a] = Compatibility Transformation Matrix
[r] = Global Displacement Column Matrix

Figure 4.62 Nodal points, nodal displacement, and global nodal displacements for finite element modeling.

is a material matrix that depends on the current state of stress and hardening of the material. If the material follows the Prandtl-Reuss theory, Eq. 4.36, and the von Mises yield criterion, Eq. 4.110, then the material matrix [D] has the form

$$[D] = [D]_e + [D]_p \quad (4.209)$$

where

$$[D]_e = 2G \begin{bmatrix} \eta & \mu & \mu & 0 & 0 & 0 \\ \mu & \eta & \mu & 0 & 0 & 0 \\ \mu & \mu & \eta & 0 & 0 & 0 \\ 0 & 0 & 0 & \tfrac{1}{2} & 0 & 0 \\ 0 & 0 & 0 & 0 & \tfrac{1}{2} & 0 \\ 0 & 0 & 0 & 0 & 0 & \tfrac{1}{2} \end{bmatrix} \quad (4.210)$$

$$[D]_p = \frac{-9G^2}{\bar{\sigma}^2(3G + H')} \begin{bmatrix} S_x S_x & S_x S_y & S_x S_z & S_x S_{xy} & S_x S_{yz} & S_x S_{zx} \\ S_x S_y & S_y S_y & S_y S_z & S_y S_{xy} & S_y S_{yz} & S_y S_{zx} \\ S_x S_z & S_y S_z & S_z S_z & S_z S_{xy} & S_z S_{yz} & S_z S_{zx} \\ S_x S_{xy} & S_y S_{xy} & S_z S_{xy} & S_{xy} S_{xy} & S_{xy} S_{yz} & S_{xy} S_{zx} \\ S_x S_{yz} & S_y S_{yz} & S_z S_{yz} & S_{xy} S_{yz} & S_{yz} S_{yz} & S_{yz} S_{zx} \\ S_x S_{zx} & S_y S_{zx} & S_z S_{zx} & S_{xy} S_{zx} & S_{yz} S_{zx} & S_{zx} S_{zx} \end{bmatrix}$$

(4.211)

$$S_x = \frac{2\sigma_x - \sigma_y - \sigma_z}{3} \quad S_y = \frac{2\sigma_y - \sigma_x - \sigma_z}{3} \quad S_z = \frac{2\sigma_z - \sigma_y - \sigma_x}{3}$$

$$S_{xy} = \tau_{xy} \quad S_{yz} = \tau_{yz} \quad S_{zx} = \tau_{zx}$$

$$\eta = \frac{1-\nu}{1-2\nu} \quad \mu = \frac{\nu}{1-2\nu} \quad (4.212)$$

and G is given by Eq. 4.8, $\bar{\sigma}$ by Eq. 4.123, $d\bar{\epsilon}_p$ by Eq. 4.126, H' by Eq. 4.132, and $[D]_e$ is the elastic material matrix, $[D]_p$ is the plastic material matrix, and ν is Poisson's ratio.

The total incremental potential energy functional (ΔV_n) for any elastoplastic solid divided into a finite number (n) of discrete elements (V_n) is given by

$$\Delta V_p = \sum_n \left(-\int_{Sn} [\Delta u]^T [\Delta T]\, dS - \int_{Vn} [\Delta u]^T [\Delta F]\, dV \right)$$

$$+ \tfrac{1}{2} \sum_n \int_{Vn} [\Delta \sigma]^T [\Delta \epsilon]\, dV \quad (4.213)$$

where $[\Delta T]$ is the incremental traction column matrix acting on a surface S_n, and $[\Delta F]$ is the incremental body force column matrix.

Substituting Eqs. 4.206–4.208 into Eq. 4.213 leads to

$$\Delta V_p = \sum_n (-[\Delta s]^T[\Delta S] + \tfrac{1}{2}[\Delta s]^T[k][\Delta s]) \qquad (4.214)$$

where

$$[k] = \int_{Vn} ([L][A])^T[D]([L][A])\,dV \qquad (4.215)$$

and

$$[\Delta S] = \int_{Sn} [A]^T[\Delta T]\,dS + \int_{Vn} [A]^T[\Delta F]\,dV + \int_{Vn} ([L][A])^T[D][\Delta \epsilon_0]\,dV$$
$$- \int_{Vn} ([L][A])^T[\sigma_0]\,dV \qquad (4.216)$$

are the element stiffness matrix and the incremental nodal force column matrix, respectively.

The incremental nodal displacements $[\Delta s]$ for different elements are not completely independent; a transformation is needed to relate the element incremental nodal displacements to the independent incremental generalized global displacements. Then the compatibility equations of the assembled structure are written in matrix form as Fig. 4.62

$$[\Delta s] = [a][\Delta r] \qquad (4.217)$$

where $[a]$ is the compatibility transformation matrix and $[\Delta r]$ is the incremental global displacement column matrix.

Substituting Eq. 4.217 into Eq. 4.214 yields

$$\Delta V_p = -[\Delta r]^T[\Delta R] + \tfrac{1}{2}[\Delta r]^T[K][\Delta r] \qquad (4.218)$$

where

$$[K] = \sum_n [a]^T[k][a] \qquad (4.219)$$

and

$$[\Delta R] = \sum_n [a]^T[\Delta S] \qquad (4.220)$$

are the global stiffness matrix and the incremental global force column matrix, respectively.

The principle of minimum incremental potential energy requires that $\delta \Delta V_p = 0$, that is,

$$\frac{\partial \Delta V_p}{\partial [\Delta r]} \delta[\Delta r] = (-[\Delta R] + [K][\Delta r])\delta[\Delta r] = 0 \qquad (4.221)$$

For arbitrary $\delta[\Delta r]$,

$$[K][\Delta r] = [\Delta R] \qquad (4.222)$$

In view of Eqs. 4.215 and 4.219 and the definition of $[D]$, it can be seen that the global stiffness matrix $[K]$ for elastoplastic problems is itself a function of current state of stress. The incremental equilibrium equation, Eq. 4.222, is therefore nonlinear and cannot be solved by a simple matrix inversion. One of the popular techniques for solving Eq. 4.222 is combining the method of incremental loading and the modified Newton-Raphson method (for examples, see refs. [13–22]) to generate the complete nonlinear response by a sequence of piecewise linear steps.

4.6.2 Example I—stress and deflection analysis of partially routed panels for depanelization

In PCB assembly of surface mount components, in order to increase throughput, multiple board manufacturing (MBM) is a common practice. MBM is the process of assembling several arrays of smaller boards on a larger standard size panel. To facilitate depanelization (separation of the boards) after assembly, the perimeter of each board is routed except for small tabs which hold the boards in the panel. While partial routing yields excellent edge quality and simplifies depanelization, it also reduces panel rigidity and leads to a flimsy panel. Excessive deflection of partially routed board will affect component planarity and the quality of solder joints. Furthermore, it will break the connecting tab if the stresses induced reach the strength of material.

In the present study, a total of eight different structures have been designed for a standard size panel. Emphasis is placed on the characterization of deflection of the partially routed panels and on the comparison of board deflection with the flatness specification for surface mount printed circuit boards. Induced stresses will also be characterized to verify that material failure did not occur.

Figures 4.63 and 4.64 show the designs of two tab configurations, Series "A" and Series "B," respectively, of a standard size panel (18" × 12" × 0.058") [11]. There are nine equally dimensioned printed circuit boards with surface mounted components on the panel. These boards (5.933" × 3.933" × 0.058") are partially routed so that each is held in place via a thin strip of material (connecting tab). These tabs are shown in circles. For each series, four different connecting tab dimensions were considered (Figs. 4.63 and 4.64). It can be seen that Series "A" has one tab connecting adjoining boards whereas Series "B" has two. It can also be seen that Series "B" tabs are half as wide as those on Series "A." As a result, the total area connecting adjoining boards is

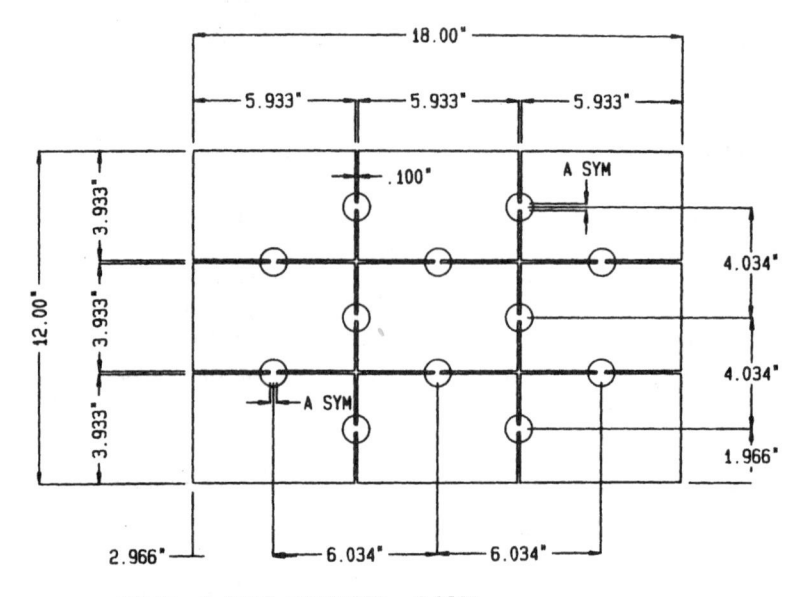

276   Chapter Four

NOTES: 1. PANEL THICKNESS = 0.058"
              2. TABS SHOWN WITH CIRCLES
              3. TAB WIDTH (A) = 0.1", 0.15", 0.2", AND 0.25"
              4. ALL TABS HAVE THE SAME DIMENSIONS (A∗0.1 OR 0.1∗A)

Figure 4.63 PCB tab configuration, Series "A."

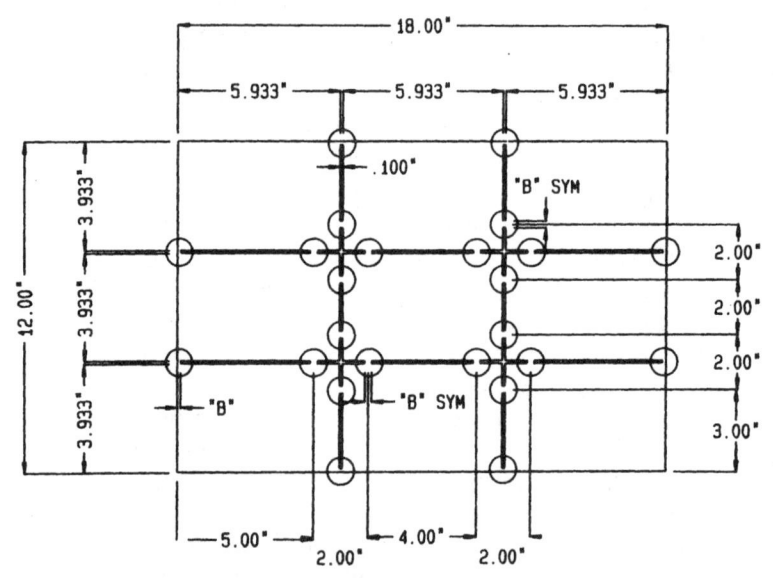

NOTES: 1. PANEL THICKNESS = 0.058"
              2. TABS SHOWN WITH CIRCLES
              3. TAB WIDTH (B) = 0.05", 0.075", 0.1", AND 0.125"
              4. ALL TABS HAVE THE SAME DIMENSIONS (B∗0.1 OR 0.1∗B)

Figure 4.64 PCB tab configuration, Series "B."

Mechanical Design/Analysis/Measurement/Reliability of Electronic Packaging 277

identical for these two series. These designs were chosen to present a worst-case scenario, while still adhering to current manufacturing practices.

A surface mount assembly line will incorporate stations for solder-paste application, component loading, solder reflow, wave solder, cleaning, inspection, testing, and depanelization. Panels will be transferred from station to station via a board handling system. While panels are being assembled in each of the process stations they will be supported from the bottom so they will not deflect. However, while panels are being transferred via the board handling system they are supported only along two opposite edges. As a result, panels will bow and components may dislodge or skew in the unhardened solder paste. Since panel deflection will occur only during board transport, the present boundary-value problem is defined to simulate this process.

Figure 4.65 defines the boundary-value problem of the panel under board transport conditions. It is simply supported along the edges $Y = 0$ and $Y = 12$, and is free to move along the edges $X = 0$ and $X = 18$. The panel is subjected to a uniform distribution load equal to 0.013 psi. (This density was established by measuring the load density of an assembled board currently being manufactured.) Because of double symmetries, only a quarter of the structure is modeled (shaded area in Fig. 4.66). In this figure, all the necessary displacement boundary conditions along the edges of the shaded area were specified. The panel is made of epoxy/glass FR-4 with a modulus of elasticity of 3100 ksi.

Noting the properties of the material to be used and industry standards for the assembly of surface mount boards, the following acceptance criteria were established for the partially routed panel under study: (1) maximum deflection of each surface mount printed circuit board is less than 0.7 percent of its longer length and (2) maximum

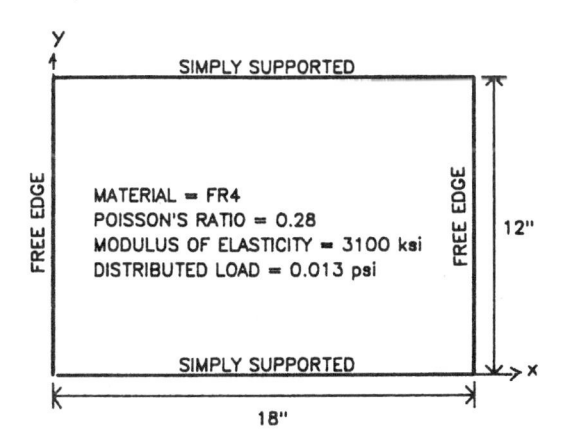

Figure 4.65 PCB boundary value problem.

278　Chapter Four

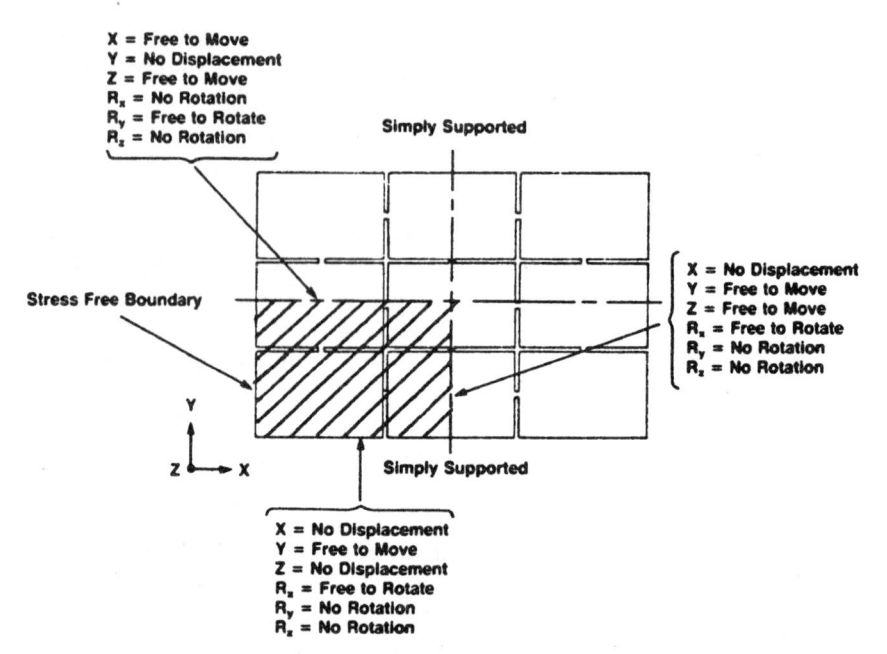

Figure 4.66　Displacement boundary conditions.

stress in tension is less than 35 ksi; maximum stress in compression is less than 35 ksi; maximum stress in shear is less than 14 ksi. These criteria will be measured against the finite element analysis results.

The finite element method was selected for the present study because the determination of a closed-form solution for the partially routed panel is very difficult, if not impossible. In order to build up confidence in and to verify the accuracy of the method, a finite element analysis was used to solve for the deflections, moments, and shear forces of an unrouted panel. The results of this analysis were compared to the closed-form solution (see Sec. 4.2.2) and were found to be in excellent agreement (see Tables 4.1 and 4.2).

The finite element models for a quarter of the panels of Series "A" and "B" are shown in Figs. 4.67 and 4.68, respectively. They were constructed by a 3-D thick-shell element. This element has eight nodal points, and each nodal point has six degrees of freedom.

The finite element analysis of Series "A" showed that the maximum deflection of the entire panel was between 0.315 inches (tab width = 0.1 inch) and 0.217 inch (tab width = 0.25 inch). While these values exceeded the allowable limit, individual deflection profiles for the boards in the panel were significantly less and within the allowed limits (as measured across each board separately). This is seen from the deflection plots shown in Figs. 4.69 and 4.70. In Fig. 4.69, the unde-

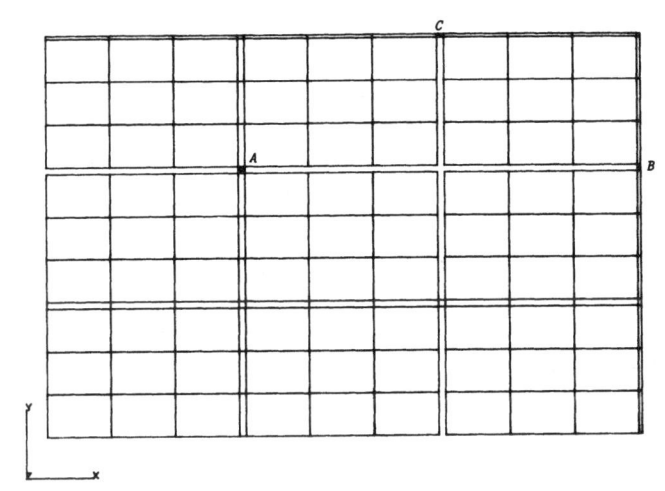

Figure 4.67 Finite element model for Series "A" (tab width = 0.1").

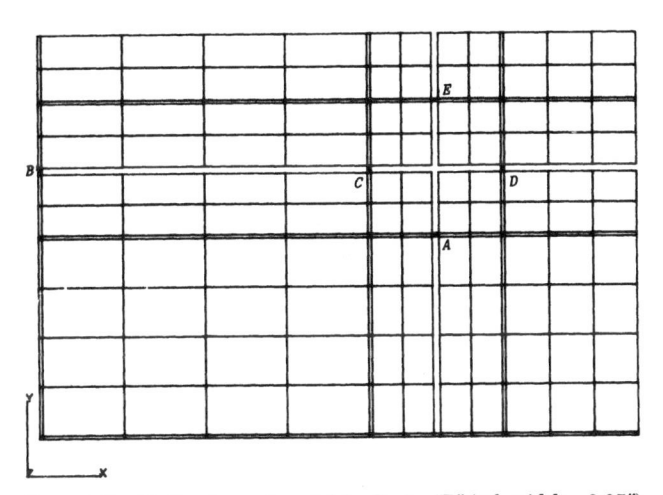

Figure 4.68 Finite element model for Series "B" (tab width = 0.05").

formed mesh is shown in the background and the deformed mesh in the foreground. While the deformed mesh shows significant panel displacement, it also shows a discontinuity in the slope of the partially routed panel. This discontinuity occurs where the connecting tabs hold the boards together. The curvature of the boards on either side of the connecting tabs appears to be insignificant. This is seen more easily when the panel is viewed from the end as in Fig. 4.70. Similar deflection plots are shown in Figs. 4.71 and 4.72 for Series "B." The maximum deflection of the entire panel was between 0.263 and 0.181 inch. As with Series "A," panel deflection is excessive while board deflection is

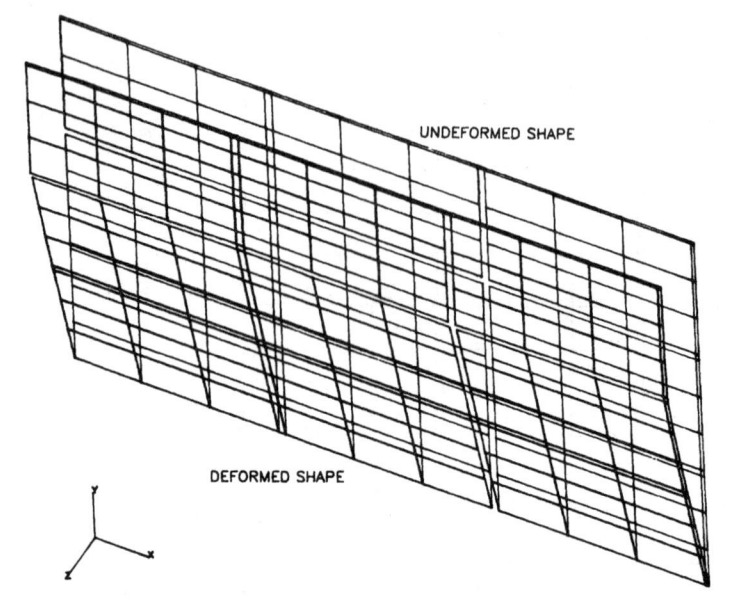

Figure 4.69 Deflection of Series "A" panel.

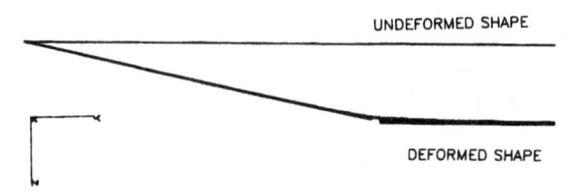

Figure 4.70 Deflection of Series "A" panel (end view).

within the allowed limit. Noting that the acceptance criteria are for board deflection (versus panel deflection), both series meet the criteria for deflection. Thus component planarity is not affected by partial routing of the panel.

Figure 4.73 shows the von Mises stress contours in a Series "A" panel. (The von Mises stress is used because it takes all the principal stresses into consideration.) As can be seen, the "higher" stresses occur near the tabs due to the effect of stress concentration. This is reasonable since the connecting tabs must support the same load as portions of the board having a larger area. Similar stress behavior has also been observed for Series "B" panels (Fig. 4.74). The maximum stresses at elements A, B, and C (Fig. 4.67) of Series "A" panels with various connecting tab widths are tabulated in Table 4.9. The maximum stresses at elements A, B, C, D, and E (Fig. 4.68) of Series "B" panels with various connecting tab widths are also tabulated in Table 4.10. A review of these tables shows that for both series, as the tab width is increased

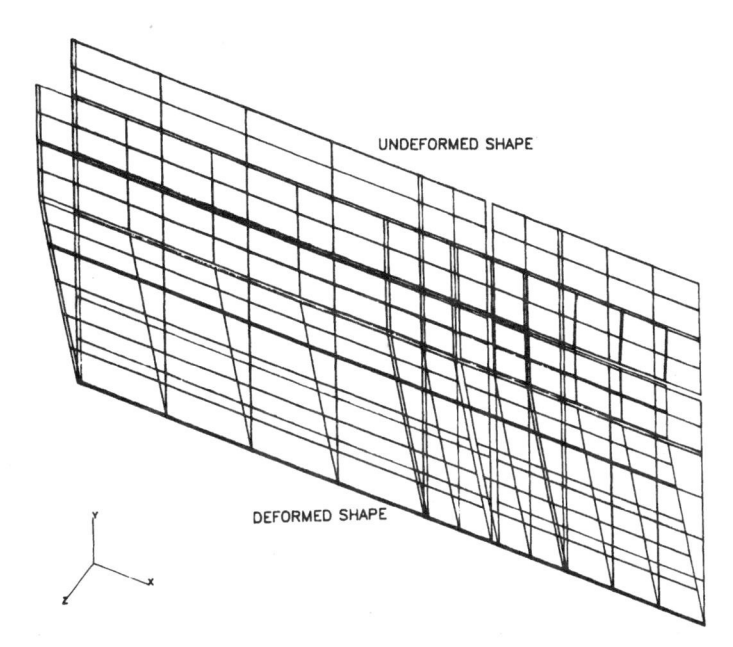

Figure 4.71 Deflection of Series "B" panel.

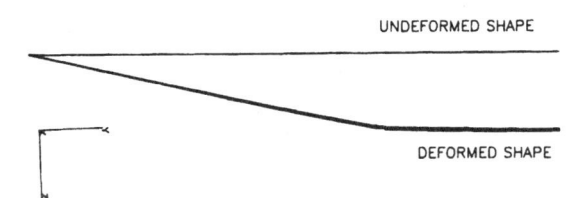

Figure 4.72 Deflection of Series "B" panel (end view).

the induced stresses are reduced. It can also be seen that in all cases, the induced stresses are less than the maximum allowable. Thus structural integrity of the panels during manufacturing is not adversely affected by partial routing.

Results of the finite element analysis indicated that partial routing of a panel would not cause excessive board bowing or material failure. To verify these results, a physical model (Series "A" with tab width = 0.1″) was created. This test panel was loaded and supported to simulate the conditions of the analysis. The experimental value for the maximum panel deflection was 0.301 inch, which is in excellent agreement with the finite element solution (0.315 inch). Furthermore, the individual board deflection was within the allowed limit of 0.7 percent of the board length and agreed with the analytical results.

The results of the present analysis indicated that panels with the prescribed dimensions could be partially routed without causing mate-

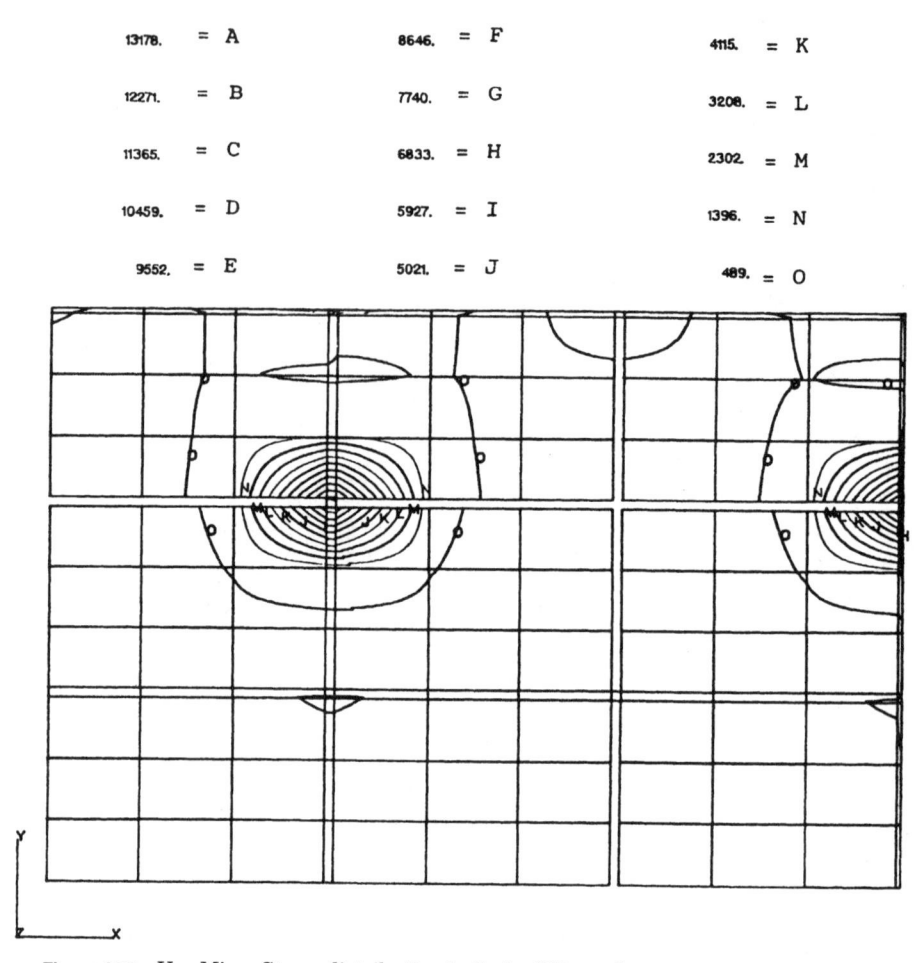

13178.	= A	8646.	= F	4115.	= K
12271.	= B	7740.	= G	3208.	= L
11365.	= C	6833.	= H	2302.	= M
10459.	= D	5927.	= I	1396.	= N
9552.	= E	5021.	= J	489.	= O

Figure 4.73 Von Mises Stress distribution in Series "A" panel.

rial failure or compromising solder joint quality. Experimental findings supported the theory that while partial routing will increase overall panel deflection, boards within the panel will not be deformed beyond the allowable limit. Since the designs used in this study represented a worst-case scenario, partial routing of surface mount boards was declared a safe practice. As a result, the need to develop a sophisticated depanelization workcell was eliminated.

4.6.3 Example II—thermal stress analysis of a plastic leaded package

Figure 4.75 shows a 3-D finite element model of a plastic leaded package [104]. The objective of this model is to determine the stresses at the lead and the entry area of the lead of the plastic package [104]. Under

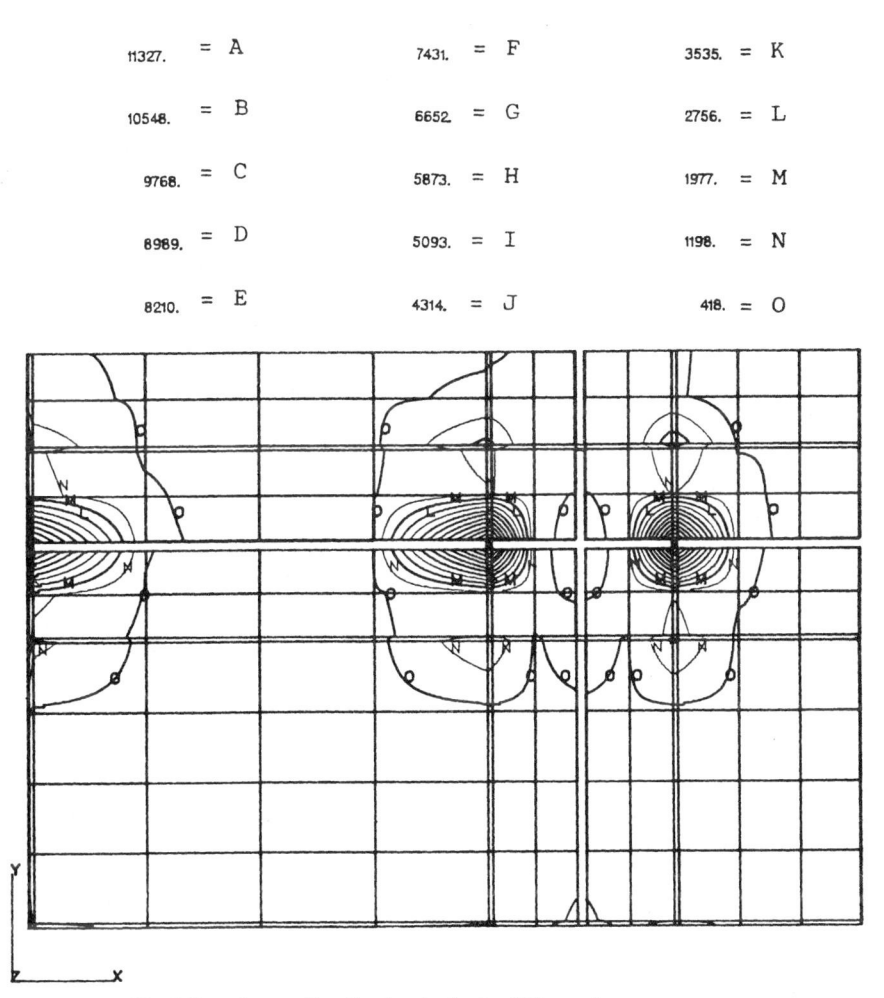

Mechanical Design/Analysis/Measurement/Reliability of Electronic Packaging   283

11327. = A	7431. = F	3535. = K
10548. = B	6652. = G	2756. = L
9768. = C	5873. = H	1977. = M
8989. = D	5093. = I	1198. = N
8210. = E	4314. = J	418. = O

Figure 4.74 Von Mises Stress distribution in Series "B" panel.

a temperature change of 180°C, the von Mises (equivalent) stress contours, Eq. 1.123, acting upon the plastic package portion of the lead entry area (darker lines around ingress point of lead indicate a cavity) are shown in Fig. 4.76. It can be seen that higher stresses occur near the outer interface between the gull-wing lead and the package. Thus, failure of the package could happen there during thermal cycling, and moisture could enter the package and affect the package reliability. Fortunately, the magnitude of stress is very small.

The von Mises stress contours acting on the gull-wing lead are shown in Fig. 4.77. It can be seen that the higher stresses occur at the inner elbows and near the interface between the package and the gull-wing. Higher stress also occurs at the outer shoulder of the gull-wing.

TABLE 4.9 Maximum Stresses at Elements A, B, and C in Series "A" Designs

	\multicolumn{12}{c}{Connecting tab width (inches)}											
	\multicolumn{3}{c}{0.10 Stress (psi) in}	\multicolumn{3}{c}{0.15 Stress (psi) in}	\multicolumn{3}{c}{0.20 Stress (psi) in}	\multicolumn{3}{c}{0.25 Stress (psi) in}
Element	x-dir.	y-dir.	z-dir.	x-dir.	y-dir.	z-dir.	x-dir.	y-dir.	z-dir.	x-dir.	y-dir.	z-dir.
A	−89.7	18830	283	−19.3	12590	188	7.46	9470	126	18	7592	86
B	369	18600	−283	95.4	12330	−188	−42.9	9213	−126	−119	7346	−86
C	1090	2213	283	792	1817	188	647	1503	126	559	1255	86

TABLE 4.10 Maximum Stresses at Elements A, B, C, D, and E in Series "B" Designs

	\multicolumn{12}{c}{Connecting tab width (inches)}
	\multicolumn{3}{c}{0.10 Stress (psi) in}	\multicolumn{3}{c}{0.15 Stress (psi) in}	\multicolumn{3}{c}{0.20 Stress (psi) in}	\multicolumn{3}{c}{0.25 Stress (psi) in}
Element	x-dir.	y-dir.	z-dir.	x-dir.	y-dir.	z-dir.	x-dir.	y-dir.	z-dir.	x-dir.	y-dir.	z-dir.
A	1108	1204	7.45	698	864	5.29	529	638	9.2	438	492	12
B	986	15160	114	725	10040	67	551	7508	49	430	6007	40
C	1016	20420	190	730	13720	119	543	10340	88	416	8292	72
D	1465	20690	76	1050	13770	52	794	10310	39	621	8233	31
E	1403	2414	76	943	1757	52	737	1364	39	619	1104	31

Due to residual manufacturing stresses, it is not uncommon to observe cracks at the shoulder of the gull-wing. In that case, further crack propagation in the gull-wing under operating conditions is likely.

Next, we would like to compare the thermal stresses in a plastic package with alloy-42 and copper leads (Fig. 4.78) [105]. The material

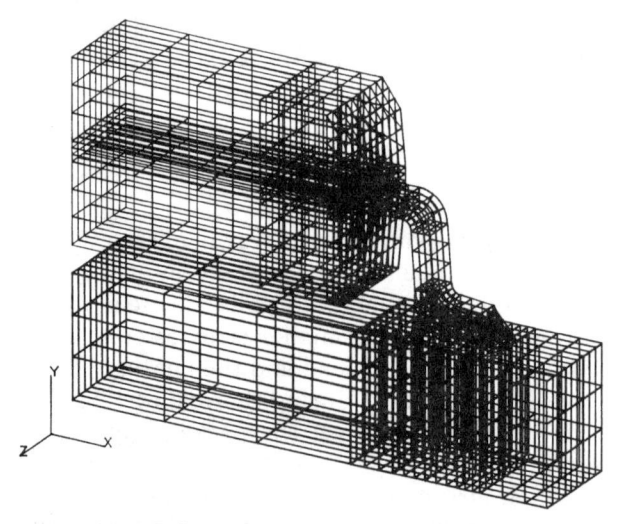

Figure 4.75 3-D finite element model for a gull-wing plastic leaded package.

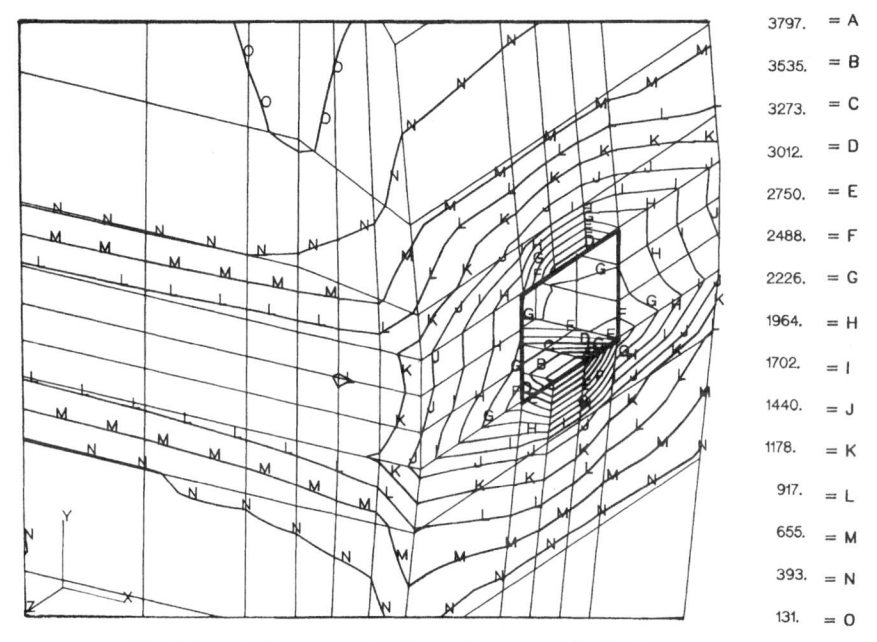

Mechanical Design/Analysis/Measurement/Reliability of Electronic Packaging 285

Figure 4.76 Von Mises (psi) contours in the package near the lead entry area.

3797.	= A
3535.	= B
3273.	= C
3012.	= D
2750.	= E
2488.	= F
2226.	= G
1964.	= H
1702.	= I
1440.	= J
1178.	= K
917.	= L
655.	= M
393.	= N
131.	= O

properties of the package are shown in Table 4.11. The resultant stresses due to uniform heating from −55 to 150°C at Locations A, B, C, D, E, F, and G are summarized in Table 4.12. Also, the maximum principal stress acting in the molding plastic of the package with alloy-42 and copper lead frames are shown, respectively, in Figs. 4.79 and 4.80.

It can be seen that for the silicon chip, the largest maximum principal stress occurs at Position B of Fig. 4.78 and any chip cracking should occur at this location. Furthermore, the silicon chip of a plastic package with copper lead frame is more likely to crack than the silicon chip of a plastic package with alloy-42 lead frame.

For the chip-attach pad, the largest maximum principal stress occurs at Position C of Fig. 4.78, and the maximum principal stresses acting at Positions C, D, and E (Fig. 4.78) are larger for the plastic package with alloy-42 material than with copper material.

For the molding plastic, the maximum principal stresses are larger for the package with alloy-42 lead frame than with copper lead frame. Therefore, the molded plastic of a package with alloy-42 lead frame is more likely to crack than the molded plastic of a package with copper lead frame. The stress acting at the molded plastic near the lower right-hand tip of the chip-attach pad (Position F of Fig. 4.78) can be reduced by rounding the corner of the chip-attach pad.

286   Chapter Four

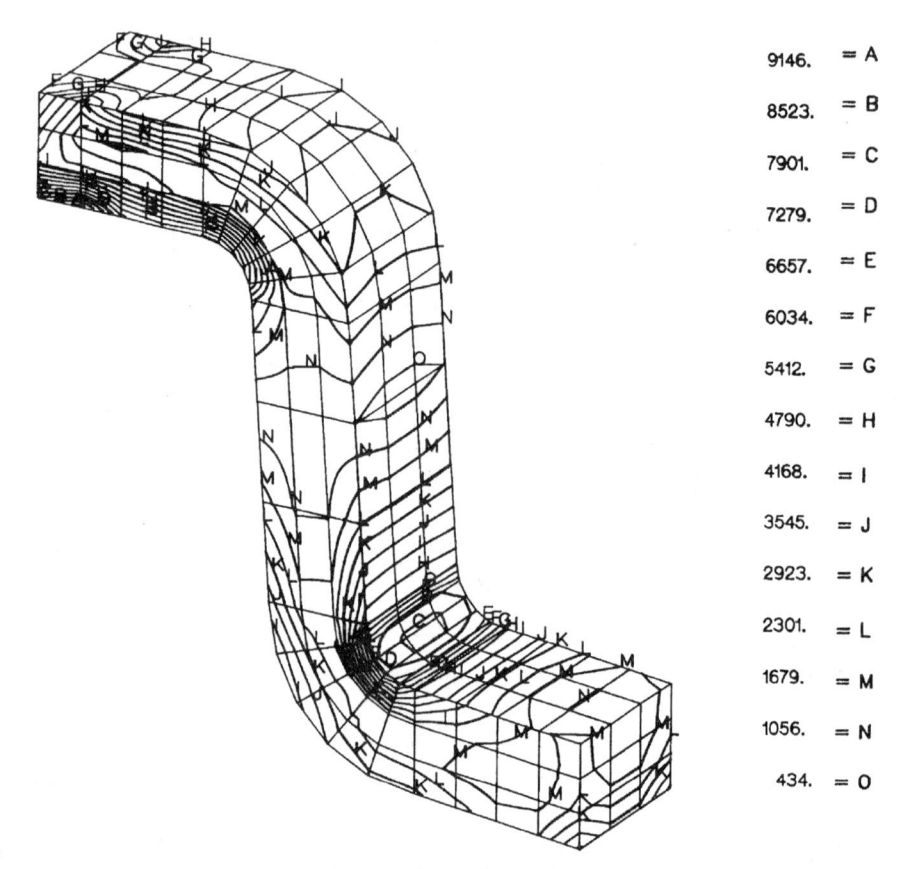

9146.	= A
8523.	= B
7901.	= C
7279.	= D
6657.	= E
6034.	= F
5412.	= G
4790.	= H
4168.	= I
3545.	= J
2923.	= K
2301.	= L
1679.	= M
1056.	= N
434.	= O

Figure 4.77 Von Mises (psi) contours in the gull-wing lead.

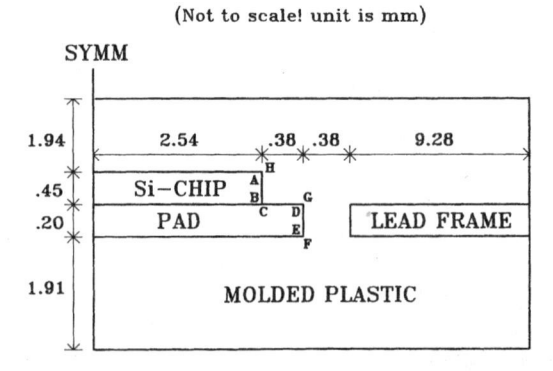

Figure 4.78 A typical cross section of a plastic leaded package.

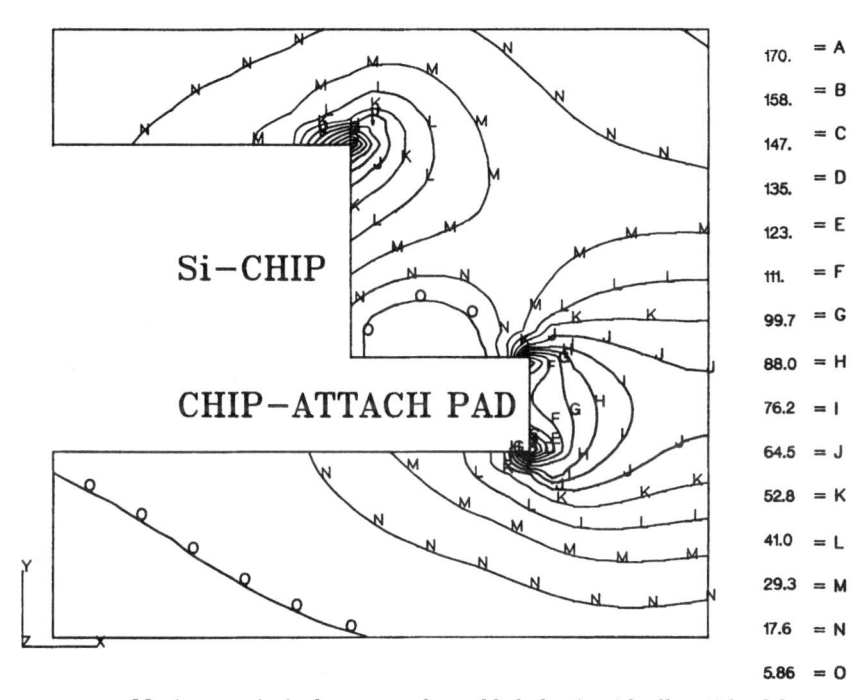

Figure 4.79 Maximum principal stress at the molded plastic with alloy-42 lead frames.

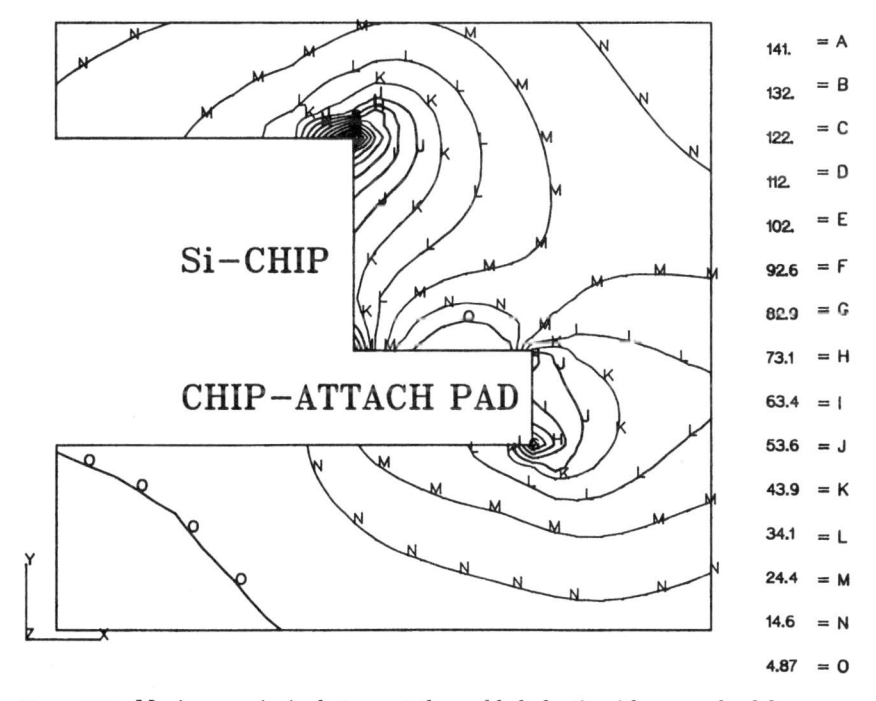

Figure 4.80 Maximum principal stress at the molded plastic with copper lead frames.

TABLE 4.11 Material Properties of the Plastic Leaded Package

	Silicon	Plastic	Alloy-42	Copper
Young's modulus (10^4 MN/m^2)	12.8	1.253	20.4	11.7
Poisson's ratio	0.28	0.3	0.29	0.35
Thermal expansion coefficient (10^{-6}/°C)	2.6	27	7	17

TABLE 4.12 Maximum Principal Stress at Various Positions of the Plastic Leaded Package with Alloy-42 and Copper Lead Frames

	Max principal stress (MN/m^2)	
Position	Alloy-42	Copper
Si-Chip		
A	290	289
B	562	708
Chip-attach pad		
C	775	751
D	534	136
E	582	250
Molded plastic		
F	170	90
G	110	45
H	135	139

4.6.4 Example III—mechanical behavior of microstrip structures made from YBa$_2$Cu$_3$O$_{7-x}$ superconducting ceramics

The composite structures under consideration are shown in Fig. 4.81a through d, where their dimensions are also specified. Figure 4.81c and d are microstrip structures where both the top conductor and the bottom ground plane are made of YBa$_2$Cu$_3$O$_{7-x}$. The boundary-value problem is to calculate the thermal stresses of these structures, while they are subjected to a temperature drop from 873 to 50°K.

The thermal expansions of YBa$_2$Cu$_3$O$_{7-x}$, GaAs, Si, Al$_2$O$_3$, and MgO have been compiled from [106–118] and are summarized in Fig. 4.82. It can be seen that for a wide range of temperatures, and the thermal coefficient of linear expansion of YBa$_2$Cu$_3$O$_{7-x}$ is almost the same as MgO, but is quite different from that of Si, GaAs, and Al$_2$O$_3$. In particular, for the temperature range, 873 to 50°K, the thermal coefficients of linear expansion for the YBa$_2$Cu$_3$O$_{7-x}$ is 11.8 ppm/°K; MgO is 11 ppm/°K; Al$_2$O$_3$ is 6.2 ppm/°K; GaAs is 5.8 ppm/°K; and Si is 2.9 ppm/°K (Table 4.13). The Young's moduli and Poisson's ratios of these materials are given by Table 4.13. It can be seen that the Poisson's ratios of these

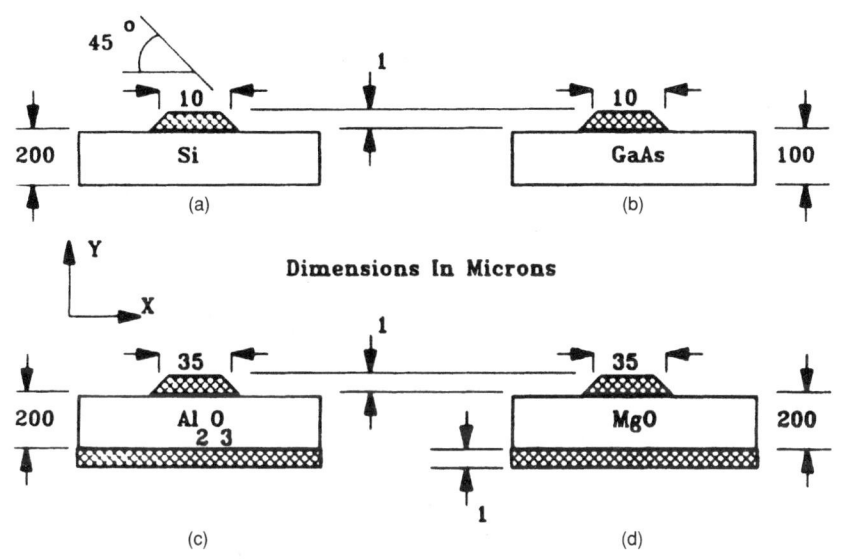

Figure 4.81 Physical structures modeled: YBa$_2$Cu$_3$O$_{7-x}$ on Si, on GaAs, on Al$_2$O$_3$, and on MgO.

materials are basically the same, but the Young's moduli vary dramatically (e.g., the Young's modulus of MgO is more than three times that of YBa$_2$Cu$_3$O$_{7-x}$).

It should be emphasized that the physical and mechanical properties of these materials may be temperature, frequency, rate, and time dependent. Furthermore, these materials may not behave linearly at

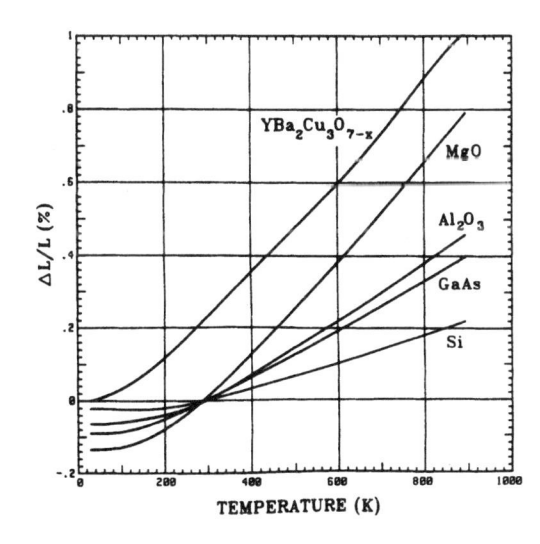

Figure 4.82 Thermal expansions of YBa$_2$Cu$_3$O$_{7-x}$, GaAs, Si, Al$_2$O$_3$, and MgO.

TABLE 4.13 Material Properties of YBa$_2$Cu$_3$O$_{7-x}$, GaAs, Si, Al$_2$O$_3$, and MgO

Material	Young's modulus (GPa)	Poisson's ratio	Thermal coefficient of linear expansion (10^{-6}/K)
YBa$_2$Cu$_3$O$_{7-x}$	96	.25	11.8
Si	131	.28	2.9
GaAs	86	.31	5.8
Al$_2$O$_3$	255	.30	6.2
MgO	317	.29	11.0

high temperatures. However, for the purpose of this study and due to the lack of material data, the material properties are assumed to be constant with a resulting linear analysis.

As the temperature drops from 873 to 50 °K, these composite structures are subjected to a very complex stress state due to the large thermal expansion mismatch and stiffness differences between the YBa$_2$Cu$_3$O$_{7-x}$ and the dielectric substrates. Furthermore, because of the geometry of the composite structures, the determination of the stresses in them is very difficult. For this reason, the finite element method was chosen. Figure 4.83, for example, shows the finite element model for YBa$_2$Cu$_3$O$_{7-x}$ on Al$_2$O$_3$ and MgO substrates [119]. Because of symmetry, only half of the structure is analyzed. Two-dimensional plane stress elements have been used for the construction of these models. This plane stress element has 8 nodal points, and each nodal point has 2 degrees of freedom. Table 4.14 summarizes [119] the maximum stresses acting within the microstrip line on various substrates. For the cases considered, MgO is structurally the best substrate match for YBa$_2$Cu$_3$O$_{7-x}$, while Si is the worst.

4.6.5 Example IV—design for plastic ball grid array solder joint reliability

Figure 4.84a shows the cross section of a full-matrix PBGA assembly [120, 121]. It is a 27 × 27-mm package with 15 × 15 = 225 solder balls on a 1.5-mm ball-pitch. The 8-node quadrilateral plane strain element is used for the finite element analysis. A uniform temperature loading from 0 to 85°C is applied to the whole assembly. Within the scope of the present study, all material properties are assumed to be temperature independent and linear elastic, Table 4.15, except the solder balls which are elastoplastic with a yield stress = 1200 psi or 8.3 MPa, yield strain = 0.0008, and strain-hardening parameter = 0.1.

The effects of chip size, chip thickness, and BT (bismaleimide triazine) thickness on the solder joint reliability are shown in Fig. 4.84b, c, and d, respectively. It can be seen that, in all cases, the maximum

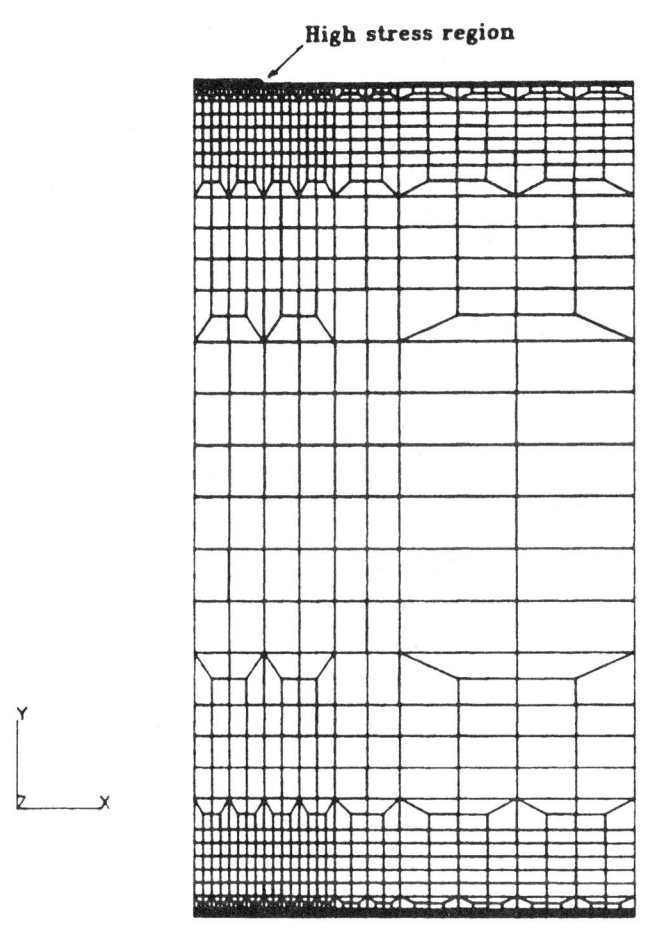

Figure 4.83 Finite element model for YBa$_2$Cu$_3$O$_{7-x}$ on Al$_2$O$_3$ and MgO substrates.

accumulated effective plastic strain (which will be used as an index for solder joint reliability in this section) occurs in the solder ball underneath the edges of the chip. This is due to the substantial local thermal expansion mismatch (Table 4.15) between the silicon chip, the BT substrate, and the FR-4 PCB. Although the BT substrate has a similar TCE to FR-4, the effect from the chip is still essential since the BT substrate is relatively thin.

In view of Fig. 4.84b, it can be seen that the larger the chip size the larger the accumulated effective plastic strain. Also, it can be seen from Fig. 4.84c that the thicker the chip, the larger the accumulated effective plastic strain; that is, because of the larger dimensions of the chip, the larger the thermal expansion mismatch. On the other hand, Fig.

TABLE 4.14 Maximum Stresses in the YBa₂Cu₃O₇₋ₓ on GaAs, Si, Al₂O₃, and MgO Substrates

Substrate materials	Maximum stress (MPa) in the YBa₂Cu₃O₇₋ₓ acting in the		
	x-direction	y-direction	xy-plane
Si	931	439	524
GaAs	774	359	418
Al₂O₃	653	306	359
MgO	95	44	51

4.84d shows that the thicker the BT substrate, the smaller the accumulated effective plastic strain; that is, because the thicker the substrate (the larger the bending stiffness), the smaller the global deflection, and the thicker BT substrate can reduce the influence of the silicon chip on the local thermal expansion mismatch.

Figure 4.85a shows a set of 1.27-mm ball-pitch, 27 × 27-mm body size perimeter-arrayed PBGAs (PA-PBGA) with and without thermal solder balls. Figures 4.86, 4.87, 4.88, and 4.89 show, respectively, the accumulated effective plastic strain of a 4-row PA-PBGA without thermal solder balls, a 4-row PA-PBGA with 16 thermal balls, a 5-row PA-PBGA without thermal balls, and a 5-row PA-PBGA with 16 thermal balls. Two different BT substrate thicknesses and three different chip sizes for each case have been studied. The important results can be summarized as following.

1. In general, for perimeter-arrayed PBGAs (Figs. 4.86 through 4.89), smaller chip size will enhance the reliability of inner perimeter solder joints but jeopardize thermal balls at the center. The outer perimeter solder joints are insensitive to the chip size.

2. For perimeter-arrayed PBGAs without thermal balls (Figs. 4.86 and 4.88), thicker BT substrates always improve the reliability of the innermost solder joints. It is especially beneficial for packages with larger chip size.

3. For perimeter-arrayed PBGAs with thermal balls (Figs. 4.87 and 4.89), thicker BT substrates always improve the reliability of thermal balls. The trend of perimeter solder joints with respect to the substrate thickness is insignificant since their plastic strain is considerably less than that in the thermal balls.

4. Without thermal balls, the innermost solder joint of 5-row perimeter-arrayed PBGA packages (Fig. 4.88) has larger plastic strain than that of 4-row ones (Fig. 4.87). However, if thermal balls exist, the difference is minimal.

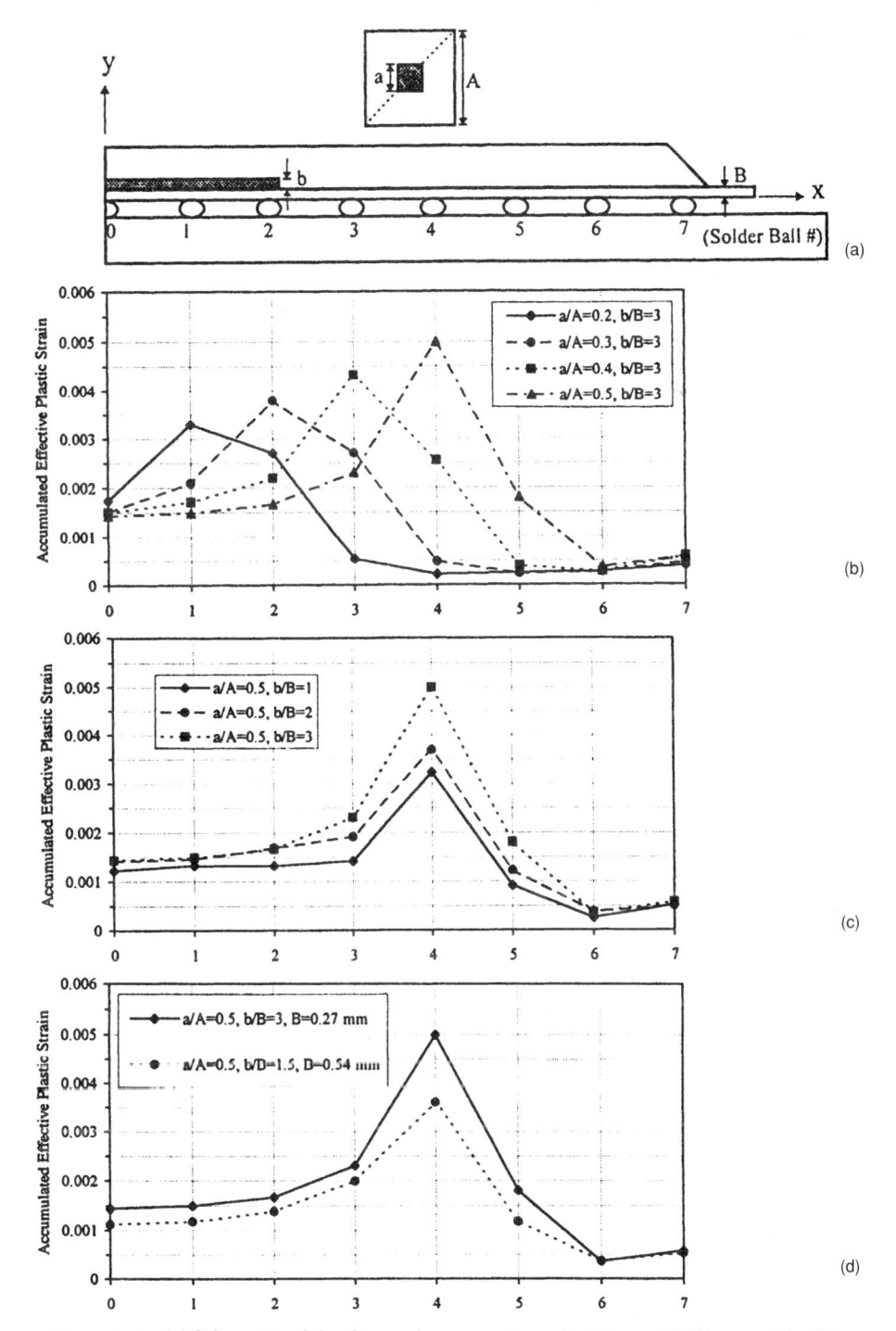

Figure 4.84 (a) Schematic of the diagonal cross section of a 225-pin PBGA assembly, (b) effect of chip size, (c) effect of chip thickness, and (d) effect of substrate thickness on accumulated effective plastic strain in solder joints.

TABLE 4.15 Material Properties for PBGAs on PCB

Materials	Young's modulus (GPa)	Poisson's ratio	CTE (10^{-6}/°C)
Silicon die	131	0.3	2.8
C4 solder (95Pb/5Sn)	8	0.4	30
Eutectic solder (63Sn/37Pb)	10	0.4	21
Underfill epoxy	6	0.35	30
BT substrate	26 (x, y) 11 (z)	0.39 (xz, yz) 0.11 (xy)	15 (x, y) 52 (z)
FR-4 PCB	22 (x, y) 10 (z)	0.28 (xz, yz) 0.11 (xy)	18 (x, y) 70 (z)

4.6.6 Example V—solder joint reliability of plastic ball grid array packages with solder-bumped flip chip

Figure 4.90 shows a set of PBGAs (27 × 27-mm body size and 1.27-mm ball-pitch) with solder-bumped flip chip on the BT substrate with underfill encapsulant (Table 4.15) [122]. They have much better electrical performance than those discussed in Example IV, which are wire bonding chip on the BT substrate with overmold encapsulant. Figure 4.91 shows the cross section for finite element modeling. The typical

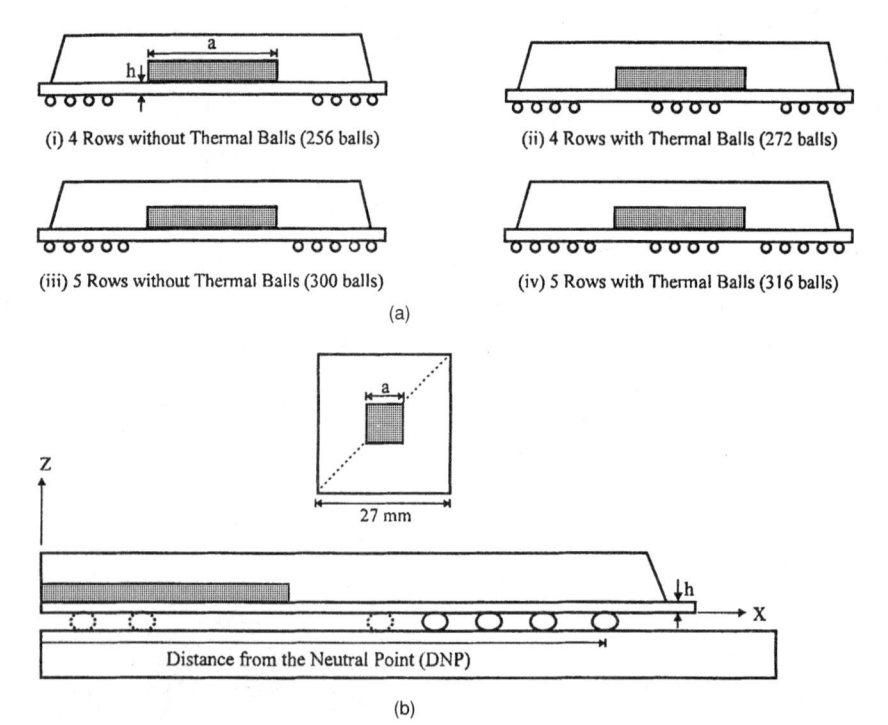

Figure 4.85 (a) Various perimeter PBGAs with and without thermal solder balls, and (b) the finite element model.

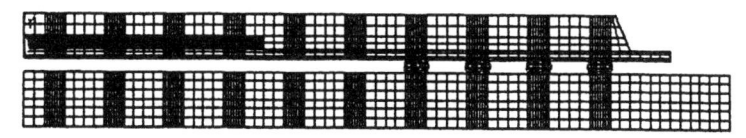

(a) Corresponding Finite Element Mesh

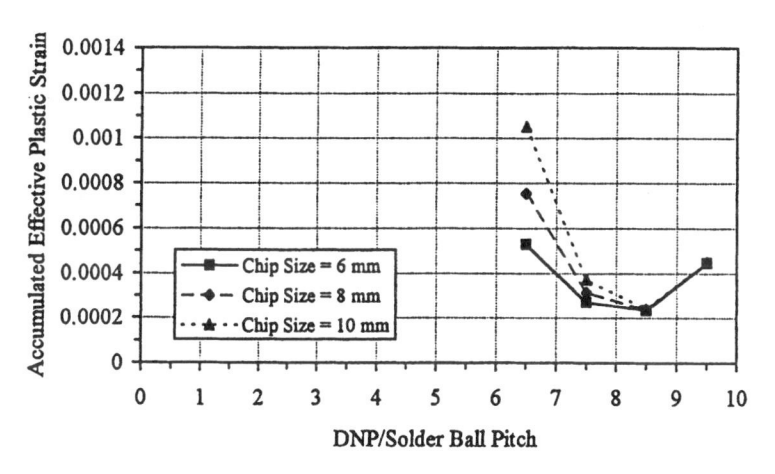

(b) BT Substrate Thickness = 0.27 mm

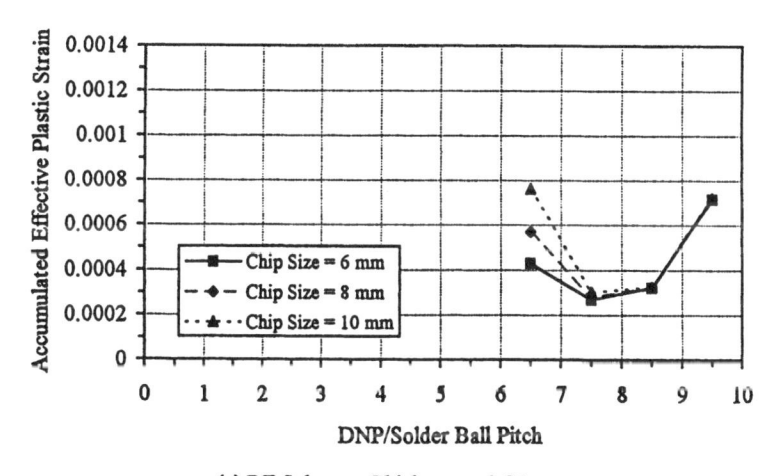

(c) BT Substrate Thickness = 0.54 mm

Figure 4.86 Numerical results for the 4-row perimeter PBGA without thermal solder balls.

von Mises stress and plastic strain contours in the second-level solder joint are shown in Fig. 4.92. It can be seen that the maximum stress and strain are acting near the corners of the solder joint. (It should be pointed out that the flip-chip solder bump is not the focus of the present study.)

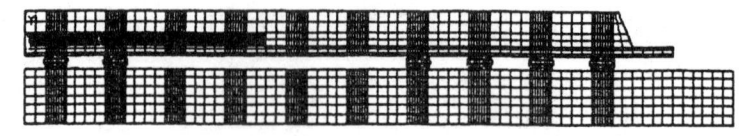

(a) Corresponding Finite Element Mesh

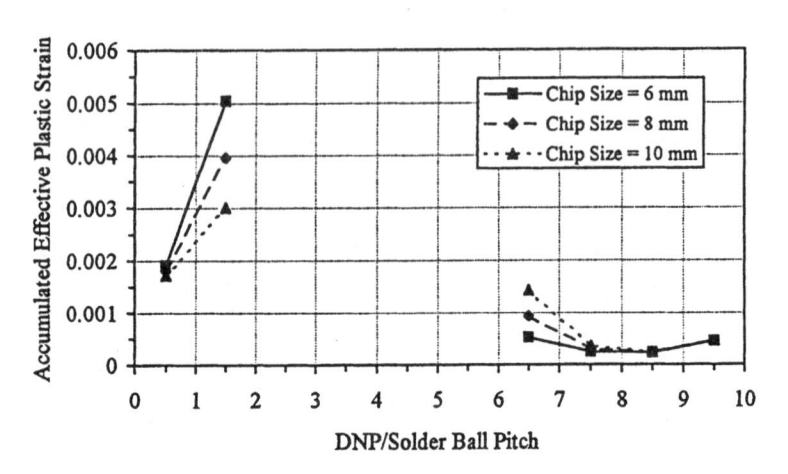

(b) BT Substrate Thickness = 0.27 mm

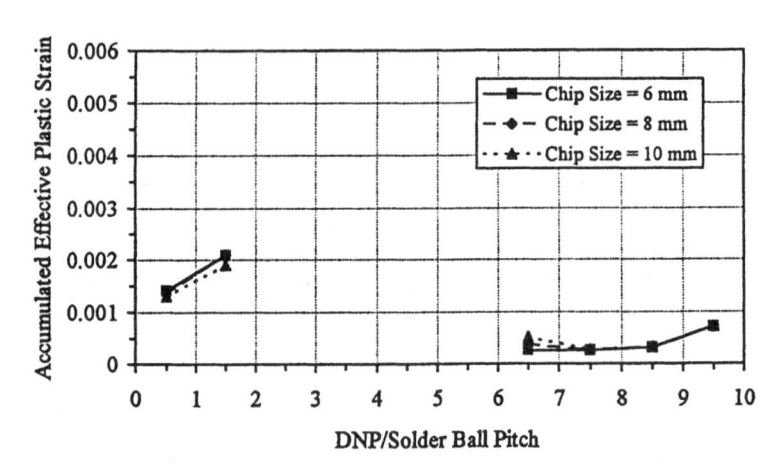

(c) BT Substrate Thickness = 0.54 mm

Figure 4.87 Numerical results for the 4-row perimeter PBGA with 16 thermal solder balls.

The effects of chip size (*a*), chip thickness (*b*), and BT substrate thickness (Fig. 4.90) on the second-level solder joint reliability are shown in Figs. 4.93 through 4.95. Some of the deformed shapes of the flip-chip solder-bumped PBGA assemblies are shown in Figs. 4.96 and 4.97. Important results are summarized as follows:

(a) Corresponding Finite Element Mesh

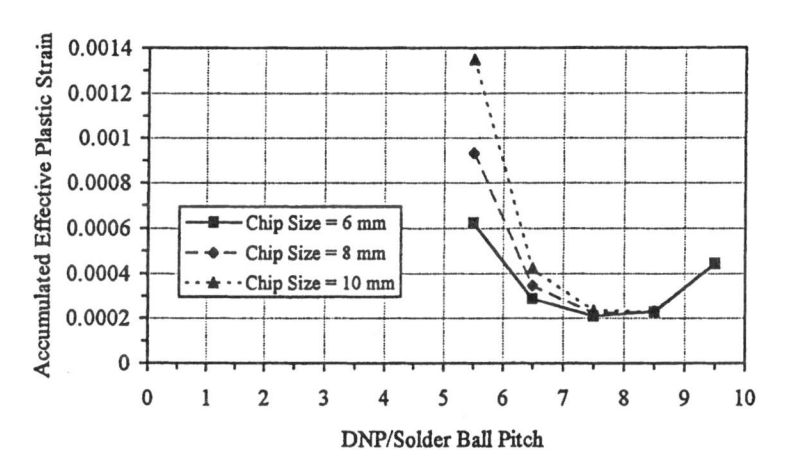

(b) BT Substrate Thickness = 0.27 mm

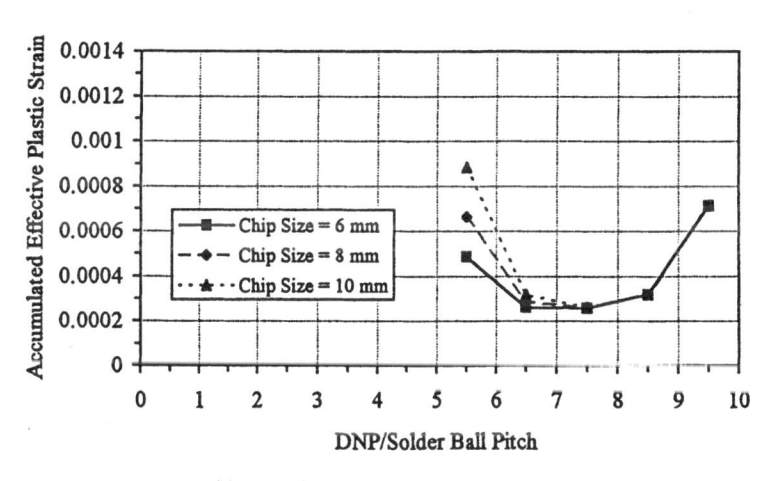

(c) BT Substrate Thickness = 0.54 mm

Figure 4.88 Numerical results for the 5-row perimeter PBGA without thermal solder balls.

1. For the perimeter-arrayed flip chip PBGA without thermal balls (Fig. 4.93), it is found that the larger chip size always induces more plastic strain in the innermost solder joint. The thickness of the chip has less effect than the planar size. The thickness of BT substrate has a substantial effect on solder-joint reliability, especially for a larger size

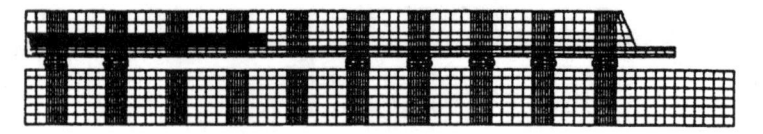

(a) Corresponding Finite Element Mesh

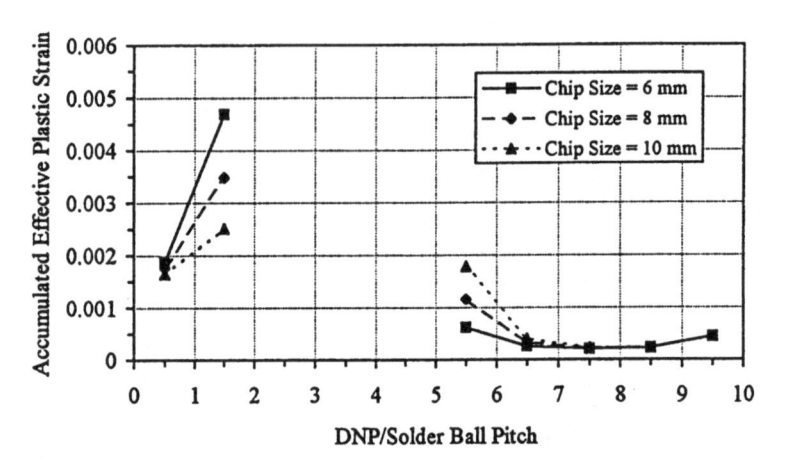

(b) BT Substrate Thickness = 0.27 mm

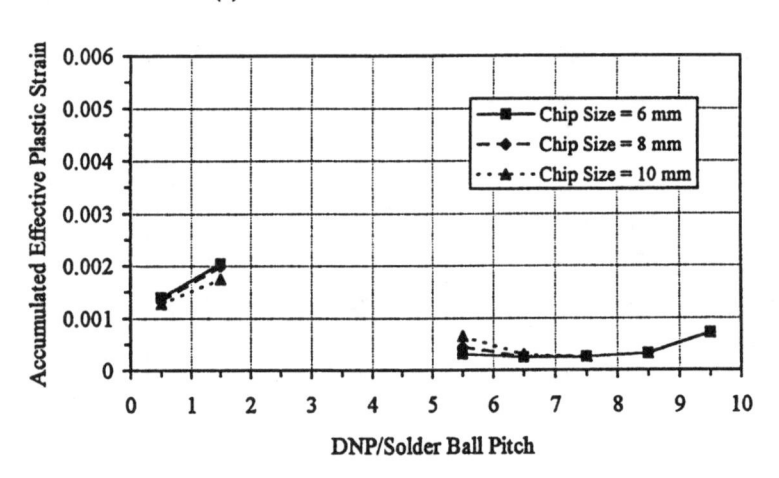

(c) BT Substrate Thickness = 0.54 mm

Figure 4.89 Numerical results for the 5-row perimeter PBGA with 16 thermal solder balls.

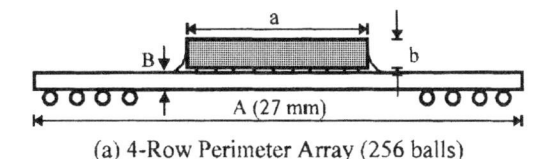

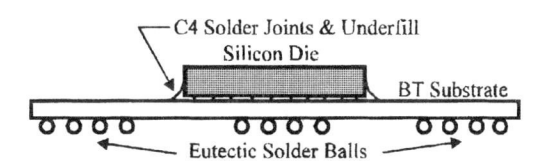

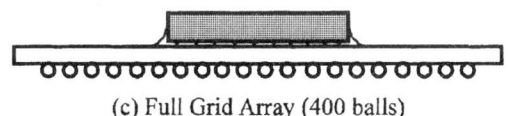

Mechanical Design/Analysis/Measurement/Reliability of Electronic Packaging 299

Figure 4.90 Various solder-bumped flip-chip PBGAs. (C4 = controlled-collapse chip connection.)

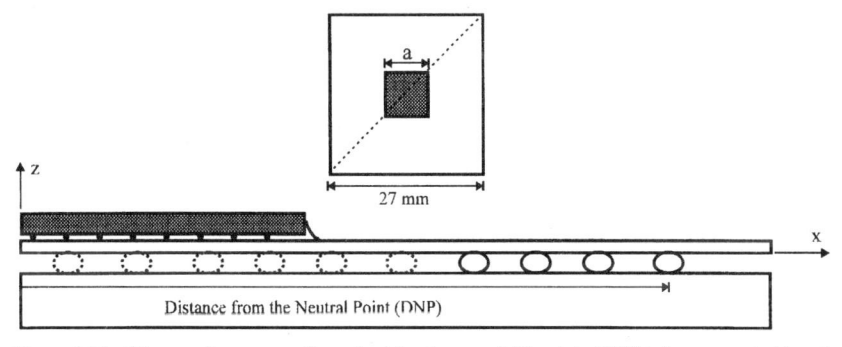

Figure 4.91 Diagonal cross section of solder-bumped flip-chip PBGA for computational modeling.

chip. The thicker BT substrate introduces more plastic strain and thus reduces solder-joint reliability. This is in contradiction to the results (Figs. 4.86 and 4.88) mentioned in Sec. 4.6.5. The reason is because of the local bending of the solder-bumped flip chip on BT substrate (Fig. 4.96a and b for smaller chips, and Fig. 4.97a and b for larger chips) and superpose on the thermal expansion ($\Delta T = 85°C$) of the solder joints in the vertical direction.

2. For the perimeter-arrayed flip chip PBGA with thermal balls (Fig. 4.94), the plastic strains in the solder joints are considerably suppressed, and the effects of chip dimensions and BT substrate thickness become insignificant. Substantial plastic strain is observed in the ther-

(a) Contour Plot of von Mises Stress

A = 16.9 MPa
B = 17.3 MPa
C = 17.7 MPa
D = 18.1 MPa
E = 18.5 MPa
F = 18.8 MPa
G = 19.2 MPa
H = 19.6 MPa
I = 20.0 MPa

(b) Contour Plot of Effective Plastic Strain

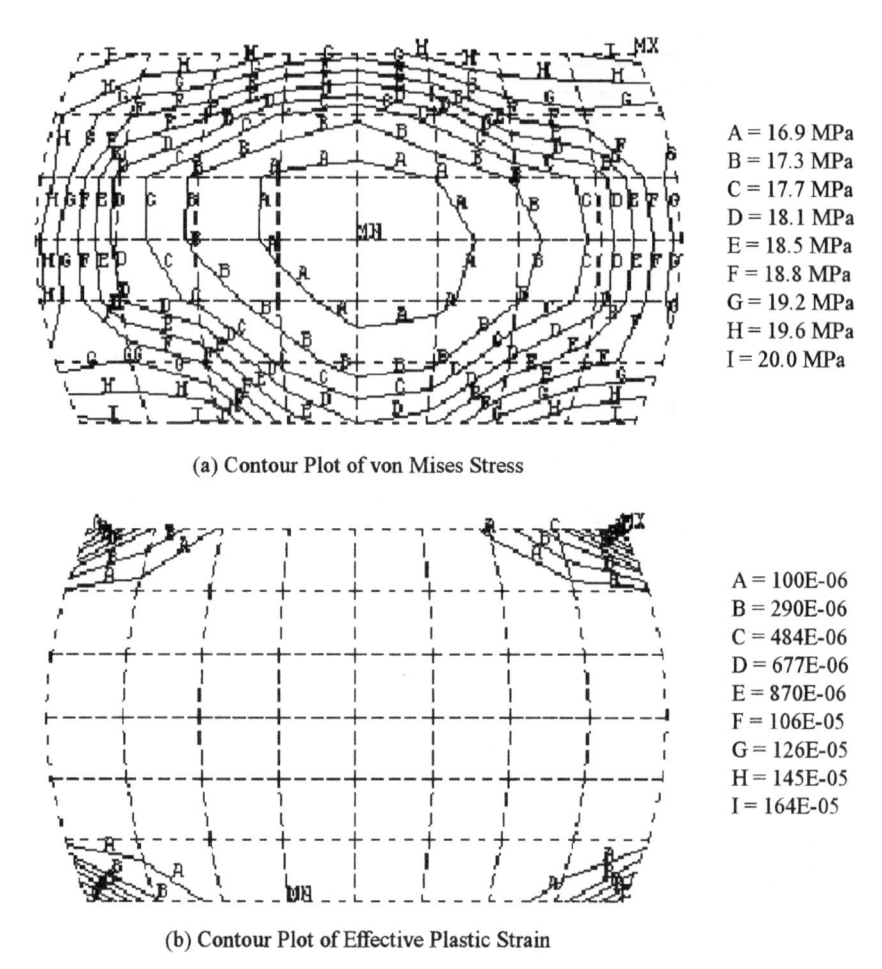

A = 100E-06
B = 290E-06
C = 484E-06
D = 677E-06
E = 870E-06
F = 106E-05
G = 126E-05
H = 145E-05
I = 164E-05

Figure 4.92 Typical stress and accumulated effective plastic strain contours in the 63Sn37Pb solder ball.

mal balls. However, thicker BT substrate can always relax the TCE mismatch directly from the silicon chip and, hence, reduce the plastic strain in the thermal balls.

3. For full-matrix flip-chip PBGA (Fig. 4.95), it is found that neither the chip size nor the chip thickness has a notable effect on solder-joint reliability. Also, the plastic strain distribution in the solder joints underneath the chip is quite uniform (Fig. 4.95c), and in some cases, the maximum plastic strain is not necessary near the chip corner (Fig. 4.95b). Besides, the plastic strain distribution is insensitive to the DNP (distance to neutral point). Also, it should be pointed out that the plastic strain in the solder joints of the full-matrix flip-chip PBGA is smaller than those in the full-matrix wire-bonded PBGA discussed in Sec. 4.6.5 (Fig. 4.84).

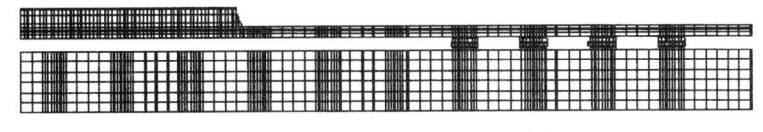

(a) Corresponding Finite Element Mesh (256 solder balls)

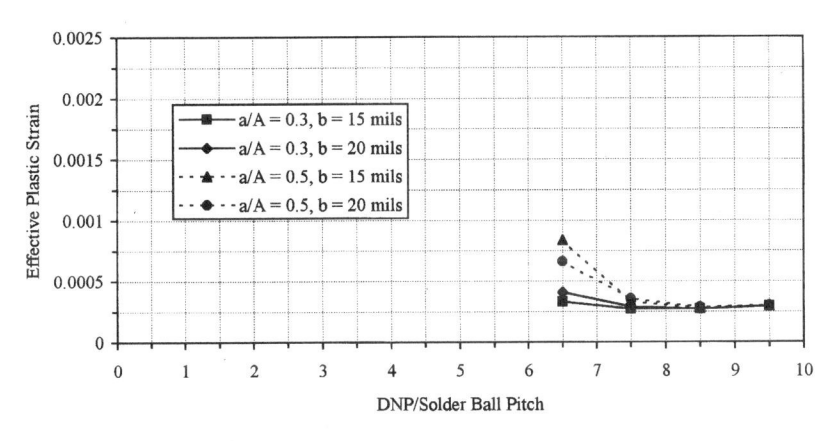

(b) BT Substrate Thickness = 0.27 mm

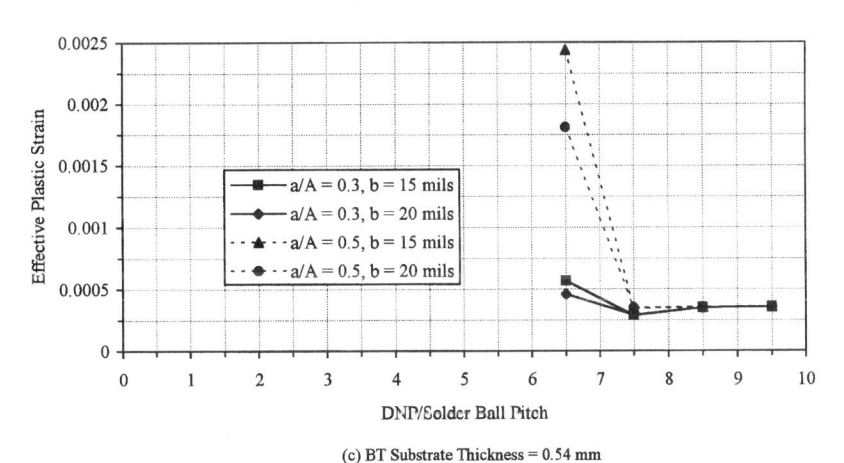

(c) BT Substrate Thickness = 0.54 mm

Figure 4.93 Distribution of maximum accumulated effective plastic strain in the (solder-bumped flip-chip) 4-row perimeter PBGA without thermal solder balls.

4.7 The Roles of DNP (Distance to Neutral Point) on Solder-Joint Reliability of Area-Array Assemblies

Distance to neutral point (DNP) is defined as the separation of a solder joint from the neutral point on a chip or on a carrier (substrate). The dimension controls the shear strain on the solder joint imposed by the thermal expansion mismatch between chip and carrier, or chip and

(a) Corresponding Finite Element Mesh (272 solder balls)

(b) BT Substrate Thickness = 0.27 mm

(c) BT Substrate Thickness = 0.54 mm

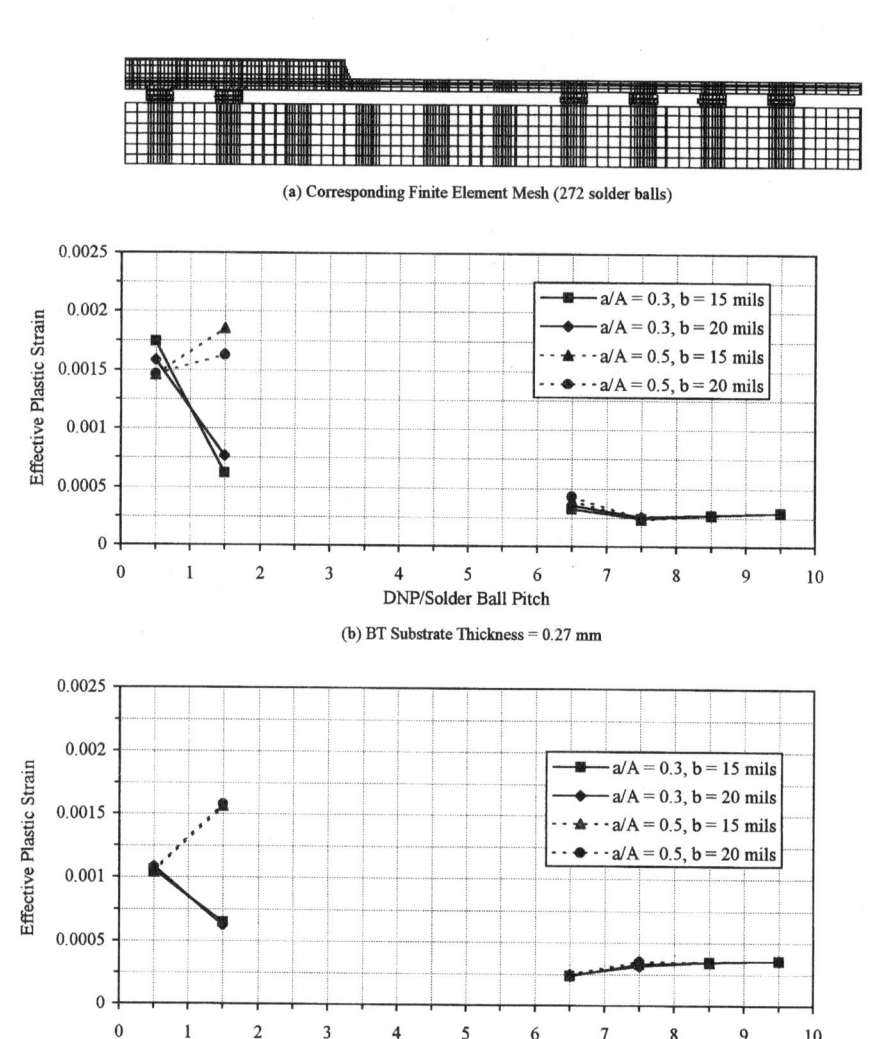

Figure 4.94 Distribution of maximum accumulated effective plastic strain in the (solder-bumped flip-chip) 4-row perimeter PBGA with 16 thermal solder balls.

PCB (printed circuit board), or carrier and PCB. The neutral point is usually the geometric center of an array of pads (the plane of symmetry) and defines the point at which there is no relative motion of chip and carrier (or chip and PCB, or carrier and PCB) in the horizontal plane during thermal loading.

The concept of DNP has been used to calculate the strain (the greater the DNP, the larger the shear strain) and then combined with the Coffin-Manson relationship (the number of cycles-to-failure depends on the

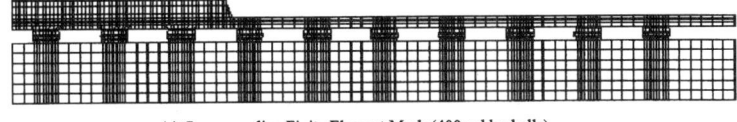

(a) Corresponding Finite Element Mesh (400 solder balls)

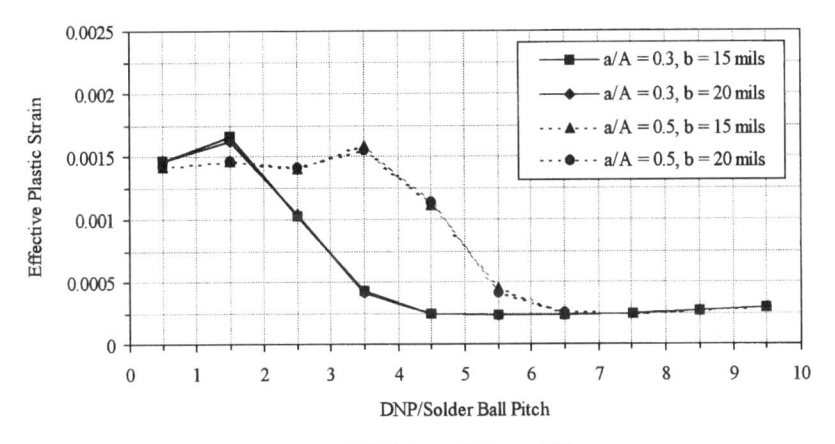

(b) BT Substrate Thickness = 0.27 mm

(c) BT Substrate Thickness = 0.54 mm

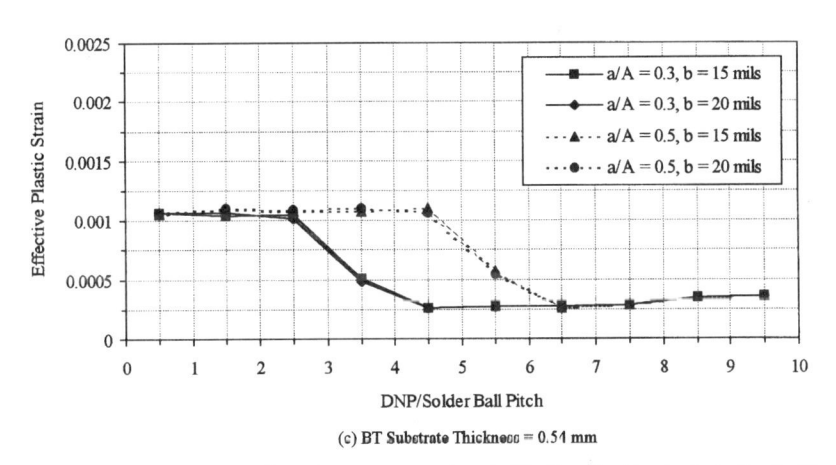

Figure 4.95 Distribution of maximum accumulated effective plastic strain in the (solder-bumped flip-chip) full-matrix PBGA.

magnitude of the shear strain raised inversely to the second power) to predict the thermal fatigue life of solder interconnects in area-array assemblies such as solder-bumped flip chip on ceramic substrate or on PCB, CBGA on PCB, PBGA on PCB, and CSP on PCB. However, the DNP concept always overestimated the value of the shear strain and thus underestimated the thermal fatigue life of the solder interconnects. Also, sometimes, the DNP concept leads to meaningless results. In this

304 Chapter Four

(a) 4-Row Perimeter FC-PBGA (256 balls), BT Substrate Thickness = 0.27 mm
(PCB is removed to avoid overlapping of deformation)

(b) 4-Row Perimeter FC-PBGA (256 balls), BT Substrate Thickness = 0.54 mm
(PCB is removed to avoid overlapping of deformation)

(c) 4-Row Perimeter FC-PBGA with Thermal Balls (272 balls), BT Substrate Thickness = 0.27 mm

(d) 4-Row Perimeter FC-PBGA with Thermal Balls (272 balls), BT Substrate Thickness = 0.54 mm

(e) FC-PBGA with Full Grid (400 balls), BT Substrate Thickness = 0.27 mm

(f) FC-PBGA with Full Grid (400 balls), BT Substrate Thickness = 0.54 mm

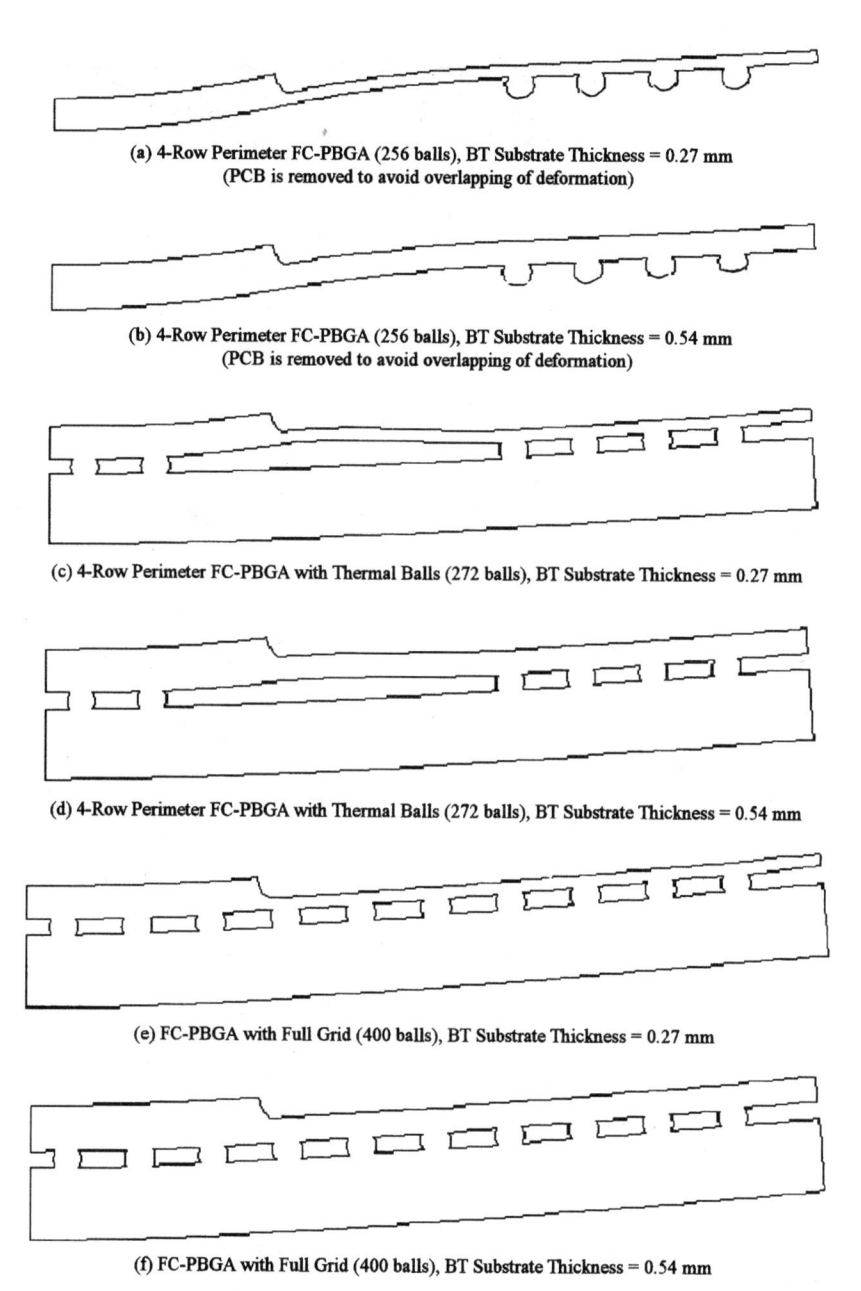

Figure 4.96 Deflection of solder-bumped flip-chip PBGA showing local bending of the BT substrate ($a/A = 0.3$). The displacement scale factor for all deformations shown is 20.

Mechanical Design/Analysis/Measurement/Reliability of Electronic Packaging 305

(a) 4-Row Perimeter FC-PBGA (256 balls), BT Substrate Thickness = 0.27 mm
(PCB is removed to avoid overlapping of deformation)

(b) 4-Row Perimeter FC-PBGA (256 balls), BT Substrate Thickness = 0.54 mm
(PCB is removed to avoid overlapping of deformation)

(c) 4-Row Perimeter FC-PBGA with Thermal Balls (272 balls), BT Substrate Thickness = 0.27 mm

(d) 4-Row Perimeter FC-PBGA with Thermal Balls (272 balls), BT Substrate Thickness = 0.54 mm

(e) FC-PBGA with Full Grid (400 balls), BT Substrate Thickness = 0.27 mm

(f) FC-PBGA with Full Grid (400 balls), BT Substrate Thickness = 0.54 mm

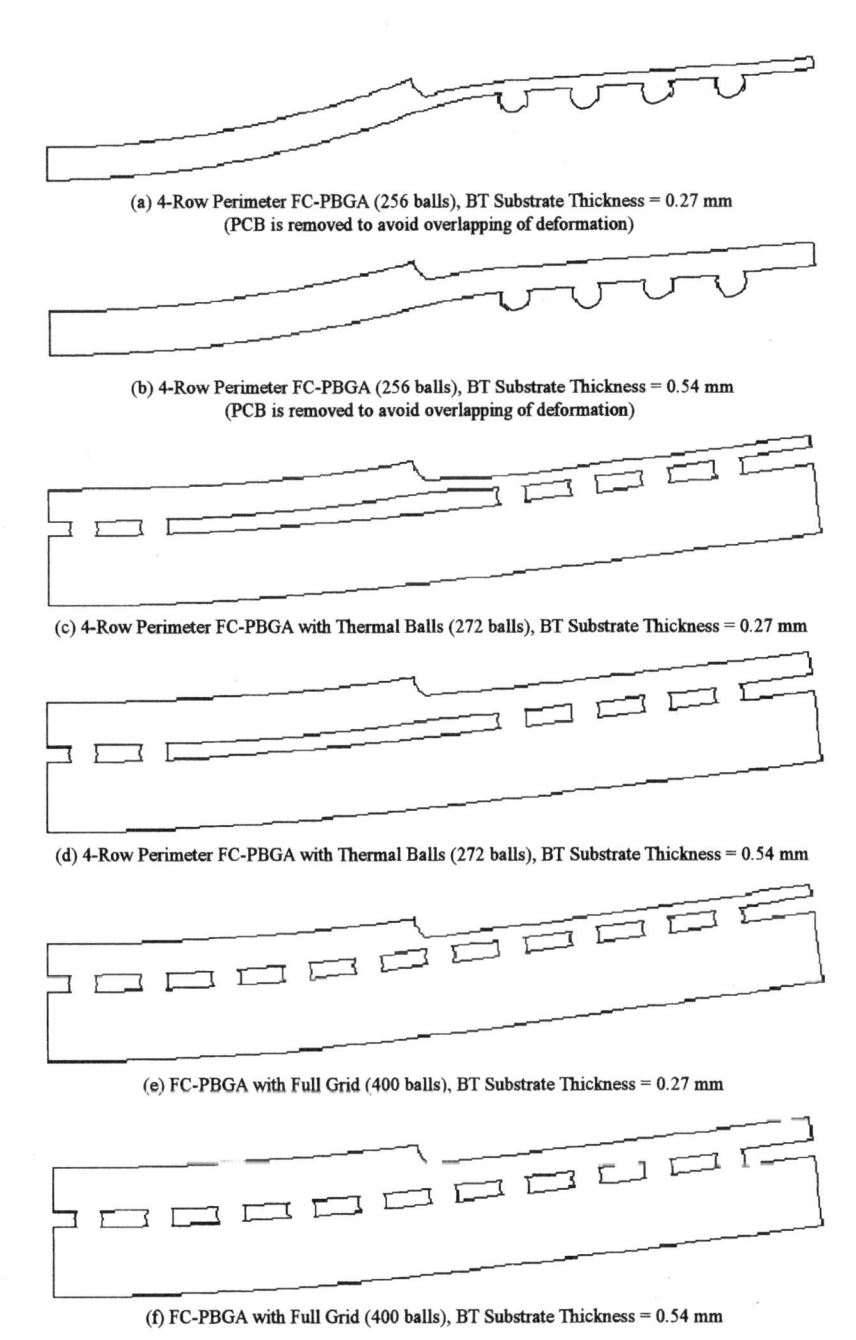

Figure 4.97 Deflection of solder-bumped flip-chip PBGA showing local bending of the BT substrate ($a/A = 0.5$). The displacement scale factor for all deformations shown is 20.

section, the limitations and roles of DNP in predicting the thermal fatigue life of area-array solder interconnects will be discussed through the following examples: (1) CSP on PCB, (2) solder-bumped flip chip on PCB, (3) CBGA on PCB, and (4) PBGA on PCB [123].

4.7.1 Example I—chip scale package on PCB

Figure 4.98 shows a CSP assembly. The chip was attached to the ceramic interposer (substrate) with gold bumps and underfill epoxy, and then soldered to the PCB. The chip size was $7 \times 7 \times 0.41$ mm with 100 peripherally distributed gold bumps on a 0.25-mm pitch. The dimensions of the gold bump were $0.1 \times 0.1 \times 0.025$ mm (Fig. 4.99). The ceramic interposer dimensions were $7.45 \times 7.45 \times 0.25$ mm (Fig. 4.100). There were 100 arrayed eutectic solder bumps (on a 0.65-mm pitch) at the bottom of the ceramic substrate. After the CSP was assembled on the PCB, the solder ball height was about 0.2 mm. The diameter of the pads on the ceramic and PCB substrates was about 0.3 mm.

Figure 4.101 shows the finite-element model (8896 2-D 8-node plane-strain elements and 27,217 nodes) for the analysis of one-half of the CSP assembly along the diagonal direction. The material properties of the silicon, gold, ceramic, FR-4, and underfill epoxy were assumed to be constant except the (63wt%Sn/37wt%Pb) eutectic solder, which was temperature dependent (Figure 4.102a and b for the stress-strain

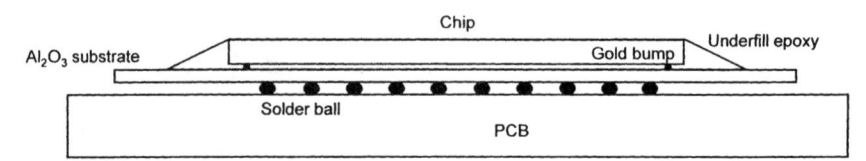

Figure 4.98 Schematic cross section of a chip scale package (CSP) assembly.

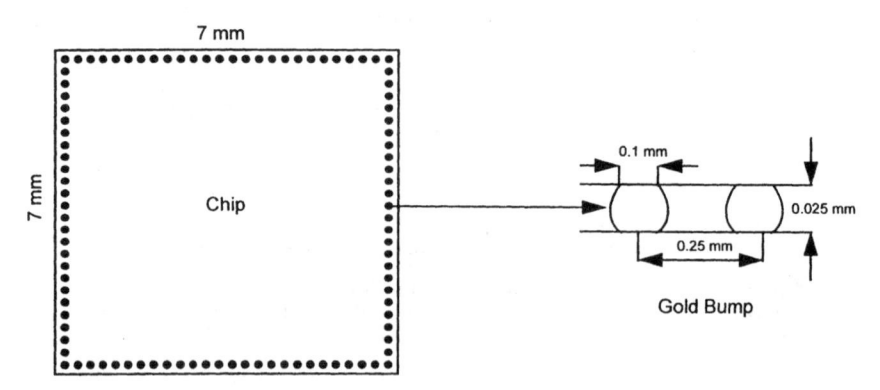

Figure 4.99 Chip size and gold bump pitch for the CSP.

Mechanical Design/Analysis/Measurement/Reliability of Electronic Packaging 307

Schematic cross section of A Chip Scale Package (CSP) assembly

$$\delta = d\Delta T\Delta\alpha$$
$$\gamma = \delta/h = d\Delta T\Delta\alpha/h$$

Ceramic substrate and eutectic solder balls

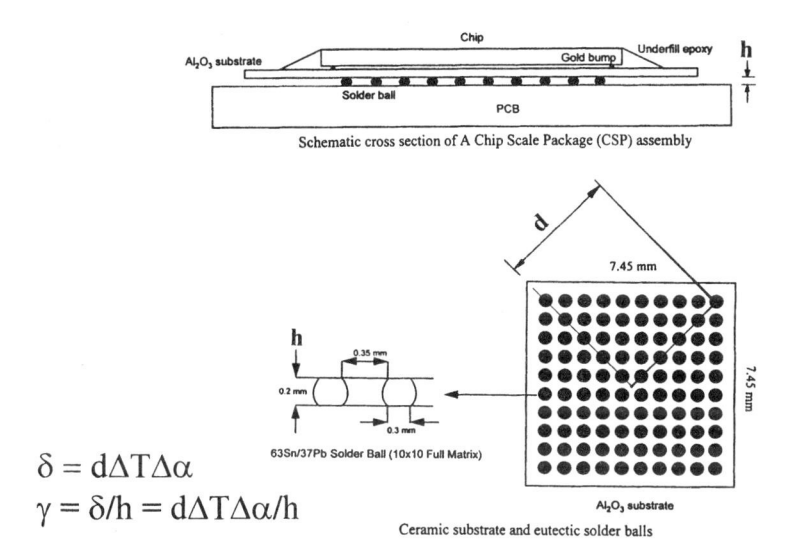

where δ is the relative displacement, d is the distance to neutral point, ΔT is the temperature change, $\Delta\alpha$ is the difference in TCE (thermal coefficient of expansion) between the substrate and the PCB, γ is the shear strain, and h is the standoff height of the solder joint.

Figure 4.100 Ceramic substrate and the 63Sn37Pb solder balls on PCB. Distance to neutral point (DNP) is defined.

curves and Fig. 4.103 for the Young's modulus). The Norton's steady-state creep relation (see, for example, Eq. 4.1 of [94]) has been used for the temperature-time-dependent elasto-plastic-creep analysis of the CSP solder joints. For 63Sn/37Pb eutectic solder the material constants of the Norton equation are given in Table 4.2 of [94].

The temperature loading imposed on the CSP assembly is shown in Fig. 4.104. It can be seen that for each cycle the temperature condition

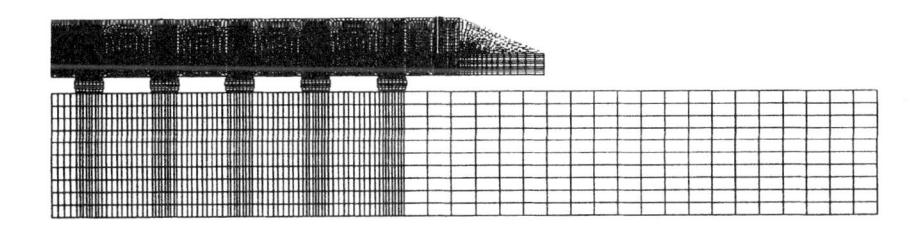

Figure 4.101 Finite element model for the nonlinear analysis.

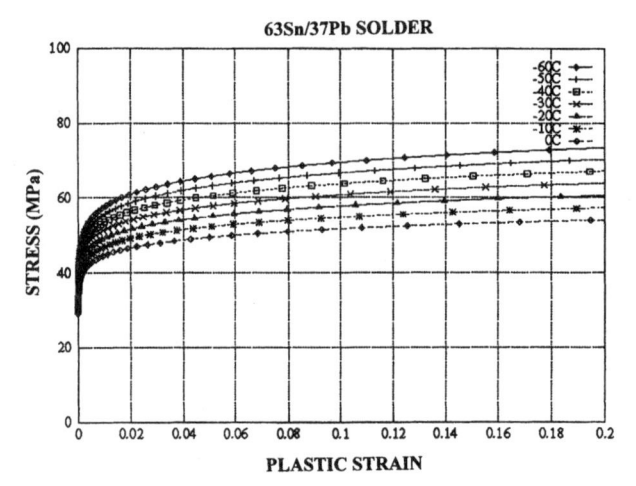

Figure 4.102 (a) Temperature-dependent stress-strain curves of 63Sn37Pb solder.

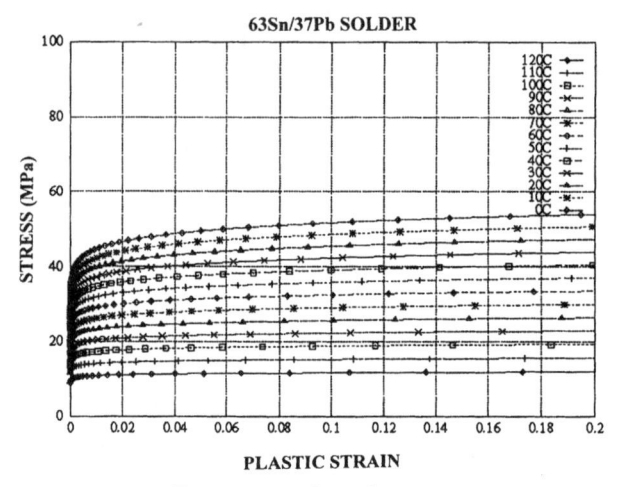

Figure 4.102 (b) Temperature-dependent stress-strain curves of 63Sn37Pb solder.

was between −40 and 125°C with 10 min ramp and 20 min hold (cycle time = 60 min). Two complete cycles were executed. The whole-field deformation is shown in Fig. 4.105. It can be seen that the maximum deflection (moving downward) is at the center of the PCB and is equal to 0.0144 mm. This is due to the thermal expansion mismatch between the ceramic substrate and FR-4 PCB and a temperature range of 165°C.

Figure 4.106 shows the solder ball shear-strain distribution (solid line) along the diagonal direction of the CSP assembly. It can be seen

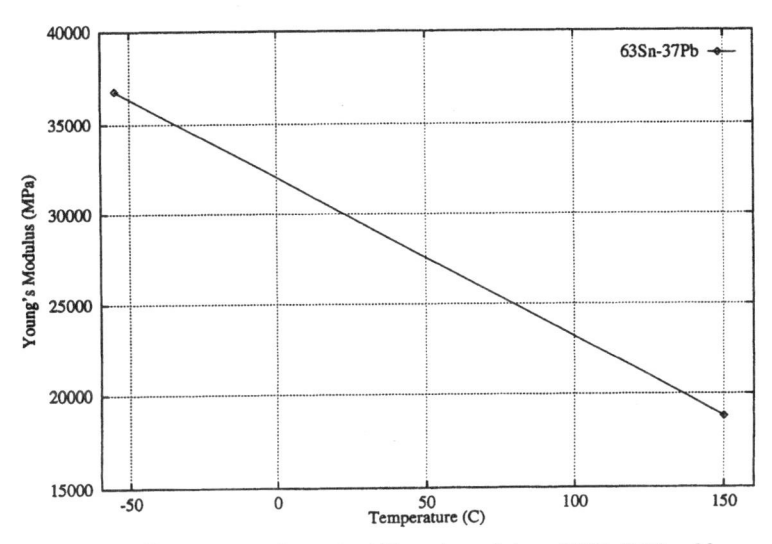

Mechanical Design/Analysis/Measurement/Reliability of Electronic Packaging 309

Figure 4.103 Temperature-dependent Young's modulus of 63Sn37Pb solder.

that the maximum shear strain is at the corner solder joint and is equal to 0.02. However, based on the concept of DNP, the calculated shear strain 4.136(165)(18 − 6.5)/0.2 = 0.0392 (dotted line), which is almost two times that calculated by the temperature-time-dependent nonlinear finite-element method. This is due, in part, to the fact that the DNP concept neglects the resistant contribution of all the interior solder balls and deflection of the whole assembly.

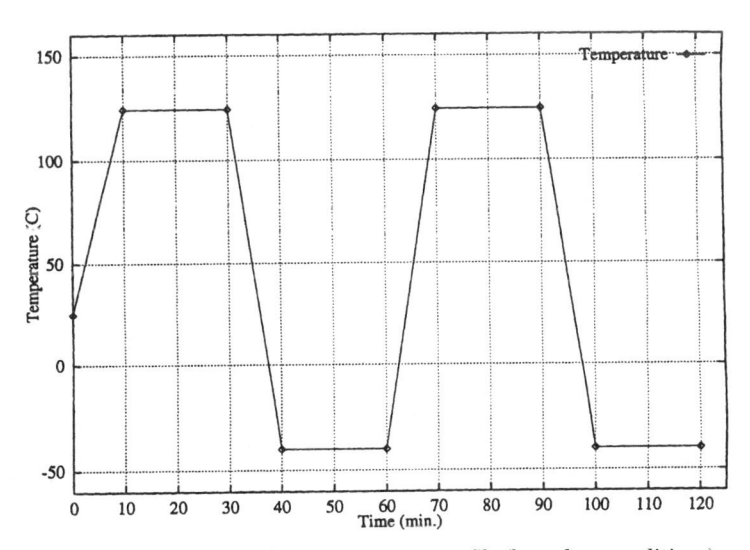

Figure 4.104 Thermal cycling temperature profile (boundary conditions).

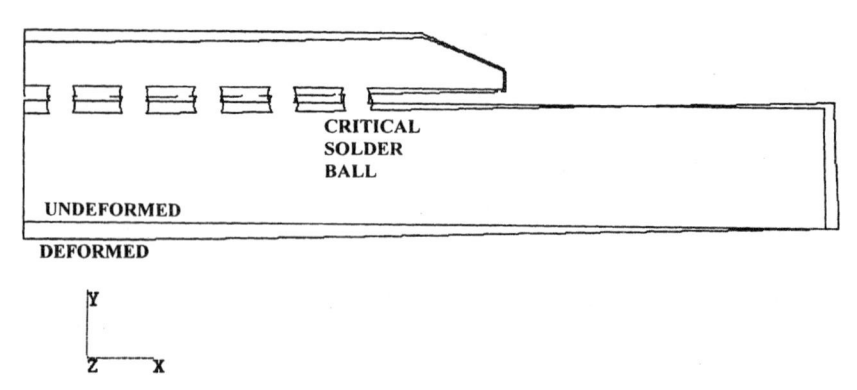

Figure 4.105 Deflection of CSP assembly from −40 to 125°C.

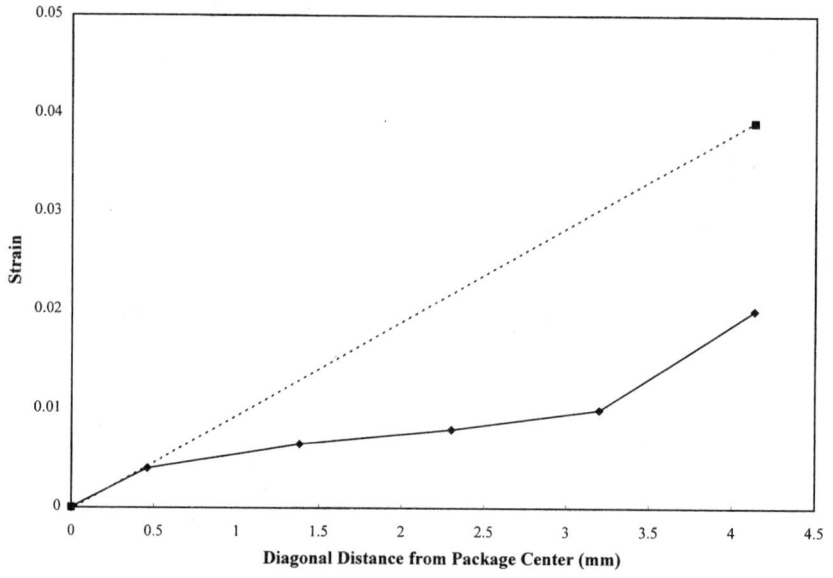

Figure 4.106 Solder ball shear strain distribution along the diagonal direction. Solid line obtained from nonlinear finite element method and dotted line from the concept of DNP.

4.7.2 Example II—solder-bumped flip chip on PCB (DCA)

Figure 4.107 shows schematically a 5wt%Sn/95wt%Pb solder-bumped flip chip on a 63wt%Sn/37wt%Pb solder-coated PCB with an underfill epoxy. The thermal coefficient of expansion of the silicon chip is 2.8 ppm/°C and of the PCB is 18 ppm/°C. The standoff height of the solder bumps is about 100 μm. For a 60°C temperature change, the U (relative displacement in the x-direction) and average shear

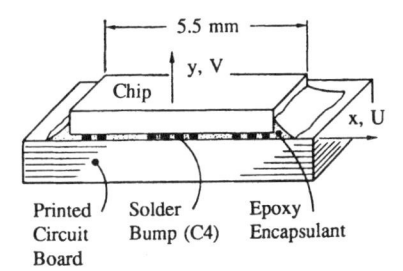

Mechanical Design/Analysis/Measurement/Reliability of Electronic Packaging 311

Figure 4.107 Schematic of a solder bumped flip chip on PCB.

strains of the whole cross section have been determined by a moire interferometry method [124], and is shown in Fig. 4.108 (solid lines). It can be seen that the chip (located at the rightmost or the leftmost solder joint) moved relative to the PCB in the x-direction about 1.2 µm and the corresponding shear strain is about 1.2 percent. However, based on the concept of DNP, the calculated relative displacement (dotted line) in the x-direction, $(2.75)(60)(18 - 2.8) = 2.5$ µm, and the shear strain, $2.5/100 = 2.5$ percent (dotted line), which are more than two times that measured by the moire interferometry method. Again, this is due, in part, to the fact that the DNP concept neglects the resistant contribution of all the interior solder bumps and deflection of the whole assembly.

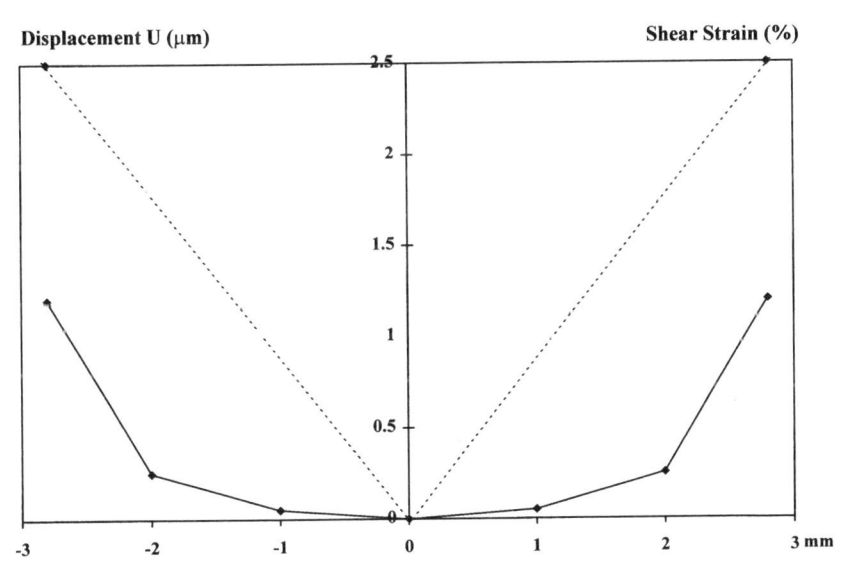

Figure 4.108 Distribution of relative displacement (µm) and shear strain (%) in the horizontal direction of the solder bumps. Solid lines obtained from measurement results, and dotted lines from the concept of DNP.

4.7.3 Example III—ceramic ball grid array (CBGA) on PCB

Figure 4.109 shows the cross section (specimen) and the moire interferometry measurement results of a 25-mm-square CBGA assembly reported by [125]. It has 19 solder balls in the xy-plane and the DNP of the rightmost or leftmost solder is 11.4 mm. The thermal coefficient of expansion of the ceramic is 6 ppm/°C and of the PCB is 18.5 ppm/°C. The CBGA assembly is subjected to a temperature change of 60°C. It can be seen that the maximum relative horizontal displacement in the x-direction is 4.75 µm and the maximum nominal shear strain is 0.17 percent. However, based on the concept of DNP, the calculated maximum relative horizontal displacement in the x-direction of the CBGA assembly is 11.4(60)(18.5 − 6) = 8.55 µm, which is almost twice that obtained by measurement.

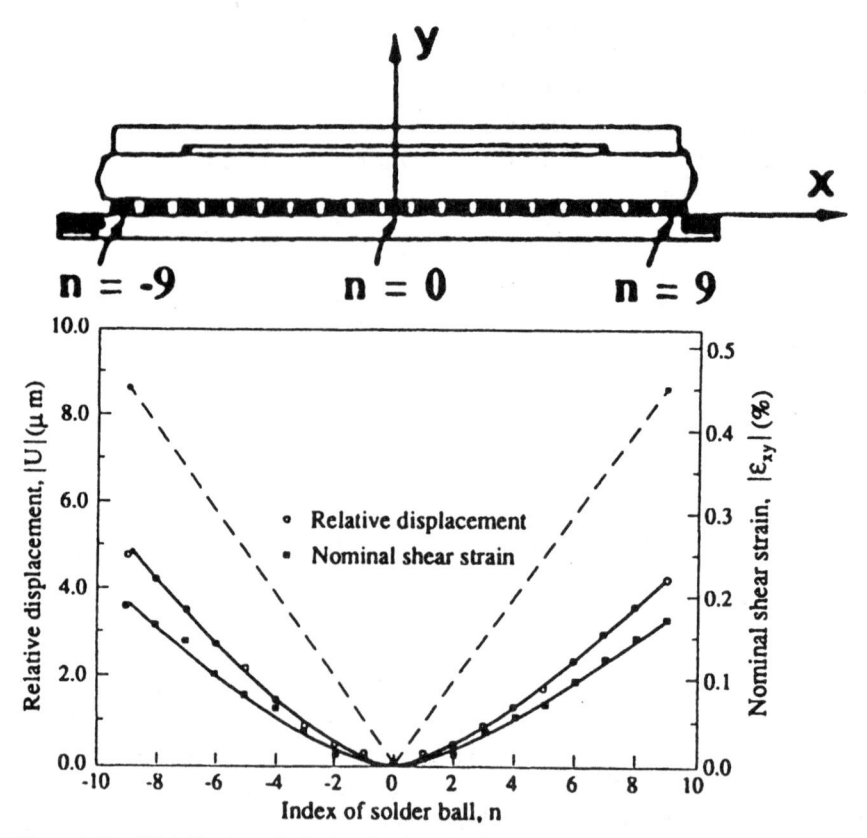

Figure 4.109 Distribution of relative displacement (µm) and shear strain (%) in the horizontal direction of the solder balls in a CBGA assembly. Solid lines obtained from measurement results, and dotted line (for displacement) from the concept of DNP.

4.7.4 Example IV—plastic ball grid array (PBGA) on PCB

Figure 4.110a and b, respectively, shows the cross section of a 225-pin (1.5-mm pitch) full-matrix PGBA and the cross section of a 256-pin (1.27-mm pitch) perimeter PBGA in the diagonal direction. The thermal coefficient of expansion of the silicon chip is 2.5 ppm/°C, of the BT (bismaleimide triazine) substrate is 15 ppm/°C, and of the FR-4 PCB is 18.5 ppm/°C. The solder is assumed to be a temperature-dependent material (Fig. 4.102a and b) and the temperature loading is shown in Fig. 4.104.

The effective plastic strain is shown in Fig. 4.111 for both the full-matrix PBGA assembly and perimeter PBGA assembly along the diagonal direction [126]. It can be seen that, in both cases, the maximum strain does not occur at the corner solder joint because the global thermal expansion mismatch between the BT substrate and FR-4 PCB is not very large. The maximum strain, in both cases, occurred at the solder joints underneath and near the chip corners due to the very large local thermal expansion mismatch between the silicon chip, BT substrate, and the FR-4 PCB. In these cases, the concept of DNP will lead to meaningless results.

In summary, the concept of DNP is a very simple way to calculate the shear strain in the solder interconnects due to the thermal expansion mismatch between chip and substrate or PCB, or substrate and PCB. Since DNP neglects the resistant contribution of all the interior solder interconnects and deflection of the whole assembly, it always overestimates the relative horizontal displacement and shear strain. Consequently, DNP always underestimates the thermal fatigue life of area-array solder joints.

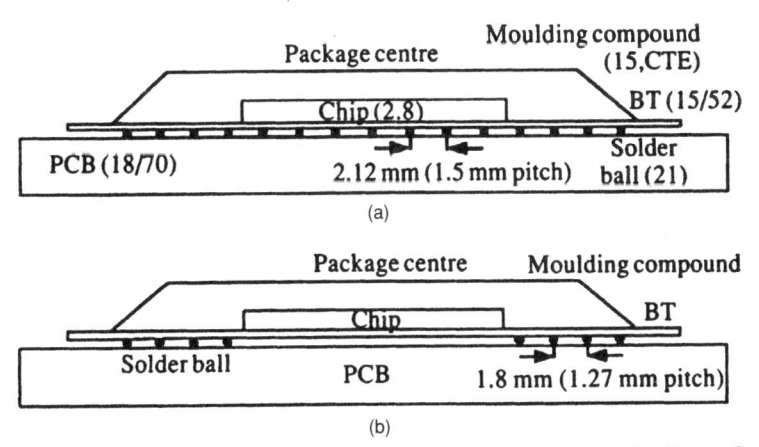

Figure 4.110 (a) Schematic of a full-matrix PBGA assembly in the diagonal direction and (b) schematic of a perimeter PBGA assembly in the diagonal direction.

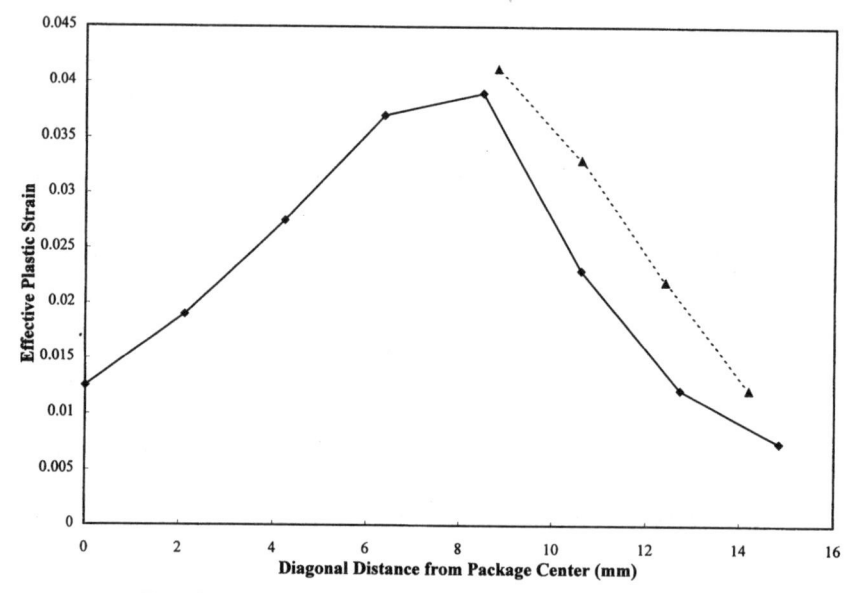

Figure 4.111 Distribution of shear strain (%) in the horizontal direction of the solder balls in a full-matrix PBGA assembly (solid line), and in a perimeter PBGA assembly (dotted line).

If the thermal fatigue life predicted by the DNP method meets the reliability requirements, then the job is done! However, if the conservative thermal fatigue life predicted by the DNP method does not meet the reliability requirements—and thus, proclaim the solder interconnects of that particular package are not reliable and abandon that package—then, it is a shame! In that case, a more precise method such as the nonlinear finite-element analysis should be used to calculate the shear strain with all the solder interconnects presented.

4.8 Mechanical Vibration of Electronic Packaging

Vibration analysis of electronic packaging consists of four major steps. First, the package of interest is identified, its boundary conditions are estimated, and its interfaces with other packages are understood. Second, the natural frequencies and mode shapes of the package are determined by analysis or direct experimental measurement. Third, the time-history of the loads on the package are estimated. Fourth, these loads are applied to an analytical model of the package to determine its response. The first two steps are called *free vibration analysis*. The last two steps are called *forced vibration analysis* and are usually executed by modal superposition, direct integration, or frequency domain methods [127–132].

Free vibration analysis is the most crucial step in package dynamics. Many vibration problems can be minimized by avoiding coincident resonances that can magnify the dynamic loads and stresses in the package.

All problems of free vibration of undamped linear systems are eigenvalue problems (see [127–132] for examples). In most of continuum mechanics the problem is to determine the natural mode of vibration for which all points in the continuum perform simple harmonic oscillations that are in phase with each other and that have the same frequency. Accordingly, all points in the continuum obtain their maximum velocities simultaneously as they pass through their neutral positions, and they all reach their extreme displacement simultaneously. A frequency for which natural vibration is possible is an eigenvalue of the boundary-value problem and is called a *natural frequency*. The number of the natural frequencies is equal to the number of degrees of freedom of the continuum. Theoretically speaking, it is infinite. In this section, we first present the free vibration of packages with heat sinks and then PCB with different supporting conditions.

4.8.1 Horizontal vibration of packages with heat sinks

Figure 4.112 shows a CBGA with solder columns and a heat sink. For high performance applications, usually the heat sink is very large and heavy. The horizontal dynamic forces introduced by the weight of the heat sink and ceramic carrier may crack the solder joints on the PCB. Figure 4.113 shows the model for determining the natural frequencies of the packaging system.

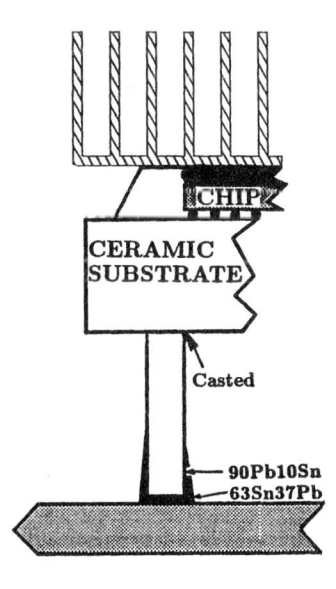

Figure 4.112 IBM's ceramic column grid array with heat sink.

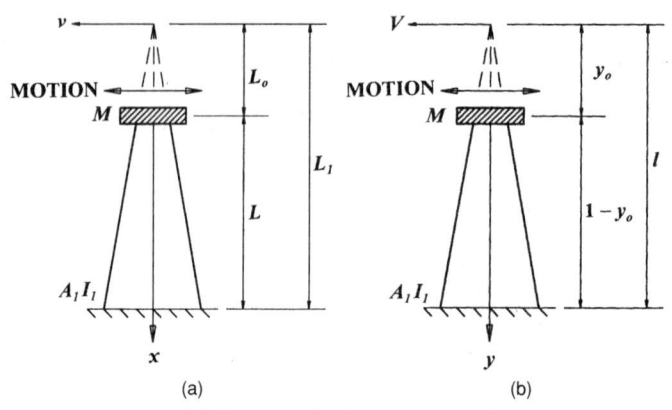

Figure 4.113 Simplify horizontal vibration model for electronic packaging with heat sink.

The partial differential equation governing the motion for small-amplitude, free transverse vibrations of the solder column shown in Figs. 4.112 and 4.113 is

$$[EI(x)v'']'' + m(x)\ddot{v} = 0 \qquad (4.223)$$

where E is Young's modulus, I is the moment of inertia, v is the deflection, m is the mass per unit length of the column, $(') = \partial/\partial x$ and $(\cdot) = \partial/\partial t$.

The boundary conditions are as follows:

At $x = L_0$

$$v'' = 0; \qquad (EI(x)v'')' = -M\ddot{v} \qquad (4.224)$$

At $x = L_1$

$$v = 0; \qquad v' = 0 \qquad (4.225)$$

where M is the concentrated mass (heat sink and ceramic carrier).

Using the standard method of separation of variables one assumes

$$v(x, t) = V(x)T(t) \qquad (4.226)$$

then Eq. 4.223 leads to two ordinary differential equations,

$$\ddot{T} + p^2 T = 0 \qquad (4.227)$$

and

$$[EI(x)V'']'' - m(x)p^2 V = 0 \qquad (4.228)$$

The boundary conditions become:

At $x = L_0$

$$V'' = 0; \qquad [EI(x)V'']' = p^2 MV \qquad (4.229)$$

At $x = L_1$

$$V = 0; \qquad V' = 0 \qquad (4.230)$$

where p is the circular frequency of vibration.

For the structural system under consideration (Fig. 4.113)

$$A(x) = \left(\frac{x}{L_1}\right)^2 A_1$$

$$I(x) = \left(\frac{x}{L_1}\right)^4 I_1 \qquad (4.231)$$

where A_1 and I_1 are, respectively, the cross-sectional area and moment of inertia at $x = L_1$. It is convenient to introduce a new variable y defined by (Fig. 4.113b)

$$y = \frac{x}{L_1} \qquad (4.232)$$

Then Eqs. 4.228–4.230 become

$$\frac{1}{y^2}\frac{d^2}{dy^2}\left(y^4 \frac{d^2 V}{dy^2}\right) - q^4 V = 0 \qquad (4.233)$$

At $y = y_0 = L_0/L_1$

$$\frac{d^2 V}{dy^2} = 0; \qquad \frac{d}{dy}\left(y^4 \frac{d^2 V}{dy^2}\right) = \overline{M} q^4 (1 - y_0) V \qquad (4.234)$$

At $y = 1$

$$V = 0; \qquad \frac{dV}{dy} = 0 \qquad (4.235)$$

where

$$q^4 = \rho A_1 L_1^4 p^2 / EI_1 \qquad (4.236)$$

and

$$\overline{M} = M/\rho A_1 L \qquad (4.237)$$

ρ is the mass density of the beam.

The general solution to Eq. 4.233 is

$$V = y^{-1}[C_1 J_2(z) + C_2 Y_2(z) + C_3 I_2(z) + C_4 K_2(z)] \quad (4.238)$$

where

$$z = 2q\sqrt{y} \quad (4.239)$$

In Eq. 4.238, J_2 and Y_2 are the second-order Bessel functions of first and second kind [133], respectively; while I_2 and K_2 are the modified Bessel functions of second order and of first and second kind, respectively.

Substituting Eq. 4.238 in Eqs. 4.234 and 4.235 results in the frequency determinant equation

$$\begin{vmatrix} J_4(a) & Y_4(a) & I_4(a) & K_4(a) \\ a_{21} & a_{22} & a_{23} & a_{24} \\ J_2(b) & Y_2(b) & I_2(b) & K_2(b) \\ J_3(b) & Y_3(b) & -I_3(b) & K_3(b) \end{vmatrix} = 0 \quad (4.240)$$

where

$$\begin{aligned} a_{21} &= J_3(a) - \overline{M}q(1-y_0)y_0^{-2.5} J_2(a) \\ a_{22} &= Y_3(a) - \overline{M}q(1-y_0)y_0^{-2.5} Y_2(a) \\ a_{23} &= I_3(a) - \overline{M}q(1-y_0)y_0^{-2.5} I_2(a) \\ a_{24} &= -K_3(a) - \overline{M}q(1-y_0)y_0^{-2.5} K_2(a) \\ a &= 2q\sqrt{y_0} \\ b &= 2q \end{aligned} \quad (4.241)$$

Equation 4.240 degenerates to the clamped-free case of [131, 132], if $M = 0$.

To find the roots of Eq. 4.240, the Bessel functions were replaced by their approximate polynomial equivalents [134], and the roots were then found by using the method of successive bisection with $\epsilon = 10^{-7}$ [135]. In order to initiate the iterative process, the roots were first bracketed by means of a straight search process [127, 129].

Once the roots have been obtained, the frequency of the structural system can be determined by Eq. 4.236 as

$$f = \frac{Q^2}{2\pi L^2} \sqrt{\frac{EI_1}{\rho A_1}} \quad (4.242)$$

where the frequency parameter $Q = q(1 - y_0)$ for a wide range of values of beam dimensions and concentrated mass are shown in Tables 4.16 through 4.25.

Mechanical Design/Analysis/Measurement/Reliability of Electronic Packaging 319

TABLE 4.16 Values of Q for $\overline{M} = 0$

y_0	Mode				
	1	2	3	4	5
0.1	2.68419	4.32206	6.09293	7.96900	9.90786
0.2	2.48926	4.28783	6.31139	8.44048	10.62205
0.3	2.34718	4.31754	6.54297	8.86120	11.22275
0.4	2.23809	4.36633	6.76301	9.23817	11.74885
0.5	2.15062	4.42127	6.96986	9.58190	12.22252
0.6	2.07817	4.47772	7.16482	9.89975	12.65692
0.7	2.01666	4.53382	7.34950	10.19682	13.06052
0.8	1.96345	4.58876	7.52531	10.47678	13.43915
0.9	1.91669	4.64222	7.69341	10.74233	13.79699
1.0	1.87510	4.69409	7.85476	10.99554	14.13717

TABLE 4.17 Values of Q for $\overline{M} = 0.2$

y_0	Mode				
	1	2	3	4	5
0.1	1.10212	3.15637	5.11296	7.10001	9.10876
0.2	1.29695	3.31638	5.45913	7.65757	9.88171
0.3	1.41336	3.46089	5.75049	8.11207	10.50084
0.4	1.48992	3.59643	6.01266	8.51076	11.03690
0.5	1.54127	3.72491	6.25635	8.87359	11.51935
0.6	1.57524	3.84671	6.48690	9.21107	11.96382
0.7	1.59681	3.96180	6.70713	9.52936	12.37965
0.8	1.60936	4.07015	6.91860	9.83234	12.77289
0.9	1.61530	4.17182	7.12218	10.12251	13.14762
1.0	1.61640	4.26706	7.31837	10.40156	13.50670

TABLE 4.18 Values of Q for $\overline{M} = 0.4$

y_0	Mode				
	1	2	3	4	5
0.1	0.92869	3.15207	5.11131	7.09909	9.10815
0.2	1.09856	3.30136	5.45278	7.65382	9.87911
0.3	1.20610	3.43098	5.73653	8.10341	10.49466
0.4	1.28282	3.54948	5.98853	8.49516	11.02548
0.5	1.34001	3.66057	6.22005	8.84917	11.50105
0.6	1.38332	3.76604	6.43711	9.17624	11.93714
0.7	1.41618	3.86680	6.64329	9.48294	12.34332
0.8	1.44093	3.96331	6.84085	9.77360	12.72590
0.9	1.45924	4.05581	7.03125	10.05118	13.08929
1.0	1.47241	4.14443	7.21549	10.31781	13.43668

TABLE 4.19 Values of Q for $\overline{M} = 0.6$

| | \multicolumn{5}{c}{Mode} |
y_0	1	2	3	4	5
0.1	0.83974	3.15064	5.11076	7.09879	9.10795
0.2	0.99509	3.29631	5.45065	7.65256	9.87824
0.3	1.09534	3.42067	5.73179	8.10049	10.49259
0.4	1.16882	3.53275	5.98021	8.48985	11.02160
0.5	1.22551	3.63669	6.20725	8.84072	11.49478
0.6	1.27031	3.73472	6.41907	9.16396	11.92787
0.7	1.30608	3.82810	6.61942	9.46622	12.33047
0.8	1.33471	3.91761	6.81076	9.75189	12.70895
0.9	1.35757	4.00369	6.99473	10.02406	13.06777
1.0	1.37567	4.08665	7.17252	10.28498	13.41021

TABLE 4.20 Values of Q for $\overline{M} = 0.8$

| | \multicolumn{5}{c}{Mode} |
y_0	1	2	3	4	5
0.1	0.78174	3.14993	5.11048	7.09864	9.10785
0.2	0.92717	3.29377	5.44958	7.65193	9.87781
0.3	1.02193	3.41545	5.72941	8.09903	10.49154
0.4	1.09233	3.52417	5.97599	8.48716	11.01966
0.5	1.14759	3.62425	6.20071	8.83643	11.49161
0.6	1.19217	3.71808	6.40976	9.15769	11.92316
0.7	1.22867	3.80712	6.60695	9.45760	12.32390
0.8	1.25876	3.89227	6.79480	9.74059	12.70022
0.9	1.28360	3.97412	6.97506	10.00979	13.05659
1.0	1.30409	4.05308	7.14898	10.26749	13.39631

TABLE 4.21 Values of Q for $\overline{M} = 1.0$

| | \multicolumn{5}{c}{Mode} |
y_0	1	2	3	4	5
0.1	0.73947	3.14950	5.11032	7.09854	9.10779
0.2	0.87751	3.29225	5.44894	7.65155	9.87755
0.3	0.96797	3.41230	5.72798	8.09815	10.49092
0.4	1.03572	3.51895	5.97345	8.48555	11.01848
0.5	1.08945	3.61661	6.19674	8.83384	11.48970
0.6	1.13335	3.70777	6.40407	9.15389	11.92031
0.7	1.16983	3.79396	6.59928	9.45234	12.31991
0.8	1.20042	3.87617	6.78492	9.73366	12.69489
0.9	1.22619	3.95507	6.96277	10.00099	13.04974
1.0	1.24792	4.03114	7.13413	10.25662	13.38776

TABLE 4.22 Values of Q for $\overline{M} = 3.0$

y_0	Mode 1	2	3	4	5
0.1	0.56219	3.14836	5.10988	7.09830	9.10763
0.2	0.66808	3.28818	5.44723	7.65054	9.87685
0.3	0.73854	3.40382	5.72413	8.09579	10.48924
0.4	0.79247	3.50478	5.96659	8.48121	11.01534
0.5	0.83643	3.59561	6.18599	8.82686	11.48456
0.6	0.87359	3.67898	6.38855	9.14358	11.91262
0.7	0.90570	3.75657	6.57814	9.43802	12.30910
0.8	0.93389	3.82956	6.75736	9.71464	12.68037
0.9	0.95889	3.89876	6.92806	9.97659	13.03093
1.0	0.98123	3.96482	7.09160	10.22621	13.36409

TABLE 4.23 Values of Q for $\overline{M} = 5.0$

y_0	Mode 1	2	3	4	5
0.1	0.49484	3.14813	5.10979	7.09825	9.10760
0.2	0.58822	3.28736	5.44688	7.65034	9.87671
0.3	0.65054	3.40212	5.72336	8.09532	10.48891
0.4	0.69844	3.50190	5.96521	8.48034	11.01471
0.5	0.73771	3.59130	6.18381	8.82546	11.48353
0.6	0.77113	3.67299	6.38538	9.14149	11.91107
0.7	0.80024	3.74867	6.57379	9.43510	12.30690
0.8	0.82603	3.81953	6.75163	9.71073	12.67741
0.9	0.84913	3.88642	6.92076	9.97155	13.02707
1.0	0.87002	3.94998	7.08254	10.21986	13.35920

TABLE 4.24 Values of Q for $\overline{M} = 7.0$

y_0	Mode 1	2	3	4	5
0.1	0.45494	3.14804	5.10975	7.09823	9.10758
0.2	0.54085	3.28702	5.44673	7.65025	9.87665
0.3	0.59827	3.40138	5.72303	8.09511	10.48876
0.4	0.64248	3.50066	5.96462	8.47997	11.01444
0.5	0.67882	3.58944	6.18287	8.82485	11.48308
0.6	0.70982	3.67040	6.38401	9.14059	11.91040
0.7	0.73693	3.74524	6.57191	9.43384	12.30596
0.8	0.76103	3.81516	6.74915	9.70905	12.67614
0.9	0.78272	3.88101	6.91759	9.96937	13.02541
1.0	0.80243	3.94344	7.07860	10.21712	13.35709

TABLE 4.25 Values of Q for $\overline{M} = 9.0$

y_0	Mode				
	1	2	3	4	5
0.1	0.42725	3.14798	5.10973	7.09822	9.10757
0.2	0.50797	3.28682	5.44665	7.65021	9.87662
0.3	0.56195	3.40098	5.72285	8.09500	10.48868
0.4	0.60356	3.49997	5.96429	8.47976	11.01429
0.5	0.63780	3.58840	6.18235	8.82451	11.48284
0.6	0.66707	3.66895	6.38325	9.14009	11.91003
0.7	0.69270	3.74332	6.57086	9.43314	12.30544
0.8	0.71555	3.81271	6.74777	9.70811	12.67543
0.9	0.73616	3.87797	6.91582	9.96815	13.02448
1.0	0.75494	3.93976	7.07639	10.21558	13.35591

The reliability of the present results have been verified by checking the clamped-free case, $\overline{M} = 0$, with [131, 132]; and the limiting case, $y_0 = 1$, with [136]. The last case corresponds to a uniform cantilever beam with end mass.

4.8.2 Vertical vibration of PCBs

The vertical (out-of-plane) vibration of rectangular PCBs can be determined by considering the potential energy of bending of a flat plate (Sec. 4.2.2 and Fig. 4.2),

$$U = \iiint dU = \frac{D}{2} \iint \left\{ \left(\frac{\partial^2 w}{\partial x^2} \right)^2 + \left(\frac{\partial^2 w}{\partial y^2} \right)^2 + 2\nu \frac{\partial^2 w}{\partial x^2} \frac{\partial^2 w}{\partial y^2} + 2(1-\nu) \left(\frac{\partial^2 w}{\partial x \partial y} \right)^2 \right\} dx\, dy \quad (4.243)$$

The kinetic energy of a transversely vibrating plate is

$$T = \frac{\rho h}{2} \iint \dot{w}^2\, dx\, dy \quad (4.244)$$

where the flexural rigidity $D = Eh/[12(1-\nu^2)]$ is given by Eq. 4.38, h is the thickness of the PCB, E is the Young's modulus, ν is the Poisson's ratio, and ρh is the mass per unit area of the PCB.

For a simply supported PCB at all four edges, the deflection (Fig. 4.114)

$$w = \sum_{m=1}^{\infty} \sum_{n=1}^{\infty} \phi_{mn} \sin \frac{m\pi x}{a} \sin \frac{n\pi y}{b} \quad (4.245)$$

will satisfy the boundary conditions:

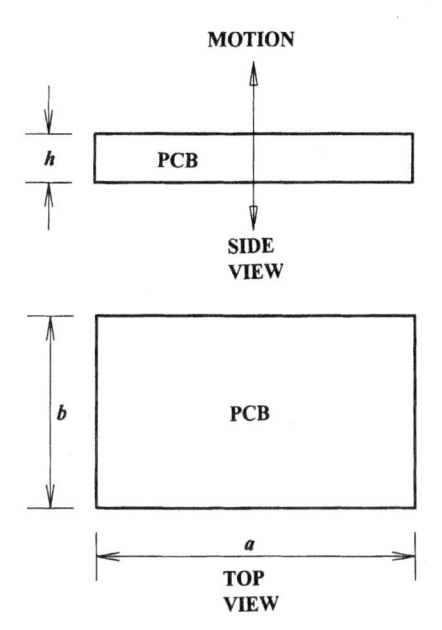

Mechanical Design/Analysis/Measurement/Reliability of Electronic Packaging 323

Figure 4.114 Vertical (out-of-plane) vibration of PCB (plate).

$$w = \partial^2 w/\partial x^2 = 0 \quad \text{at } x = 0 \text{ and } x = a$$
$$w = \partial^2 w/\partial y^2 = 0 \quad \text{at } y = 0 \text{ and } y = b \quad (4.246)$$

Substituting Eq. 4.245 into Eqs. 4.243 and 4.244, we have

$$U = \frac{\pi^4 ab}{8} D \sum_{m=1}^{\infty} \sum_{n=1}^{\infty} \phi_{mn}^2 \left(\frac{m^2}{a^2} + \frac{n^2}{b^2} \right)^2 \quad (4.247)$$

$$T = \frac{\rho h a b}{8} \sum_{m=1}^{\infty} \sum_{n=1}^{\infty} \dot{\phi}_{mn}^2 \quad (4.248)$$

The inertial force for a typical element of the PCB is $-\rho h \ddot{w} dx dy$ and considering a virtual displacement

$$\delta w_{mn} = \delta \phi_{mn} \sin \frac{m \pi x}{a} \sin \frac{n \pi y}{b} \quad (4.249)$$

Then we have the equation of motion of the plate in principal coordinates

$$\rho h \ddot{\phi}_{mn} + \pi^4 D \phi_{mn} \left(\frac{m^2}{a^2} + \frac{n^2}{b^2} \right) = 0 \quad (4.250)$$

The solution of Eq. 4.250 is

$$\phi_{mn} = c_1 \cos \omega t + c_2 \sin \omega t$$

where

$$\omega = \pi^2 \sqrt{\frac{D}{\rho h}} \left(\frac{m^2}{a^2} + \frac{n^2}{b^2} \right) \qquad (4.251)$$

and the frequencies of the simply supported PCB are

$$f_{mn} = \frac{\omega}{2\pi} = \frac{\pi}{2} \sqrt{\frac{D}{\rho h}} \left(\frac{m^2}{a^2} + \frac{n^2}{b^2} \right) \qquad (4.252)$$

When $m = n = 1$, then we have the first (fundamental) natural frequency

$$f = \frac{\pi}{2} \sqrt{\frac{D}{\rho h}} \left(\frac{1}{a^2} + \frac{1}{b^2} \right) \qquad (4.253)$$

For other boundary conditions of PCB, we can obtain the frequency by similar procedures or by Rayleigh-Ritz's approximate method. Some useful results for the fundamental natural frequency of PCB with various boundary conditions are summarized in the following (see Fig. 4.114 for notations).

1. Two long edges (a) free and two short edges (b) free:

$$f = \frac{\pi}{2} \left[\frac{D}{\rho h} \left(\frac{2.08}{a^2 b^2} \right) \right]^{1/2} \qquad (4.254)$$

2. Two long edges (a) free, one short edge (b) fixed, and one short edge (b) free:

$$f = \frac{0.56}{a^2} \left(\frac{D}{\rho h} \right)^{1/2} \qquad (4.255)$$

3. Two long edges (a) free and two short edges (b) fixed:

$$f = \frac{3.55}{a^2} \left(\frac{D}{\rho h} \right)^{1/2} \qquad (4.256)$$

4. Two long edges (a) free, one short edge (b) fixed, and one short edge (b) supported:

$$f = \frac{0.78\pi}{a^2} \left(\frac{D}{\rho h} \right)^{1/2} \qquad (4.257)$$

5. Two long edges (a) free and two short edges (b) supported:

$$f = \frac{\pi}{2a^2} \left(\frac{D}{\rho h} \right)^{1/2} \qquad (4.258)$$

Mechanical Design/Analysis/Measurement/Reliability of Electronic Packaging 325

6. One long edge (a) free, one long edge (a) supported, one short edge (b) free, and one short edge (b) supported:

$$f = \frac{\pi}{11} \left(\frac{D}{\rho h}\right)^{1/2} \left[\frac{1}{a^2} + \frac{1}{b^2}\right] \qquad (4.259)$$

7. Two long edges (a) supported, one short edge (b) free, and one short edge (b) supported:

$$f = \frac{\pi}{2} \left(\frac{D}{\rho h}\right)^{1/2} \left[\frac{1}{4a^2} + \frac{1}{b^2}\right] \qquad (4.260)$$

8. One long edge (a) free, one long edge (a) supported, one short edge (b) free, and one short edge (b) fixed:

$$f = \frac{\pi}{2} \left[\frac{D}{\rho h} \left(\frac{0.127}{a^4} + \frac{0.2}{a^2 b^2}\right)\right]^{1/2} \qquad (4.261)$$

9. One long edge (a) free, one long edge (a) fixed, one short edge (b) free, and one short edge (b) fixed:

$$f = \frac{\pi}{5.42} \left[\frac{D}{\rho h} \left(\frac{1}{a^4} + \frac{3.2}{a^2 b^2} + \frac{1}{b^4}\right)\right]^{1/2} \qquad (4.262)$$

10. Two long edges (a) fixed, one short edge (b) free, and one short edge (b) fixed:

$$f = \frac{\pi}{3} \left[\frac{D}{\rho h} \left(\frac{0.75}{a^4} + \frac{2}{a^2 b^2} + \frac{12}{b^4}\right)\right]^{1/2} \qquad (4.263)$$

11. Two long edges (a) fixed and two short edges (b) fixed

$$f = \frac{\pi}{1.5} \left[\frac{D}{\rho h} \left(\frac{3}{a^4} + \frac{2}{a^2 b^2} + \frac{3}{b^4}\right)\right]^{1/2} \qquad (4.264)$$

12. Two long edges (a) supported and two short edges (b) fixed

$$f = \frac{\pi}{3.46} \left[\frac{D}{\rho h} \left(\frac{16}{a^4} + \frac{8}{a^2 b^2} + \frac{3}{b^4}\right)\right]^{1/2} \qquad (4.265)$$

13. One long edge (a) free, one long edge (a) supported, and two short edges (b) fixed:

$$f = \frac{\pi}{1.74} \left[\frac{D}{\rho h} \left(\frac{4}{a^4} + \frac{1}{2a^2 b^2} + \frac{1}{64 b^4}\right)\right]^{1/2} \qquad (4.266)$$

14. Two long edges (*a*) fixed, one short edge (*b*) fixed, and one short edge (*b*) supported:

$$f = \frac{\pi}{2}\left[\frac{D}{\rho h}\left(\frac{2.45}{a^4} + \frac{2.90}{a^2 b^2} + \frac{5.13}{b^4}\right)\right]^{1/2} \quad (4.267)$$

15. One long edge (*a*) fixed, one long edge (*a*) supported, one short edge (*b*) fixed, and one short edge (*b*) free:

$$f = \frac{\pi}{2}\left[\frac{D}{\rho h}\left(\frac{0.127}{a^4} + \frac{0.707}{a^2 b^2} + \frac{2.44}{b^4}\right)\right]^{1/2} \quad (4.268)$$

16. One long edge (*a*) fixed, one long edge (*a*) supported, one short edge (*b*) fixed, and one short edge (*b*) supported:

$$f = \frac{\pi}{2}\left[\frac{D}{\rho h}\left(\frac{2.45}{a^4} + \frac{2.68}{a^2 b^2} + \frac{2.45}{b^4}\right)\right]^{1/2} \quad (4.269)$$

17. Two long edges (*a*) supported, one short edge (*b*) fixed, and one short edge (*b*) supported:

$$f = \frac{\pi}{2}\left[\frac{D}{\rho h}\left(\frac{2.45}{a^4} + \frac{2.32}{a^2 b^2} + \frac{1}{b^4}\right)\right]^{1/2} \quad (4.270)$$

18. One long edge (*a*) free, one long edge (*a*) fixed, and two short edges (*b*) supported:

$$f = \frac{\pi}{2}\left[\frac{D}{\rho h}\left(\frac{1}{a^4} + \frac{0.608}{a^2 b^2} + \frac{0.126}{b^4}\right)\right]^{1/2} \quad (4.271)$$

4.9 Design, Analysis, and Measurement of Hermetic Microwave Packages

In the present study, three very small microwave packages have been designed and analyzed. Emphasis is placed on the material construction, fabrication, analysis, and qualification testing of these packages. The results show that the packages, designed with closely matched coefficient of linear thermal expansion materials, have excellent reliability (by finite-element analysis) and consistent hermeticity yield (by testing).

The objective of this work is to develop reliable, small, and hermetic microwave packages. These packages must meet the required microwave electrical performance parameters and the stringent requirements of the Military Standard (MIL-STD) specifications (particularly the JAN-S environmental testing), as well as the Hewlett-Packard General Semiconductor Specifications (GSS).

Mechanical Design/Analysis/Measurement/Reliability of Electronic Packaging 327

The reliability of a hermetic package depends on the process, the material construction, and the mechanical design of the package. In the present study, three different microwave packages have been designed, analyzed, fabricated, and tested for reliability and hermeticity (fine and gross leak tests) [137, 138].

The first part of this section describes the design and fabrication of these packages, and the second part discusses the package hermeticity performance. Finally, the thermal and mechanical performance of these packages is provided. Since the microwave electrical performance parameters of these packages are comparable, the package design consideration to meet this requirement is not discussed in this section.

4.9.1 Design and fabrication of packages

All three package designs have the same external package dimensions (see Fig. 4.115). However, each one of the packages has a different internal material construction and design. The packages consist of ceramic bases and ceramic lids, hermetically sealed together with solder glass seal (Kyocera solder glass KC-402). The leads for all three packages are made of Kovar. The physical and mechanical properties of these packages are shown in Table 4.26. After sealing, the packages were split into two categories: one category has packages with straight leads, and the other category has packages with bent leads (for surface-mount applications).

4.9.1.1 Package "A." Package "A" has metallized areas deposited on the top side, to which the leads are brazed (tungsten is used for metallization and copper/silver alloy is used for braze material). The contacts of the device are mounted directly on the leads (see Fig. 4.116a). The lid has a cavity and is sealed to the base with a solder glass seal over the brazed leads (see Fig. 4.116b).

A more common structure of this type of package has the leads brazed either at the sides of the base or at the top or bottom of the base. This type of package is more commonly known as a side-braze, a top-braze, or a bottom-braze package [80]. However, the side-braze and bottom-braze package designs are not suitable for the required microwave performance of the package. By the same token, the standard top-braze design does not provide sufficient space for device attachments (due to the small size of the package).

4.9.1.2 Package "B." Package "B" has no metallization on top of the base. Leads were attached to the base using low temperature solder glass (KC-402). Again, the contacts of the device are mounted directly

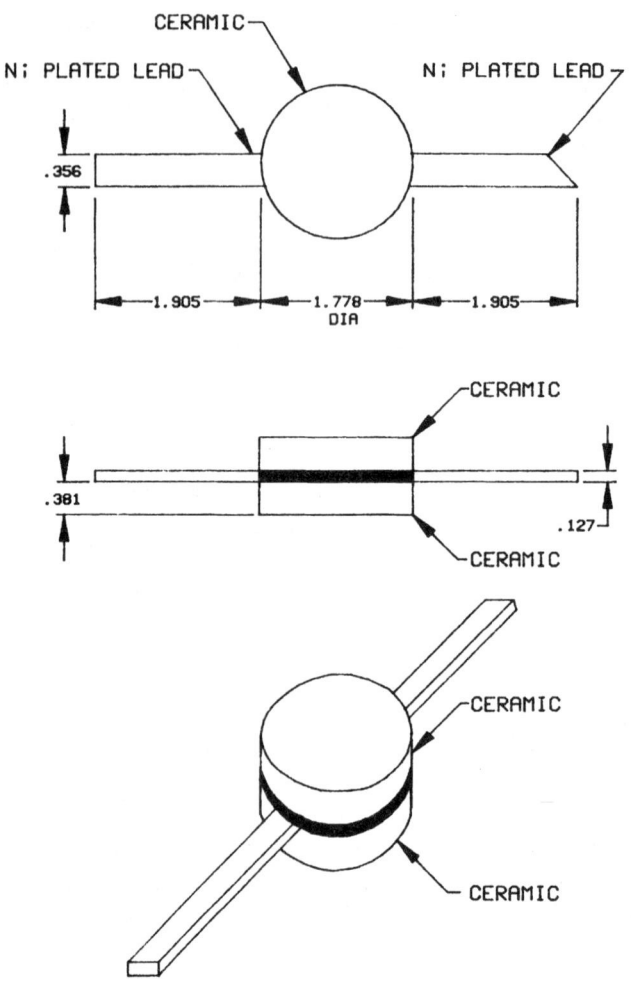

Figure 4.115 Top and side views of the hermetically sealed microwave package (units in mm).

on the leads (see Fig. 4.117a). The lid (which also has a cavity) is sealed to the base with the same solder glass seal over the leads (see Fig. 4.117b). This package structure is similar to the "cerdip" package, except the seal area of package "B" is much smaller than the seal area of the smallest "cerdip" package available.

4.9.1.3 Package "C." Package "C" has metallized areas (tungsten) deposited on the top side of the base, to which small portions of the leads are brazed (copper/silver alloy braze material). However, the contacts of the device are mounted directly on the metallization instead of on the

TABLE 4.26 Material Properties of the Microwave Packages

	Ceramic (92% Al_2O_3)	Kovar (17% Co, 29% Ni, 54% Fe)	Tungsten (W)	Copper/silver (Cu-Ag)	Sealing glass (K/C-402)
Young's modulus (10^6 psi)	39	20	50	14	9
Poisson's ratio	0.23	0.29	0.28	0.37	0.22
Coefficient of thermal expansion (10^{-6} in/in/°C)	7.2	6.2	4.7	19.3	6.7

leads (see Fig. 4.118a). The lid is then sealed to the base with solder glass seal (KC-402) over the brazed leads and metallization (see Fig. 4.118b). The effective sealing area is the area where the glass flows over the metallization (see Fig. 4.118b). The package structure is a combination of the "middle"-braze and the side-braze package structures.

4.9.2 Hermeticity performance of packages

All three packages were assembled, fabricated, and tested for hermeticity using both the helium (He) fine-leak test and the dye-penetration gross-leak test [138–140]. The bubble test (a more common gross leak test method) is not an effective test for these packages, due to their minute size. The dye penetration test was chosen (destructive test) to analyze the package hermeticity and reliability [138 and 139]. Parts were considered bad if dye was observed (by using a microscope) at the center portion of the package (Fig. 4.119). The experimental results of these three packages are shown in Table 4.27.

4.9.2.1 Hermeticity performance of package "A."
Package "A" showed poor (inconsistent) hermeticity yield (both fine and gross leaks). There seems to be no difference whether the leads were straight or bent (Table 4.27). The inconsistency in the hermeticity yield can be traced to the mismatched thermal coefficient of expansion of the copper/silver braze material compared with the rest of materials in the sealing area (see Table 4.27). In most cases, the dye was observed along the brazed material. Furthermore, a finite-element analysis of the thermal stresses in this package design showed much higher stress in this type of package than the packages "B" and "C" designed with closely matched coefficient of thermal expansion materials [141].

4.9.2.2 Hermeticity performance of package "B."
It can be seen from Table 4.27 that the straight lead version of package "B" showed good and consistent hermeticity yield (both fine and gross leaks). This could be

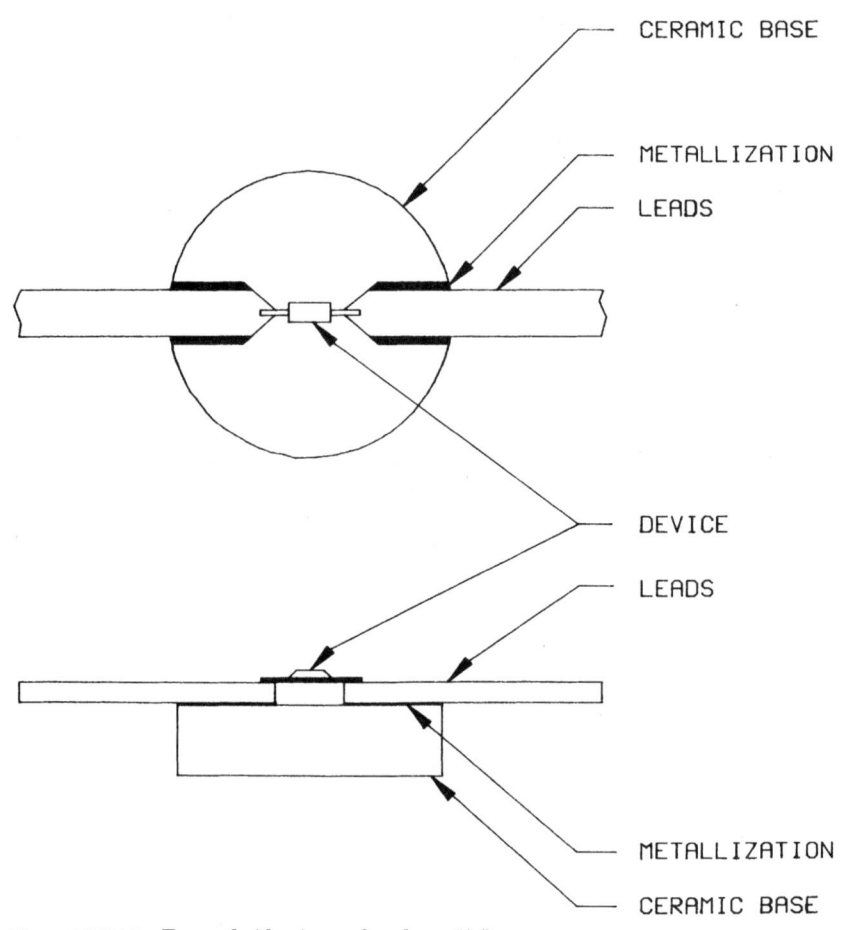

Figure 4.116 (a) Top and side views of package "A."

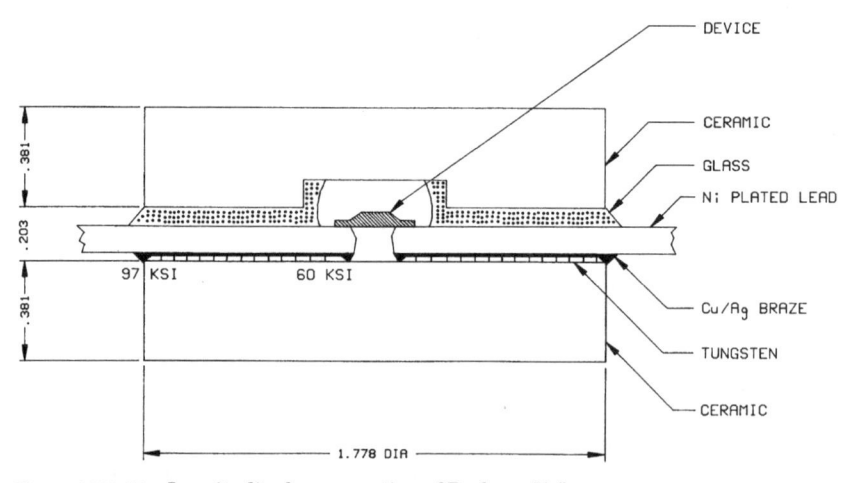

Figure 4.116 (b) Longitudinal cross section of Package "A."

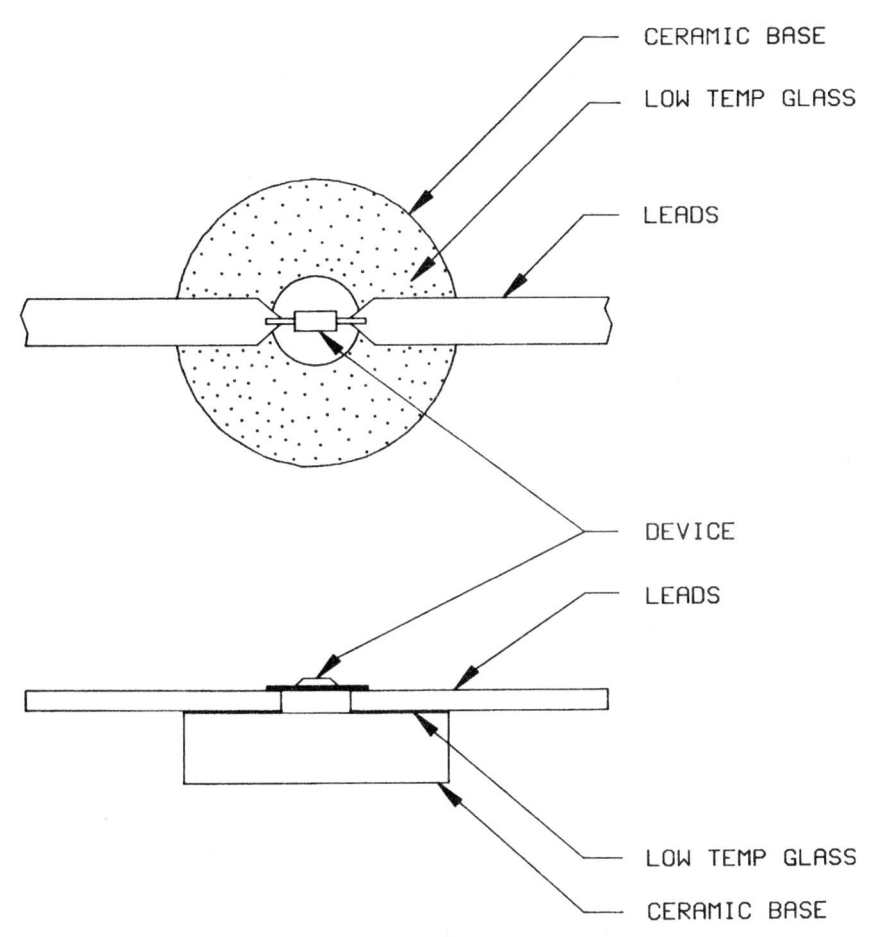

Figure 4.117 (a) Top and side views of package "B."

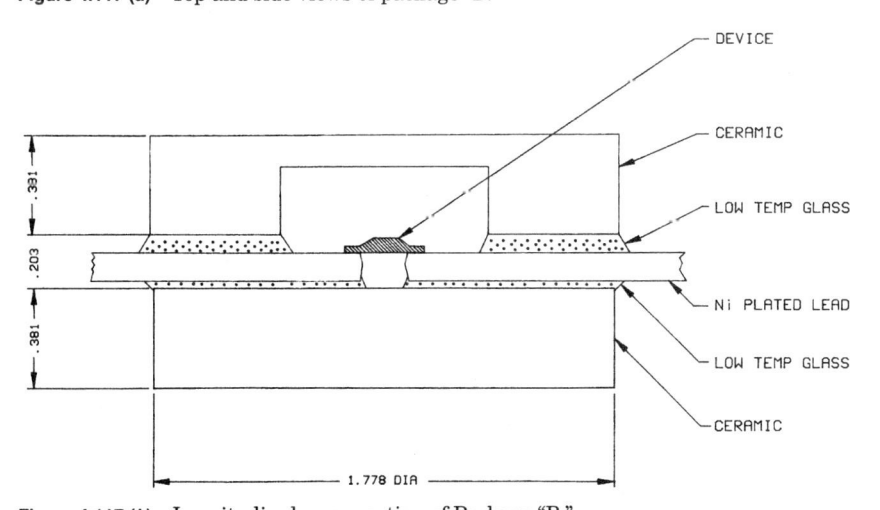

Figure 4.117 (b) Longitudinal cross section of Package "B."

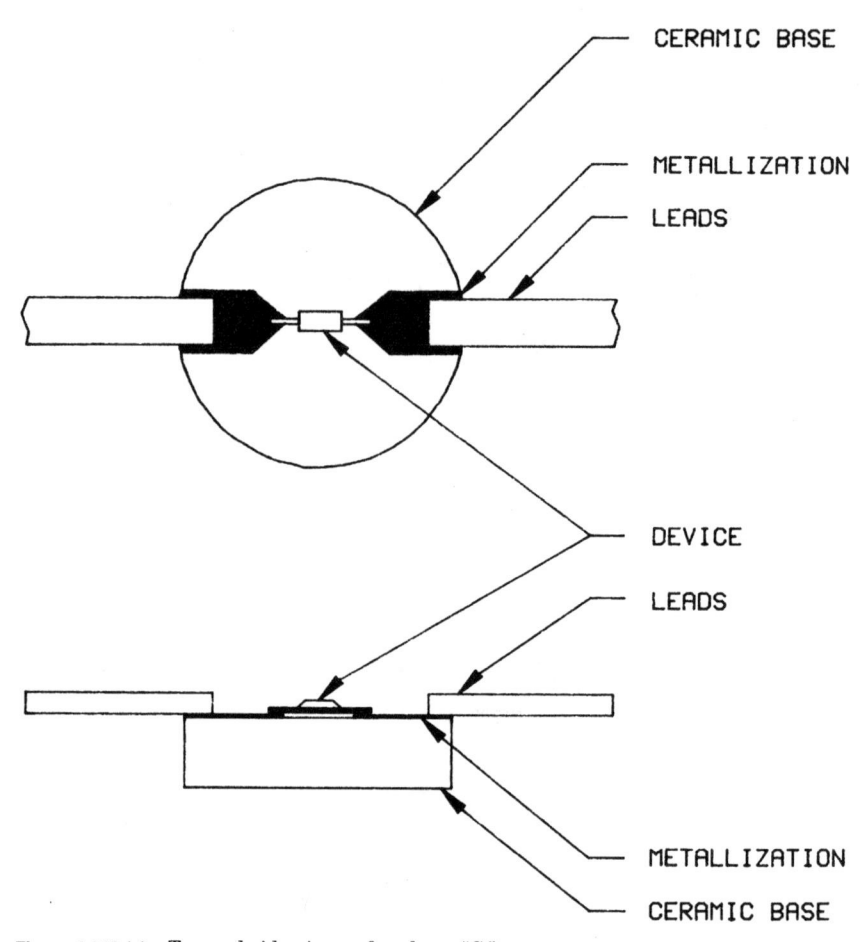

Figure 4.118 (a) Top and side views of package "C."

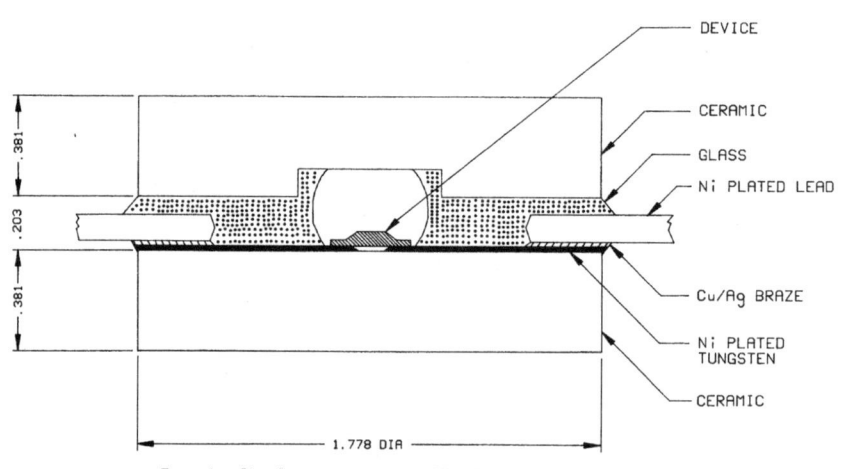

Figure 4.118 (b) Longitudinal cross section of Package "C."

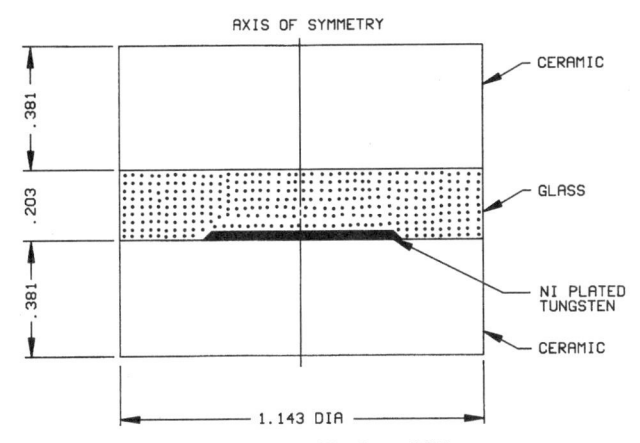

Figure 4.118 (c) Cross section of Package "C."

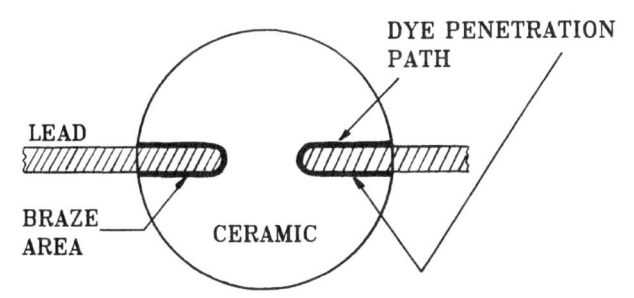

Figure 4.119 Dye penetration test (dye is usually observed along the braze area).

explained by the presence of the closely matched materials in the sealing area (ceramic bases and lids, Kovar leads, and solder glass/KC-402). A finite-element analysis of this type of package showed [141] the thermal stresses in this type of package to be much less than that of package "A." However, the bent-lead version of this package showed a significant number of fine leak rejects (see Table 4.27). The increase in fine leak rejects of the bent-lead version of package "B" is due to microcracks generated along the sealing glass during the lead-bending process.

4.9.2.3 Hermeticity performance of package "C." Package "C" showed very good hermeticity yields for both the straight and bent-lead versions of the package (see Table 4.27). In this package design, the glass provides a hermetic seal over the metallization areas, because all of the materials have closely matched thermal coefficient of expansions. The brazed leads are necessary to add sturdiness to help meet the environmental requirements of the JAN-S testing and to allow the bending of the leads for surface-mount applications. Since the leads were "anchored down" to the

TABLE 4.27 Experimental Results

Package	Percent of acceptable parts after*	
	He leak test	Dye-penetrant test
"A" Straight leads	57%	69%
Bent leads	79%	57%
"B" Straight leads	99%	99%
Bent leads	85%	99%
"C" Straight leads	99%	99%
Bent leads	99%	99%

* He leak test was performed first, followed by dye-penetrant test.

base of the package (brazed leads), no microcracks were generated during the lead-bending process. This resulted in excellent fine leak yield results.

4.9.3 Thermal stress performance of packages

Finite-element analyses of these three packages have been performed and reported in [141]. In this section, only the results of package "C" are briefly discussed.

The finite-element models for the cross and longitudinal sections of package "C" are shown in Figs. 4.120 and 4.121. Two-dimensional 8-node plane stress elements have been used for the construction of these models. Because of symmetry, only half of the package was analyzed. The statement of the present boundary-value problem is to calculate the thermal stresses of these models, while they are subjected to a temperature drop from 440 to 20°C.

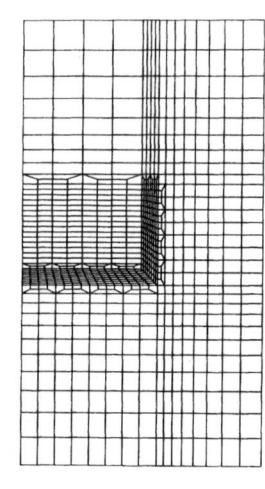

Figure 4.120 Finite element modeling of package "C" (cross-sectional model).

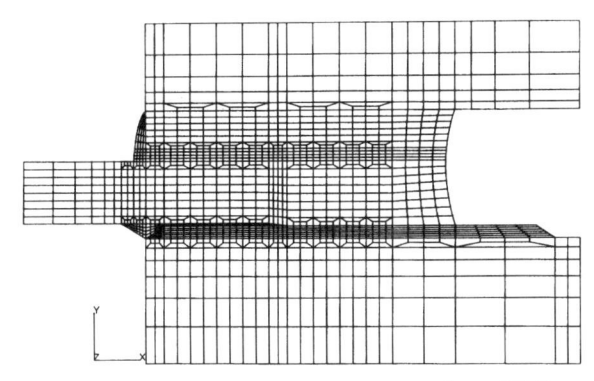

Mechanical Design/Analysis/Measurement/Reliability of Electronic Packaging   335

Figure 4.121 Finite element modeling of package "C" (longitudinal model).

The finite-element results of the cross-sectional model (sealing glass, Ni-plated tungsten, and ceramic body) are shown in Figs. 4.122 through 4.124, respectively. It can be seen that except at the stress concentration areas, the stress value at the sealing glass is about 13 ksi (90 kPa; Fig. 4.122); the stress value at the Ni-plated tungsten is about 30 ksi (207 kPa; Fig. 4.123); and the stress value at the ceramic body is about 15 ksi (103 kPa; Fig. 4.124). These stress values are larger than those of package "B" but are smaller than those of package "A" [141].

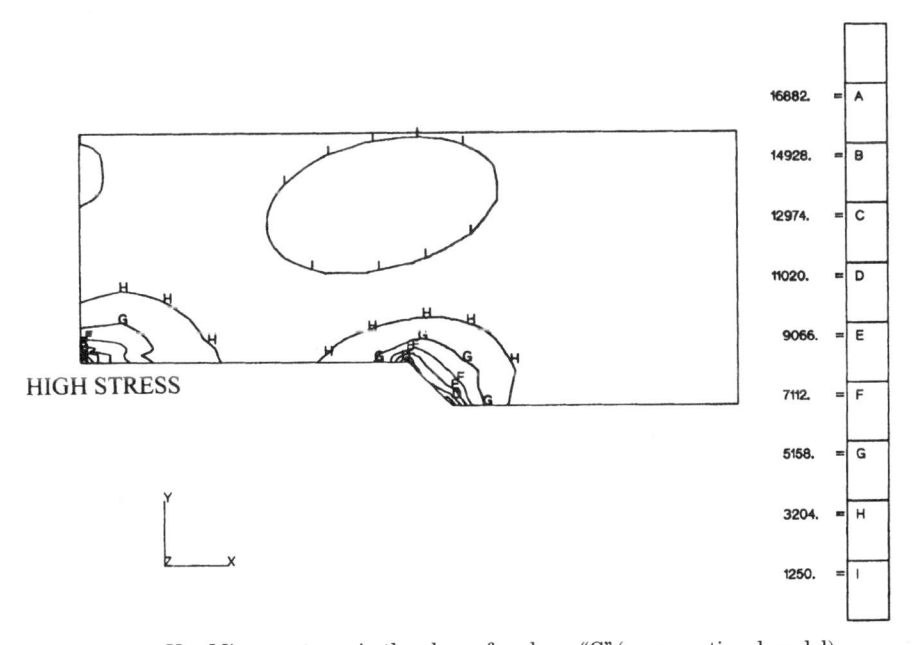

Figure 4.122 Von Mises contours in the glass of package "C" (cross-sectional model).

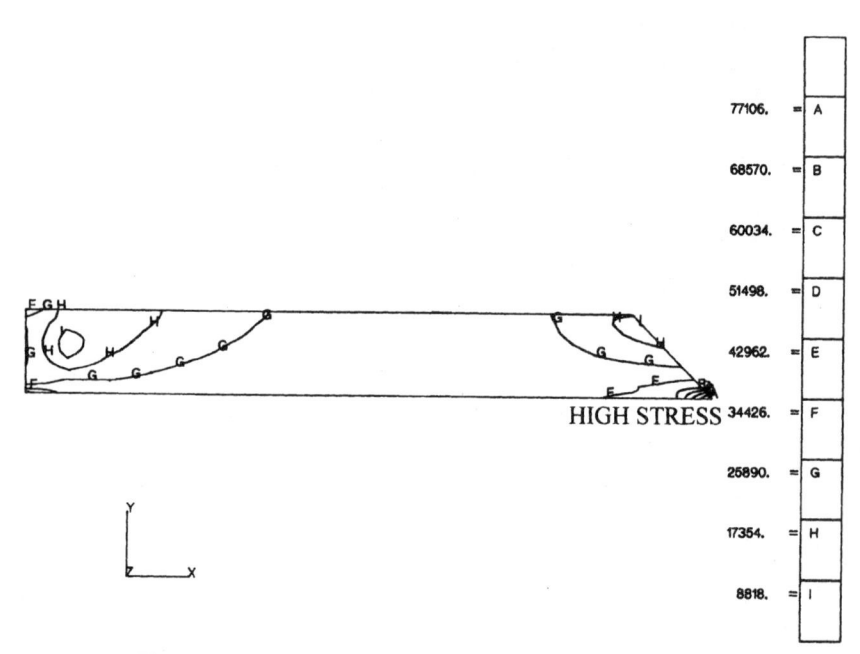

Figure 4.123 Von Mises contours in the tungsten of package "C" (cross-sectional model).

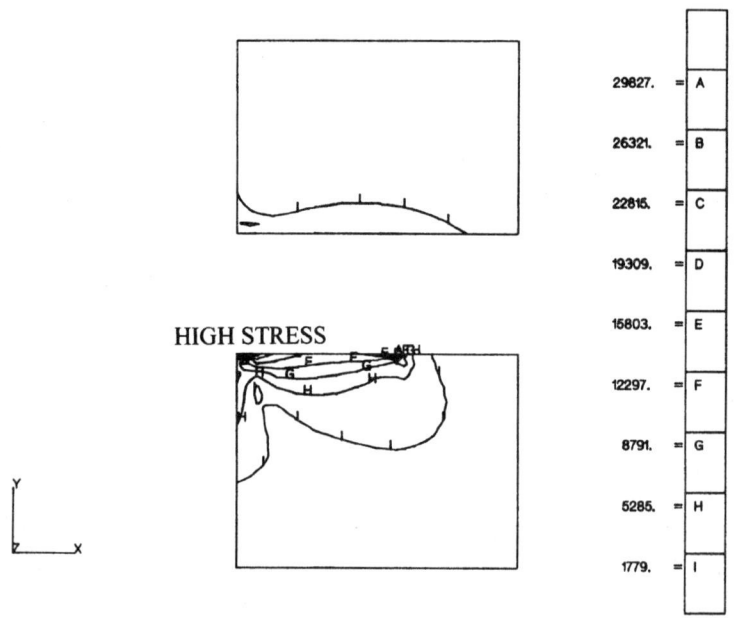

Figure 4.124 Von Mises contours in the ceramic of package "C" (cross-sectional model).

Since the stress values in package "C" are smaller than the strength of the materials, this package is very good for both insertion-hole mounting and surface mounting.

4.9.4 Qualification tests of packages

Since package "C" (both the bent- and straight-lead versions) showed the best hermeticity result for fine and gross leak tests and very good thermal-stress performance by finite-element analysis, this package was selected for the standard qualification tests.

The most demanding qualification tests performed on these packages are the environmental tests. In these tests, 78 packages (from two different lots) were subjected to the thermal shock test, the mechanical shock test, the vibration variable frequency test, the acceleration test, and the lead-fatigue test, respectively [142–146].

For the thermal shock test, the Hewlett-Packard General Semiconductor Specification of 200 cycles was followed (more stringent than the JAN-S specification) [145, 146]. The lead-fatigue test was required after the thermal-shock test (per JAN-S specification) [145]. However, in this evaluation, the lead-fatigue test was not performed after the thermal-shock test; instead, it was performed at the very end of the test sequence (see Table 4.28).

The mechanical-shock test, the mechanical-vibration test, the acceleration test, and the lead-fatigue test were performed per the Military Standard Mil-Spec-750C [142–144]. Fine and gross leak tests were performed after each test and before proceeding to the next test. Table 4.28 shows the environmental test results. It can be seen that they all passed the tests. For the environmental tests, LTPD (maximum percent defective) of 5 was chosen for the sample size [148].

TABLE 4.28 Environmental Test Results

Test*	Test conditions	Sample size	Status
Thermal shock	−65°C/+175°C 200 cycles	78 LTPD = 5	Passed
Mechanical shock	1500 G's; 0.5 msec. pulse each/all axis	78 LTPD = 5	Passed
Vibration	20–2000 Hz; 20 G's 4 min. each/all axis	78 LTPD = 5	Passed
Acceleration	20,000 g. for 1 min. each/all axis	78 LTPD = 5	Passed
Lead fatigue	90 degree lead bend (3×)	78 LTPD = 5	Passed

* All tests were done sequentially. He and dye-penetrant tests were performed after each stress group.

In addition to the tests performed in Table 4.28, other qualification tests were applied to these packages: 1000 hours each of high-temperature operating-life test [149] and high-temperature reverse-bias test [150] were performed on two lots of package "C." The moisture-resistance test [147] was also performed on the packages (JAN-S requirement). Again, packages were tested for He fine-leak and dye-penetration leak tests after each stress (test).

Table 4.29 shows the qualification test results of the package "C" design [147, 149, 150]. It can be seen that all the packages passed the tests. Each test in Table 4.29 was performed in parallel. New packages were used for each test.

4.9.5 Summary

Three very small microwave packages have been designed and fabricated. The hermeticity performance and reliability of these packages were evaluated. Package "A" has brazed leads, hence superior lead strength (lead tension and lead fatigue), but poor and inconsistent hermeticity yields.

The leads of package "B" are sandwiched between the base and lid using the solder glass, hence the lead strength is not as good as the brazed leads. Furthermore, the sealed package is more susceptible to handling related problems. The bending process caused microcracks along the sealing glass, which deteriorate the hermeticity of the package.

The package "C" design incorporated the package "A" design and the side-braze package design. The reliability of this package meets the JAN-S requirements as well as other military standard specifications.

TABLE 4.29 Additional Qualification Test Results

Test*	Test conditions	Sample size	Status
High temp. op. life	$T_a = 155°C$ $lf = 10$ mA D.C. $t = 1,000$ hours	78 LTPD = 5	Passed
High temp. reverse bias	$T_a = 150°C$ VBR = 3.2 V $t = 1000$ hours	78 LTPD = 5	Passed
Moisture resistance	85/85% R.H. VBR = 3.2 V $t = 1000$ hours	22 LTPD = 10	Passed
Lead fatigue	90 degree lead bend (3×)	78 LTPD = 5	Passed

* All tests were done in parallel. He and dye penetrant tests were performed after each stress group.

The performance of the package is also acceptable for high-frequency microwave applications.

The results further indicated that packages designed with closely matched coefficient of linear thermal expansion materials provide reliable hermetic packages. These results were confirmed by finite-element analysis.

4.10 Design, Analysis, and Measurement of a Low-Cost Chip Size Package—NuCSP

A new wire-bondable land grid array (LGA) chip size package called *NuCSP* is presented in this section. NuCSP is a minimized body size wire-bondable package with rigid substrate interposer. The design concept is to utilize the plating bars on the edges of the package substrate as the wire-bond fingers. Bond fingers are redistributed inward to an array of plated through-hole vias underneath the chip, then are connected to copper pads on the bottom of the package. NuCSP package size is about equal to die size + 2.5 mm. Using conventional PCB substrate manufacturing with 4/4 mils routing width/space and wire-bonding process, NuCSP offers a very low cost package suitable for memory chips and low-pin-count ASIC applications. The other advantages are that the use of wirebonding allows NuCSP to be applicable for die size, pad count, and pitch variations. Because it is wire bondable, NuCSP may be as generic as plastic quad flat pack (PQFP), yet provides smaller body size, lower cost, and smaller package electrical parasitic parameters.

The first part of this section describes the design concept and applications of NuCSP. The package family of NuCSP and design examples of a 64-pin standard NuCSP and a 104-pin thermally enhanced NuCSP follow. The electrical parasitic parameters of the NuCSP package family and the comparisons with corresponding pin-count PQFPs are discussed next. Thermal performance of NuCSP is also discussed in this section. Finally, the solder joint reliability of NuCSP on PCB is presented.

4.10.1 NuCSP design concept

Figure 4.125 schematically shows the cross section of a generic wire bondable chip size package—NuCSP. It leverages the PBGA design into a minimized package body size for lower power and pin-count applications. The design concept utilizes the plating bars on the edge of the package substrate as the wire-bond fingers. The bond fingers are then redistributed (fanning) inward on the top layer of the substrate, then dropped using vias to connect to the pads on the bottom side of the package substrate, as illustrated in Fig. 4.125. NuCSP presents the following specific design features [151, 152]:

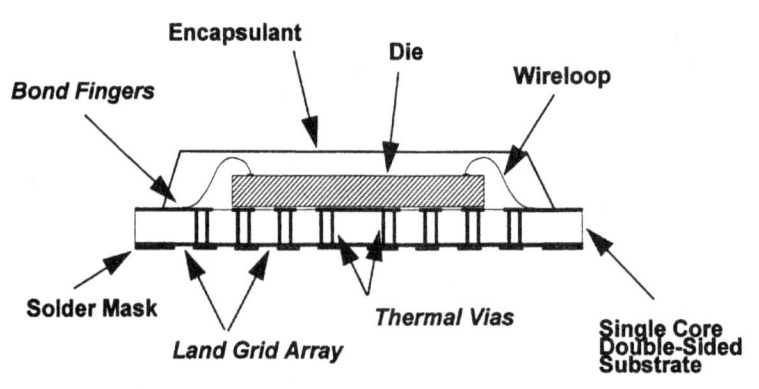

Figure 4.125 Cross section of the NuCSP.

- It is an LGA package that can be soldered onto PCB through a 0.15-mm (6-mil)-thick solder paste (Fig. 4.126), and formed 3-mil (0.08-mm)-height solder joints.
- Plating bars are used as bond fingers that locate at the edges of the package substrate.
- Bond fingers are redistributed (fanning) inward under the die.
- Redistribution traces are connected to the copper pads on the bottom side of the package substrate through vias.
- Package copper pads are area-array at 0.5-, 0.75-, and 1-mm pitch.
- Package body size is about equal to die size + 2.5 mm.
- It is wire bondable such that it allows die size and pitch variations.
- Its substrate manufacturing is conventional PCB process with 4/4-mil (.1/.1-mm) trace width/space routing.

4.10.2 NuCSP design examples

A preliminary family of NuCSP designs for pin-count ranges from 32 to 104 are listed in Table 4.30. Note that the package size primarily depends on the location of the bond fingers (2 sides or 4 sides) and solder land pitch. This section presents two design examples: one is a standard 64-pin NuCSP and the other is a thermally enhanced 104-pin NuCSP.

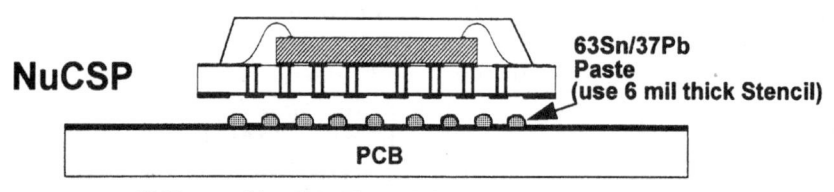

Figure 4.126 SMT assembly of NuCSP on PCB with 6-mil-thick solder paste.

TABLE 4.30 Wire-Bonding NuCSP Package Family

Ball count	Ball pitch (mm)	Ball distribution	Bond finger distribution	Trace width/ space (mm)	Via diameter (mm)	Via land diameter (mm)	Package size (mm)	Die size (mm)	Applications
32	1	6 rows	4 sides	0.1/0.1	0.3	0.5	7.5 × 7.5	Up to 4 mm	Memory
32	1	4 × 8	2 sides	0.1/0.1	0.3	0.5	6.5 × 9.5	Up to 3.6 × 6.4	Memory
48	1	6 × 8	2 sides	0.1/0.1	0.3	0.5	7.8 × 9.5	Up to 5.5 × 6.5	Memory
64	1	8 × 8	4 sides	0.1/0.1	0.3	0.5	8.8 × 8.8	Up to 5.8 mm	Memory/ASIC
80	0.812	10 rows	4 sides	0.1/0.1	0.3	0.5	10.3 × 10.3	Up to 7.3 mm	Memory/ASIC
104	0.864	12 rows	4 sides	0.1/0.1	0.3	0.5	12.75 × 12.75	Up to 8 mm	Memory/ASIC

4.10.2.1 A 64-pin standard NuCSP. A standard 64-pin NuCSP is shown in Fig. 4.127. The substrate is a 22-mil-thick high-T_g organic material. This design uses 4/4-mil trace width/space routing pitch. Through-hole via diameter is 12 mils; via land diameter is 20 mils. Solder land diameter is 20 mils and land pitch is 30 mils. There is an extra 20 mils space from the edge of the encapsulant to the edge of the substrate. The resulting package body size is 8×8 mm.

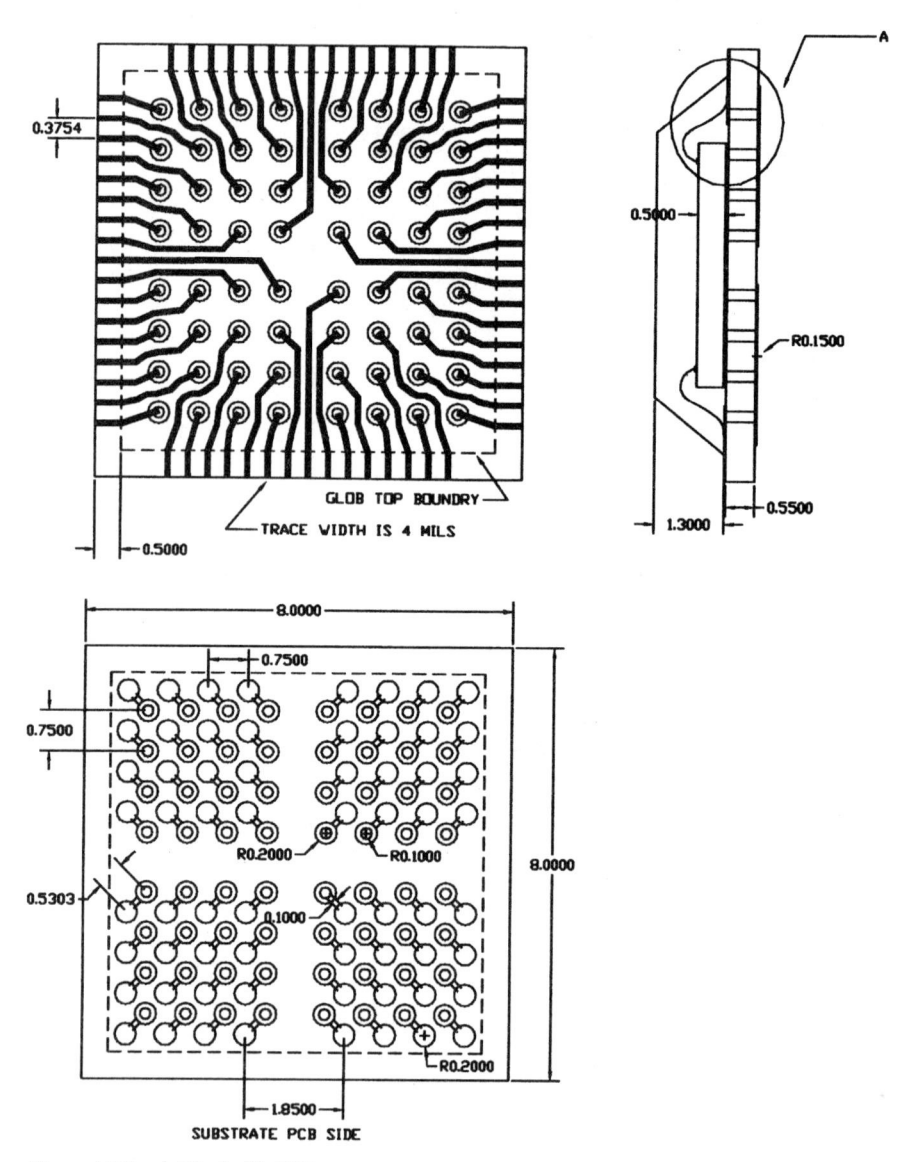

Figure 4.127 A 64-pin NuCSP.

Mechanical Design/Analysis/Measurement/Reliability of Electronic Packaging   343

4.10.2.2 A thermally enhanced 104-pin NuCSP. Due to small package size and lack of metal planes on the substrate, most CSP presents poor thermal performance as compared to low-pin-count leadframe package PQFP and TSOP. Figure 4.128 shows a 104-pin NuCSP with thermal enhancement using 16 thermal vias connecting metal planes on both

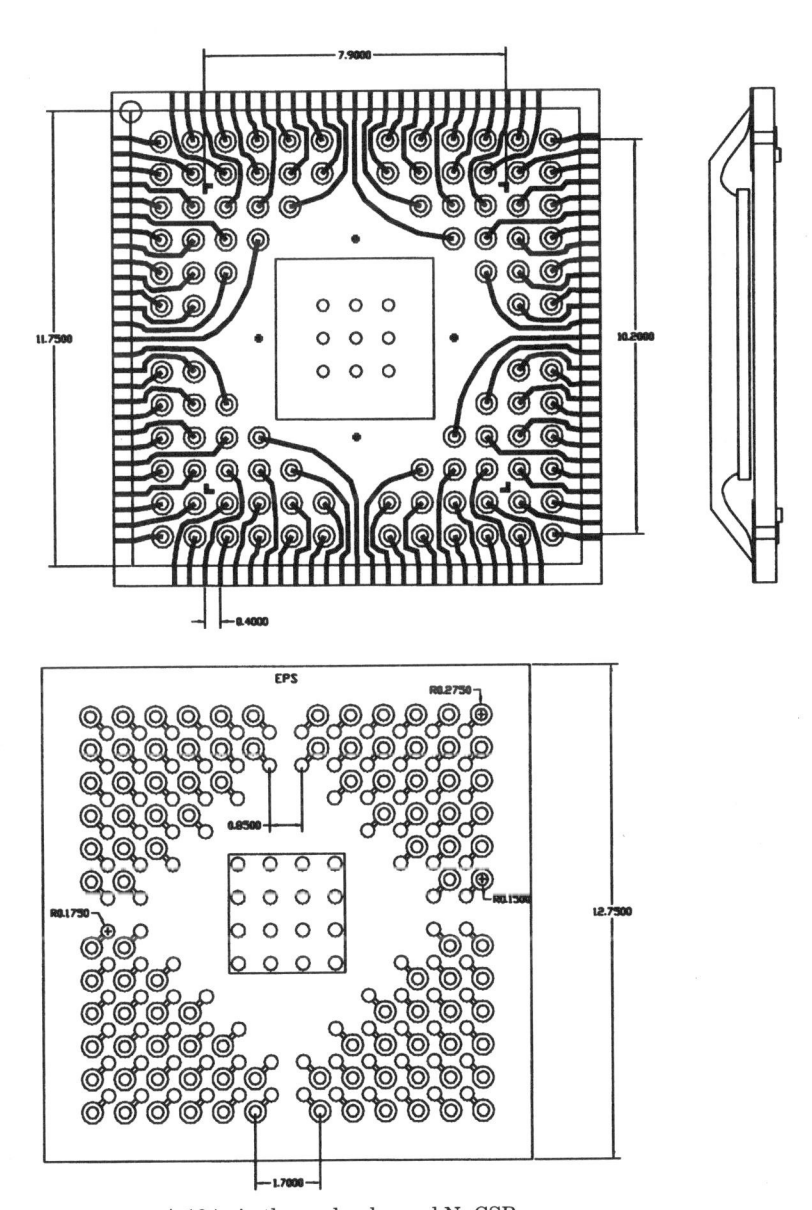

Figure 4.128 A 104-pin thermal-enhanced NuCSP.

sides of the substrate under the die. The trace routing pitch is 4/4 mils width/space. Through-hole via diameter is 12 mils and via land diameter is 20 mils. Solder land pitch is 40 mils. The resulting package size is 12.75 × 12.75 mm.

4.10.3 Electrical performance of NuCSP

Electrical parameters: resistance (R), inductance (L) and capacitance (C) of the NuCSP family are listed in Table 4.31. Comparisons of RLC values with a 64-Pin thin quad flat packs (TQFP) are shown in Table 4.32.

It is clearly demonstrated in Table 4.32 that NuCSP-64 has smaller parasitic parameters R, L, and C than TQFP-64. Smaller parasitic parameters lead to reduced cross talk, ground bounce, and power drop. NuCSP has smaller package size and shorter trace length and that is why its electrical performance is superior to that of TQFP.

4.10.4 Thermal performance of NuCSP

Thermal resistance of the 64-pin and 104-pin NuCSP is predicted by using finite-element analysis. Figures 4.129 and 4.130 show the models of 64-pin and 104-pin NuCSP, respectively. Each CSP is mounted on a 25.4 × 25.4 mm, two-metal layer FR-4 PCB. Figures 4.131 and 4.132 show the typical temperature distribution of the NuCSPs. The thermal resistances under various air velocity conditions are summarized in Figs. 4.133 and 4.134. They compared very well with measurement results (within 3%).

Since the CSPs are small packages and no copper leadframe is used, the thermal resistance is relatively high. One can conclude from Fig.

TABLE 4.31 R, L, and C Values of NuCSP Family*

Package types	Package size (mm × mm)	R (mΩ)	L (nH)	K_{Lm}	C (pF)	C_m (pF)	K_b
32 NuCSP	7.5 × 7.5	29–34	2.1–4	0.39	0.42–0.63	0.12–0.2	0.16
64 NuCSP	8 × 8	30–35	2.5–4.4	0.39	0.49–0.7	0.15–0.22	0.16
80 NuCSP	10.3 × 10.3	33–40	2.8–5.2	0.39	0.63–0.9	0.19–0.28	0.16
104 NuCSP	12.75 × 12.75	35–42	3.5–5.8	0.39	0.7–0.98	0.21–0.34	0.16

* The values assume 1-mm long bondwire.

TABLE 4.32 Comparisons of Package Electrical Parameters of a 64 NuCSP with TQFPs

Package types	Package size (mm × mm)	R (mΩ)	L (nH)	K_{Lm}	C (pF)	C_m (pF)	K_b
64 NuCSP	8 × 8	30–35	2.5–4.4	0.39	0.49–0.7	0.15	0.16
64 TQFP	14 × 14	47–70	5.2–6.9	0.49	0.4–0.62	0.18–0.3	0.24
64 TQFP	12 × 12	40–60	4.6–5.9	0.51	0.34–0.5	0.2–0.26	0.25

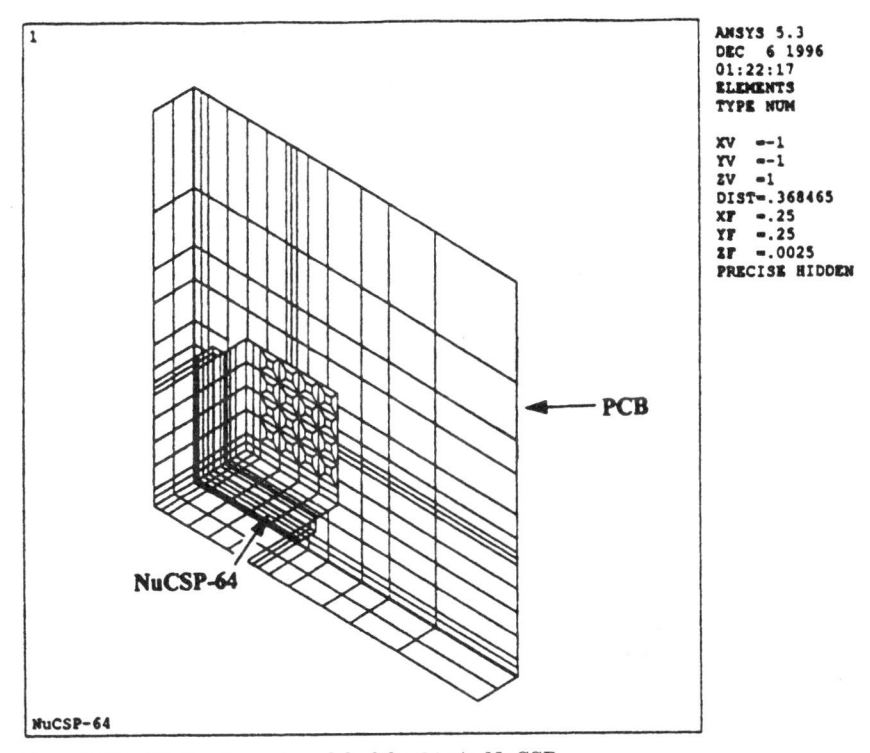

Figure 4.129 Finite element model of the 64-pin NuCSP.

4.133 that NuCSP-64 can dissipate 0.28 watt of power under still air condition and 0.4 watt of power under 1 m/s air flow velocity condition.

By comparing Figs. 4.133 and 4.134, one can conclude that thermal resistance of NuCSP is improved by 5 percent by increasing package size (from 8 × 8 mm to 12.75 × 12.75 mm) and ball count (from 64 to 104). However, thermal resistance can be further reduced by applying ground planes at the substrate, for example, if the ground planes are half-ounce copper planes (0.018 mm thick) on the top and the bottom surfaces of the BT substrate. The thermal resistance of NuCSP-104 with ground planes is 20 percent lower than that of NuCSP-64. Ground planes are very important for thermally enhanced CSP without heat sink.

4.10.5 Solder-joint reliability of NuCSP on PCB

The solder-joint reliability of a low-cost wire-bondable LGA NuCSP on FR-4 PCB assembly under thermal cycling loading is considered in this section. In Figs. 4.125 and 4.126, it can be seen that the chip was attached to the single-core double-sided BT interposer, and then sol-

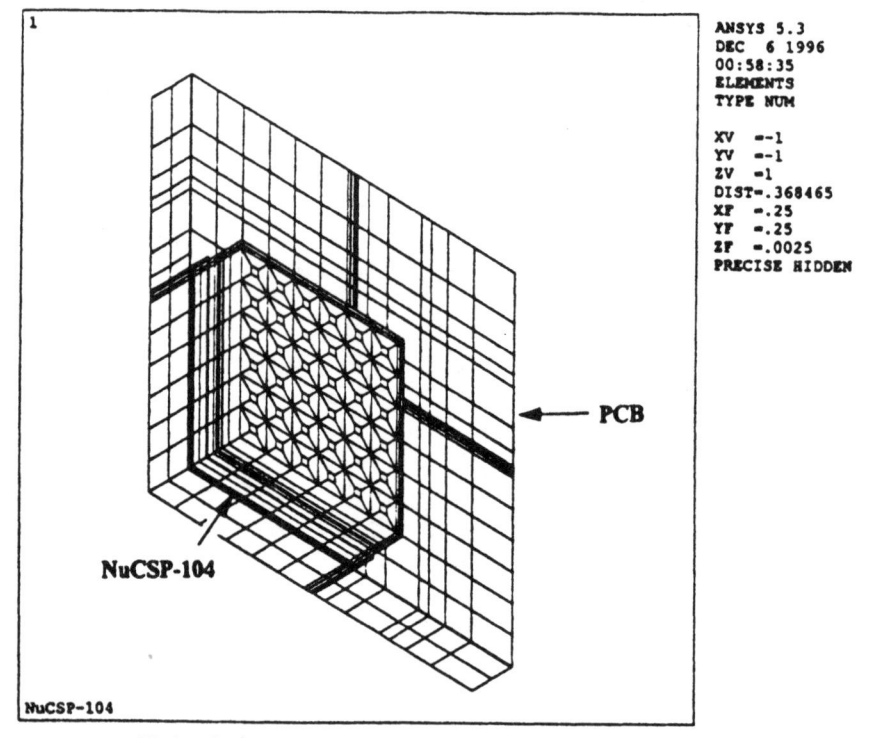

Figure 4.130 Finite element model of the 104-pin thermal-enhanced NuCSP.

dered to the PCB. The chip size was 7.2 × 4 × 0.38 mm with 32 pads; the BT interposer dimensions were 11 × 7.2 × 0.56 mm. After the NuCSP was assembled on the PCB, the solder-bump height was about 0.08 mm, and the solder bumps were in area-array format with 1-mm pitch.

Figure 4.135 shows one-half of the assembly (finite-element model) along the diagonal direction for the analysis. There were 8896 2-D 8-node plane-strain elements and 27,217 nodes. The Norton's steady-state creep relation (Eqs. 4.145 and 4.154)

$$d\gamma_{crp}/dt = B^* \exp\{-\Delta H/kT\}\tau^n \qquad (4.272)$$

has been used for the time-dependent elasto-plastic-creep analysis of the NuCSP solder joints. In this equation, $d\gamma_{crp}/dt$ is the creep shear strain rate, τ is the shear stress, ΔH is the activation energy (= 0.49 eV), T is the absolute temperature, k is the Boltzmann's constant (= 8.617 × 10^{-5} eV/K), and n is the stress exponent. For 63Sn/37Pb eutectic solder, the material constants of the Norton equation are given in Table 4.2 of [94] as $n = 5.25$ and $B^* = 0.205$ MPa$^{-5.25}$ sec^{-1}. The temperature-

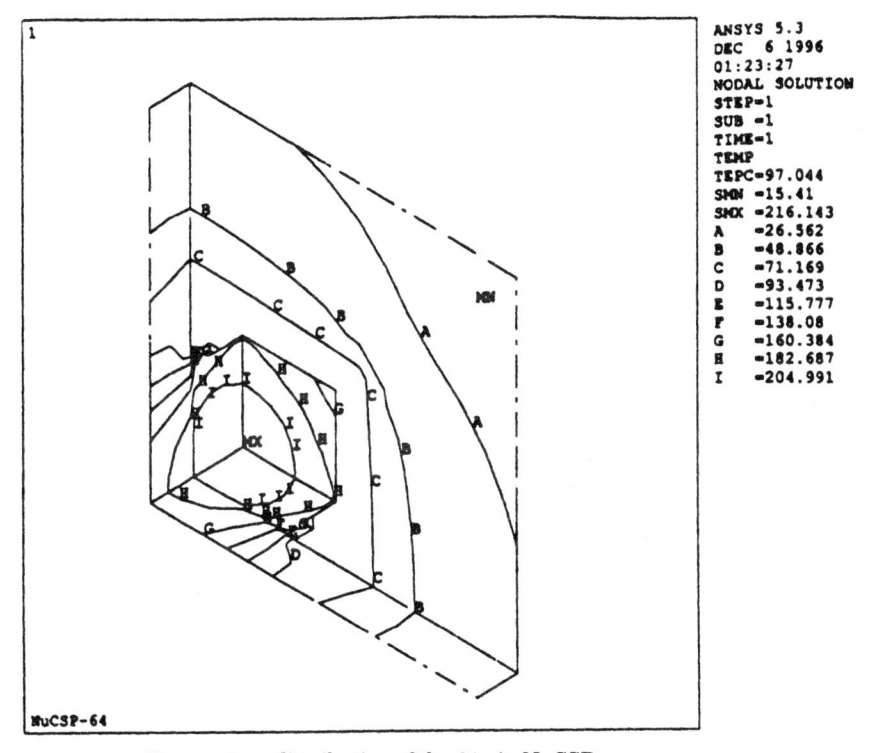

Mechanical Design/Analysis/Measurement/Reliability of Electronic Packaging 347

Figure 4.131 Temperature distribution of the 64-pin NuCSP.

dependent stress-strain curves and Young's modulus for the solder have already been given in Figs. 4.102a and b, and 4.103.

The temperature loading imposed on the NuCSP assembly is shown in Fig. 4.104. It can be seen that for each cycle (60 min) the temperature condition was between −40 and 125°C with 10 min ramp and 20 min hold. Two complete cycles were executed.

The location of the maximum stress and strain hysteresis responses was at the solder joint right underneath the corners of the chip. Figure 4.136 shows the notations in the critical solder joint. The hysteresis loops of the shear stress and shear strain at various locations (NODE-A, B, C, D, E, F, G, H, and I on the upper and lower interfaces) in that solder joint are shown in Fig. 4.137 (upper interface between BT substrate and solder) and Fig. 4.138 (lower interface between solder and PCB). It can be seen that the maximum shear stress and shear strain location occurs at the lower right-hand corner (NODE-I on interface 2) of the solder joint.

The viscoplastic strain energy density per cycle can be determined by using the average of the two cycles from the hysteresis loop plot.

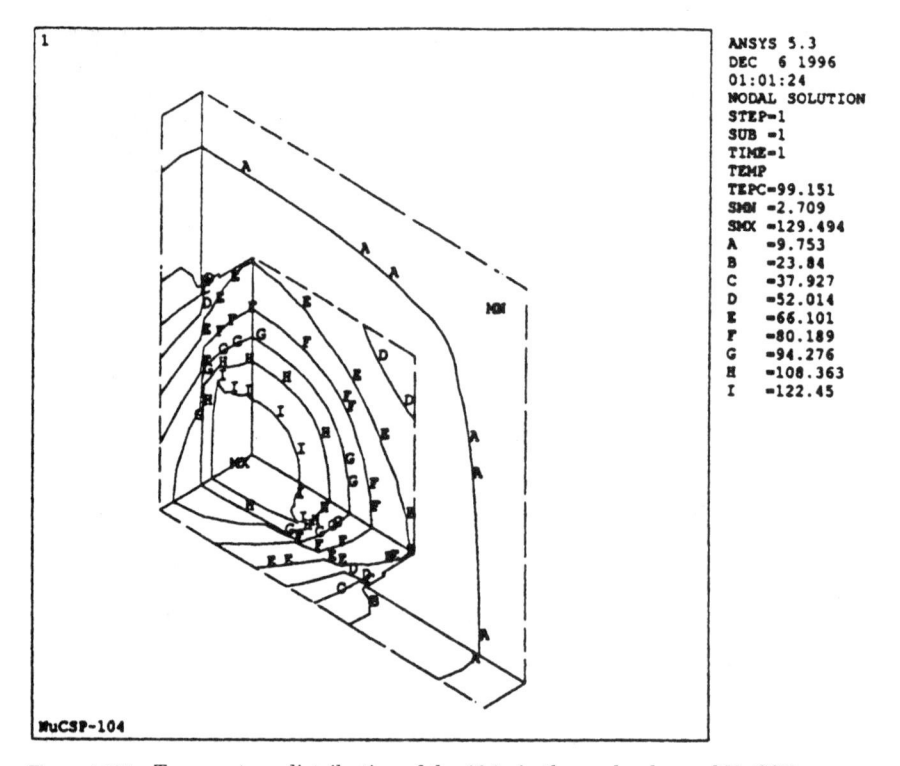

Figure 4.132 Temperature distribution of the 104-pin thermal-enhanced NuCSP.

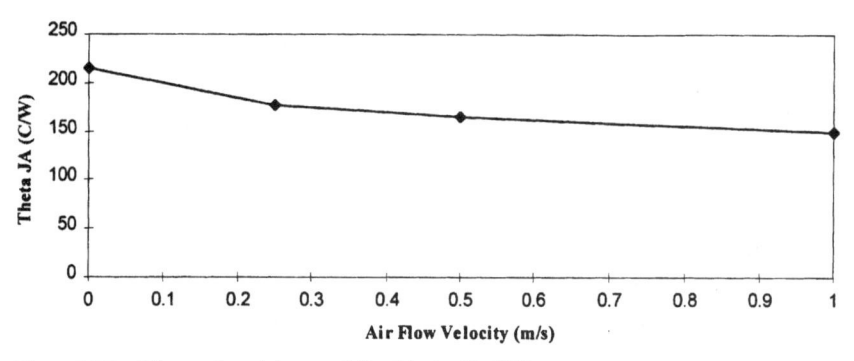

Figure 4.133 Thermal resistance of the 64-pin NuCSP.

These are: 0.14 MPa (20 psi) at NODE-A on interface 1; 0.35 MPa (51 psi) at NODE-I on interface 1; 0.09 MPa (12.6 psi) at NODE-A on interface 2; and 0.52 MPa (76 psi) at NODE-I on interface 2. The typical size of element width and height was 2 mil (0.05 mm) and 0.44 mil (0.011 mm), respectively.

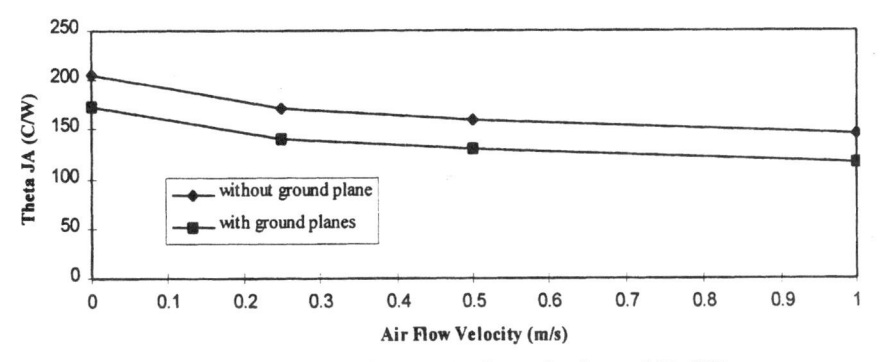

Figure 4.134 Thermal resistance of the 104-pin thermal-enhanced NuCSP.

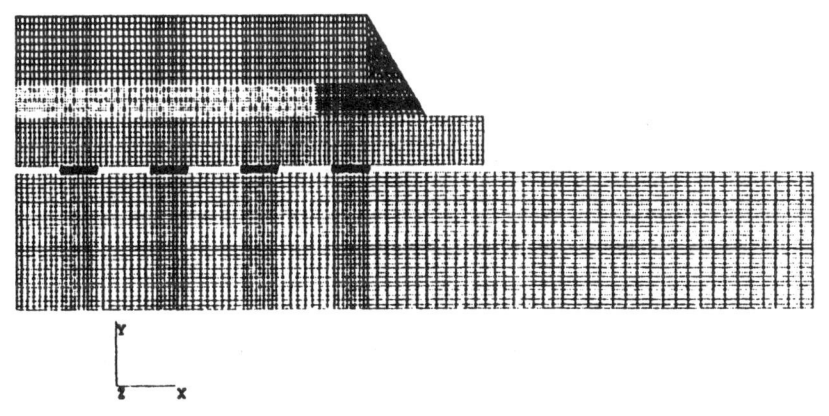

Figure 4.135 Finite element model for temperature-time-dependent analysis of the NuCSP's solder joints.

Once we have the average viscoplastic strain energy density per cycle (ΔW), the thermal fatigue crack initiation life (N_0) can be estimated from Eq. 13.35 of [95] (see Table 4.33), that is,

$$N_0 = 7860 \, \Delta W^{-1} \qquad (4.273)$$

and the thermal fatigue crack propagation life (N) based on the linear fatigue crack growth rate theory can be estimated by Eq. 13.36 of [95], that is,

$$da/dN = 4.96 \times 10^{-8} \, \Delta W^{1.13}$$

or

$$\frac{a_f - a_0}{N - N_0} = 4.96 \times 10^{-8} \, \Delta W^{1.13}$$

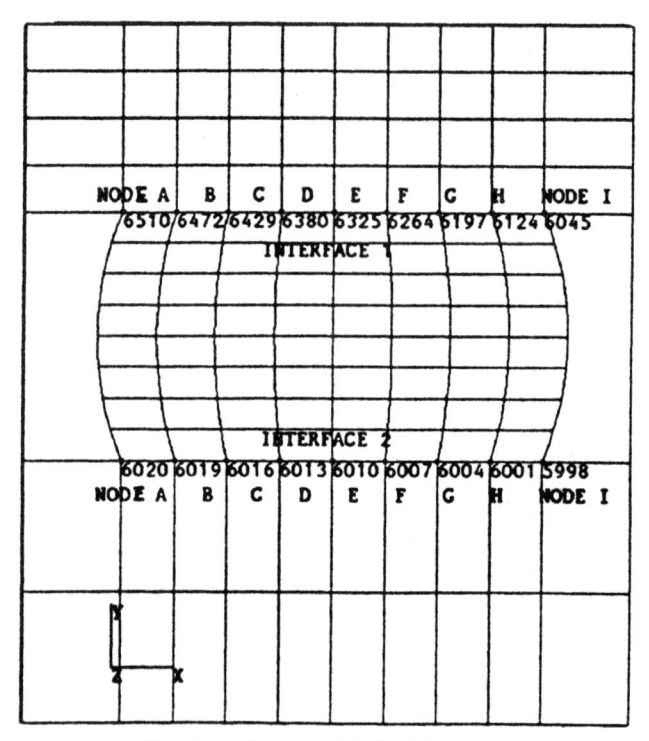

Figure 4.136 Notations in the critical solder joint in temperature-time-dependent analysis.

or

$$N = N_0 + a_f/4.96 \times 10^{-8} \Delta W^{1.13} \qquad (4.274)$$

where $a_0 = 0$ (the initial crack length) and a_f is the final crack length (the width = 0.0124 in) of the solder joint. The total thermal fatigue life of the solder joint is $N_0 + N$. For our case, the N_0, N, and total thermal fatigue life are, respectively, 397 cycles, 8167 cycles, and 8564 cycles at NODE-A on interface 1; 154 cycles, 2810 cycles, and 2964 cycles at NODE-I on interface 1; 624 cycles, 13,611 cycles, and 14,235 cycles at NODE-A on interface 2; and 103 cycles, 1784 cycles, and 1887 cycles at NODE-I on interface 2 (Table 4.33).

4.10.6 Summary

A low-cost wire-bondable LGA chip size package called NuCSP has been presented. NuCSP is a minimized body size wire-bondable package with rigid substrate interposer. The design concept of NuCSP is to utilize the plating bars on the edge of the package substrate as the

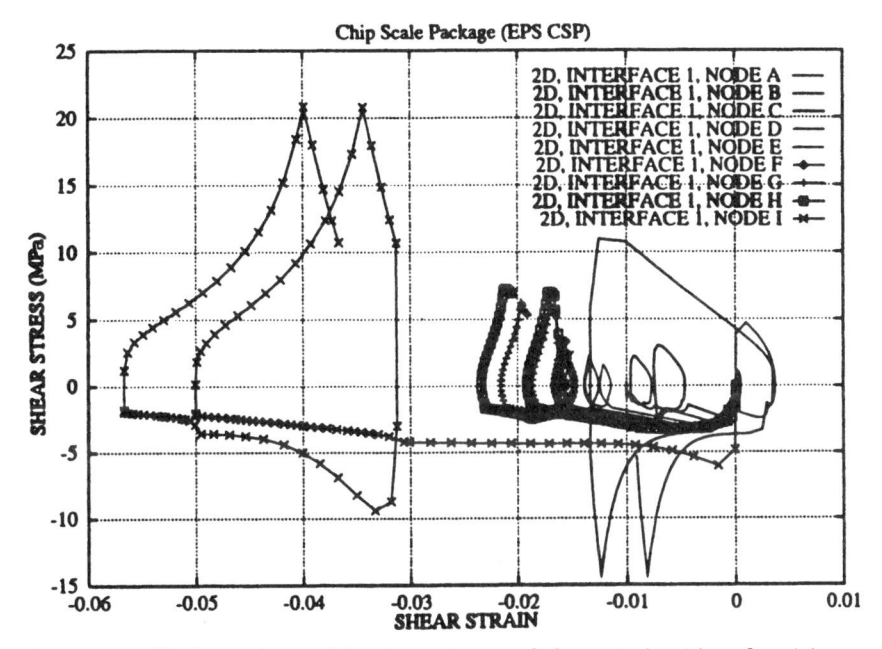

Figure 4.137 Hysteresis loops of the shear stress and shear strain at interface 1 (near the carrier).

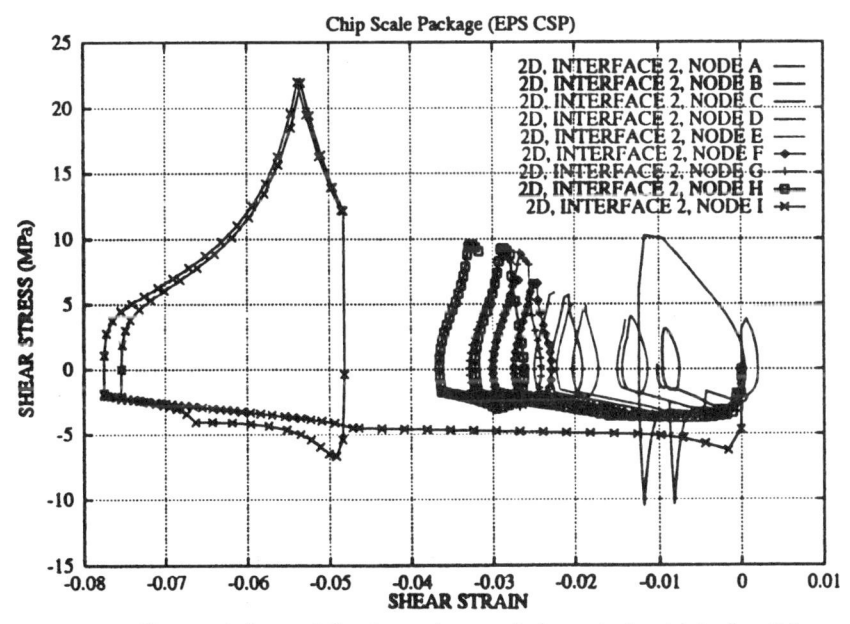

Figure 4.138 Hysteresis loops of the shear stress and shear strain at interface 2 (near the PCB).

TABLE 4.33 Thermal Fatigue Life Prediction of the NuCSP Solder Joints

Nodes	Viscoplastic strain energy density per cycle (ΔW) MPa	psi	Crack initiation $N_0 = 7860\Delta W^{-1}$	Fatigue life prediction (cycles) Crack propagation $N = (0.0124/4.96) \times 10^{-8} \Delta W^{1.13} + N_0$	Total life $N_0 + N$
Node 1-a	.1364	19.78	397	8167	8564
Node 1-i	.3507	50.85	154	2810	2964
Node 2-a	.0868	12.59	624	13611	14235
Node 2-i	.5243	76.03	103	1784	1887

Linear fracture mechanics has been assumed. Initial crack length $a_0 = 0$ and final crack length $a_f = 0.0124$ inches is the width of the solder joint.

wire-bond fingers. Bond fingers are redistributed inward to an array of through-hole vias underneath the die, then are connected to solder lands on the bottom of the package.

NuCSP package size is about equal to die size + 2.5 mm. Using conventional PCB substrate manufacturing and wire-bonding process, NuCSP offers a low-cost chip size package suitable for memory chips and low-pin-count ASIC applications. The NuCSP may be as generic as PQFP, yet provides smaller body size, potentially lower cost, and smaller package electrical parasitic parameters. Also, the solder joints on the PCB with the LGA NuCSP are reliable.

Another version of the NuCSP is shown in Fig. 4.139. It can be seen that instead of wire bonding the chip on the substrate, solder-bumped flip-chip technique is used. In this case, the NuCSP size is even smaller and the electrical performance is much better [153].

4.11 Design of Electrical and Thermal Enhanced Plastic Ball Grid Array Packages—NuBGA

Driven by high-performance applications, the electronics industry is devoting unprecedented effort to develop cost-effective, high-density,

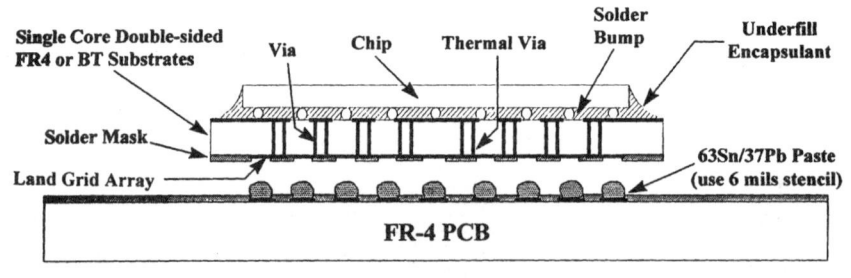

Figure 4.139 NuCSP with solder bumped flip chip [153].

high-I/O, and highly reliable electrical and thermal enhanced packages. NuBGA, a single-core, two-metal layer, cavity-down plastic ball grid array (PBGA) package, is presented to fulfill the preceding requirements.

In this section, the design concept of a low-cost, single-core with two-metal-layers package is described, which provides an optimization design from the aspect of electrical and thermal performance. The electrical performance of the package is studied by both numerical simulation and experimental measurements. Parasitic parameters are extracted from time-domain reflectometer (TDR) and time-domain transmission (TDT) measurement. The package resistance, inductance, and capacitance of the NuBGA are compared with that of the PQFP, conventional single-core PBGA, and enhanced multilayer PBGA (EPBGA). Cross talk and simultaneous switch output (SSO) noise of the package are also investigated.

The thermal performances (thermal resistance, temperature, and heat-dissipation distributions) of NuBGA with different sizes of heat spreader are studied by finite-element simulations and experimental measurements. Measurements were also made of the thermal performance of NuBGA with different heat sinks attached onto a full-size heat spreader under various air flow and power.

4.11.1 NuBGA design concepts

To achieve good thermal performance, NuBGA adopts the cavity-down design configuration with the back of the die attached directly to a heat spreader covering the back surface of the package. Standard NuBGA uses a 22-mil-thick high-T_g organic substrate as the core layer, which is compatible with low-cost PCB substrate manufacturing process. Optimized electrical performance on the single-core and double-sided substrate is achieved by applying the concept of split-wrap-around (SWA) or split-via-connection (SVC). With SWA or SVC design technique, programmable power/ground split ring segments for wire bonding, shorter wire-bond length, and controlled impedance µ-stripline and coplanar stripline signal traces can be obtained. The heat spreader is attached to the core substrate through standard PCB pre-preg lamination. Figure 4.140 illustrates the cross section of standard NuBGA design.

4.11.1.1 Programmable VDD/VSS split-wrap-around (SWA) and µ-stripline and coplanar stripline traces.
Figure 4.141 illustrates the design concept of SWA [154, 155]. Unlike conventional PBGA design, NuBGA design has only one ring for power and ground wire bonding. The ring is split or segmented to support at least 4-to-1 (signal-to-power/ground) pad ratio. To provide transmission-line structures for signal traces on

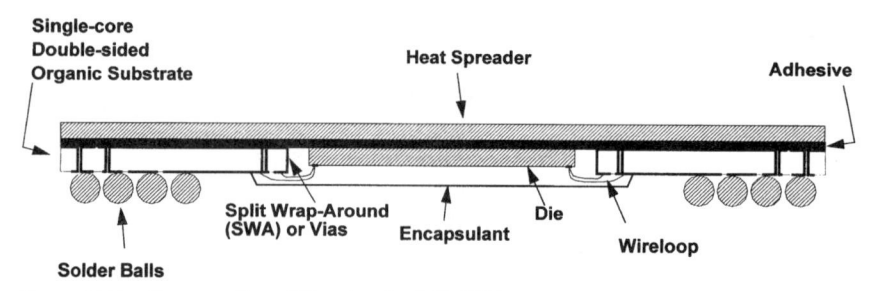

Figure 4.140 Cross section of the standard NuBGA.

the top side of the substrate, split planes on the bottom side of the substrate are designed to match the configuration of the split-ring segments. Split-ring segments and split planes are then connected through split-wrap-around (SWA) on the side wall of the cavity of the substrate. Signal traces on the top side of the substrate are covered by the split power/ground planes on the bottom side of the substrate to form the µ-stripline structure. Signal traces routed on the bottom side of the substrate are sandwiched by the split power/ground planes to form coplanar stripline structures. The µ-stripline and coplanar stripline structures provide controlled impedance transmission line characteristics and smaller cross talk for signal traces. The transmission-line structures also provide stronger coupling between signal traces and power/ground planes that reduce the effective power/ground inductance and SSO noise for better signal and power-loop integrity.

4.11.1.2 Programmable VDD/VSS split-via-connections (SVC) and µ-stripline and coplanar stripline traces. From a substrate manufacturing point of view, SWA requires more tooling processes, thus, a variation of SWA called split-via-connections (SVC) was developed. Instead of using SWA, via is applied to connect the split-ring segments and planes as illustrated in Fig. 4.142.

4.11.2 NuBGA design examples

Layout designs of a 35 × 35-mm body size, 352-solder-balls SWA NuBGA are illustrated in Fig. 4.143a and b for electroless Ni/Au bonding fingers, and in Fig. 4.144a and b for electroplated Ni/Au bonding fingers. (Figures 4.143a and b show an example of SWA NuBGA, while Fig. 4.144a and b show an example of SVC NuBGA.) Figures 4.143a and 4.144a show the top-side layout of split power/ground ring segments, bonding fingers for the µ-stripline and coplanar stripline traces, and solder balls. Figures 4.143b and 4.144b show the bottom-side layout of split power/ground planes and coplanar stripline traces. From the previous discussion, NuBGA designs can be summarized as follows [154, 155]:

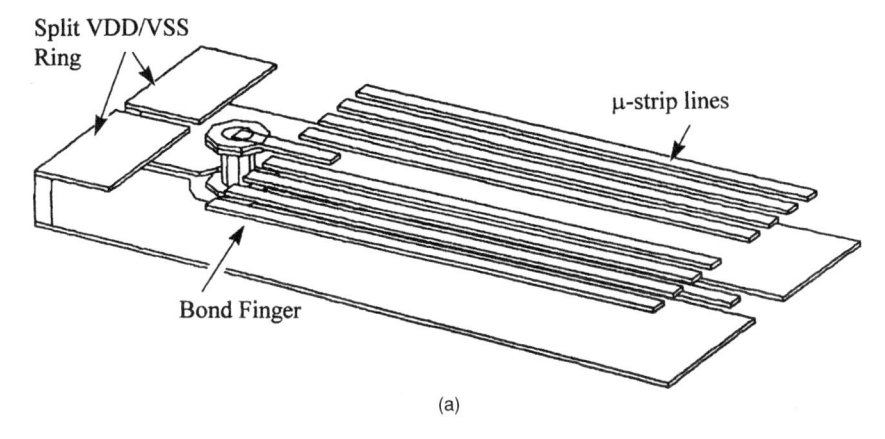

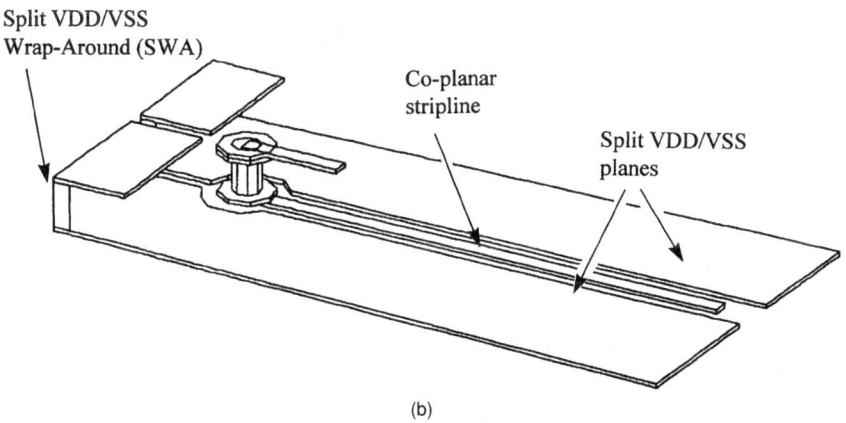

Mechanical Design/Analysis/Measurement/Reliability of Electronic Packaging   355

Figure 4.141 (*a*) Split power/ground ring (into segments) on the top side of the substrate connected to (*b*) split power/ground planes through split wrap around (SWA). Signal traces are routed in either μ-stripline or coplanar stripline structures.

4.11.2.1   Conventional PCB design rules and process

- Package body size: <50 mm
- Chip size: <20 mm
- Number of solder balls: <600
- Solder ball distributions: 4 or 5 rows
- Solder ball pitch: 50 or 40 mils (1.27 or 1 mm)
- Bond-finger pitch: 4/4 or 5/3 mils (.1/.1 or .125/.075 mm)
- Trace width/space: 4/4 mils (.1/.1 mm)
- Through-hole via diameters: 12 mils (.3 mm)
- Through-hole land diameter: 22 mils (.56 mm)
- Solder ball land diameter: 30 mils (.76 mm)

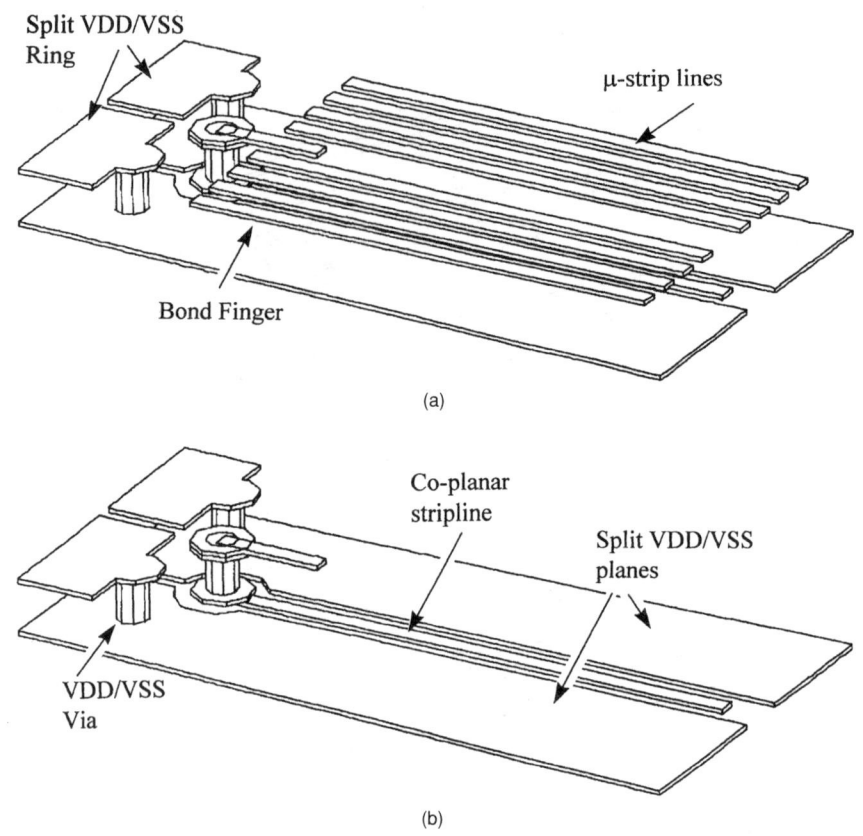

Figure 4.142 In contrast to SWA. (*a*) Split ring (in segments) on the top side of the substrate and (*b*) the split planes on the bottom side of the substrate are connected through via (SVC).

- Solder mask opening for solder ball: .65 mm
- Single-core two-metal layer substrate
- Cavity-down design with heat spreader

4.11.2.2 Electrical optimized and thermally enhanced low-cost package

- Single split-ring segments for power/ground wire-bonding
- Power/ground split-wrap-around (SWA) or split-via-connections (SVC)
- Flexible I/O to power/ground pad ratio
- Flexible power/ground supporting multiple power supply and noise decoupling
- Controlled impedance µ-stripline and coplanar stripline traces
- Strong coupling between signal to both power/ground to reduce simultaneous switch output (SSO) noise

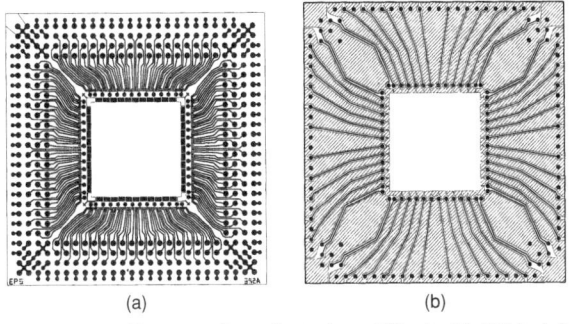

Figure 4.143 Layout of an electroless 352-pin NuBGA: (a) top side, and (b) bottom side.

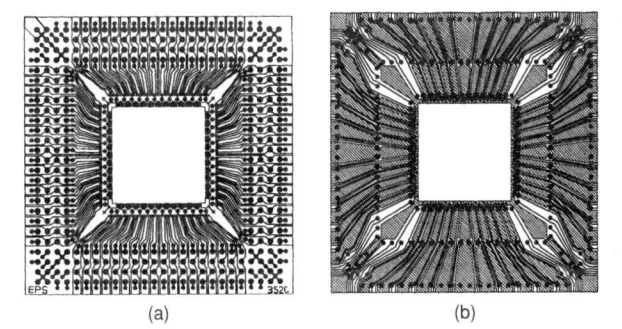

Figure 4.144 Layout of an electroplated 352-pin NuBGA: (a) top side, and (b) bottom side.

- Low thermal resistance
- Low package warpage

4.11.3 NuBGA package family

As described earlier, with the special design concept, NuBGA provides outstanding electrical and thermal performance. Table 4.34 shows a NuBGA package family (1.27-mm ball-pitch) with different body sizes and ball counts. Further thermal and electrical enhanced NuBGA may be achieved by applying additional metal stiffeners and thinner core substrates.

4.12 Analysis and Measurement of NuBGA— Electrical Performance

Parasitic parameters (R, L, C) are extracted from TDR and TDT measurements [156]. Package resistance (R), inductance (L), and capacitance (C) are compared with the competitive designs: PQFP, conventional single-core double-sided cavity-up PBGA and four-metal-layer

TABLE 4.34 NuBGA Package Family (1.27-mm Ball-Pitch Only)

Package size (mm × mm)	Ball pitch (mm)	Ball matrix	Ball footprint (mm)	Ball count 4 rows	Ball count 5 rows
27 × 27	1.27	20 × 20	24.13	256	300
31 × 31	1.27	23 × 23	27.94	304	360
35 × 35	1.27	26 × 26	31.75	352	420
37.5 × 37.5	1.27	29 × 29	35.56	400	480
40 × 40	1.27	31 × 31	38.1	432	520

EPBGA. To evaluate the SSO noise performance, system circuits representing a generic point-to-point system net were created. System simulations were performed to compare the system performance of NuBGA with the said packages. (For more information about R, L, C, and SSO noise, please refer to Chap. 2 of this book.)

4.12.1 NuBGA package parasitic parameters (R, L, C)

Electrical parameters (R, L, C) of a 35 × 35-mm body size, 352-ball NuBGA is measured from TDR and TDT measurements. Given V_{step} = 250 mV, cable and fixture Z_0 = 50 ohms, electrical parameters from TDR/TDT measurements can be extracted from the following equations.

Inductance (L) measurement:

$$L = \frac{Z_0}{2V_{step}} \int_0^\infty (W_{pkg} - W_{short})\, dt \tag{4.275}$$

Mutual inductance (L_m) measurement:

$$L_m = \frac{Z_0}{2V_{step}} \int_0^\infty W_{xt}\, dt \tag{4.276}$$

Capacitance (C) measurement:

$$C = \frac{1}{2Z_0 V_{step}} \int_0^\infty (W_{open} - W_{pkg})\, dt \tag{4.277}$$

Mutual capacitance measurement:

$$C_m = \frac{1}{2Z_0 V_{step}} \int_0^\infty W_{xt}\, dt \tag{4.278}$$

where V_{step} is the absolute value of amplitude of the reflected step determined from W_{short} and Z_0 is the impedance of the TDR fixture lead.

Mechanical Design/Analysis/Measurement/Reliability of Electronic Packaging 359

W_{xt} is the signal wave form induced in the adjacent lead. W_{pkg} and W_{open} are the wave forms shown in Figs. 4.145 and 4.146.

4.12.1.1 Measurement results of traces with coplanar stripline, A12 (S-G measurement). Figure 4.145 shows the measured waveforms for inductance and capacitance extractions of a long coplanar stripline trace, A12. The flat TDR waveform of the package trace shows that the trace has a well-controlled impedance. The measured inductance $L = 5.7$ nH, and capacitance $C = 1.16$ pF.

4.12.1.2 Measurement results of a µ-stripline, D13. Figure 4.146 shows the measured waveforms for inductance and capacitance extractions of a long µ-stripline trace, D13. The measured inductance $L = 4.25$ nH, and capacitance $C = 0.84$ pF.

4.12.1.3 Cross-talk measurement results between two adjacent µ-stripline traces, D13 and B13. Figure 4.147 shows the measured cross-talk waveforms between two adjacent µ-stripline traces, D13 and B13, for mutual inductance and capacitance extractions. The measured mutual inductance is $L_m = 0.6$ nH, and mutual capacitance is $C_m = 0.23$ pF.

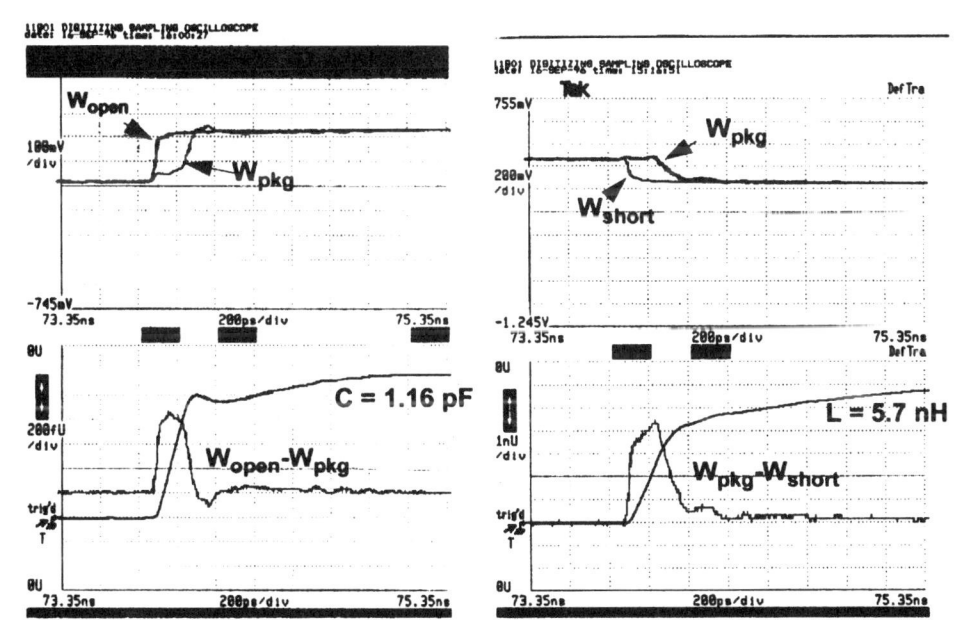

Figure 4.145 Measured TDR wave forms for L and C extraction for a stripline trace A12. Notice the constant impedance profile of the trace.

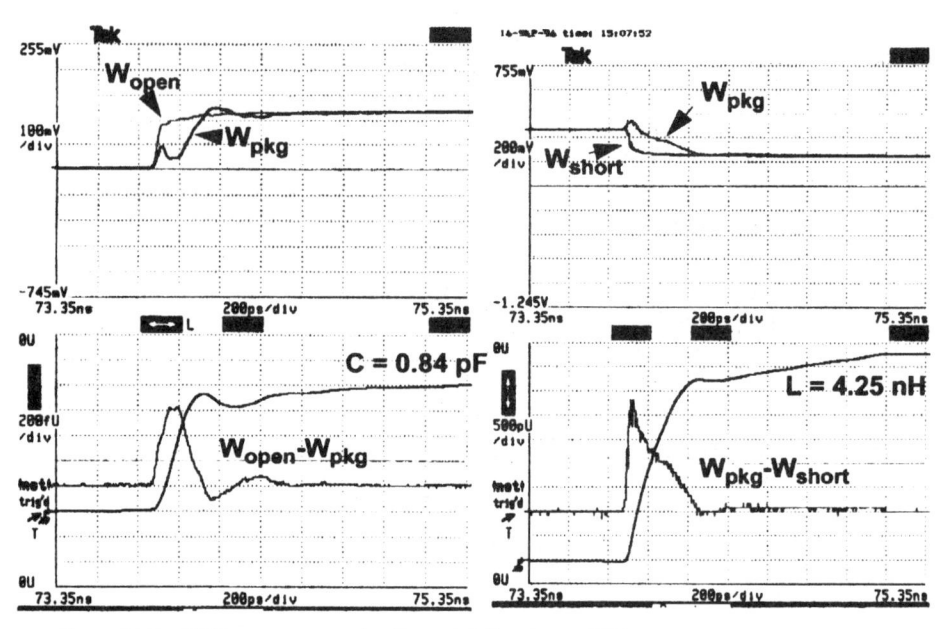

Figure 4.146 TDR measurements of a µ-stripline trace, D13.

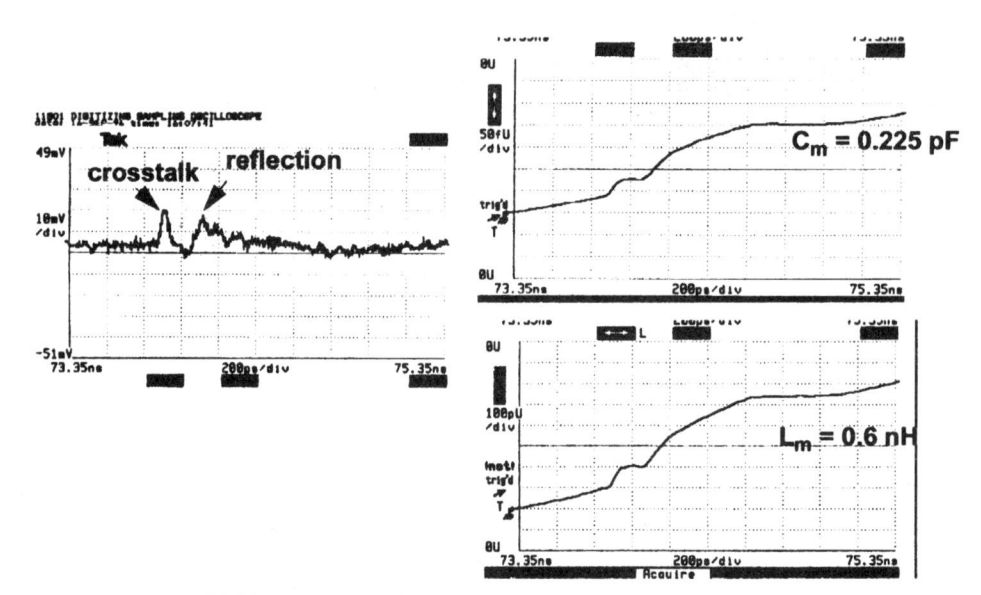

Figure 4.147 TDT measurement for crosstalk between two adjacent µ-striplines (D13 and B13).

Mechanical Design/Analysis/Measurement/Reliability of Electronic Packaging   361

4.12.2   NuBGA electrical analysis and comparison with other packages

4.12.2.1   NuBGA package parasitic parameters' performance. Table 4.35 lists the electrical parameters of a 352 NuBGA and Table 4.36 shows the comparisons of these parameters with those of a PQFP, a conventional cavity-up PBGA and a four-metal-layer EPBGA. As shown in Tables 4.35 and 4.36, the inductances (L) of both center trace (A12) and corner trace (B4) are reduced by about 30 percent due to the coplanar striplines' structure, which is achieved by SWA or SVC design concepts for the NuBGA package.

As shown in Table 4.36, the inductive coupling coefficient, K_{Lm}, between signal traces ranges from 0.14 to 0.15. It is not only relatively small when compared to other packages, it is also more uniformly distributed and less dependent on the traces' location. This more evenly distributed phenomenon also applies to other parameters, such as capacitive coupling coefficient K_{Cm} and backward coupling coefficient K_b. Since NuBGA provides more evenly distributed electrical parasitic parameters, a controlled impedance ball grid array package is hence obtained.

Furthermore, the coupling coefficients between signal traces and power/ground planes $K_{sig-VDD}$ and $K_{sig-VSS}$ for NuBGA are also enhanced due to the SWA or SVC design concepts, as shown in Table 4.36. By increasing the magnitude of $K_{sig-VDD}$ and $K_{sig-VSS}$ the ground-bounce effect will be reduced. Therefore, NuBGA provides better overall electrical performance for system applications, such as less cross talk and lower SSO noise, which is discussed next.

4.12.2.2   NuBGA SSO noise on the quiet low and high lines. SSO noise performance of the packages discussed in Table 4.36 are compared through system simulations. Figure 4.148 illustrates the simulated SSO noise on quiet low and high lines at 100-MHz clock frequency. Figure 4.149 compares the noise values among different packages. Figure 4.150 shows the relative performance index of different packages by normalizing the noise values with respect to PQFP. In Figs. 4.148 to 4.150, SSO noise on the quiet low lines results from a ground bounce, which is indicative of package signal and ground performance. SSO noise on the quiet high line results from a power drop, which is indicative of package signal and power performance. Figures 4.148 to 4.150 show that NuBGA provides up to 3× SSO noise performance over PQFP, and 2.5× over conventional PBGA. NuBGA provides comparative performance with EPBGA in noise performance on the quiet low line and outperforms EPBGA in noise performance on the quiet high line due to stronger coupling between signal traces and power plane. In

TABLE 4.35 TDR/TDT Measurement of an Electroplated 352-Pin NuBGA*

Parameters	Pin name	L (nH)	L_m (nH)	K_{Lm}	C (pF)	C_m (pF)	K_{Cm}	K_b	Z_0 (ohms)
Center traces	D13 (μ-strip)	4.25	0.6 (D13–B13)	0.14	0.84	0.225 (D13–B13)	0.27	0.1	98
	A12 stripline	5.7	—	—	1.16	—	—	—	65
Corner traces	B5 μ-strip	6.8	1.0 (B5–C6)	0.15	1.3	0.35 (B5–C6)	0.27	0.12	98
	B4 stripline	6.4	—	—	1.5	—	—	—	65
Summary	μ-strip	4.25–6.8	0.6–1	0.14–0.15	0.84–1.3	0.225–0.35	0.27	0.1–0.13	98
	stripline	4.3–4.8	—	—	1.16–1.5	—	—	—	65

*Measurement data includes package substrate and solder balls.

TABLE 4.36 Comparisons of the 352-Pin NuBGA Electrical Parameters (Assuming 2-mm Signal Wirelength)

Package types	L (nH)	K_{Lm}	C (pF)	K_{Cm}	K_b	$K_{sig\text{-}VDD}$	$K_{sig\text{-}VSS}$	R (mOhm)	L_{VDD}	L_{VSS}
NuBGA	4.25–6.8	0.14–0.15	0.8–1.3	0.27	0.1–0.13	0.1–0.2	0.1–0.2	91–125	0.06–2.5	0.06–2.5
PQFP	7.3–9.8	0.3–0.5	0.7–1.1	0.3–0.5	0.15–0.25	0	0	80–90	7.3–9.8	7.3–9.8
PBGA 2-metal, 1 dielectric	4–11	0.2–0.45	1–1.5	0.2–0.4	0.1–0.21	0	0	120–178	0.2–3.5	0.08–2.5
PBGA 4-metal, 3 dielectric	3.6–9.0	0.1–0.35	1.2–1.8	0.1–0.33	0.05–0.17	0	0.2	132–190	0.1–3.0	0.06–2.0

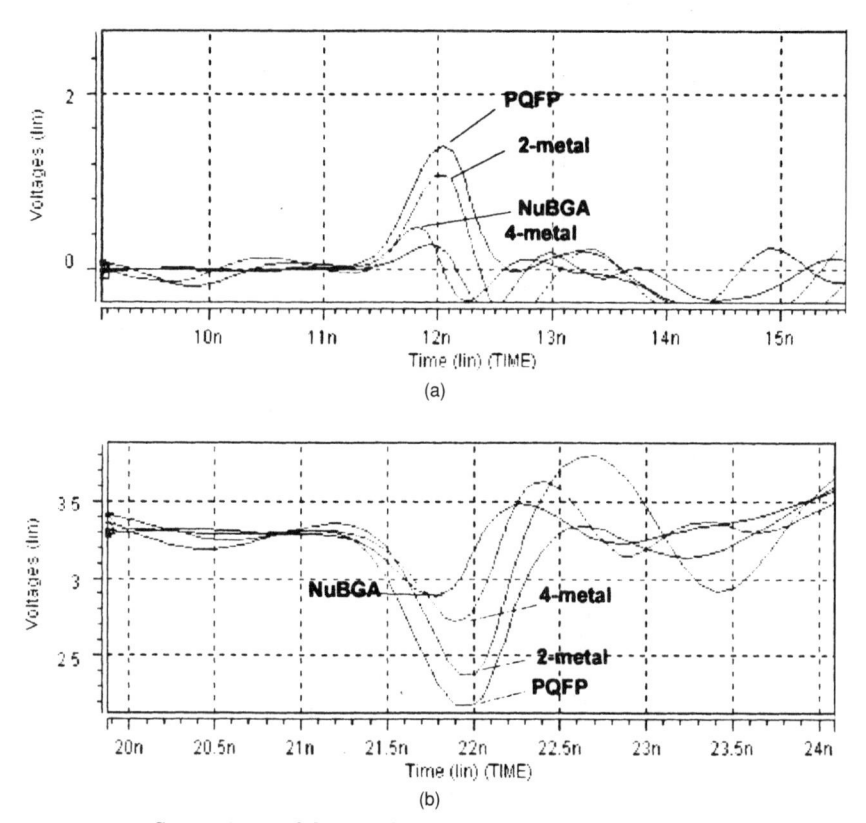

Figure 4.148 Comparisons of the simulated SSO noise on quiet low and high lines. (*a*) SSO noise on quiet low lines and (*b*) SSO noise on quiet high lines.

summary, NuBGA's low-cost single-core double-sided substrate design has the same cost structure as PBGA, yet it has similar performance as multilayer PBGA.

4.13 Analysis and Measurement of NuBGA—Thermal Performance

In this section, the thermal performance of a 35 × 35-mm NuBGA with various sizes of heat spreader are presented. First, we use the finite-element method to determine the temperature and heat-dissipation distributions, and thermal resistance of the NuBGA with a partial (28 × 28-mm) heat spreader under various air flow rates. Next, we use a wind tunnel to measure the thermal resistance of the NuBGA. The confidence of the finite-element modeling is gained by comparing with the experimental results. Finally, the finite-element method is used to

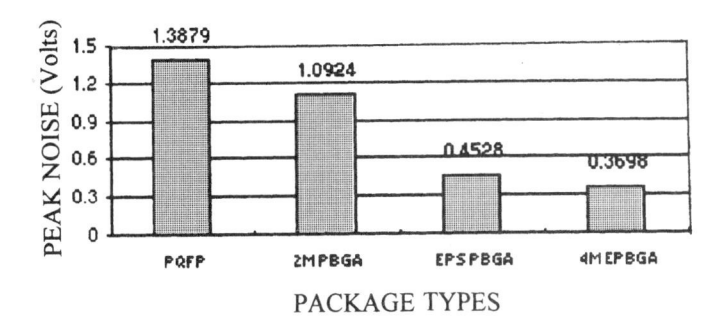

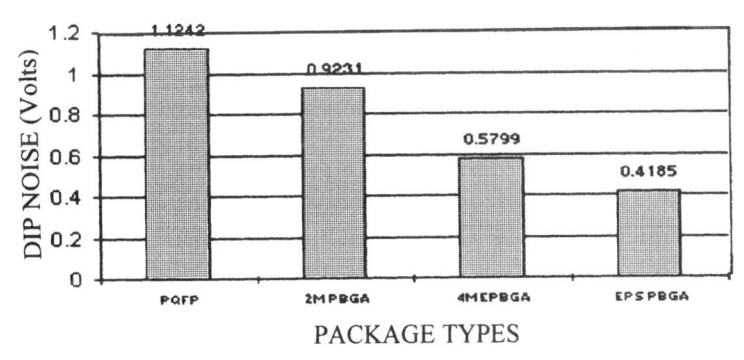

Mechanical Design/Analysis/Measurement/Reliability of Electronic Packaging 365

Figure 4.149 Comparisons of SSO values.

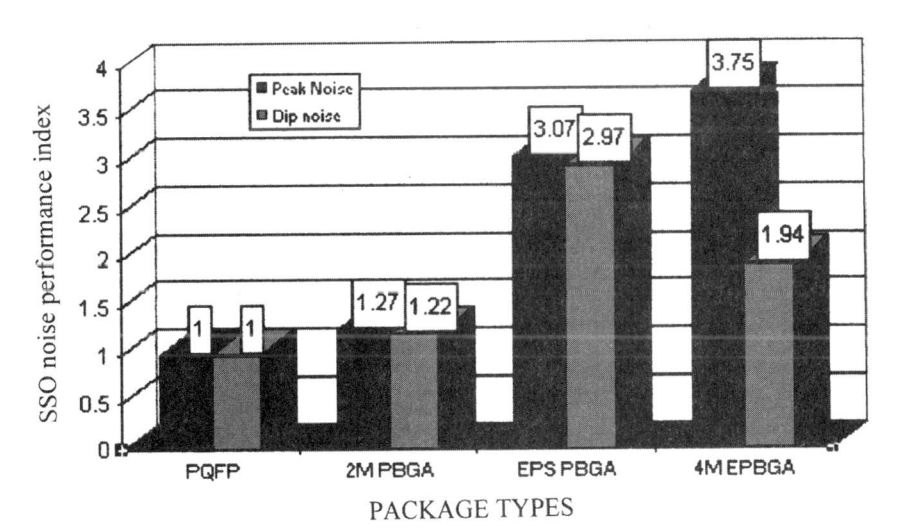

Figure 4.150 Comparisons of relative performance index.

determine the heat-dissipation distribution and thermal resistance of the NuBGA with six other sizes of heat spreader.

Also, in this section, the thermal performance of NuBGA with a couple of heat sinks is measured. Finally, the thermal-impedance measurement of NuBGA is presented.

4.13.1 Finite-element analysis of NuBGA with a partial heat spreader

The NuBGA shown in Figs. 4.140 and 4.144 has four arrays of peripheral solder balls with 1.27-mm pitch. The solder material is 63wt%Sn/37wt%Pb. The package is square, with body size of 35 × 35 mm. There is a 12 × 12-mm cavity in the single-core BT substrate with two metal layers to accommodate the silicon chip. A 10 × 10-mm silicon chip is bonded with 0.025-mm-thick thermally conductive adhesive to a 28 × 28-mm copper heat spreader 20 mils (0.5 mm) thick. The cavity is then filled by encapsulant to protect the silicon chip and bond wires.

The NuBGA is mounted to a multilayer PCB. The board dimensions are 10.54 × 10.16 × 1.57 mm and is an FR-4 epoxy glass with four half-ounce (0.018-mm-thick) copper layers. The top and bottom copper layers are for signal and the two embedded layers are for power and ground, respectively.

In this study, all the finite-element models include most of the key elements of overall package systems such as die, copper heat spreader, die attach, BT substrate, encapsulant, solder balls, via, and PCB. A typical finite-element model for the NuBGA package with a multilayer PCB is shown in Fig. 4.151. Due to the symmetry of the package, only one quarter of the package is modeled using the eight-node brick elements. Table 4.37 lists the material properties used in the simulations, where k is thermal conductivity, E is Young's modulus, α is coefficient of thermal expansion, and ν is Poisson's ratio. Table 4.38 lists the coefficients of convective heat transfer h for the NuBGA and PCB surfaces calculated by the flat-plate correlation [155, 157]. The main purposes of thermal analyses are to determine the temperature distribution and to predict the junction-to-ambient thermal resistance, which will be compared to experimental results discussed in a later section.

4.13.1.1 Temperature distribution.
A typical temperature distribution of the NuBGA and PCB, chip and copper heat spreader, substrate and solder ball, and PCB only are shown in Fig. 4.152 for an air flow rate of 1 m/s. While the distribution is very much geometrically dependent, it does in general provide a primary cooling path for the overall package. For the 35 × 35-mm NuBGA package with a 28 × 28-mm heat spreader, the calculation shows that 37 percent of the heat generated by the chip is dissipated through the copper heat spreader; 45 percent of heat goes

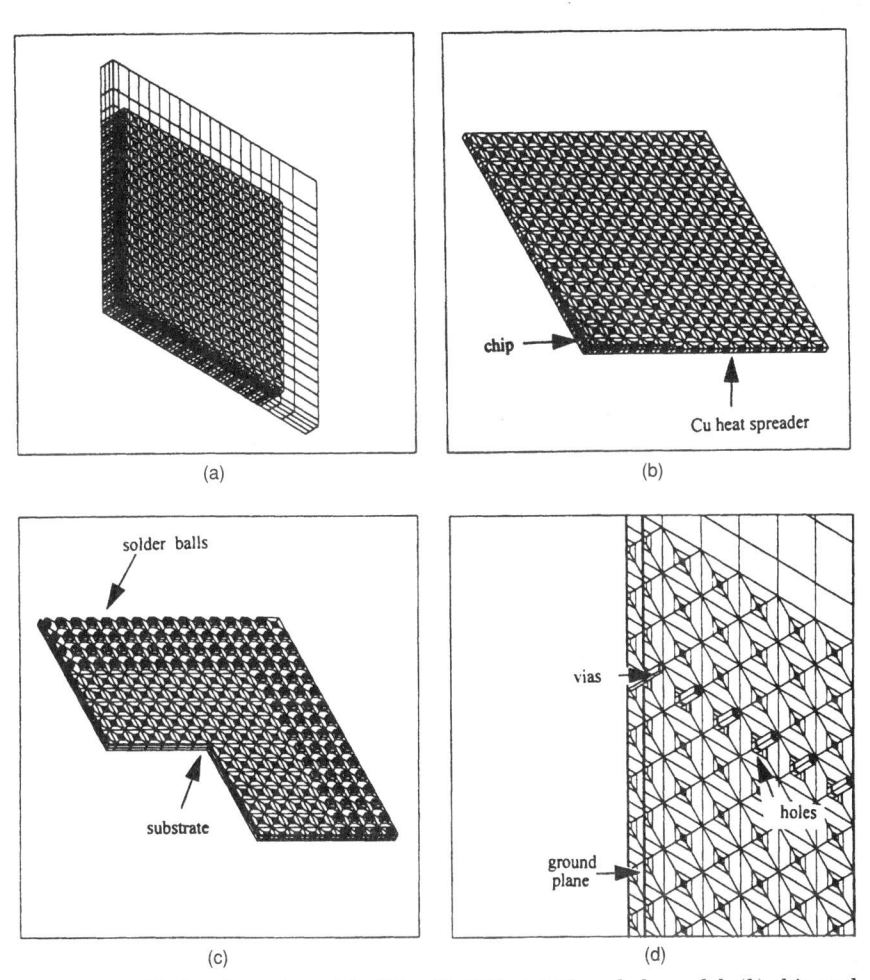

Mechanical Design/Analysis/Measurement/Reliability of Electronic Packaging 367

Figure 4.151 Finite element model of the NuBGA. (*a*) The whole model, (*b*) chip and heat spreader, (*c*) substrate with solder balls, and (*d*) power and ground planes with via in the PCB.

through the copper heat spreader, BT substrate, solder ball, and dissipates through PCB; 1.5 percent is through four sides of the copper heat spreader; 11 percent of heat transfers to PCB through encapsulant (in this case, the gap between the encapsulant and PCB is 0.2032 mm), and 5.5 percent transfers to PCB through the air gap between solder balls, as shown in Fig. 4.153 (air flow rate = 1 m/s).

4.13.1.2 Thermal Resistance. The calculated maximum temperatures from the simulations are used to calculate the thermal resistance of the NuBGA. With the known device power, the specified ambient tempera-

TABLE 4.37 Material Properties of the 352-Pin NuBGA

Material	K (W/m°C)	E (GPa)	α (ppm/°C)	ν
Silicon	150	128.2	3.6	.3
Encapsulant	65	11.8	30	.3
Die attach	1	.7	40	.3
Cu heat spreader	260	130.3	17	.35
Organic substrate	.3	18.6 (x)	19.1 (x)	.16 (xy)
		18.6 (y)	19.1 (y)	.42 (yz)
		7.6 (z)	77.3 (z)	.42 (xz)
63/37 solder	50	10.3	21	.4
FR-4 PCB	.3	—	—	—
Power/ground planes	390	—	—	—
Air	.024	—	—	—

TABLE 4.38 Convective Heat Transfer Coefficients of the NuBGA

Air flow rate (m/sec)	h at heat spreader (W/m²–°C)	h at four sides of the package (W/m²–°C)	h at PCB (W/m²–°C)
0	11.0	7.7	20.5
0.25	24.5	11.0	20.5
0.5	31.1	14.9	20.5
1.0	42.4	21.5	21.5
2.0	52.9	27.6	27.6

ture, and the calculated device junction temperature, the thermal resistance of the NuBGA with a 28 × 28-mm heat spreader is calculated and is shown in Fig. 4.154 for different air flow rates. It shows that the thermal resistance of the NuBGA is about 15°C/W at zero air flow rate. It also shows that the thermal resistance of the NuBGA drops significantly with small air flow (12°C/W at 0.25 m/s) and approaches to an asymptotic value with increasing air rates (e.g., 10°C/W at 1 m/s).

4.13.1.3 Cooling power. Assuming the device maximum junction temperature = 115°C and the ambient temperature = 55°C, then the cooling power of the NuBGA for different air flow rates can be obtained, and is shown in Fig. 4.155. It can be seen that the 35 × 35-mm NuBGA with a 28 × 28-mm heat spreader can dissipate 4 watts of heat at zero air flow and 6 watts of heat at 1 m/s.

4.13.2 Experimental measurement of NuBGA with a partial heat spreader

Thermal resistance of the 35 × 35-mm NuBGA with a 28 × 28-mm heat spreader determined by the finite-element method is verified herein by experimental measurements.

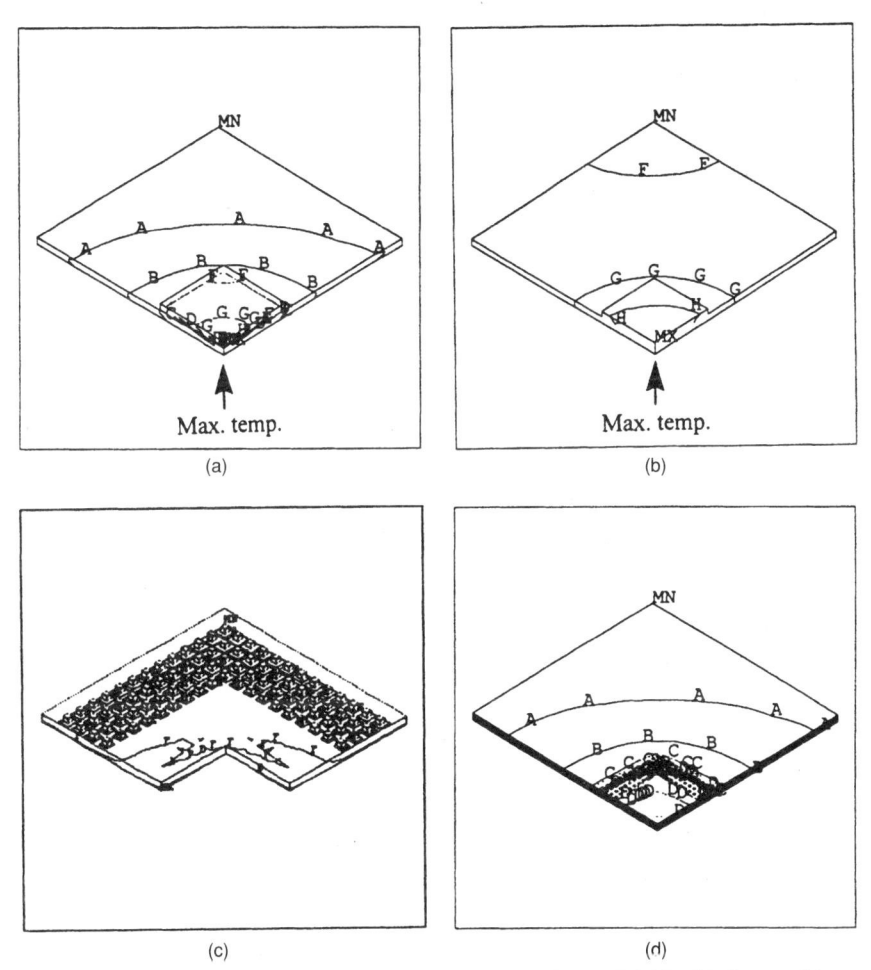

Mechanical Design/Analysis/Measurement/Reliability of Electronic Packaging 369

Figure 4.152 Temperature distribution in the (*a*) NuBGA and PCB, (*b*) chip and heat spreader, (*c*) substrate and solder balls, and (*d*) PCB only (10.54 × 10.16 × 1.57 mm). Note (for 1 watt of heat source): A = 0.9°C, B = 2°C, C = 3.5°C, D = 5°C, E = 6.5°C, F = 7.7°C, G = 8.2°C, H = 8.8°C.

4.13.2.1 Test board. The PCB dimensions are 10.54 × 10.16 × 1.57 mm with four half-ounce copper layers. There are six pairs of solder-ball pads (Fig. 4.156) for thermal test (accessibility) purpose; for example, the resistor pads on the chip can be connected to an external power supply, and the diode pads on the chip can be connected to an external sense current circuit.

4.13.2.2 Test die. The test die (PST6; Fig. 4.157) used in this study is designed and manufactured by Delco and is 10 × 10 mm. It provides one pair of bridge diodes and one pair of heating resistors. The test die is wire-bonded (six pairs) on the bonding fingers of the 325-pin NuBGA

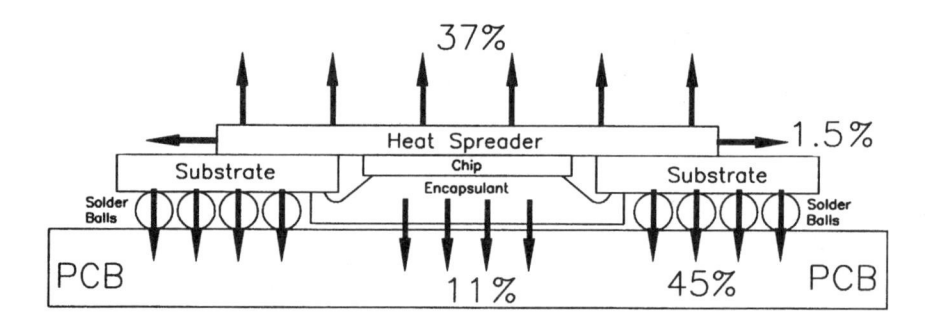

Figure 4.153 Heat dissipation distribution in the (35 × 35 mm) 352-pin NuBGA with a 28 × 28-mm heat spreader (PCB dimensions: 10.54 × 10.16 × 1.57 mm).

Thermal Resistance of 352-pin NuBGA

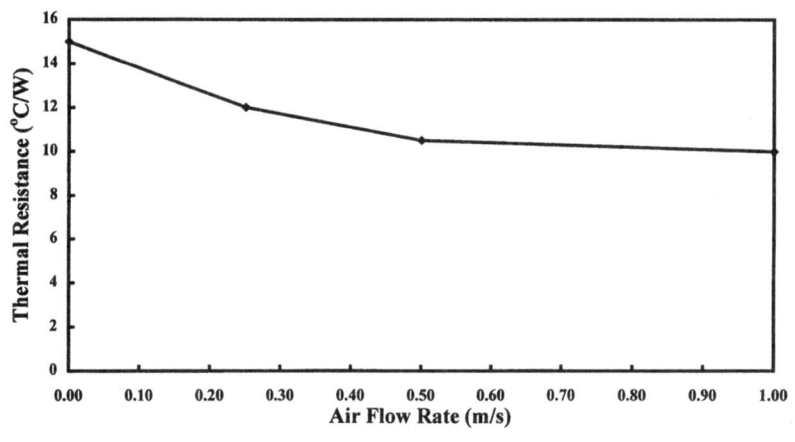

Figure 4.154 Thermal resistance of the 352-pin NuBGA with a 28 × 28-mm heat spreader (PCB dimensions: 10.54 × 10.16 × 1.57 mm).

substrate. These six pairs of bonding wires correspond to those six pairs of solder-ball pads as shown in Fig. 4.156 on the test board, which are routed out for accessibility purpose.

4.13.2.3 Device calibration. The junction temperature in the NuBGA package measured is based on temperature and voltage dependency exhibited by semiconductor diode junctions. The voltage-temperature relationships are an intrinsic electrothermal property of semiconductor junctions. These relationships are characterized by a nearly linear relationship between forward-biased voltage drop and junction temperature, when a constant forward-biased (sense) current is applied.

Cooling Capability of 352-pin NuBGA

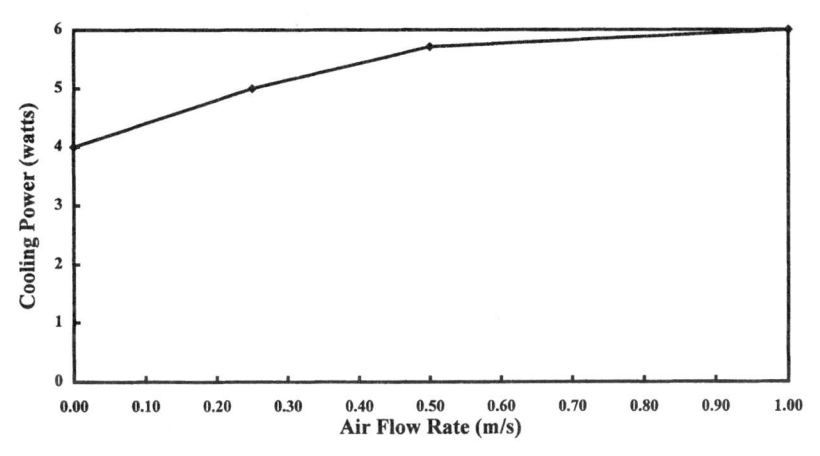

Figure 4.155 Cooling power of the 352-pin NuBGA with a 28 × 28-mm heat spreader (PCB dimensions: 10.54 × 10.16 × 1.57 mm).

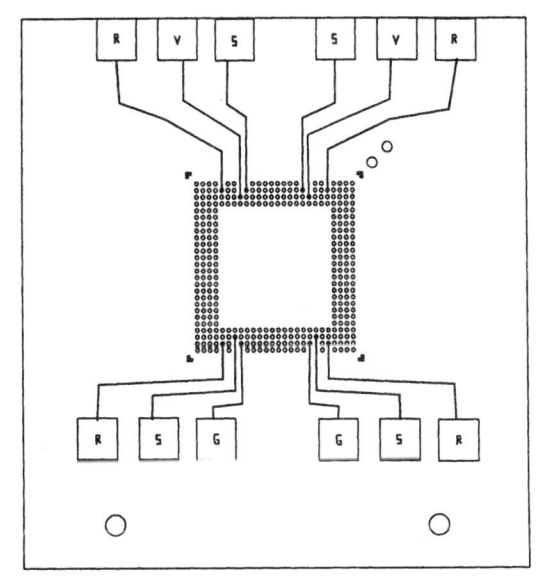

Figure 4.156 Thermal test board (10.54 × 10.16 × 1.57 mm) for measurements.

Figure 4.158 shows the voltage-temperature relationship of the present test die under constant sense current. It can be seen that the slope is equal to 528°C/V and the temperature ordinate-intercept is equal to 415°C. T_j is the junction temperature and V_f is the forward-biased voltage drop.

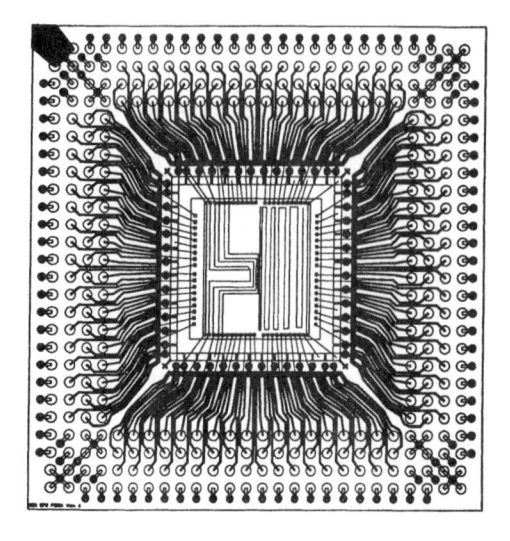

Figure 4.157 Thermal test die (PST6) wire bonded on substrate.

4.13.2.4 Thermal resistance measurement results. Steady-state thermal resistance is measured for the 35 × 35-mm NuBGA with a 28 × 28-mm heat spreader at several power levels of 1, 2, 5, and 7 watts and different air flow rates, 0, 0.12, 0.25, 0.5, 0.75, and 1 m/s. The results are shown in Fig. 4.159. It can be seen that the thermal resistance ranges from 14.5°C/W under still air to 10.8°C/W under 1 m/s air flow rate. The thermal resistance is also slightly reduced under higher power operation, especially when air flow rate is less than 0.25 m/s. The predictions from simulation are also shown in the same figure. The

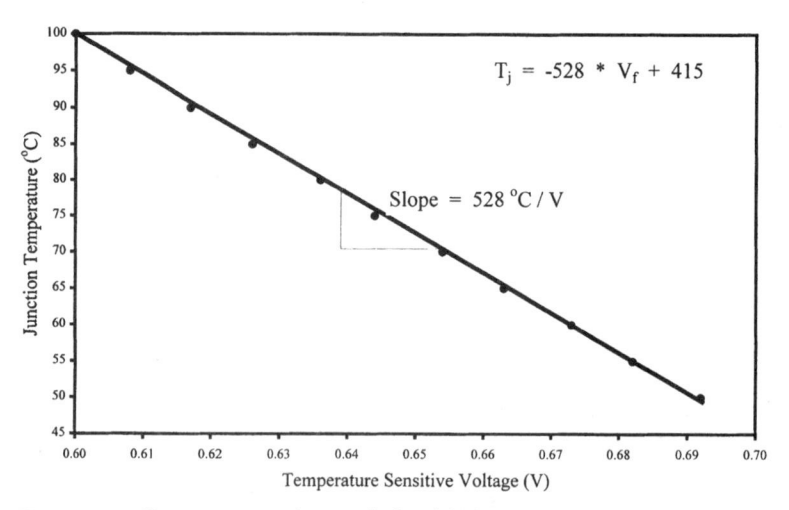

$T_j = -528 * V_f + 415$

Slope = 528 °C / V

Figure 4.158 Temperature-voltage relationship (temperature-sensitive parameter plot).

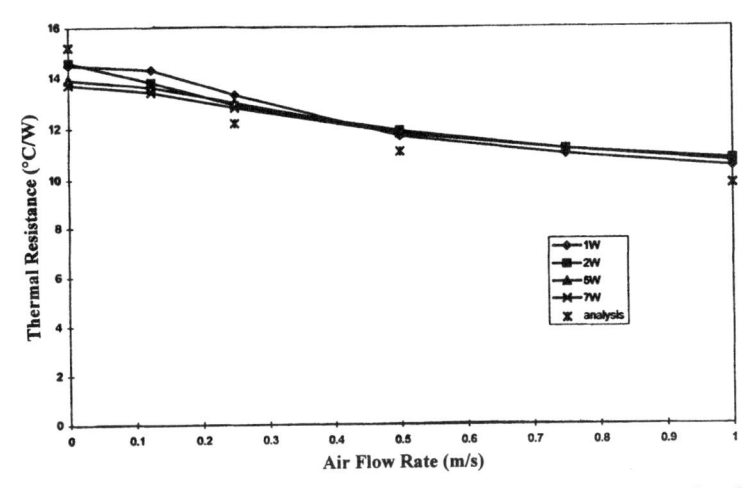

Mechanical Design/Analysis/Measurement/Reliability of Electronic Packaging 373

Figure 4.159 Thermal resistance measurement and finite element results of the 352-pin NuBGA with a 28 × 28-mm heat spreader (PCB dimensions: 10.54 × 10.16 × 1.57 mm).

discrepancy between the measurement and simulation results is within 5 percent.

4.13.3 Experimental measurement of NuBGA with a full-size heat spreader

Thermal resistances of the 35 × 35-mm NuBGA with a full-size (35 × 35-mm) heat spreader at several air flow rates and different power levels are measured. The results are shown in Fig. 4.160. It can be seen that the thermal resistance is about 11.5°C/W at still air and 9°C/W at 1 m/s air flow rate.

4.13.4 Finite-element analysis of NuBGA with various sizes of heat spreaders

In addition to determining the heat-dissipation distribution, thermal resistance and cooling power of the NuBGA with a 28 × 28-mm heat spreader, other (six different) sizes of heat spreader (Fig. 4.161) are considered by the finite-element method in this section. The PCB, die, and the BT substrate of the package are exactly the same as those discussed earlier.

4.13.4.1 Thermal resistance. Figure 4.162 shows the thermal resistance of the NuBGA with different sizes of heat spreader at 1 m/s air flow rate. It can be seen that the thermal resistance of the 35 × 35-mm NuBGA with a full-size (35 × 35-mm) heat spreader is 9°C/W at an air flow rate equal to 1 m/s. (This result agrees very well with the measurement result shown in Fig. 4.160.) Also, it can be seen that the ther-

352-pin NuBGA Thermal Resistance

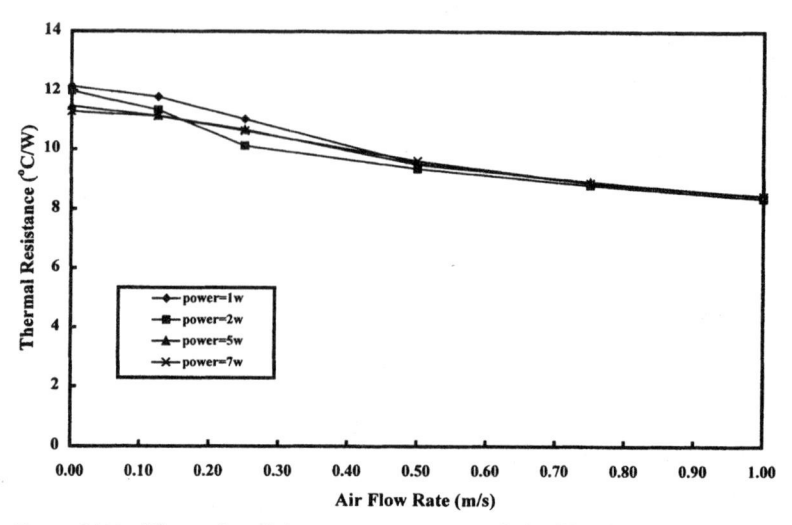

Figure 4.160 Thermal resistance measurement of the NuBGA with full-size (35 × 35 mm) heat spreader (PCB dimensions: 10.54 × 10.16 × 1.57 mm).

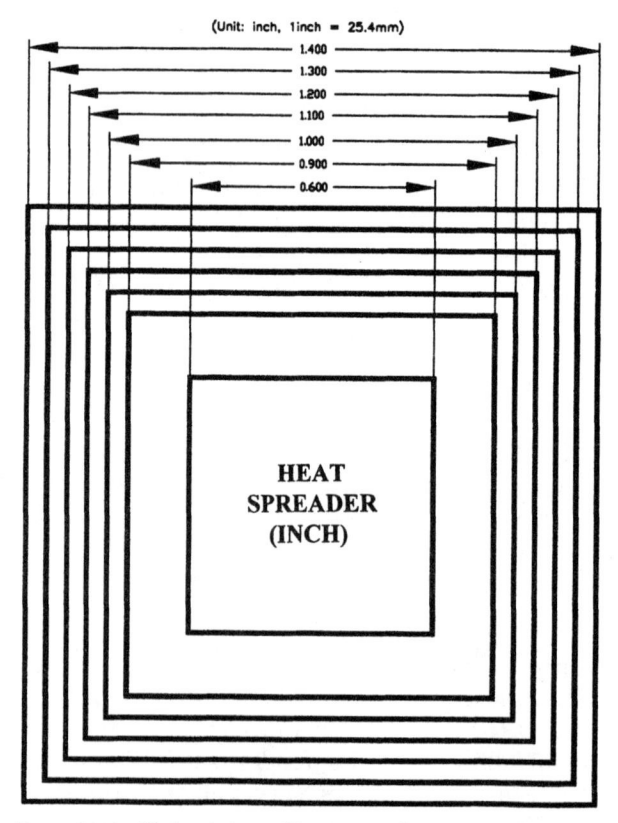

Figure 4.161 Various sizes of heat spreader.

374

Thermal Resistance of 352-Pin NuBGA (1m/sec)

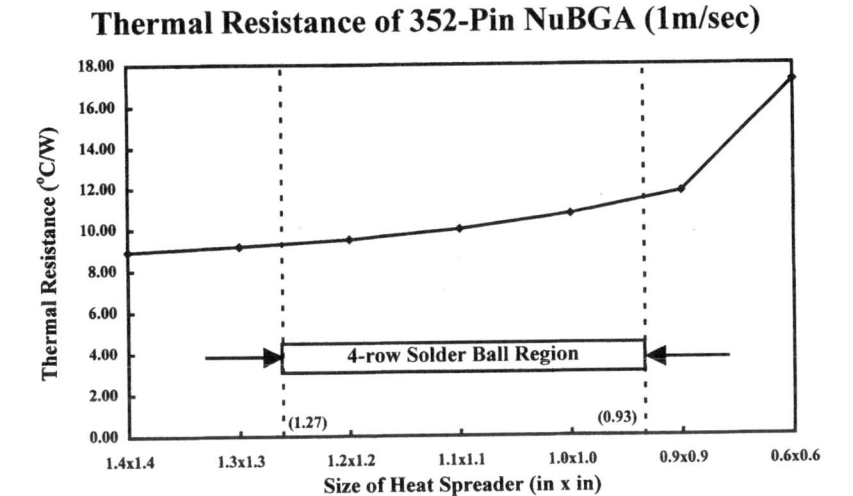

Figure 4.162 Thermal resistance of the NuBGA with various sizes of heat spreader (1 m/s, PCB dimensions: 10.54 × 10.16 × 1.57 mm).

mal resistance does not increase much with the reduction of the size of heat spreader as long as it covers most of the solder balls. However, the thermal resistance increases significantly when the heat spreader is not covering the solder balls. These phenomena can be explained by examining the heat-dissipation distributions.

4.13.4.2 Heat-dissipation distributions.
Heat dissipation from the die to the back of the heat spreader, the die to four sides of the heat spreader, the die to heat spreader to BT substrate to solder balls and then to PCB, and the die to encapsulant to air and then to PCB for different sizes of heat spreader are shown in Fig. 4.163. It can be seen that for all sizes of heat spreader under consideration, the heat dissipation from the die to four sides of the heat spreader is very small (~1.5%).

The heat dissipation from the die to encapsulant to air and then to PCB (~10%) does not increase much as long as the heat spreader is covering most of the solder balls. However, it could become a major heat path (as much as 28%) if the heat spreader (15 × 15 mm) barely covers the die and the cavity opening (12 × 12 mm).

The heat dissipation from the die to heat spreader to BT substrate to solder ball and then to PCB is a major heat path. It increases with a decrease in the size of the heat spreader as long as the heat spreader covers most of the solder balls. However, it decreases when the heat spreader no longer covers the solder balls.

The heat dissipation from the die to the back of the heat spreader is another major heat path. However, it decreases as the heat spreader becomes smaller and has a sharp drop when the heat spreader (15 × 15

Heat Dissipation Distributions of 352-Pin NuBGA (1m/sec)

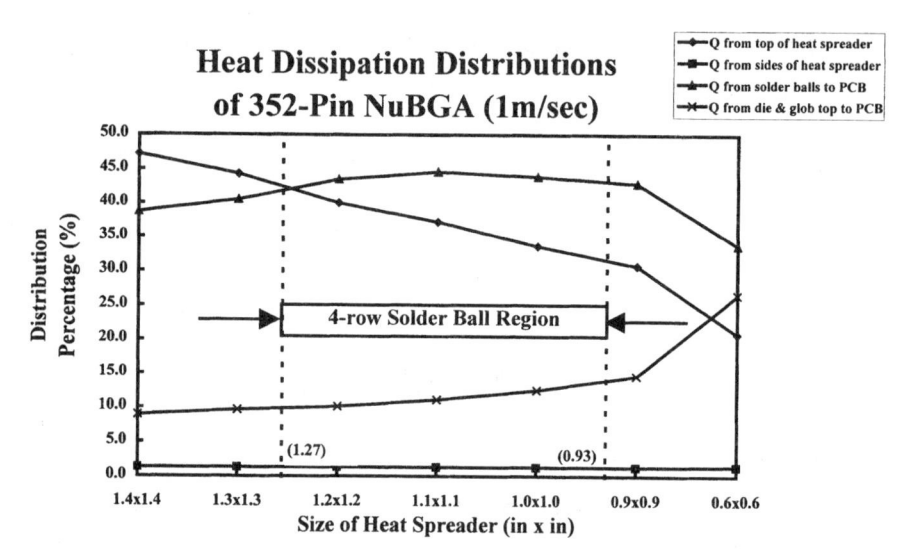

Figure 4.163 Heat dissipation distribution of the NuBGA with various sizes of heat spreader (1 m/s, PCB dimensions: 10.54 × 10.16 × 1.57 mm).

mm) barely covers the die and the cavity opening (12 × 12 mm). As a matter of fact, in this case, the heat dissipation from the die to encapsulant to air and then to PCB (27%) is more than that from the die to the back of the heat spreader (21%).

4.13.4.3 Cooling power. Assuming the device maximum junction temperature = 115°C and the ambient temperature = 55°C, then the cooling power (at 1 m/s) of the NuBGA for different sizes of heat spreader is shown in Fig. 4.164. It can be seen that the cooling power of the NuBGA with a full-size heat spreader at an air flow rate equal to 1 m/s is 6.7 watts. Also, it can be seen that the cooling power of the NuBGA decreases as the size of the heat spreader decreases, especially when the heat spreader barely covers the die and the cavity opening.

4.13.4.4 Solder ball temperature. In predicting the solder-joint reliability of PBGA assemblies under thermal conditions [94], we always assume a temperature for the solder balls. Since it is very difficult to measure or even model the solder ball temperature (thus little available data), we usually assume a very large (conservative) temperature acting at the solder balls. If the predicted thermal fatigue life meets the reliability requirements, then the job is done! However, if the conservatively predicted thermal fatigue life does not meet the reliability requirements—and thus proclaim the solder balls of the package are not reliable and abandon that package—then it is a shame!

In the present study, since we use very detailed finite-element models to calculate the temperature distribution in the assemblies, we are

Cooling Capability of 352-pin NuBGA (1m/sec)

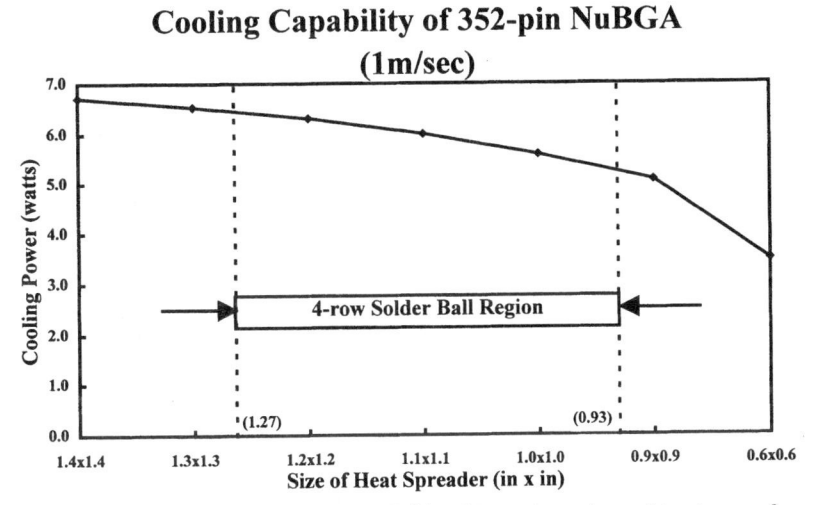

Figure 4.164 Cooling power of the NuBGA with various sizes of heat spreader (1 m/s, PCB dimensions: 10.54 × 10.16 × 1.57 mm).

able to determine the maximum solder ball temperature, which is shown in Fig. 4.165. It can be seen that for the 35 × 35-mm NuBGA with various sizes of heat spreader, the maximum solder ball temperature is between 6 and 7.5°C (this is under 1 watt of heat dissipated from the chip and 1 m/s air flow velocity). By adding this temperature to the ambient temperature, we then obtain the maximum solder-ball temperature.

4.13.5 NuBGA with different sizes of heat sinks

In this study, we measure the thermal resistance of a unidirectional finned heat sink attached on the 35 × 35-mm NuBGA with a full-size (35 × 35-mm) heat spreader. In addition, a bidirectionally cut finned heat sink made by AAVID is also considered.

4.13.5.1 Heat sinks.
Two different heat sinks are studied, as shown in Figs. 4.166 and 4.167. These heat sinks are attached to the NuBGA with a full-size heat spreader by a double-sided tape. The first one has eight 25.5-mm-high fins, with a 50 × 50-mm mounting surface area (Fig. 4.166). The AAVID heat sink has 169 (13 × 13) 8.6-mm-high (0.95 × 0.95-mm) vertical fins, with a 40 × 40-mm mounting surface area (Fig. 4.167).

4.13.5.2 Thermal performance of NuBGA with heat sinks.
Under still air, the larger heat sink (shown in Fig. 4.166) reduces the thermal resistance from 12°C/W to 5.7°C/W (Fig. 4.168), while the smaller AAVID heat sink (shown in Fig. 4.167) reduces it to 8°C/W (Fig. 4.169). At 1 m/s

Figure 4.165 Maximum solder ball temperature in the NuBGA with various sizes of heat spreader (1 watt, 1 m/s, PCB dimensions: 10.54 × 10.16 × 1.57 mm).

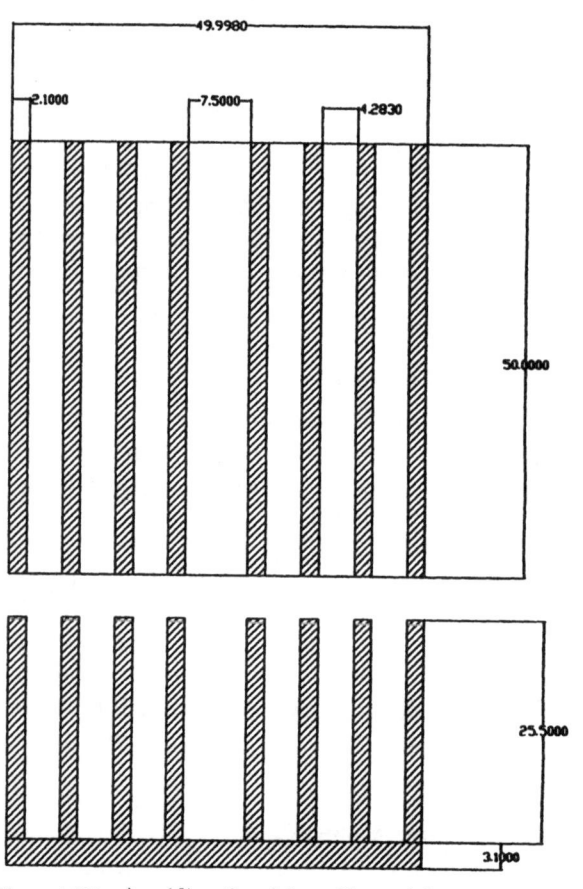

Figure 4.166 A unidirectional finned heat sink.

air flow rate, they are 4°C/W and 6°C/W, respectively. It is noted that, with the larger heat sink (Fig. 4.166), the effect of the size (35 × 35 mm versus 28 × 28 mm) of the heat spreader on the thermal resistance is very small.

Assuming the device maximum junction temperature = 115°C and the ambient temperature = 55°C, then the cooling power of the NuBGA with the larger heat sink is 10.5 watts at still air and 15 watts at 1 m/s, and with the AAVID heat sink is 7.5 watts at still air and 10 watts at 1 m/s (Fig. 4.170).

4.13.6 Thermal impedance measurement of NuBGA

Thermal impedance is to characterize the package temperature ranging from transient to steady state. Thermal impedance at thermal equilibrium is identical with thermal resistance. The thermal impedance transient response to a power step-change is often called *heat characterization* or *heating curve*. One of the insightful data that can be obtained from the heating curve is that it shows the major

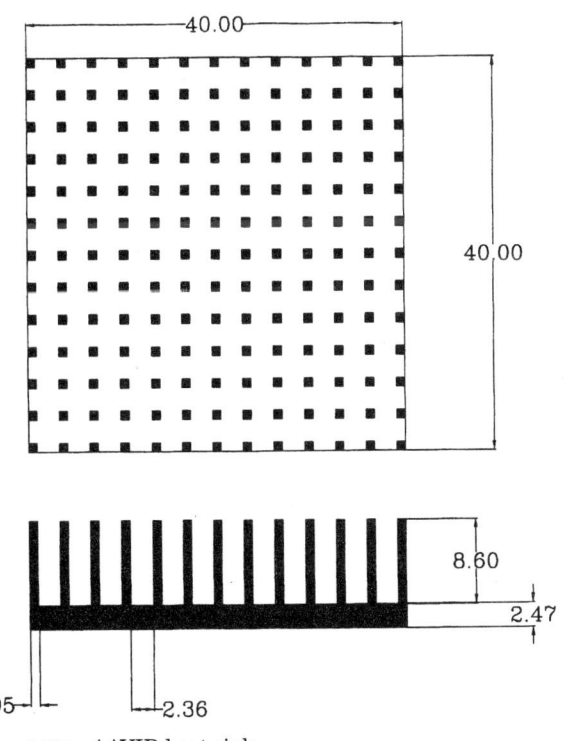

Figure 4.167 AAVID heat sink.

352-pin NuBGA Thermal Resistance with Large Heat Sink

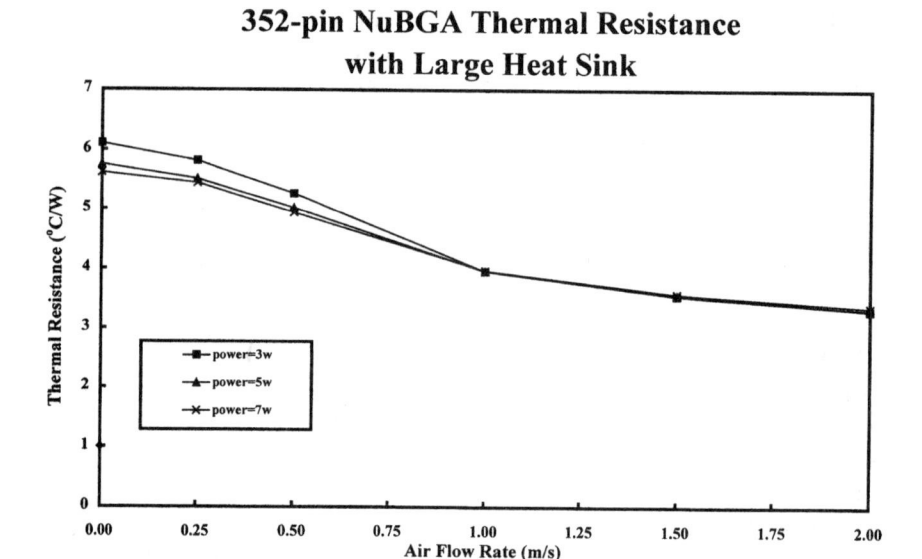

Figure 4.168 Thermal resistance measurement of the full-size heat spreader NuBGA with a unidirectional finned heat sink (PCB dimensions: $10.54 \times 10.16 \times 1.57$ mm).

352-pin NuBGA Thermal Resistance with AAVID Heat Sink

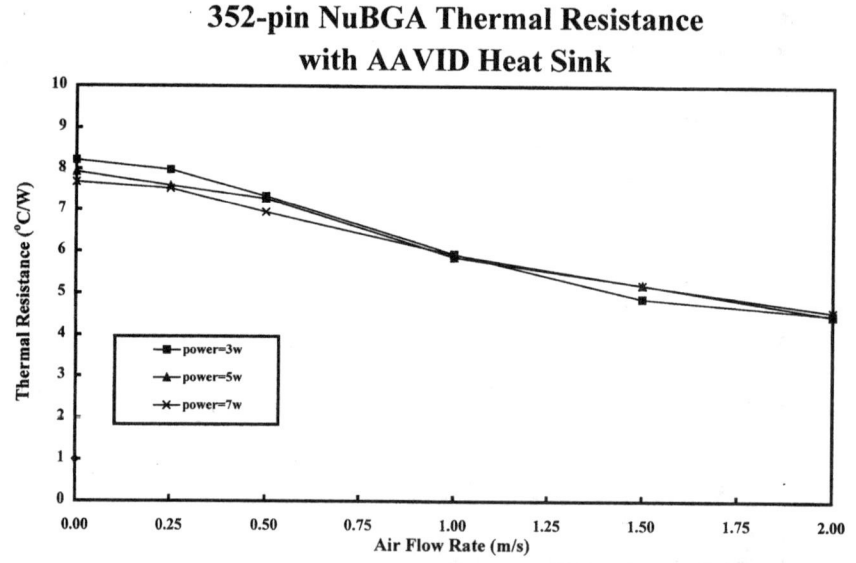

Figure 4.169 Thermal resistance of the full-size heat spreader NuBGA with an AAVID heat sink (PCB dimensions: $10.54 \times 10.16 \times 1.57$ mm).

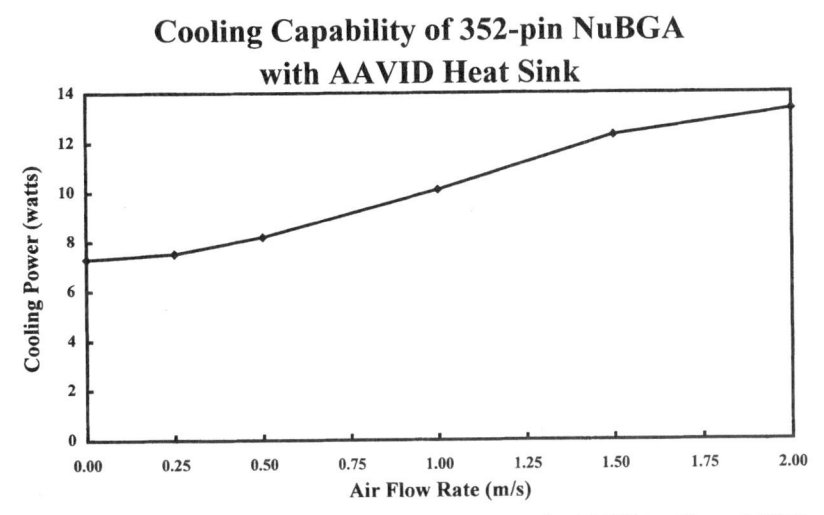

Cooling Capability of 352-pin NuBGA with AAVID Heat Sink

Figure 4.170 Cooling power of the full-size heat spreader NuBGA with an AAVID heat sink (PCB dimensions: 10.54 × 10.16 × 1.57 mm).

obstacle of the thermal cooling path. The heating curve of the 35 × 35-mm NuBGA with a 28 × 28-mm heat spreader under 5 watts and 1000 second duration period is shown in Fig. 4.171.

Figure 4.171 shows that the heat dissipates from the chip within 0.01 ~ 0.1 seconds. At this stage, the thermal impedance is low and flat due to high thermal conductivity of silicon. In between 0.1 and 1 second, the heat is transferred to the copper heat spreader. After the first second,

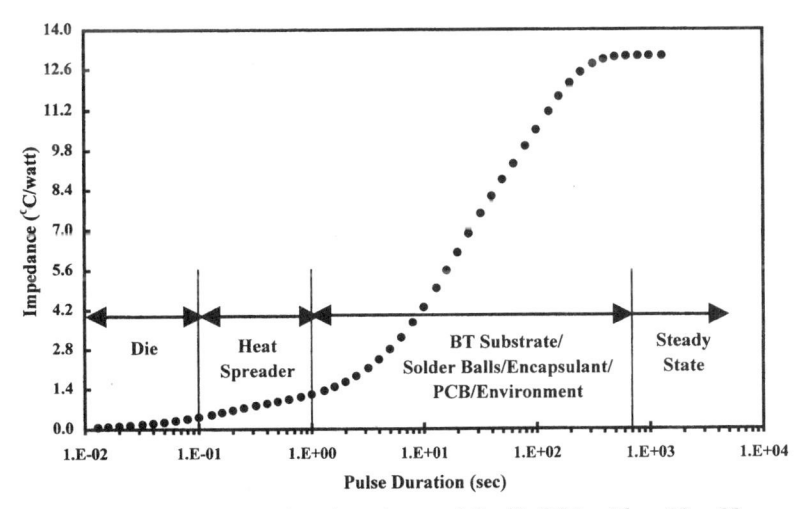

Figure 4.171 Heating transient impedance of the NuBGA with a 28 × 28-mm heat spreader (PCB dimensions: 10.54 × 10.16 × 1.57 mm).

the heat transfers to the BT substrate, solder balls, encapsulant, PCB, and environment, and then reaches a steady state at around 700 seconds. The thermal impedance increases significantly during this period due to low thermal conductivity of the BT, FR-4, and resin materials.

4.13.7 Solder-joint reliability of NuBGA on PCB

The solder-joint reliability of NuBGA on PCB is studied from the cross section shown in Fig. 4.172. The deformation is shown in Fig. 4.173. It can be seen that there is a very large local deflection at the center portion of the assembly, which is due to the thermal expansion mismatch between the silicon chip and the copper heat spreader.

The accumulated effective plastic strain in the solder joints is shown in Fig. 4.174 [158]. It can be seen that the values are very small (less than 0.1 percent) and are smaller than those for the face-up PBGA assemblies discussed in Sec. 4.6.5, and those for the solder-bumped flip-chip PBGA assemblies discussed in Sec. 4.6.6. The solder joints of NuBGA are reliable for use in most of the applications.

4.13.8 Summary

A low-cost and high (electrical and thermal)-performance PBGA package called NuBGA has be designed, analyzed, and measured. It is low

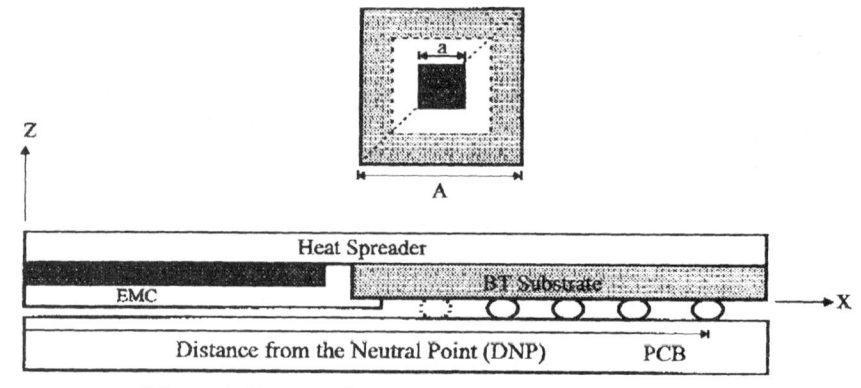

Figure 4.172 Schematic diagram of a cross section of the NuBGA for modeling.

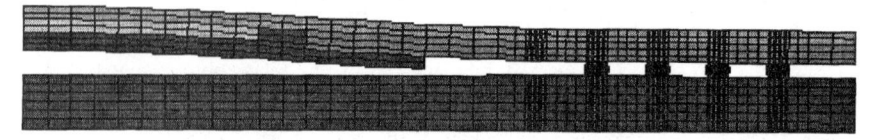

Figure 4.173 Deformation of the NuBGA assembly.

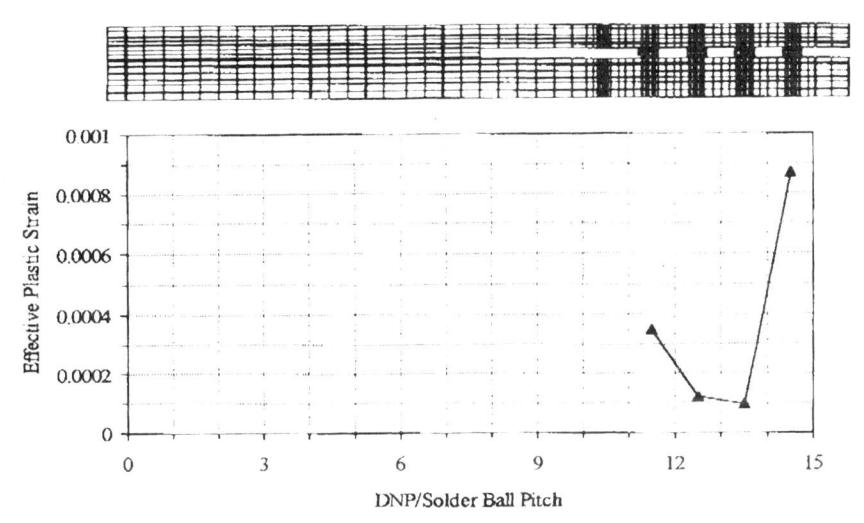

Figure 4.174 Maximum accumulated effective plastic strain in the solder joints.

cost because its substrate uses the conventional PCB design rules and manufacturing process. It has super electrical performance because it uses the concepts of SWA and SVC, such that programmable power/ground ring wire-bonding, shorter wire-bond length, and controlled impedance μ-stripline and coplanar stripline signal traces can be obtained. It has better thermal performance because it allows the back of the chip to attach to a heat spreader. Further enhanced thermal and electrical performance NuBGA may be achieved by applying an additional metal stiffener/heat sink and thinner-core substrate.

4.14 Summary

The thermomechanical governing equations for electronic packaging have been briefly presented and discussed in this chapter. The limitation and assumptions of these equations for electronic packaging applications have also been highlighted. Furthermore, a few closed-form solutions for some simple electronic packaging problems have been obtained.

The definition of stress and strain and the constitutive (stress-strain) relations for elastic, plastic, and creep materials have also been given, especially for the eutectic solder. The TMA, DMA, DSC, and TGA of the 63Sn37Pb solder and four different underfill materials are provided. The effect of these underfill materials on the mechanical and electrical performance of a functional solder-bumped flip chip on PCB has been presented.

Finite-element formulation of electronic packaging with elastic-plastic materials has been provided. Numerical results such as the determination of the deflection and stress of partially routed panels for depanelization, the thermal stress of a plastic leaded package, the mechanical behavior of microstrip structures made from $YBa_2Cu_3O_{7-x}$ superconducting ceramics, the solder joint reliability of PBGAs with either wire-bonding face-up chip or solder-bumped flip chip on BT substrate.

The limitations and roles of DNP in predicting the thermal fatigue life of area-array solder joints has been presented. Four examples: chip scale package on PCB, solder-bumped flip chip on PCB, ceramic ball grid array on PCB, and plastic ball grid array on PCB have been discussed.

The mechanical vibration of electronic packaging has been briefly discussed. In particular, the natural frequencies of packages with heat sink vibrating in the horizontal direction have been provided for a wide range of material and geometry of packaging systems. Also, the fundamental frequency of PCB vibrating in the vertical (out-of-plane) direction is given for PCBs with various boundary conditions.

Finally, a family of hermetic microwave packages, a family of low-cost chip size package—NuCSP—and a family of low-cost, thermal and electrical enhanced plastic ball grid array package—NuBGA—have been designed, analyzed, measured, and qualified.

With the advance of computers and the rapid development of commercially available finite element codes, boundary-value problems in electronic packaging with nonlinear temperature-dependent effects can be solved. However, it should be pointed out that no boundary-value problem can completely describe the complex world around us. Every boundary-value problem aims at a certain class of phenomena, formulates their essential features, and disregards what is of minor importance. The finite-element method is merely a technique to obtain approximate solutions of a boundary value problem. Thus the physical insights, basic assumptions, governing equations, and special features of the boundary-value problem should be borne in mind while constructing and analyzing a finite-element model of the problem. Verify the finite-element results by actual tests, if possible. If not, then verify the analysis procedures and the accuracy of the finite-element codes by solving similar types of problems that have experimental results or closed-form solutions. As a matter of fact, this last step is vital and should be done first.

4.15 Acknowledgments

The author (Lau) thanks G. Barrett for his contribution on the stress and deflection analysis of partially routed panels for depanelization; C. Chang for his contribution on the TMA, DMA, DSC, and TGA of solders; C. Chang and R. Chen for their contribution on the effects of underfill encapsulant on the mechanical and electrical performance of a functional flip-chip device; L. Moresco for his contribution on mechanical behavior of microstrip structures made from $YBa_2Cu_3O_{7-x}$ superconducting ceramics; R. Lee for his contribution on design for plastic ball grid array solder joint reliability; L. Lian-Mueller for her contribution on finite element modeling for optimizing hermetic package reliability; T. Chou, K. Chen, F. Wu, Y. Chen, T. Chen, C. Chang, R. Chen, L. Hu, and W. Koh for their contribution on the low-cost chip size package - NuCSP; and T. Chen, T. Chou, K. Chen, F. Wu, Y. Chen, and W. Koh for their contribution on the low-cost thermal and electrical enhanced plastic ball grid array package - NuBGA. Useful and constructive input from C. H. Cheng, S. W. Lao, T. J. Tseng, J. C. Deng, Hermen Liu, and C. T. Lin are greatly appreciated.

4.16 References

1. Suhir, S. "Interfacial Stresses in Bimetal Thermostats," *ASME Journal of Applied Mechanics*, vol. 55, 1989, pp. 595–600.
2. Engel, P. A. "Structural Analysis for Circuit Card Systems Subjected to Bending," *Journal of Electronic Packaging, Transactions of ASME*, vol. 112, March 1990, pp. 2–10.
3. Chen, W. T., and C. W. Nelson. "Thermal Stresses in Bonded Joints," *IBM Journal of Research and Development*, 23(2), 1979, pp. 179–187.
4. Lau, J. H. "Thermoelastic Solutions for a Semi-Infinite Substrate with an Electronic Device," *Journal of Electronic Packaging, Transactions of ASME*, vol. 114, September 1992, pp. 353–358.
5. Lau, J. H. "Thermoelastic Solutions for a Finite Substrate with an Electronic Device," *Journal of Electronic Packaging, Transactions of ASME*, vol. 113, March 1991, pp. 84–88.
6. Hu, S. M. "Film-Edge Induced Edge Stress in Substrates," *Journal of Applied Physics*, 50(7), 1979, pp. 4661–4666.
7. Isomae, S. "Stress Distributions in Silicon Crystal Substrates with Thin Films," *Journal of Applied Physics*, 52(4), 1981, pp. 2782–2791.
8. Kuo, A. "Thermal Stresses at the Edge of a Bimetallic Thermostat," *Journal of Applied Mechanics, Transactions of ASME*, vol. 56, 1989, pp. 585–589.
9. Eischen, J. W., C. Chung, and J. H. Kim. "Realistic Modeling of Edge Effect Stresses in Bimaterial Elements," *Journal of Electronic Packaging, Transactions of ASME*, vol. 112, March 1990, pp. 16–23.
10. Lau, J. H. "Closed-Form Solutions for the Large Deflection of Curved Optical Glass Fibers Under Combined Loads," *Journal of Electronic Packaging*, vol. 115, September 1993, pp. 337–330.
11. Lau, J. H., and G. E. Barrett. "Stress and Deflection Analysis of Partially Routed Panel for Depanelization," *IEEE Transactions on CHMT*, 10(3), September 1987, pp. 411–419.
12. Lau, J. H. "Temperature and Stress Time History Responses in Electronic Packaging," *IEEE Transactions on CPMT, Part B*, 19(1), February 1996, pp. 248–254.
13. Zienkiewicz, O. C. *The Finite Element Method in Engineering Science.* London, England: McGraw-Hill, 1971.
14. Silvester, P., and R. L. Ferrari. *Finite Element for Electrical Engineers.* Cambridge, England: Cambridge University Press, 1983.
15. Strang, G., and G. J. Fix. *An Analysis of the Finite Element Method.* Englewood Cliffs, N.J.: Prentice-Hall, 1973.
16. Bathe, K. J. *Finite Element Procedures in Engineering Analysis.* Englewood Cliffs, N.J.: Prentice-Hall, 1973.
17. Weaver, W., and P. R. Johnston. *Finite Elements for Structural Analysis.* Englewood Cliffs, N.J.: Prentice-Hall, 1984.

18. Oden, J. T. *Finite Elements of Nonlinear Continua.* New York: McGraw-Hill, 1973.
19. Tong, P., and J. N. Rossettos. *Finite-Element Method: Basic Technique and Implementation.* Cambridge, Mass.: MIT Press, 1977.
20. Cheung, Y. K., and M. F. Yeo. *A Practical Introduction to Finite Element Analysis.* Marshfield, Mass.: Pitman, 1979.
21. Kikuchi, N. *Finite Element Methods in Mechanics.* Cambridge, England: Cambridge University Press, 1986.
22. Hughes, T. *The Finite Element Method.* Englewood Cliffs, N.J.: Prentice-Hall, 1987.
23. Gatewood, B. E. *Thermal Stresses.* New York: McGraw-Hill, 1957.
24. Boley, B. A., and J. H. Weiner. *Theory of Thermal Stresses.* New York: Wiley, 1960.
25. Nowacke, W. *Dynamic Problems of Thermoelasticity.* Leyden, The Netherlands: Noordhoff International Publishing, 1975.
26. Nowacke, W. *Progress in Thermoelasticity.* Warsaw: Panstwowe Wydawnictwo Naukowe, 1969.
27. Nowacki, W. *Thermoelasticity.* 2d ed. New York: Pergamon Press, 1986.
28. Kupradze, V. D. *Three-Dimensional Problems of the Mathematical Theory of Elasticity and Thermoelasticity.* New York: North-Holland Publishing Company, 1976.
29. Nowinski, J. L. *Theory of Thermoelasticity with Applications.* Sephin AAN Rijn, The Netherlands: Sijthoff & Noordhoff International Publishers, 1978.
30. Kovalenko, A. D. *Thermoelasticity.* Groningen, The Netherlands: Wolters-Noordhoff Publishing, 1969.
31. Parkus, H. *Thermoelasticity.* London: Blaisdell Publishing Company, 1968.
32. Boley, B. A. *Thermoinelasticity.* New York: Springer-Verlag, 1970.
33. Fung, Y. C. *Foundations of Solid Mechanics.* Englewood Cliffs, N.J.: Prentice-Hall, 1965.
34. Boresi, A. P., and O. M. Sidebottom. *Advanced Mechanics of Materials.* New York: John Wiley & Sons, 1984.
35. Gere, J. M., and S. P. Timoshenko. *Mechanics of Materials,* 2d ed. Boston, Mass.: PWS Engineering, 1984.
36. Timoshenko, S. P., and J. N. Goodier. *Theory of Elasticity.* New York: McGraw-Hill Book Company, 1970.
37. Polukhin, P., S. Gorelik, and V. Vorontsov. *Physical Principles of Plastic Deformation.* Moscow: Mir, 1983.
38. Atkins, A. G., and Y. W. Mai. *Elastic and Plastic Fractures.* Chichester: Horwood, 1985.
39. Hill, R. *The Mathematical Theory of Plasticity.* Oxford: Clarendon Press, 1983.
40. Thomas, T. Y. *Plastic Flow and Fracture in Solids.* New York: Academic Press, 1961.
41. Johnson, W., and P. B. Mellor. *Engineering Plasticity.* Chichester: Horwood, 1983.
42. Washizu, K. *Variational Methods in Elasticity and Plasticity.* Oxford: Pergamon Press, 1982.
43. Gopinathan, V. *Plasticity Theory and Its Application in Metal Forming.* New York: Wiley, 1982.
44. Mendelson, A. *Plasticity, Theory and Application.* Malabar, Fla.: Krieger, 1983.
45. Lau, J. H., and T. T. Lau. "Bending and Twisting of Pipes with Strain Hardening," *Journal of Pressure Vessel Technology, Transactions of ASME,* vol. 106, May 1984, pp. 188–195.
46. Lau, J. H., and C. K. Hu. "Nonlinear Stress Analysis of Curved Bars," *Proceedings of the 5th ASCE Engineering Mechanics Conferences,* 1984, pp. 917–920.
47. Fung, Y. C. *A First Course in Continuum Mechanics.* Englewood Cliffs, N.J.: Prentice-Hall, 1969.
48. Malvern, L. E. *Introduction to the Mechanics of a Continuous Medium.* Englewood Cliffs, N.J.: Prentice-Hall, 1969.
49. Kennedy, A. J. *Processes of Creep and Fatigue in Metals.* New York: Oliver & Boyd, 1962.
50. Gittus, J. *Viscoelasticity and Creep Fracture in Solids.* New York: John Wiley–Halsted Press, 1975.
51. Evans, H. E. *Mechanics of Creep Fracture.* New York: Elsevier Applied Science Publishers, 1984.
52. Evans, H. E., and B. Wilshire. *Creep of Metals and Alloys.* London, U.K.: The Institute of Metals, 1985.

Mechanical Design/Analysis/Measurement/Reliability of Electronic Packaging 387

53. Conway, S. B. *Numerical Methods for Creep and Rupture.* New York: Gordon and Breach, Science Publishers, Inc., 1967.
54. Garofalo, F. *Fundamentals of Creep and Creep-Rupture in Metals.* New York: Macmillan, 1965.
55. Clauss, F. J. *Engineer's Guide to High Temperature Materials.* Reading, Mass.: Addison-Wesley, 1969.
56. Wilshire, B., and R. J. Owen. *Recent Advances in Creep and Fracture of Engineering Materials and Structures.* Swansea, U.K.: Pineridge Press, 1982.
57. Lubahn, J. D., and R. P. Felgar. *Plasticity and Creep of Metals.* New York: John Wiley & Sons, 1961.
58. Sully, A. H. *Metallic Creep and Creep Resistant Alloys.* New York: Interscience Publishers, 1949.
59. Ponter, A. R. S., and F. A. Leckie. "Constitutive Relationships for the Time Dependent Deformation of Metals," *Journal of Engineering Materials and Technology, Transactions of ASME,* vol. 98, 1976, pp. 47–51.
60. Miller, A. K. "An Inelastic Constitutive Model for Monotonic, Cyclic, and Creep Deformation," *Journal of Engineering Materials and Technology, Transactions of ASME,* vol. 98, 1976, pp. 97–105.
61. Hart, E. W. "Constitutive Relations for the Non-elastic Deformation of Metals," *Journal of Engineering Materials and Technology, Transactions of ASME,* vol. 98, 1976, pp. 193–202.
62. Perzyna, P. "The Constitutive Equations for Rate Sensitive Plastic Materials," *Quarterly Journal of Mechanics and Applied Mathematics,* vol. XX, 1963, pp. 321–332.
63. Lau, J. H., and G. K. Listvinsky. "Bending and Twisting of Internally Pressurized Thin-Walled Cylinder With Creep," *Journal of Applied Mechanics, Transactions of ASME,* vol. 48, June 1981, pp. 439–441.
64. Lau, J. H. "Bending of Circular Cylinder with Creep," *Journal of Engineering Mechanics Division, Proceedings of ASCE,* vol. 107, 1981, pp. 265–270.
65. Lau, J. H. "Bending and Twisting of Pipe with Creep," *International Journal of Nuclear Engineering and Design,* June 1981, pp. 367–374.
66. Lau J. H., and T. T. Lau. "Creep of Pipes Under Axial Force and Bending Moment," *Journal of Engineering Mechanics Division, Proceedings of ASCE,* vol. 108, 1982, pp. 190–195.
67. Lau, J. H., and T. T. Lau. "Deformation of Elbows With Creep," *Proceedings of the 4th ASME National Congress on Pressure Vessel and Piping Technology,* June 1983.
68. Lau, J. H., S. S. Jung, and T. T. Lau. "Creep of Thin-Wall Cylinder Under Axial Force, Bending, and Twisting Moments," *Journal of Engineering for Power, Transactions of ASME,* vol. 106, 1984, pp. 79–83.
69. Lau, J. H., and C. K. Hu. "Creep of Thin-Wall Cylinder Under Combined Loads," *Proceedings of the 5th ASCE Engineering Mechanics Conferences,* 1984, pp. 917–920.
70. Lau, J. H. "Bending and Twisting of 96.5Sn3.5Ag and 97.5Pb2.5Sn Solder Interconnects with Creep," *IEEE ECTC Proceedings,* May 1994, pp. 1108–1114.
71. Lau, J. H., and C. K. Hu. "Deformation of Curved Bars With Creep," *Journal of Engineering for Power, Transactions of ASME,* vol. 107, 1985, pp. 225–230.
72. Lau, J. H., and T. T. Lau. "Creep of Pipes Under Axial Force and Twisting Moment," *Journal of Engineering Mechanics, Proceedings of ASCE,* vol. 108, 1982, pp. 174–179.
73. Lau, J. H. "Creep of Solder Interconnects under Combined Loads," *IEEE Transactions on CHMT,* **16**(8), December 1993, pp. 794–798.
74. Lau, J. H. "Bending and Twisting of 63Sn37Pb Solder Interconnects with Creep," *J. of Electronic Packaging, Trans. of ASME,* vol. 116, June 1994, pp. 154–157.
75. Lau, J. H. "Creep of 96.5Sn3.5Ag Solder Interconnects," *Soldering & Surface Mount Technology,* no. 15, September 1993, pp. 45–49.
76. Darveaux, R., and K. Banerji. "Constitutive Relations for Tin-Based Solder Joints," *IEEE Transactions on CHMT,* **15**(6), December 1992, pp. 1013–1024.
77. Schroeder, S. A., and M. R. Mitchell. "Torsional Creep Behavior of 63Sn-37Pb Solder," *ASME Proceedings of Advances in Electronic Packaging,* April 1992, pp. 649–653.
78. Cortez, R., E. Cutiongeo, M. Fine, and D. Jeannotte. "Correlation of Uniadial Tension-Tension, Torsion, and Multiaxial Tension-Torsion Fatigue Failure in a 63Sn-37Pb

Solder Alloy," *Proceedings of the 42nd IEEE Electronic Components & Technology Conference,* May 1992, pp. 354–359.
79. Smith, J. O., and O. M. Sidebottom. *Inelastic Behavior of Load-Carry Members.* New York: Wiley, 1965.
80. Tummala, R. R., and E. J. Rymaszewski. *Microelectronics Packaging Handbook,* New York: Van Nostrand Reinhold, 1989.
81. Lau, J. H. *Flip Chip Technologies.* New York: McGraw-Hill, 1996.
82. Lau, J. H. "Thermal Fatigue Life Prediction of Flip Chip Solder Joints by Fracture Mechanics Method," *International Journal of Engineering Fracture Mechanics,* vol. 45, July 1993, pp. 643–654.
83. Lau, J. H. "Thermomechanical Characterization of Flip Chip Solder Bumps for Multichip Module Applications," *Proceedings of IEEE International Electronics Manufacturing Technology Symposium,* September 1992, pp. 293–299.
84. Lau, J. H., and D. Rice. "Thermal Fatigue Life Prediction of Flip Chip Solder Joints by Fracture Mechanics Methods," *Proceedings of the 1st ASME/JSME Electronic Packaging Conference,* April 1992, pp. 385–392.
85. Peterson, D., J. Sweet, S. Burchett, and A. Hsia. "Stresses from Flip-Chip Assembly and Underfill: Measurements with the ATC4.1 Assembly Test Chip and Analysis by Finite Element Method," *Proceedings of IEEE Electronic Components & Technology Conference,* May 1997, pp. 134–143.
86. Lau, J. H., C. Chang, and R. Chen. "Effects of Underfill Encapsulant on the Mechanical and Electrical Performance of a Functional Flip Chip Device," *Proceedings of IEEE Int. Symposium on Polymeric Electronics Packaging,* October 1997, pp. 265–272.
87. Wong, C. P., S. H. Shi, and G. Jefferson. "High Performance No Flow Underfills for Low-Cost Flip-Chip Applications," *Proceedings of IEEE Electronic Components & Technology Conference,* May 1997, pp. 850–858.
88. Wong, C. P., J. M. Segelken, and C. N. Robinson. "Chip On Board Encapsulation," in *Chip On Board Technologies for Multichip Modules,* J. H. Lau, ed. New York: Van Nostrand Reinhold, 1994, pp. 470–503.
89. Wu, T. Y., Y. Tsukada, and W. T. Chen. "Materials and Mechanics Issues in Flip-Chip Organic Packaging," *Proceedings of IEEE Electronic Components & Technology Conference,* May 1996, pp. 524–534.
90. Tsukada, Y. "Solder Bumped Flip Chip Attach on SLC Board and Multichip Module," in *Chip On Board Technologies for Multichip Modules,* J. H. Lau, ed. New York: Van Nostrand Reinhold, 1994, pp. 410–443.
91. Shi, D. M., and J. W. Carbin. "Advances in Flip-Chip Underfill Flow and Cure Rates and Their Enhancement of Manufacturing Processes and Component Reliability," *Proceedings of IEEE Electronic Components & Technology Conference,* May 1996, pp. 1025–1031.
92. Suryanarayana, D., J. Varcoe, and J. Ellerson. "Repairability of Underfill Encapsulated Flip Chip Packages," *Proceedings of IEEE Electronic Components & Technology Conference,* May 1995, pp. 524–528.
93. Lau, J. H. *Chip On Board Technologies for Multichip Modules.* New York: Van Nostrand Reinhold, 1994.
94. Lau, J. H., and Y.-H. Pao. *Solder Joint Reliability of BGA, CSP, Flip Chip, and Fine Pitch SMT Assemblies.* New York: McGraw-Hill, 1997.
95. Lau, J. H. *Ball Grid Array Technology,* New York: McGraw-Hill, 1995.
96. Lau, J. H. "Thermal Fatigue Life Prediction of Encapsulated Flip Chip Solder Joints for Surface Laminar Circuit Packaging," ASME Paper No. 92W/EEP-34, *1992 ASME Winter Annual Meeting.*
97. Lau, J. H., M. Heydinger, J. Glazer, and D. Uno. "Design and Procurement of Eutectic Sn/Pb Solder-Bumped Flip Chip Test Die and Organic Substrates," *Circuit World,* vol. 21, March 1995, pp. 20–24.
98. Wun, W., and J. H. Lau. "Characterization and Evaluation of the Underfill Encapsulants for Flip Chip Assembly," *Circuit World,* vol. 21, March 1995, pp. 25–32.
99. Kelly, M., and J. H. Lau. "Low Cost Solder Bumped Flip Chip MCM-L Demonstration," *Circuit World,* vol. 21, July 1995, pp. 14–17.

100. Lau, J. H., T. Krulevitch, W. Schar, M. Heydinger, S. Erasmus, and J. Gleason. "Experimental and Analytical Studies of Encapsulated Flip Chip Solder Bumps on Surface Laminar Circuit Boards," *Circuit World,* **19**(3), March 1993, pp. 18–24.
101. Lau, J. H. "Solder Joint Reliability of Flip Chip and Plastic Ball Grid Array Assemblies Under Thermal, Mechanical, and Vibration Conditions," *IEEE Transactions on CPMT, Part B,* **19**(4), November 1996, pp. 728–735.
102. Lau, J. H., E. Schneider, and T. Baker. "Shock and Vibration of Solder Bumped Flip Chip on Organic Coated Copper Boards," *ASME Transactions, Journal of Electronic Packaging,* June 1996, pp. 101–104.
103. Zakel, E., and H. Reichl. "Flip Chip Assembly Using the Gold, Gold-Tin and Nickel-Gold Metallurgy," in *Flip Chip Technologies,* J. H. Lau, ed. New York: McGraw-Hill, 1996, pp. 415–490.
104. Lau, J. H., and C. Harkins. "Thermal-Stress Analysis of SOIC Packages and Interconnections," *IEEE Transactions on CHMT,* **11**(4), December 1998, pp. 380–389.
105. Lau, J. H. "Thermal Stress Analysis of Plastic Leaded Chip Carriers," *IEEE Proceedings of InterSociety Conference on Thermal Phenomena,* 1990, pp. 57–66.
106. Kwon, O. K., B. W. Langley, R. F. W. Pease, and M. R. Beasley. "Superconductors as Very High-Speed System-Level Interconnects," *IEEE Electron Device Letters,* **EDL-8**(12), Dec. 1987, pp. 582–585.
107. Henry, R. L., H. Lessoff, E. M. Swiggard, and S. B. Qadri. "Thin Film Growth of $YBa_2Cu_3O_{7-x}$ From Nitrate Solutions," *Journal of Crystal Growth,* vol. 85, 1987, pp. 615–618.
108. Edelstein, A. S., S. B. Qadri, R. L. Holtz, P. R. Broussard, J. H. Claassen, T. L. Francavilla, D. U. Gubser, P. Lubitz, E. F. Skelton, and S. A. Wolf. "Formation of the Structure of the Superconducting Phase of La-Sr-Cu-O by DC Sputtering," *Journal of Crystal Growth,* vol. 85, 1987, pp. 619–622.
109. Salomons, E., H. Hemmes, J. J. Scholtz, N. Koeman, R. Brouwer, A. Griessen, D. G. De Groot, and R. Griessen. "Thermal Expansion, Compressibility and Gruneisen Parameter of $YBa_2Cu_3O_{7-x}$," *Journal of Physica,* 145B, 1987, pp. 253–259.
110. Bayot, V., C. Dewitte, J.-P. Erauw, X. Gonze, M. Lambricht, and J.-P. Michenaud. "Thermal Expansion of a $YBa_2Cu_3O_{7-x}$ Superconducting Ceramic," *Solid State Communications,* **64**(3), 1987, pp. 327–328.
111. Touloukian, Y. S., ed. *Thermalphysical Properties of Matter,* vol. 12, New York: IFI/Plenum, 1970–1979.
112. Soma, Y., J. Satoh, and H. Matsuo. "Thermal Expansion Coefficient of GaAs and InP," *Solid State Communications,* vol. 42, 1982, pp. 889–892.
113. Blakemore, J. S. "Semiconducting and Other Major Properties of Gallium Arsenide," *Journal of Applied Physics,* **53**(10), October 1982, pp. 123–159.
114. Conwell, E. M. "Properties of Silicon and Germanium," *Proceedings of the I.R.E.,* 1952, pp. 1327–1337.
115. Ledbetter, H. M., M. W. Austin, S. A. Kim, and M. Lei. "Elastic Constants and Debye Temperature of Polycrystalline $YBa_2Cu_3O_{7-x}$," *Journal of Material Research,* **2**(6), Nov/Dec 1987, pp. 786–789.
116. Davidge, R. W. *Mechanical Behavior of Ceramics.* Cambridge: Cambridge University Press, 1979.
117. Bever, M. B., ed. *Encyclopedia of Materials Science and Engineering,* vol. 4, Cambridge, Mass.: J-N/MIT Press, 1986.
118. Nakahara, S., G. J. Fisanick, M. F. Yan, R. B. Van Dover, T. Boone, and R. Moore. "On the Defect Structure of Grain Boundaries in $YBa_2Cu_3O_{7-x}$," *Journal of Crystal Growth,* vol. 85, 1987, pp. 639–651.
119. Lau, J. H., and L. Moresco. "Mechanical Behavior of Microstrip Structures Made from $YBa_2Cu_3O_{7-x}$ Superconducting Ceramics," *IEEE Transactions on CHMT,* **11**(4), December 1988, pp. 419–426.
120. Lee, S. W., and J. H. Lau. "Effect of Chip Dimension and Substrate Thickness on the Solder Joint Reliability of Plastic Ball Grid Array Packages," *Circuit World,* **23**(1), 1996, pp. 16–19.
121. Lee, S. W., and J. H. Lau. "Design for Plastic Ball Grid Array Solder Joint Reliability," *Circuit World,* **23**(2), 1997, pp. 11–13.

122. Lee, S. W., and J. H. Lau. "Solder Joint Reliability of Plastic Ball Grid Array with Solder Bumped Flip Chip," *ASME Paper No.* 97WA/EEP-6, November 1997.
123. Lau, J. H. "The Role of DNP (Distance to Neutral Point) on Solder Joint Reliability of Area Array Assemblies," *Soldering & Surface Mount Technology*, no. 26, July 1997, pp. 58–60.
124. Guo, Y., W. Chen, and C. Lim. "Experimental Determinations of Thermal Strain in Semiconductor Packaging Using Moire Interferometry," *Proceedings of the 1st Joint ASME / JSME Conference on Electronic Packaging*, April 1992, pp. 779–784.
125. Guo, Y., C. Lim, W. Chen, and C. Woychik. "Solder Ball Connect (SBC) Assemblies Under Thermal Loading: I. Deformation Measurement via Moire Interferometry, and Its Interpretation," *IBM Journal of Research and Development*, 37(5), September 1993, pp. 635–647.
126. Jung, W., J. H. Lau, and Y. Pao. "Nonlinear Analysis of Full-Matrix and Perimeter Plastic Ball Grid Array Solder Joints," *ASME Paper No. 96WA/EEP-9*, November 1996.
127. Lau, J. H. "Vibration Frequencies of Tapered Bars With End Mass," *Trans. of ASME, J. of Applied Mech.*, vol. 51, March 1984, pp. 179–181.
128. Lau, J. H. "Vibration Frequencies and Mode Shape for a Constrained Cantilever," *Trans. of ASME, J. of Applied Mech.*, vol. 51, March 1984, pp. 182–187.
129. Lau, J. H. "Vibration Frequencies for a Non-Uniform Beam With End Mass," *J. of Sound and Vibration*, vol. 97, December 1984, pp. 513–521.
130. Lau, J. H., and C. A. Keely. "Dynamic Characterization of Surface Mount Component Leads for Solder Joint Inspection," *IEEE Trans. on CHMT*, 12(4), December 1989, pp. 594–602.
131. Conway, H. D., E. Becker, and J. Dubil. "Vibration Frequencies of Tapered Bars and Circular Plates," *ASME Transactions, J. of Applied Mechanics*, vol. 31, 1964, pp. 329–331.
132. Conway, H. D., and J. Dubil. "Vibration Frequencies of Truncated-Cone and Wedge Beam," *ASME Transactions, J. of Applied Mechanics*, vol. 32, 1965, pp. 932–934.
133. Watson, G. *A Treatise on the Theory of Bessel Functions*. Cambridge, England: Cambridge University Press, 1952.
134. Abramowitz, M., and I. Stegun. *Handbook of Mathematical Functions*. New York: Dover, 1970.
135. McCalla, T. *Introduction to Numerical Methods and FORTRAN Programming*. New York: Wiley, 1967.
136. Laura, P., J. Pombo, and E. Susemihl. "A Note on the Vibrations of a Clamped-Free Beam with a Mass at the Free End," *Journal of Sound and Vibration*, vol. 37, 1974, pp. 161–168.
137. Zakraysek, L., and H. Lin. "Glass to Metal Seal Quality," US Air Force Rome Air Development Center, Griffiss AFB, N.Y., 1987.
138. McCormick, J., H. Lin, and L. Zakraysek. "Liquid Penetrant Testing for Microelectronic Package Hermeticity," US Air Force Rome Air Development Center, Griffiss AFB, N.Y., 1987.
139. Ebel, G., and R. De Cristofaro. "A Method of Lead Detection and Location for Conformally Coated Packages," *Proceedings of the International Symposium for Testing and Failure Analysis*, 1980, pp. 26–29.
140. Rafanelli, A. "Failure Analysis of 132-Pin Lead Chip Carrier Gate Arrays," *ASME Paper No. 86-WA/EEP-6*, 1986.
141. Lau, J. H., and L. B. Lian-Mueller. "Finite Element Modeling for Optimizing Hermetic Package Reliability," *ASME Transactions, J. of Electronic Packaging*, vol. 111, December 1989, pp. 255–260.
142. MIL-STD-750C, "Lead Fatigue, Test Condition B2," Method 2004.4, 25 December 1983.
143. MIL-STD-750C, "Mechanical Shock," Method 2016-2, 17 September 1987.
144. MIL-STD-750C, "Acceleration," Method 2006, 17 September 1987.
145. MIL-S-19500G, "Amendment 1," 16 February 1984.
146. Hewlett-Packard Company, "General Semiconductor Specification, Method 12-106, Thermal Shock Test," Revision N, 21 November 1988.

147. Hewlett-Packard Company, "General Semiconductor Specification, Method 12-106, Moisture Resistance Test," Revision N, 21 November 1988.
148. MIL-S-19500 and MIL-M-38510, "LTPD Sampling Plans Chart," 28 November 1978.
149. MIL-STD-750C, "Steady State Operation Life," Method 1026, 17 September 1987.
150. MIL-STD-750C, "Burn-In (For Diodes and Rectifiers)," Method 1038, 17 September 1987.
151. Lau, J. H. "Solder Joint Reliability of a Low Cost Chip Size Package - NuCSP," *ISHM Microelectronics Symposium Proceedings,* October 1997, pp. 691–696.
152. Chou, T., and J. H. Lau. "A Low Cost Chip Size Package - NuCSP," *Circuit World,* 24(1), November 1997, pp. 34–38.
153. Lau, J. H., C. Chang, T. Chen, D. Cheng, and E. Lao. "A Low-Cost Solder-Bumped Chip Scale Package—NuCSP," to be published in the *Proceedings of NEPCON West,* March 1998.
154. Lau, J. H., and T. Chou. "Electrical Design of a Cost-Effective Plastic Ball Grid Array Package - NuBGA," *IEEE Transactions on CPMT, Part B,* 21(1), February 1998.
155. Lau, J. H., and T. Chen. "Cooling Assessment and Distribution of Heat Dissipation of a Cavity Down Plastic Ball Grid Array Package-NuBGA," *ISHM Microelectronics Symposium Proceedings,* October 1997, pp. 482–493.
156. Tektronix, "TDR Tools in Modeling Interconnects and Packages," Applications Note, Tektronix, Beaverton, Oregon, 1993.
157. Lau, J. H., and K. Chen. "Thermal and Mechanical Evaluations of a Cost-Effective Plastic Ball Grid Array Package," *ASME Transactions, J. of Electronic Packaging,* vol. 119, September 1997, pp. 208–212.
158. Lee, S. W., and J. H. Lau. "Solder Joint Reliability of Cavity-Down Plastic Ball Grid Array Assemblies," to be published in the *Journal of Soldering & Surface Mount Technology,* 1998.

Chapter 5

Polymers for Electronic Packaging: Materials, Processes, and Reliability

5.1 Introduction

The advances of semiconductor technology are largely attributed to the advances of materials, and are due in particular, to the advances of polymeric materials. The deep submicron device feature size (<0.25 µm) requires ultrasensitive and high-resolution photoresists for the electron beam or X-ray lithography; the ultrafast device with operating frequency in excess of 500 MHz requires low dielectric constant material ($E_1 < 3.0$) such as fluorinated polyimides, Teflon and aerogel-type materials to reduce the signal propagation delays. Furthermore, the ever increasing integrated circuit (IC) dimension requires low-stress encapsulants and molding compounds to reduce the thermomechanical stress of the large IC package. The high-performance microprocessor and application-specific IC (ASIC) devices require a huge (>1000) input/output (I/O) signal for on-chip and off-chip interconnects; consequently, the area array flip-chip interconnect assembly is gaining more acceptance by the industry—flip-chip underfill encapsulant thereby plays a critical role for the reliability enhancement of temperature cycling flip-chip devices. Conductive adhesives with low-process temperature (less residue stress), better temperature cycle performance, fine-pitch capability, and "green" (lead-free) environmental properties are getting more attention in microelectronic assembly and packaging. These are some of the attributes that make advanced materials as the hearts and soul of modern electronic technology [1]. This chapter describes and addresses some of these materials, their processes, and reliability aspects as they are related to the next generation of electronic packaging in the 21st century.

Generally speaking, an advanced ultra-large-scale-integration (ULSI) device is a very complex and delicate, three-dimensional structure. It consists of millions of components in a single IC chip. These components are densely packaged in a multilayer structure with different metallized conductor lines (Al or Cu) separated with dielectric (organic and inorganic) insulating layers. The advances of IC technology have had great technological and economic impact on the electronic industry throughout the world. The rapid growth of the number of components per chip, the rapid decrease of device dimensions, the steady increase in IC chip size, and the ever increasing input/output (I/O) interconnects have imposed stringent requirements, not only on the IC physical design and fabrication, but also on the IC encapsulants and packaging.

For the past couple decades, we have seen the number of components per chip double for every 18-month period. This type of advance in semiconductor technology follows precisely with Moore's law, which was predicted by Gordon Moore at Fairchild (now Texas Instruments) almost 25 years ago [2] (Fig. 5.1). The increase of integration in ULSI

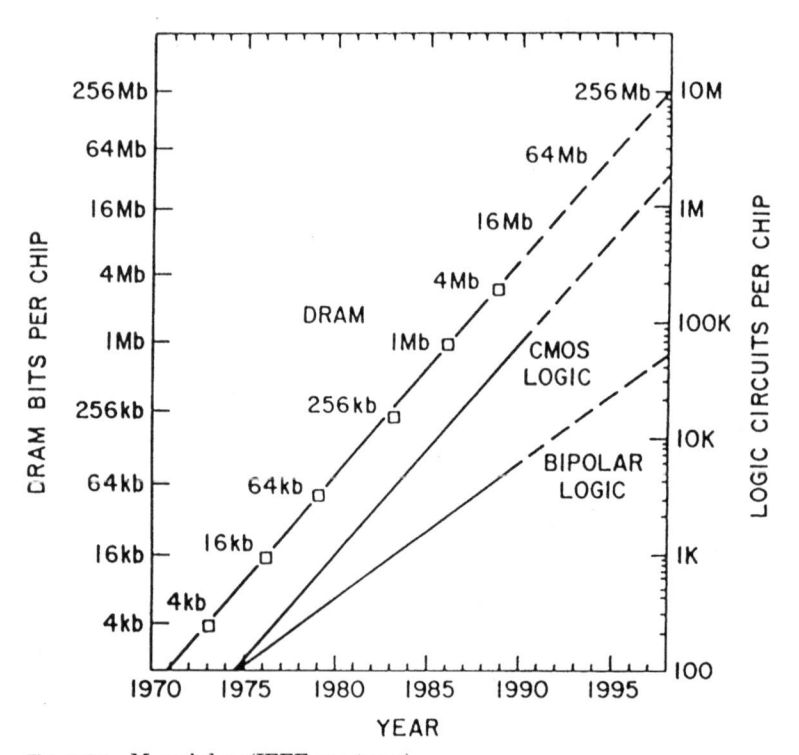

Figure 5.1 Moore's law (IEEE spectrum).

technology has resulted in the miniaturization of the device size, which has reduced the propagation delay due to higher density packaging and interconnection. As a result, a modern advanced device operates at a faster speed (>500 MHz), consumes less power, and consequently dissipates less heat during operation. In addition, ULSI technology has increased the reliability of microelectronic devices due to the elimination of poor interconnections and decreased the cost per function of the devices, which has had a profound impact on the modern electronics industry. However, due to the high-density packaging of these devices, the power consumption per package—as well as the physical dimension of the chip size and packages—are also increased. These advances provide challenges and opportunities to the materials and processes engineering communities. Table 5.1 shows the Semiconductor Industry Association (SIA) road map that clearly maps out the semiconductor technological advancement.

The various materials used to fabricate integrated circuits and their related interconnects are generally not capable of surviving their end-use operating, or field, environment without some level of additional protection. Metals susceptible to corrosion, dissimilar metals forming galvanic cells, and structures that are stress- or heat-sensitive, and so forth, can all lead to performance degradation or failure in the field without proper protection. With the latest advances in IC metallization, copper low-k material will rapidly replace aluminum. The low resistivity of copper is the main driving force for this metallization

TABLE 5.1 Semiconductor Industry Association (SIA)'s IC Technology Road Map

	1995	1998	2001	2004	2007
Feature size (μm)	0.35	0.25	0.18	0.12	<0.10
Gates/chip	800K	2M	5M	10M	20M
Bits/chip					
■ DRAM	64M	256M	1G	4G	16G
■ SRAM	16M	64M	256M	1G	4G
Chip size (mm^2)					
■ Logic/Microprocessor	400	600	800	1000	1250
■ DRAM	200	320	500	700	1000
Wafer diameter (mm)	200	200–400	200–400	200–400	200–400
Number of interconnect levels (logic)	4–5	5	5–6	6	6–7
Max. power (W/Die)					
■ High-performance	15	30	40	40–120	40–200
Power supply (V)					
■ Portable	2.2	2.2	1.5	1.5	1.5
Number of I/Os	750	1500	2000	3500	5000
Performance (MHz)					
■ Off-chip	100	175	250	350	500
■ On-chip	200	350	500	700	1000

change. Silver will be the ultimate low resistivity metallization for conductor line; however, there are many problems such as migration and oxidation that are associated with silver metallization. Furthermore, field environments can vary greatly for any one device depending on the product in which it is incorporated, or the customer's application of the product. These environmental stresses include not only the thermomechanical stresses associated with device operation, but also the contaminants present in the operating environment. These pollutants and contaminants probably have the greatest variation of all the stresses, varying widely, dependent upon location and over the lifetime of the product.

Consumer products may be appropriate for many uses—even a mobile one such as cellular phone, PCMCIA—so that a single product may see several environments in its lifetime. When factors related to the device are coupled in, such as the cost of manufacture, production volume, expected use environments or markets, and so forth, packaging issues and decisions become very complex. These complexities require the packaging engineer to make critical decisions on how to most cost-effectively package the device while obtaining acceptable reliability in the field. Packaging selection can be seen to be a compromise between cost and performance, where performance includes reliability issues as well as other design considerations.

Early hermetically sealed metal cases such as transistor outline (TO) packages were used to protect sensitive chips from environmental stresses. Corrosion problems prompted the military and aerospace industries to require hermetic packages in order to achieve high reliability and long life. Hermetic packages utilize a sealed environment that is impervious to gases and moisture to protect the devices. Final sealing is accomplished with caps or lids using glass or metal seals. Moisture content within the package is standardly limited to a volumetric maximum of 5000 parts per million (MIL-TD-883C), which was chosen to eliminate the possibility of condensation occurring inside the enclosure. For extremely sensitive devices, however, moisture limits are lowered even further, increasing both the time and cost to manufacture and test these hermetically sealed packages.

Hermetic packaging proved to be such a robust method of obtaining a long field life that little has changed in hermetic packaging technology in recent years. After a slow development and initial reliability problems, hermetically packaged devices gave way to lower-cost nonhermetic plastic molded packaging, especially in electronics designed for commercial and industrial markets. Today, however, nonhermetic does not imply nonreliable. Plastic packages are not hermetic, yet demonstrate admirable reliability and now account for approximately 90 to 95 percent of all worldwide device packaging (>60 billions per year).

Many obstacles had to be overcome with respect to understanding materials and processes associated with plastic packaging before their use was widespread. Organic materials are not hermetic and, therefore, allow moisture to penetrate and be absorbed. Different organic material families have greatly varying material properties related to their moisture permeation rates and moisture absorption percentages. In the first-generation plastic packages, corrosion was found to be the primary cause of failure due to poor adhesion and high levels of mobile ionic contaminants, such as sodium and chloride ions in the materials. This acquired understanding identified the presence of moisture along with mobile ions as the major contributors to corrosion, among other failure mechanisms. Improvements in plastic packaging materials through the advances of materials and processes have all but removed corrosion as a source of failure, and have led to reliability that approaches that of hermetic packages. Now, stress-related failures from increasing die sizes in high-speed, heat-flux devices present the largest obstacle to be solved by the plastic packaging engineer. However, the use of the low-stress and high-thermal conductive molding compound has alleviated some of this problem.

These major improvements in plastic packaging were derived from the effective use of environmental test data—amassing large volumes of information and tracing field failure over long periods of time—that assisted engineers in understanding factors governing the principal failure modes associated with protection through organic materials. Accelerated environmental stress tests on packaged devices were the primary source of this data. In some cases, combinations of accelerated stress, along with corrosive ambients, were used to predict reliability behavior under severe use conditions, and were further extended as a means of assessing the effects of impurities on device lifetimes. Accelerated stress tests—such as the 85°C/85%RH (85/85) humidity test—became useful tools for screening and assessing the coupling between the processes and material systems. In the beginning of plastic packaging, 1000 hours of 85/85 was not only difficult to obtain with existing materials and processes, but was considered sufficient for many commercial markets. It is now quite common for plastic packages to exceed 10,000 hours of 85/85 without failure.

For the sake of saving development time, more severe environmental tests have been devised to shorten test and evaluation times, decreasing the cost of the final product. The highly accelerated stress test (HAST), which is usually tested at 121°C, 100 percent relative humidity, 2 atmosphere (30 psi), with or without bias, at the test duration of 2- to 168-h intervals, is the most prominent example, and expands upon the 85/85 approach. HAST utilizes a pressurized chamber to allow the temperature to be elevated while maintaining humidity control, to produce the additional activation energy for more accelerated testing. Quite a bit is still unknown with regard to accelerated testing and its correlation to field life. However, data show good correlation between 85/85 versus HAST testing. This chapter considers these issues and their relationship in the environmental protection of integrated circuits and electronic components.

In both hermetic and nonhermetic packages, it is assumed that passivation is the first line of protection. This first line of defense is what is customarily termed the *passivation layer* and is typically applied at the back end of the circuit fabrication process. The passivation layer covers the entire integrated circuit, with the exception of the bond pads that remain exposed to facilitate signal and power interconnect to the device. The most prevalent circuit passivations consist of chemically vapor-deposited phosphorous silicate glass (P glass), silicon nitride (sin cap) or silicon oxynitride, or photo-definable polyimides or benzocyclobutenes that are first dispensed in the spinning process then exposed, etched, and cured. The passivation layer on the devices, although very important in providing device reliability, is considered a given input to any sealing and encapsulation packaging processes and is an art to itself. It is discussed later on in this chapter.

5.2 Purpose of Electronic Packaging

Electronic packaging is defined as: (1) protection/encapsulation; (2) power and ground electrical signals; (3) thermal management; and (4) interconnects from discrete devices to the entire electronic system (see Fig. 5.2). It is an important part in system integration. In this chapter, we focus on electronic protection and encapsulation.

5.2.1 Protection and encapsulation

The purpose of encapsulation is to protect electronic devices from an adverse environment and increase their long-term reliability. However, the ultimate goal of encapsulation is to ensure the device's reliability and increase the production yield with the lowest cost. Some of the possible contributors to degradation that could negatively affect electronic device performance or lifetime of the IC are moisture; contaminants; mobile ions; ultraviolet, visible, and alpha-particle radiation; and hostile environmental conditions such as low- and high-temperature cycling temperature ranges (−65 to 175°C). Passivating layers on IC devices are not 100 percent free of cracks and pinholes. Furthermore, the IC bond pad areas are etched out for interconnect and need protection as well. That is why passivated ICs need further coating materials to enhance their reliability. The following sections define and discuss adverse elements/conditions that could affect IC performance.

Definition of Electronic Packaging

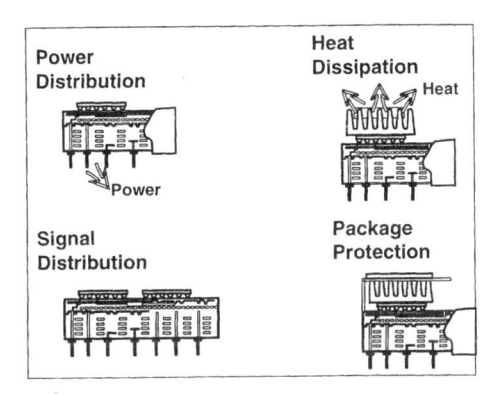

Figure 5.2 Four functions of electronic packaging (Ref. 1).

5.2.2 Moisture permeability

Moisture is one of the major sources of corrosion for IC devices. Electrooxidation and metal migration are associated with the presence of moisture. The diffusion rate of moisture depends on the encapsulant material and is a function of the diffusive encapsulant thickness and exposure time. Pure crystals and metals are the best materials as a barrier against moisture. Glass (silicon dioxide) is an excellent moisture barrier, but is slightly inferior to pure crystals and metals. Organic polymers—such as fluorocarbons, epoxies, and silicones—are a few orders of magnitude more permeable to moisture as compared with glass. Silicone materials, which have the highest moisture permeability in most polymers, nevertheless are one of the best device encapsulants. The reason for this is the subject of our discussion in a later section. Obviously, of all materials, gases are the most permeable to moisture. In general, for each particular material the moisture diffusion rate is proportional to the water vapor partial pressure and inversely proportional to the material thickness. This is accurate when moisture diffusion rates are in steady-state permeation; however, moisture transient penetration rates (perhaps more important because they determine the time it takes for moisture to break through) are inversely proportional to the "square" of material thickness.

5.2.3 Mobile ion contaminants

Mobile ions, such as sodium or potassium, tend to migrate to the p-n junction of the IC device where they acquire an electron and deposit as the corresponding metal on the *p-n* junction—consequently destroying the device [3]. Furthermore, mobile ions will also support leakage currents between biased device features, which degrade device performance and ultimately destroy the devices by electrochemical processes such as metal conductor dissolution. For example, even in trace amounts (in ppm level), chloride and fluoride ions could cause the dissolution of aluminum metallization of complementary metal oxide semiconductor (CMOS) devices. Unfortunately, CMOS is likely to be the trend of the ULSI technology. Hence, contaminants must be kept away from these sensitive devices. Furthermore, due to its high conductivity, copper is emerging as the next generation IC metallized material. However, its ease of oxide formation still needs further study. The protection of these devices from the effects of these mobile ions is an absolute requirement. The use of an ultra-high-purity encapsulant to encapsulate the passivated IC is the answer to some of these mobile ion contaminant problems.

5.2.4 Light and alpha-particle radiations

Opto-electronic devices are sensitive to ultraviolet or visible (UV-VIS) light radiation and can be damaged by this type of radiation. However, UV-VIS protection can be achieved by choosing an opaque encapsulant. Furthermore, impurity in an encapsulant, such as low levels of uranium in the ceramic or plastic package, can cause appreciable alpha-particle radiation. So can cosmic radiation in the atmosphere. The alpha radiation can generate a temporary "soft error" in operating dynamic random access memory (DRAM) devices. This type of alpha-particle radiation has become a major concern, especially in high-density memory devices. Good encapsulants must have alpha-radiation levels less than 0.001 alpha particles/cm^2/hour, and opaque in nature to protect UV-VIS radiation. Since the alpha particle is a weak radiation, a few micrometers thickness of the high-purity encapsulant usually will prevent this radiation damage to the DRAM devices.

5.2.5 Hostile environments

Hostile device operational environments, such as extreme cycling temperatures (values from −65 to +150°C in MIL-STD-883), high relative humidity (85 to 100%), shock and vibration, and high-temperature operating bias are part of the real-life operation. It is critical for the device to survive these operational life cycles. In addition, encapsulants must also have suitable mechanical, electrical, and physical properties (such as minimal stress and matching thermal expansion coefficient, etc.) that are compatible to the IC devices. In addition to the preceding requirements, the encapsulant, as mentioned earlier, must have a low dielectric constant to reduce the device propagation delay, excellent thermal conductivity to dissipate those power-hungry, high-speed (>500 MHz) IC and high-density multichip (MCM) packages. Furthermore, the encapsulant must be of ultrapure material, with extremely low amounts of ionic contaminants. Since the encapsulation is the final process step and some of the devices are expensive, particularly in the high-density MCM, it must be easy to apply and repair in production and service. With the proper choice of encapsulant and process, the encapsulation could enhance the reliability of the fragile IC device, improve its mechanical, physical properties, and its manufacturing yields, yet lower the overall production cost. These are the ultimate goals of the encapsulation and protection of the electronics.

5.3 Prepackaging Cleaning

Prior to IC device encapsulation, the preencapsulation cleaning is the most critical step to ensure the long-term reliability of the device. Encapsulating a dirty device is a guarantee of device failure [4]. It is imperative that we remove even trace amounts of the contaminant from the IC device surface prior to the encapsulation process. There are three main cleaning processes: (1) conventional cleaning; (2) reactive oxygen cleaning; and (3) reactive DC-hydrogen plasma cleaning. These are described as follows.

5.3.1 Conventional solvent and aqueous cleaning

The conventional cleaning process includes organics, such as detergents and solvents (i.e., chlorofluorohydrocarbon (CFC)-Freons, and chlorohydrocarbon-trichloroethane, methylene chloride, etc.), to remove organic contaminants. However, due to environmental concerns of CFC, terpenes (orange peel extracts) are currently gaining acceptance to replace CFCs [5]. Inorganic ionic salts are most effectively removed with high-quality aqueous water or polar solvents, such as alcohols.

5.3.2 Reactive oxygen cleaning

In addition to the conventional cleaning process, reactive oxygen is very effective in removing low-level organic contaminants. There are three types of reactive oxygen gas in cleaning:

1. *UV-ozone.* UV-ozone is very effective in removing a few monolayers of organics from the substrate surface. However, the device under cleaning must be placed directly under the UV source.

2. *Plasma oxygen.* Plasma oxygen operates at 13.6 MHz radio frequency (RF), is fast and effective in cleaning metal oxide semiconductor (MOS) devices, and preserves the aluminum metallization of the devices. However, the thermal stress associated with the plasma process may damage some device structures.
3. *Microwave discharge of oxygen.* Carried out at 2.5 GHz RF, this is also a powerful technique in the device cleaning process. This process is similar to the oxygen plasma process, except the microwave frequency is used.

5.3.3 Reactive DC-hydrogen plasma cleaning

Recently, it has been reported that DC-hydrogen plasma cleaning is an alternative cleaning process for high-performance IC packages [6]. The plasma process is based on an argon-hydrogen discharge that generates between the heated filament (cathode) and the reactor wall (anode). The discharge is based on the current density from 10 to 100 amps and low voltage of 20 to 30 volts. As such, it will only mildly clean the interfacial contaminant and eliminate the sputtering and damage to the IC packages. The process is simple and environmentally friendly as it reduces organic and inorganic contaminants. Furthermore, the hydrogen cleaning process eliminates the oxide formations generated by UV-ozone.

To clean the IC device effectively, the combination of conventional cleaning solvents, highly purified deionized water (with 18-megaohm/cm resistivity), and reactive atomic oxygen or hydrogen gas cleaning processes are ideal in providing a thorough cleaning of the IC devices. In addition, the cleaning process should be performed in a clean room or clean hood environment where the encapsulation process could proceed as soon as cleaning is completed to minimize the contamination to the precleaned devices.

5.4 Polymeric Materials for Microelectronic Packaging

Polymers are used both as interlayer dielectrics and encapsulants. There are numerous organic polymeric materials that are used as electronic encapsulants. These materials are typically used as on-chip as well as off-chip encapsulation and packaging. However, some of these materials are also used for interlayer dielectrics and passivation of the IC devices. These materials are described in the following sections.

5.4.1 Interlayer dielectrics

A modern ULSI device is a complex three-dimensional structure with metallized conductor lines separated (insulated) by the interlayer dielectrics. Conventionally, inorganic materials such as silicon dioxides and nitrides (as mentioned in the previous sections) are used for this purpose. However, the advances of polymeric materials have made polyimides, silicone-polyimides, and benzocyclobutenes—in particular, the photo-definable derivatives of these materials—to become widely used interlayer dielectrics. These materials tend to have lower dielectric constants (versus their inorganic interlayer dielectrics), which reduce the signal propagation delays and enhance the system performance. As an example, Teflon is one of the polymers that has the lowest dielectric constant ($E_1 = 2.0$) of all known polymers, which is a potentially good interlayer dielectric material. However, the adhesive property, thermal stability, and process issues of this material still need to be addressed and defined prior to its common use. Furthermore, "Zero gel," a microscopic or nanostructure pore dielectric, is being investigated for its potential application in this low-k dielectric area. In addition to polyimides, the recently modified benzocyclobutenes by Dow Chemical and polycyclic olefins by BF Goodrich are becoming important materials as interlayer dielectrics. We discuss these materials in the coming sections.

5.4.2 Passivation materials for IC device packaging

Passivating materials are deposited on devices while they are still in the wafer form, usually done at the completion of the IC fabrication process. These types of materials are mainly used for the mechanical protection of the IC devices during dicing of the wafer (*singulation process*). Furthermore, the passivation layer also serves as corrosion protection of the device. Inorganic polymers such as silicon dioxide, silicon nitride, and silicon-oxynitride are usually used by the semiconductor industry. Although silicon dioxide and silicon nitride are both excellent moisture barriers, silicon dioxide is still permeable to mobile ions such as sodium—in particular, it is under bias condition. The use of "getter" such as phosphite-doped silicon dioxide (P glass) by either CVD, plasma, or spin-on types, is readily available. However, the use of silicon nitride eliminates the mobile-ion diffusion problem. Recently, organic polymers such as polyimides, benzocyclobutenes, and silicone-polyimides (in particular, the photo-definable derivatives of this class of materials) are increasingly being used as passivating materials. These materials usually deposit a thickness of 0.25 to 2 μm, or with multilevel coatings for better reliability protection of the ICs. Since bond pad areas of the IC are etched out for further interconnections, and passivation layers are not entirely free of pinholes or cracks, corrosion of devices still occurs. As such, a second layer of the high-performance organic encapsulant is still needed for their protection. This layer can also act as a buffer coating of the large IC as a stress-relief coating prior to transfer molding of these plastic packages. Nevertheless, we address some organic materials for this type of application.

5.4.3 Organic encapsulants

There are numerous organic polymeric materials that are used as electronic encapsulants. These materials are typically used as on-chip as well as off-chip encapsulation and packaging. However, some of these materials are also used for passivation and interlayer dielectrics of the IC devices (as mentioned in the previous sections). These materials may be divided into three categories: (1) thermoplastic polymers; (2) thermosetting polymers; and (3) elastomers (see Table 5.2).

1. *Thermoplastic polymers* are materials that, when subjected to heat, will flow and solidify upon cooling without cross-linking. These thermoplastic processes are reversible (recyclable) and the polymers become suitable engineering plastic materials. Examples of high-performance thermoplastic polymers include: polyvinyl chloride, polystyrene, polyethylene, fluorocarbon polymers, asphalt, acrylics, tars, Parylenes (Union Carbide's poly-para-xylylene), and preimidized silicone-modified polyimides, which were originally developed by General Electric and subsequently developed by M&T Chemicals, National Starch and Chemical, Occidental Chemicals, Sumitomo Blakite Plastics, and Hitachi Chemical. Furthermore, the recently developed liquid-crystal-type thermoplastics are potential device encapsulants.

2. *Thermosetting polymers* are cross-linking polymers that cannot be reversed to the original polymer after curing. Examples of electronic thermosetting encapsulants include: silicones, polyimides, epoxies, silicone-modified polyimides, silicone-epoxies, polyesters, butadiene-styrenes, alkyd resins, allyl esters, silicon-carbons by Hercules, and polycyclicolefins by BF Goodrich.

3. *Elastomers* are thermosetting materials that have high elongation or elasticity. These types of materials consist of a long, linear, flexible molecular chain that is joined by internal covalent chemical cross-linking. Examples include silicone rubbers, silicone gels, natural rubbers, and polyurethanes. For IC applications, however, only a few of the materials in the preceding three groups that can be made ultrapure—such as epoxies, silicones, polyurethanes, polyimides, silicone-polyimides, Parylenes, cyclicolefins, silicon-carbons, and benzocyclobutenes—have been shown to be acceptable IC encapsulants

[6]. In addition, the recently developed high-performance liquid crystal materials (high-performance engineering plastic materials) are also potential organic polymeric materials for electronic applications. We now discuss some of these in detail.

5.4.3.1 Silicones (polyorganosiloxanes). Although polyorganosiloxanes (silicone) compounds have been known since the 1900s, it was not until the 1940s that research directed toward obtaining heat-resistant electrically insulating materials led Dow Corning (formed by the merger of Dow Chemical and Corning Glass Works silicone divisions) and General Electric into the manufacturing of silicone polymers.

General chemical properties of silicone. Silicone, with a repeating unit of alternating silicon-oxygen (Si-O) siloxane backbone, has some unique chemistry [7–13]. The siloxane backbone, with a partial double-bond Si-O character, derived from orbital interaction, provides a very thermally stable and flexible polymer. Two monovalent organic radicals are attached to a silicon atom with the general structure of $(R_2Si)_n$. The Si-O bond has a bonding energy of 110 kcal/mole [10], and the C-C bond has an energy of 82.6 kcal/mole. This bonding energy of silicon-oxygen could attribute to silicone's thermal stability. Furthermore, substituting the methyl group(s) with phenyl group(s), one could further increase its thermal stability by increasing electron delocalization from the electron-rich phenyl group to the siloxane backbone. In addition to the increase in thermal stability, the electron delocalization effect from the phenyl substitution also increases the refractive index of the siloxane, so phenyl-substituted siloxanes can be used in optical fiber index matching coating applications [14]. The phenyl-substituted groups disrupt the orderly conformation of the dimethylsiloxane and eliminate or suppress the low-temperature crystallization or melting temperature ($T_m = -45°C$) to form excellent low-temperature cycling materials. Furthermore, a 1,1,1-trifluoropropyl-substituted group on the siloxane backbone could generate a low dielectric constant and more solvent-resistant silicone. In many cases, silicone can be tailored for specific needs by changing the substitution groups. The most common use of polydimethylsiloxane elastomer with a large molar volume (77.5 cm^3/mole) and a low cohesive energy density is to cause the methyl group to rotate easily around the siloxane backbone. These freely rotating methyl groups provide a low T_g (glass transition temperature) = $-125°C$ polymer, which is extremely useful for low-temperature electronic applications. In addition, one or more of the substituted methyl groups could be replaced with a reactive functional group such as hydroxyl, alkoxyl, hydride, vinyl, amino, or chloro functional group to generate cross-linking. A thorough review by Yilg and

406 Chapter Five

TABLE 5.2 Typical Properties of Some Selected Electronic Encapsulants

	Dielectric strength (V/mil)	Resistivity (ohm-cm)	Water absorption (%[b])	Dielectric constant at 10^{10} Hz	Loss tangent at 10^{10} Hz	Relative arc track resistance[c]	Distortion temp. (°C)
Thermoplastics							
Asphalt and tars[k]	300	10^{10}	0.06	3.5	0.04	5	55
Fluorocarbon	450	10^{18}	0.00	2.1	0.0003	1	120
Polyethylene[k]	500	10^{16}	<0.01	2.3	0.0005	3	—
Polystyrene[k]	550	10^{18}	0.04	2.5	0.0003	3	80
Polyvinyl chloride	400	10^{15}	0.15	2.8	0.006	3	65
Wax[k]	400	10^{17}	0.02	2.6	0.001	3	25
Silicone-polyimide[a]	1500–2800	10^{15-17}	<1	3.0	0.007	2	150–240
Parylene[d]	500–7000	10^{13-16}	0.03	2.8	0.01–0.003	2	280–400
Polyketone	550	10^{15}		3.4	0.003	2	186
Polyetherketone	750	10^{15}		3.3	—	2	160
Polyetheretherketone	750	10^{14}		3.4	0.0015	2	160
Polyaryletherketone	590	10^{13}	0.5	3.5	0.001	2	245
Polysulfone[j]	600	10^{13}	0.5	3.2	0.001	2	200
Polyethersulfone[i]	500	10^{13}	0.5	3.5	0.001	2	205
Liquid crystal polymers	1220–1700	10^{14}		3.0–4.5	0.005	2	250–280
Thermosets							
Alkyd[k]	350	10^{14}	0.4	3.8	0.025	2	105
Allylester[k]	400	10^{14}	0.7	—	—	3	>90
Butadiene styrene[k]	600	10^{16}	0.03	2.4	0.006	3	125
Epoxode[k]	450	10^{14}	0.20	2.9	0.018	2	200
Phenolaldehyde[k]	350	10^{12}	0.3	4.7	0.04	4	80
Polyester[k]	350	10^{13}	0.4	3.5	0.05	3	90
Silicones[k]	600	10^{15}	0.03	2.8	0.002	2	40
Polyimides[e]	3400	10^{16}	—	3.6	0.002	2	>310
Silicone-Epoxy[f]	246–338	10^{15}	0.1	3.6	0.004	2	—
Benzocyclobutene[e]	10,000	10^{19}	0.2	2.6	0.0008	2	300(N_2)
Elastomers							
Buna-S rubber	500	10^{14}	—	2.5	0.01	4	—
Chloro rubber	400	10^{12}	—	2.7	0.05	3	—
Natural rubber	500	10^{16}	—	2.1	0.03	4	—
Silicone rubber[k]	600	10^{13}	—	3.0	0.05	2	>230
Thioplast[k]	150	10^{11}	—	14	0.15	4	—
Urethane[k]	350	10^{11}	0.4	3.5	0.04	4	>65
Inorganics							
SiO_2	5000	>10^{16}		3.5		1	760
Si_3N_4	5000	10^{12}	0	7.10		1	760
SiON	5000	10^{12}	0	7		1	760

[a] Ablestiks.
[b] In 24 hours. ⅛ inch thick.
[c] 1 = best; 5 = poorest.
[d] Union Carbide.
[e] Du Pont.
[f] Dow Corning.
[g] M = Rockwell M; SA = Shore Durometer A; SD = Shore Durometer D.
[h] At 21–32°C.
[i] ICI.
[j] Amoco.
[k] Unfilled.
[l] Dow Chemicals.

Thermal		Physical[h]					
Safe use temp. (°C)	Linear expansion × 10⁵/°C	Ultimate tensile strength (kpsi)	Ultimate tensile strength (MPa)	Ultimate elongation (%)	Hardness[g]	Relative adhesion[c]	General comments
70	8	0.6	4	5	SD 60	4	Lowest cost
260	5.5	3	20	200	SD 60	none	Good solvent resistance
115	9.5	4.4	30	1000	SD 65	5	Flexible
85	4	7.3	50	1.5	M 80	4	Rigid
100	3	3	20	100	SD 80	3	Costing material
55	11	0.3	2	5	SD 30	4	Melt, pour, and chill
400	3–10	2	14	200	—	1	Good solv. resist. high temp.
120	3.5–4.0	10.2	70	200	—	4	Conformal costing
260	5.7	13.8	95	150	—	2	
—	4.4	15.2	105	50	M 105	2	
250	4.7	14.5	100	5	M 99	2	High temperature
250	4.0	10	69	5	M 86	2	Crystalline material
200	3.0	12.2	84	6	M 69	2	
235	3.0	13	89	7	M 88	2	
180–240	0–2.5	20.3–34.8	140–240	1–7	M 60–M 100	2	High performance materials
120	4	8	55		SD 90	2	
100	4	5.8	40		M 70	3	
245	5	4.4	30	4	SD 80	4	
230	4–8	10.2	70	<1	M 90	1	Excellent solvent resistance
80	4	7.3	50	1.5	M 126	2	
165	6	8.0	55	<5	M 100	3	
260	13–100	2.5	17	8		4	Excellent THB performance
<430	0.3–80	14–20	100–140	10–80	—	3	Good solv. resist. high temp
<200	3–6	8	55		SD 60	3	Good for molding
350(N₂)	3.4	10.3	71	<1	3H	1	High temperature
120	6	0.3	2	400	SA 50	2	Flexible
—	9	2.5	17	500	SA 70	3	Flexible
65	4	3	20	700	SA 50	2	Flexible
260	—	0.58	4	100	SA 60	4	Flexible
120	10	0.3	2	400	SA 40	2	Poor elec. props high temp.
95	10	5.1	35	400	SA 60	1	Poor elec. props. high temp.
760	0.05–0.09	14–56	100–390	0		4	Excellent passivation
370–700	0.45–0.54	14–140	100–1,000	0		4	Excellent passivation
500	0.05–0.35	14–140	100–1,000	0			

McGrath describes all these siloxanes and their latest development and syntheses in detail [13].

The basis of commercial production of silicones is that chlorosilanes are readily hydrolyzed to give disilanols, which are unstable and condense to form siloxane oligomers and polymers. Depending on the reaction conditions, a mixture of linear polymers and cyclic oligomers is produced. The cyclic components can be ring opened by either acid or base to become linear polymers; it is these linear polymers that are of commercial importance. The *linear polymers* are typically liquids of low viscosity and, as such, are not suited for use as encapsulants. These must be cross-linked (or vulcanized) in order to increase the molecular weight to a sufficient level where the properties are useful. Two methods of cross-linking are used: those which can be classified as condensation cures (see Fig. 5.3) and those which are addition-cure systems (see Fig. 5.4). For electronic applications, only the high-purity room-temperature vulcanized (RTV) condensation cure silicone, which used the alkoxide cure system with noncorrosive alcohol by-products, and the platinum-catalyzed addition heat cure vinyl and hydride (hydrosilation) silicone systems are suitable for device encapsulation. Since these RTV condensation and heat hydrosilation cure systems are of interest to us in electronic and photonic applications, we discuss these in the following sections.

Room-temperature vulcanized (RTV) silicones. RTV silicone is a typical condensation cure system material. The moisture initiated catalyst (such as organotitanate, tin, dibutyldilaurate, etc.)-assisted process generates water or alcohol by-products, which could cause outgassing and voids. However, by careful control of the curing process, one could achieve a very reliable encapsulant. Since the silicone has a low surface tension, it tends to creep and run over the encapsulated IC circuits. To better control the rheological properties of the material, thixotropic agents (such as fumed silica) are usually added to the formulation. The thixotropic agent provides a yield stress, increases the suitable G' (storage modulus), G'' (loss modulus), and η^* (dynamic viscosity) of the

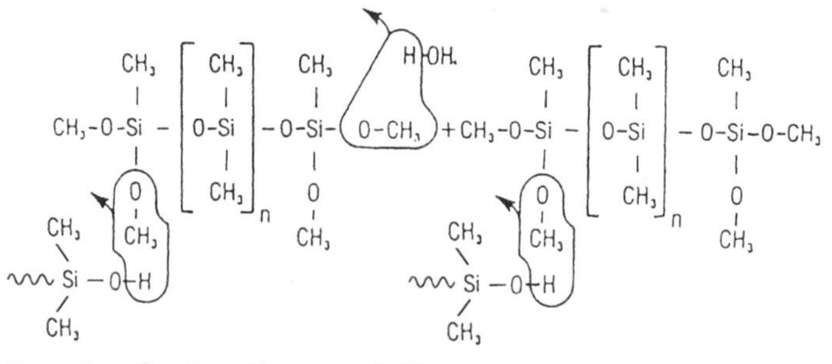

Figure 5.3 Condensation cure of silicone.

CHEMICAL REACTION MECHANISM OF SILICONE GEL

(A) Silicone Gel Additional Cure Mechanism:

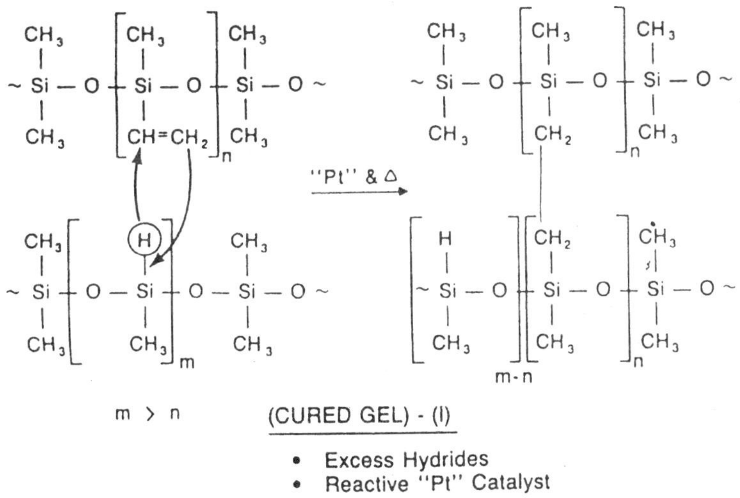

Figure 5.4 Addition cure of silicone.

encapsulant. Filler-resin and filler-filler interactions are important in obtaining a well-balanced and well-controlled encapsulant. This rheological controlled material tends to flow evenly in all circuit edges, covers all the underchip area, and prevents wicking and run-over of the circuits, which is a critical parameter in coating production (see Fig. 5.5) [15]. In addition, pigments such as carbon-black and titanium dioxide are usually added as opacifier to protect light-sensitive devices. Organic solvents such as xylenes and Freons are incorporated into the formulation to reduce the encapsulant viscosity. Table 5.3 shows a typical RTV silicone system that AT&T has been using in electronic encapsulation. This RTV silicone has been used to protect bipolar, MOS, and hybrid ICs (HICs) for over 20 years [16]. The ability of the RTV silicone to form chemical bonds with the coated substrate is one of the key reasons the material achieves excellent electrical performance. The reactive alkoxy functional groups of the silicone react with the surface hydroxyl groups to form a stable inert silicon-oxygen-substrate bond. In addition, this chemical reaction consumes the substrate surface hydrophilic hydroxyl groups that would hydrogen (H)-bond with diffused moisture. When

RHEOLOGY OF RTV SILICONES

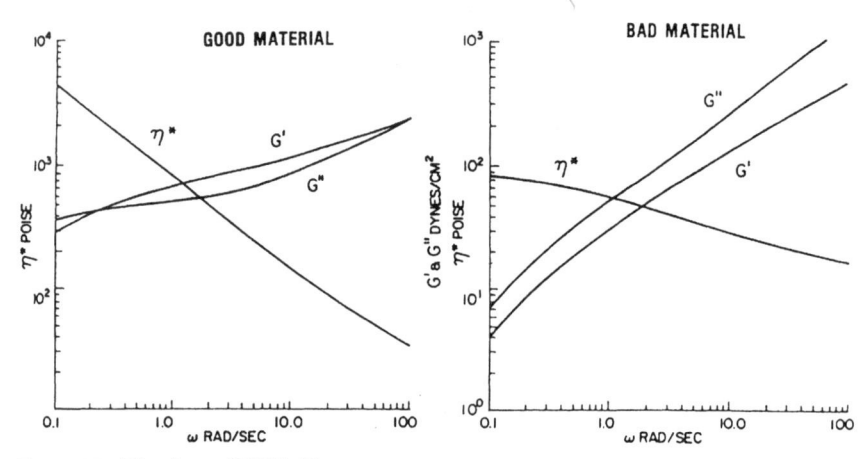

Figure 5.5 Rheology of RTV silicone.

sufficient diffused moisture is H-bonded with the surface hydroxyl groups, surface conduction probably takes place by hopping of protons from hydroxyl group to hydroxyl group. Furthermore, diffused moisture could form a continuous path of thin water layers. Under such circumstances, and in the presence of those contaminant mobile ions and applied electrical bias, corrosion of the IC metallization will result. However, when all surface hydroxyl groups are reacted with the silicone alkoxy groups, even though moisture continues to diffuse through the silicone matrix, as long as no continuous water path is formed on the interface that could result in a surface conduction and/or "H-bonding" of the diffused moisture, the diffused moisture will only diffuse in and out of the siloxane matrix polymer in an equilibrium fashion, which does not cause electrical corrosion of the encapsulated devices. That is probably one of the reasons why RTV silicone is capable of achieving superior performance in temperature-humidity-bias (THB) accelerated electrical tests (see Fig. 5.6).

TABLE 5.3 Typical RTV Silicon Formulations

Ingredients	Concentration (phr)	Impact on properties
Base polymer (OH-terminated polydimethylsiloxane)	100	Physical and mechanical
Crosslinker: $(OMe)_3SiMe$	8–12	Physical and mechanical
Catalyst (titanate)	0.5	Cure rate
Inhibitor (isopropanol)	0.25	Longer shelf life and pot life
Filler(s):SiO_2	10–12	Physical and mechanical
TiO_2	2–3	Opacifier
C-black	0.05	Pigment
Solvent (xylenes)	0–50	Rheological control diluent

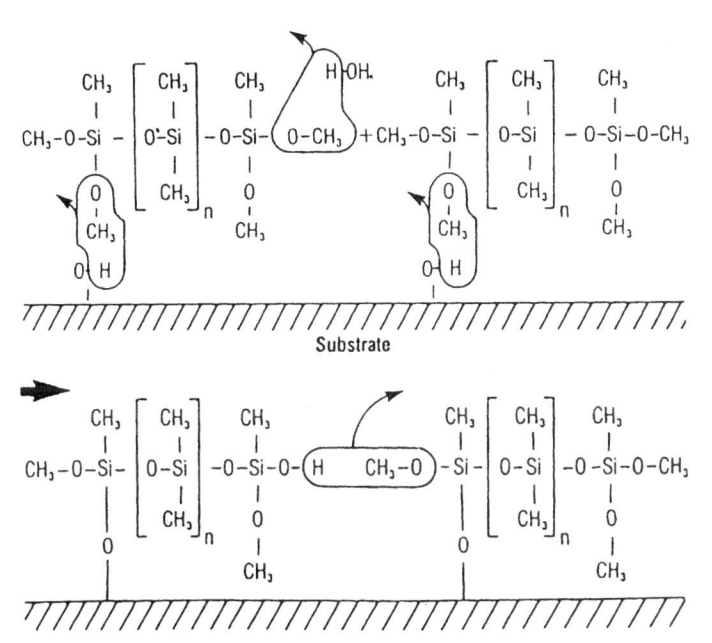

Polymers for Electronic Packaging: Materials, Processes, and Reliability 411

Figure 5.6 Mechanism of RTV silicone: condensation cure mechanism reaction with coated substrate.

Heat-curable hydrosilation silicones (elastomers and gels). Heat-curable hydrosilation silicone (either elastomer or gel) has become an attractive device encapsulant. Its curing time is much shorter than the RTV-type silicone. Heat-curable silicones also tend to have slightly better stability at elevated temperatures than the conventional RTV silicone. With its jellylike (very low modulus) intrinsic softness, silicone gel is a very attractive encapsulant in wire-bonded large-chip-size IC devices. The two-part heat-curable system, which consists of the vinyl and hydride reactive functional groups, and the platinum catalyst hydrosilation addition cure system provides a fast-cure system without any by-product. (See Fig. 5.3 for cure mechanism.) To formulate a low-modulus silicone gel, a vinyl-terminated polydimethylsiloxane with a moderate low viscosity range from 200 to a few thousand centipoise (cps), and a low viscosity (range from a few cps to ~100 cps) di- or multifunctional hydride-terminated polydimethylsiloxane are used in the formulation. The low-viscosity hydride resin usually blends in with the higher-viscosity vinyl resin to achieve an easier mixing ratio of part A (only vinyl portion) and part B (hydride plus some vinyl portion for ease of mixing). The key to formulate a low-modulus silicone is the deliberate under-cross-linking of the silicone system. A few ppm platinum catalyst, such as chloroplatanic acid or organoplatinum, is used in this system. This catalyst is usually incorporated in the part A vinyl portion of the resin. (For typical silicone gel formulation, see Table 5.4.)

TABLE 5.4 Typical Silicone Gel Formulations

Ingredients	Concentration (phr)	Impact on properties
Base Polymer (vinyl-terminated polydimethylsiloxane)	100	Physical and mechanical
Crosslinker: tri- or tetrafunctional silicon hydride	10–15	Physical and mechanical
Catalyst: chloroplantanic acid or organoplatinum	5–10 ppm	Cure rate

However, a highly deactivated platinum catalyst system (by premixing a chelating compound such as 2-methyl-3-butyn-2-ol to coordinate the reactive platinum catalyst) is used to formulate a one-component system. This one-component silicone gel system provides less mixing and a problem-free production material. This solventless type of heat-curable silicone gel will have increased use in electronic applications.

Applications of silicones in electronic coatings. Since World War II, silicones have been used in a variety of applications where high thermal stability, hydrophobicity, and a low dielectric constant are necessary, for example, as encapsulants or conformal coating for integrated circuits. In 1969, it was demonstrated that room-temperature vulcanized (RTV) silicones exhibited excellent performance as moisture-protection barriers for IC devices, and a number of different RTV silicones have been adapted for use in the electronics industry [16–38]. In addition to their superior electrical, chemical, and physical properties, the main reasons for their use in electronic applications are mainly due to the noncorrosive by-products (alcohols) generated and the low level of ionic contaminants in the silicone resin.

Ionic contaminants, whether from the device surface, encapsulation materials, or the environment, affect the electrical reliability of encapsulated IC devices. For this reason, the silicones are subjected to intense purification. The concentration of Na^+, K^+, F^- and Cl^- mobile ions is less than a few ppm, and alpha-particle emission is less than 0.001 alpha/cm^2. Thus, silicones also offer excellent alpha-particle shielding for eliminating soft-error in dynamic random access memory (DRAM) devices, such as megabit DRAM chips [26]. The drawbacks of RTV and heat-curable silicone as an IC encapsulant are both its poor solvent resistance and weak mechanical properties. However, a recently developed silicone material with well-controlled cross-linking density and high filler loading system have significantly improved the solvent resistance and mechanical properties, yet maintains its excellent temperature cycling performance of the silicone encapsulant [39–40]. (See Fig. 5.7.)

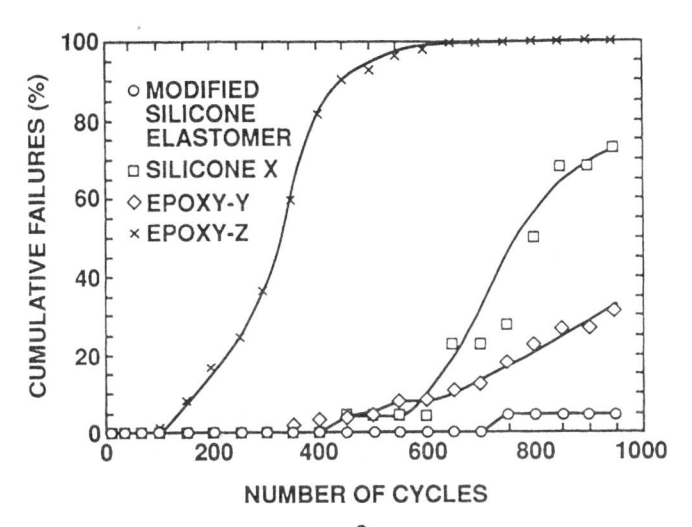

Figure 5.7 Temperature cycling test results of modified silicone.

Chart legend: MODIFIED SILICONE ELASTOMER, SILICONE X, EPOXY-Y, EPOXY-Z. Axes: CUMULATIVE FAILURES (%) vs NUMBER OF CYCLES. TEST CONDITION: -40 TO 130 °C, 20 MINUTES DWELL TIME.

5.4.3.2 Epoxies. Epoxies are one of the most utilized polymeric materials in electronics. This class of materials were first prepared in early 1930. Their unique chemical and physical properties—such as excellent chemical and corrosion resistances, electrical and physical properties, excellent adhesion, thermal insulation, low shrinkage, and reasonable material cost—have made epoxy resins very attractive in electronic applications [41–43]. The commercial preparation of epoxies is based on bisphenol A, which, upon reaction with epichlorohydrin, produces diglycidyl ethers. The repetition number n varies from zero (liquid) to approximately 30 (hard solid). The reactants' ratio (bisphenol A versus epichlorohydrin) determines the final viscosity of the epoxies. In addition to the bisphenol A resins, the Novolac resins with multifunctional groups and the recent biphenyl epoxies, which lead to higher cross-link density and better thermal and chemical resistance, have gained increasing acceptance in electronic applications. Typical epoxy curing agents are amines, anhydrides, dicyanodiamides, melamine/formaldehydes, urea/formaldehydes, phenol/formaldehydes, imidazoles, and so forth. Anhydrides and amines are two of the most frequently used curing agents.

Selecting the proper curing agents is dependent on application techniques, curing conditions, pot-life required, and the desired physical properties. Besides affecting viscosity and reactivities of the epoxy formulations, curing agents determine the degree of cross-linking and the formulation of chemical bonds in the cured epoxy system. The reactivity of some anhydrides with epoxies is slow; therefore, an accelerator, usually a tertiary amine, is used to assist the cure. "Novolacs" and "Resole" are two major commonly used phenolformaldehyde epoxies. A Novolac is a phenol-formaldehyde, acid-catalyzed epoxy polymer. The phenolic groups in the polymer are linked by a methylene bridge, which provides highly cross-linked systems and a high temperature and excellent chemical-resistant polymer. "Resole" is a base-catalyzed phenol-formaldehyde epoxy polymer. In most phenolic resins, the phenolic group is converted into an ether to give improved base resistance. Phenolic resins are cured through the secondary hydroxyl group on the epoxy backbone. High-temperature curing is required in this system and it provides excellent chemical resistance.

Recently developed high-purity epoxies have become very attractive encapsulants for electronics [41–46]. These new types of resin contain greatly reduced amounts of chloride and other mobile ions, such as sodium and potassium, and have become widely used in device encapsulation and molding compounds. The incorporation of a high level of the well-controlled spherical silica particles, with bimodal distribution as filler in the epoxy systems, has drastically reduced the thermal coefficient of expansion of these materials and makes them more compatible with the IC die-attached substrate materials. The incorporation of a small amount of an elastomeric material (such as silicone elastomeric domain particles) to the rigid epoxy has drastically reduced the elastic modulus, reduced the thermal stress, and increased the toughness of the epoxy material [45–46]. This new type of low-stress epoxy encapsulant has great potential application in molding large IC devices. In addition, the newly developed glob-top-type (a glob of polymeric resin that covers the entire IC device, including the wire bonding, able to meet all the requirements as a device encapsulant) epoxy material, which is applied as a drop on top of the IC chip, is becoming increasingly more acceptable as an encapsulant for higher-reliability chip-on-board and flip-chip-type electronic devices and systems [47–48]. (See Fig. 5.8.) When the epoxy materials are properly formulated and applied, and their stress-related issues (such as reduced stress and reduced thermal coefficient of expansion) have been properly considered and resolved, they could become a very attractive high-performance encapsulant. The continuous advancements in epoxy material development will have a great impact in device packaging. Table 5.5 shows various structure-property relationships of epoxies.

5.4.3.3 Polyurethanes. Polyurethane was first made available by Otto Bayer in the late 1920s in Germany [49]. The early study of polyurethane was simply based on di-isocyanates and diols or polyols. However, recent work is focused on the use of *intermediates*, which are low

ENCAPSULATED FLIP-CHIP DEVICE

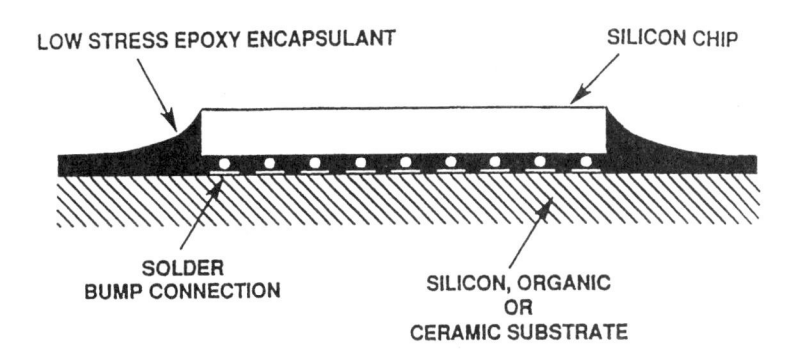

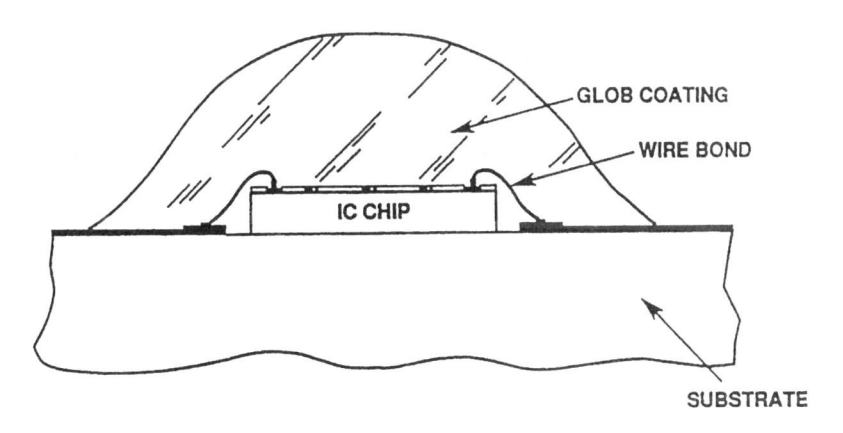

DIRECT CHIP ATTACHMENT WITH PROTECTIVE POLYMER OVERCOAT ("GLOB COATING")

Figure 5.8 Epoxy flip-chip and glob-top structures.

molecular weight polyethers with reactive functional groups such as hydroxyl or isocyanate groups able to further cross-link, chain extend, or branch with other chain extenders to become higher molecular weight polyurethanes. Diamine and diol are chain extended with the prepolymer (either polyester or polyether) to form polyurethanes with urea or urethane linkages, respectively [50]. The morphology of polyurethane is well characterized. Hard and soft segments from diisocyanates and polyols, respectively, are the key to excellent physical properties of this material (see Fig. 5.9).

Bases are more widely used than acids as catalysts for polyurethane polymerization. The catalytic activity increases with the basicity. Amines such as tertiary alkylamines and organic metal salts, such as

TABLE 5.5 Specialty Epoxy Resins

$$CH_2\text{-}CH\text{-}CH_2\text{-}[\text{-}R\text{-}]\text{-}CH_2\text{-}CH\text{-}CH_2$$
(with epoxide groups on terminal CH₂–CH)

Type	Structure	Function
Tetrabromo-BPA	—O—(Br,Br-C₆H₂)—C(CH₃)₂—(Br,Br-C₆H₂)—O—	Fire Retardant
Dimer acid	—O—C(=O)—C₃₄—C(=O)—O—	Flexibility
Polyglycols	—O—[CH₂—CH₂]ₙ—O—CH₂—CH₂—O—	Flexibility
Novolac	—O—(C₆H₄)—[CH₂—(C₆H₃(O-))—CH₂]ₙ—(C₆H₄)—O—	High performance resins
Tetraphenol Ethane	—O—(C₆H₄)—CH—(C₆H₄)—O— / —O—(C₆H₄)—CH—(C₆H₄)—O—	High performance resins
Methylene Dianiline	>N—(C₆H₄)—CH₂—(C₆H₄)—N<	High performance resins
Biphenyl	—[O—(CH₃,CH₃-C₆H₂)—(C₆H₂-CH₃,CH₃)—O]—	Low viscosity

tin or lead octoates, promote the reaction of isocyanate and hydroxyl functional groups in the polyurethane system and accelerate the crosslinking. However, the hydrolytic stability of the polyurethane can be affected by the catalyst used. UV stabilizers are usually added to reduce the radiation sensitivity of the material. In addition, polyurethane has unique high strength, high modulus, high hardness, and high elongation. It is one of the toughest elastomers used today. High-performance polyurethane elastomers are used in conformal coating, potting, and in reactive injection molding of IC devices.

Polymers for Electronic Packaging: Materials, Processes, and Reliability   417

SYNTHESIS OF POLYURETHANE ELASTOMER

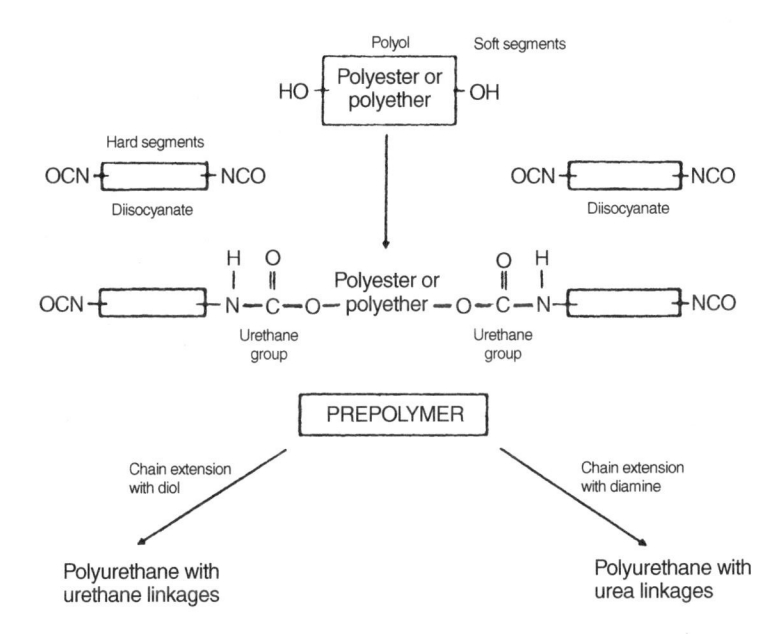

Polyol — Soft segments
HO—[Polyester or polyether]—OH

Hard segments
OCN—[   ]—NCO   OCN—[   ]—NCO
Diisocyanate   Diisocyanate

OCN—[   ]—N—C—O—Polyester or polyether—O—C—N—[   ]—NCO
           H O   O H
           | ‖   ‖ |
         Urethane   Urethane
          group   group

PREPOLYMER

Chain extension with diol ← → Chain extension with diamine

Polyurethane with urethane linkages   Polyurethane with urea linkages

HARD & SOFT SEGMENTS OF POLYURETHANE

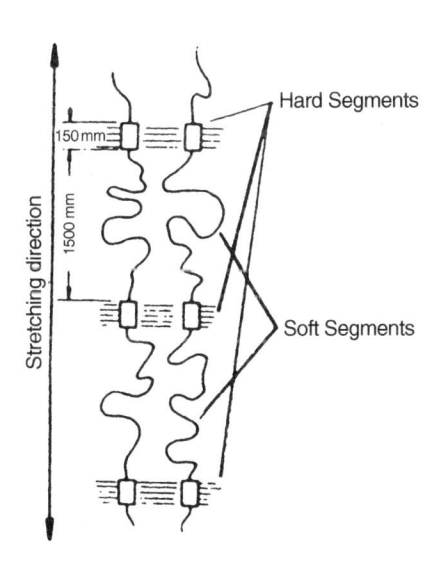

Stretching direction
150 mm
1500 mm

Hard Segments
Soft Segments

morphology of polyurethanes.

Figure 5.9 Polyurethane—synthesis and structures.

5.4.3.4 Polyimides. Polyimides are widely used high-performance electronic polymers, first developed at DuPont in the 1950s. During the past couple of decades, there has been tremendous interest in this material for electronic applications [51]. The superior thermal (stability up to 500°C), mechanical, and electrical properties of polyimide have made its use possible in many high-performance applications, from aerospace to microelectronics. In addition, polyimides show very low electrical leakage in surface or bulk. They form excellent interlayer dielectric insulators and also provide excellent step coverage, which is very important in fabrication of the multilayer IC structures. They have excellent solvent resistance and ease of application. They can be easily sprayed or spun-on and imaged by a conventional photolithography and etch process.

Most polyimides are aromatic diamine and dianhydride compositions (see Fig. 5.10). However, by changing the diamine and dianhydride substitutes, one will derive a variety of high-performance polyimides (see Fig. 5.11). Polyamic acids are precursors of the polyimides. Thermal cyclization of polyamic acid is a simple curing mechanism for this material. Siemens of Germany developed the first photo-definable polyimide material [52]. However, Ciba Geigy developed the first new type of photo-definable polyimide that does not

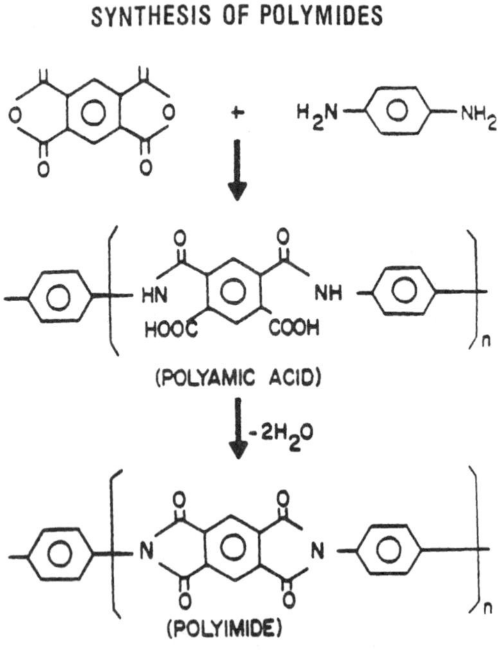

Figure 5.10 Polyimides—synthesis and structures.

Polymers for Electronic Packaging: Materials, Processes, and Reliability   419

Polymer Type	Group	Anhydride	Diamine	Modulus of Elasticity kg/cm²	Elongation at Break	T_g
PMDA-benzidine	1	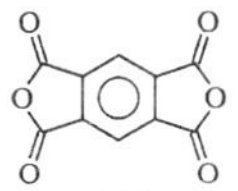 rigid	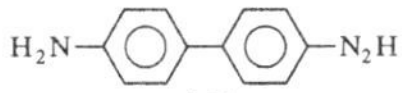 rigid	120 K	2	none
ODPA-PPDA	2	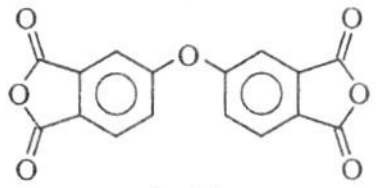 flexible	 rigid	65 K	5	none
PMDA-ODA	3	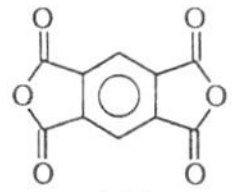 rigid	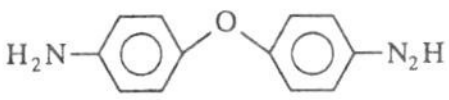 flexible	35 K	100	"crosslinks"
ODPA-ODA	4	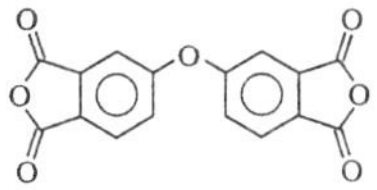 flexible	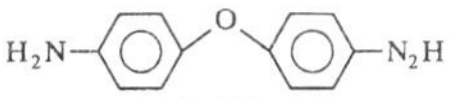 flexible	30 K	100	270°C fusible

Figure 5.11   Various structures of polyimides.

require a photoinitiator [53]. Both of these photo-definable materials are negative-resist-type polyimides. A positive-resist-type polyimide that reduces the processing step in IC fabrication has recently been reported by Sumitomo Bakelite in Japan [54]. An interpenetration network (IPN) of two types of polyimides is used to achieve the positive-tone material. Hitachi has developed an ultralow thermal coefficient of expansion (TCE) polyimide that has some potential in reducing the thermal stress of the silicon chip and the polyimide encapsulant. The rigid rodlike structure of the polyimide backbone structure is the key in preparing the low-TCE polyimide [55]. By simply blending a high- and low-TCE polyimide, one will be able to achieve a desirable TCE encapsulant that could match the TCE on the substrate and reduce the thermal stress problem in encapsulated device temperature cycling testing. However, the affinity for moisture absorption due to the carbonyl polar groups of the polyimide, a high-temperature cure, and high cost of the polyimide are the only drawbacks that prevent its use in low-cost consumer electronic application. Preimidized polyimides that cure by evaporation of solvent may reduce the drawback of the high-temperature cure of the material. Advances in polyimide syntheses have reduced the material's moisture absorption and dielectric constant by the incorporation of siloxane segments and fluorinated substitution groups into the polyimide backbone, respectively. However, the affinity of moisture to the polyimide chemical structure is still a concern in its use in electronics. Nevertheless, polyimides are widely used as IC encapsulant, interlayer dielectrics, ion implant masks, and alpha-particle getters/passivations.

5.4.3.5 Silicone-polyimide (new modified polyimides). Combining the low modulus of the siloxane (silicone), and the high thermal stability of polyimide, the siloxane-polyimide (SPI) copolymers were first developed at General Electric [56]. SPI copolymers have become very attractive IC device encapsulants [57–60]. Silicone-polyimides are fully imidized copolymers and are soluble in low-boiling solvents such as diglyme, which reduces the high processing temperature and eliminates the outgassing of water during normal polyimide imidization (cure) process. The high processing temperature and outgassing of water are main drawbacks of the polyimides. Besides, the SPI has good adhesion to many materials due to the siloxane property and eliminates the need for an adhesion promoter. Polycondensation and polyaddition processes are used to synthesize these materials (see Fig. 5.12). Thermoplastic and thermosetting SPI materials can be obtained by these processes. In addition, photo- and thermocurable SPIs are also obtainable by incorporating photo-reactive functional groups in these types of materials. One can control the imide and siloxane blocks within the copolymer matrix to tailor the SPI properties. Since most of these are preimidized thermoplastic materials, their shelf life is very stable. These materials will have potential as IC device encapsulants, interlayer dielectrics, and passivation in microelectronic applications.

5.4.3.6 Parylenes. Parylene, a poly-(para-xylylene), was first developed by Union Carbide Corporation [61]. The process uses a thermal

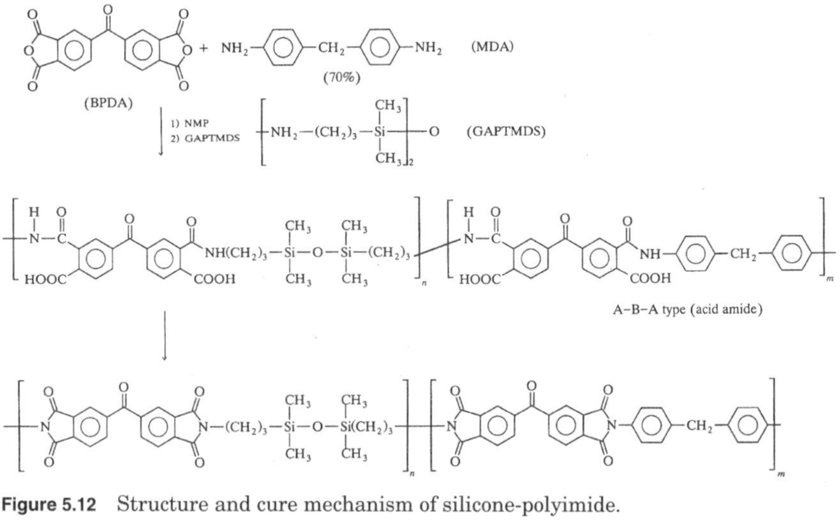

Figure 5.12 Structure and cure mechanism of silicone-polyimide.

reactor to first vaporize (at 150°C, 1 torr pressure), and pyrolyze (at 680°C, 0.5 torr pressure) the di-para-xylene, then polymerize the dimer into polymer at room temperature (see Fig. 5.13). This room-temperature deposition is a very attractive encapsulation process, especially for temperature-sensitive, low glass transition substrate materials. Parylene deposition provides an excellent conformal step-coverage and conformal film, with thickness ranging from 2 to 50 µm. Since it is a room-temperature, spontaneous, and gas-to-solid deposition process, encapsulated electronic parts experience only minimal stress; it is an excellent, reliable, and labor-saving process. In addition, the deposition rate of Parylene depends on the types of Parylenes. The dichloro- and monochloro-substituted on the benzene ring of the

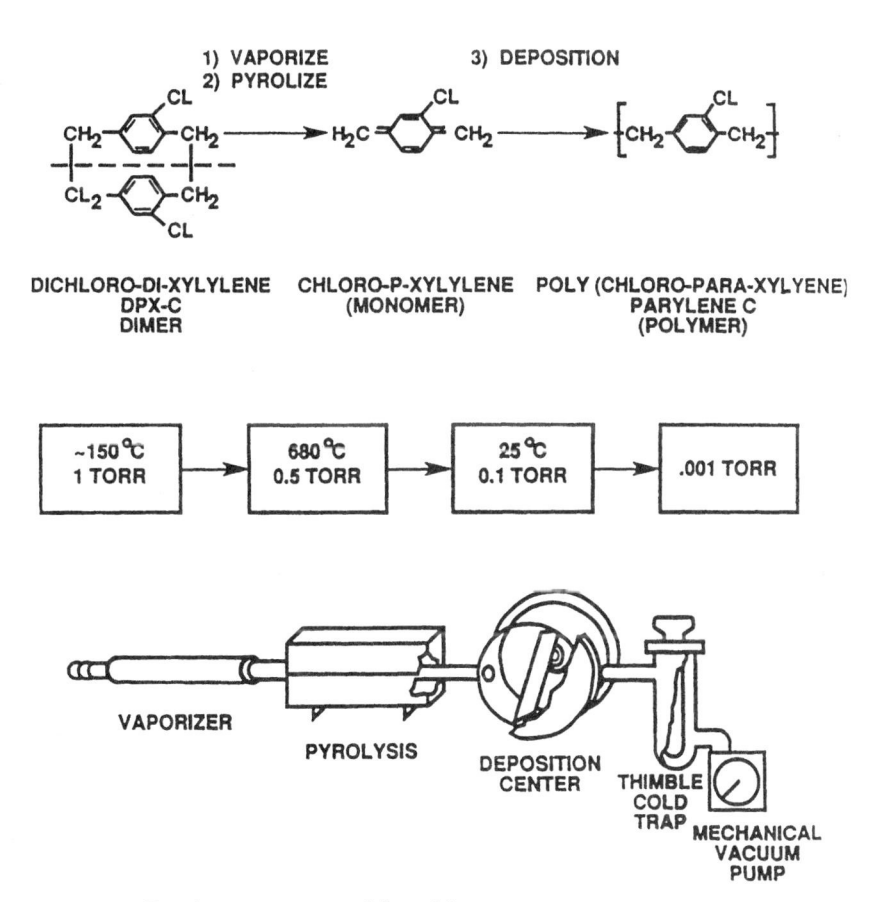

Figure 5.13 Panylene structure and deposition process.

Parylene D and C, respectively, and tetrafluoro-methylene substituted Parylene F, have a faster deposition rate than the normal unsubstituted Parylene N. The deposited Parylene film has excellent chemical resistance and electrical properties (see Table 5.6). This process pyrolyzes only the dimer to a tough polymer. There are no solvents, catalysts, and so forth, to create impurity in the formed film. This is a vapor-phase deposition, so it is suitable for depositing continuous, pinhole-free films in hard-to-reach areas, such as underchips on flip-chip or beam-leaded devices. Currently, this process is widely used in conformal coating of military circuit boards for protection of electronic parts subjected to severe abrasion. However, the high cost of the starting dimer and the deposition equipment may prohibit its wide use in consumer electronic applications. Nevertheless, it is a unique conformal coating material with potential coating applications [62]. However, adhesion of Parylene to substrate is only fair, so a silane coupling agent treatment is needed for good adhesion between these surfaces.

5.4.3.7 Benzocyclobutenes. The high-performance benzocyclobutene (BCB) polymers were recently developed by Dow Chemical Company [63–64] from the structure shown in Fig. 5.14. The cross-linking process is carried out by the thermal rearrangement of the dicyclobutyl monomer to form the reactive intermediate orthoquinodimethane, which can polymerize with the unsaturated functional group. Since it

TABLE 5.6 Types of Parylenes (Poly-para-xylylenes)

Type of parylenes	Deposition rate (μm/hr)	Dissipation factor	Dielectric constant	Water absorption (%)
N type: $-[CH_2-C_6H_4-CH_2]_n-$	1	0.0002	2.6	0.02
C type: $-[CH_2-C_6H_3Cl-CH_2]_n-$	3–5	0.02	3.1	0.06
D type: $-[CH_2-C_6H_2Cl_2-CH_2]_2-$	10–15	0.01	3.6	0.07
F type: $-[CF_2-C_6H_4-CF_2]_2-$	—	—	—	—

Benzocyclobutene Chemistry

Benzocyclobutene $\xrightarrow{>180°C}$ Orthoquinodimethane
- latent diene
- very reactive

$\xrightarrow{>180°C}$ Thermoset polymers

1

Polymerization of Benzocyclobutenes

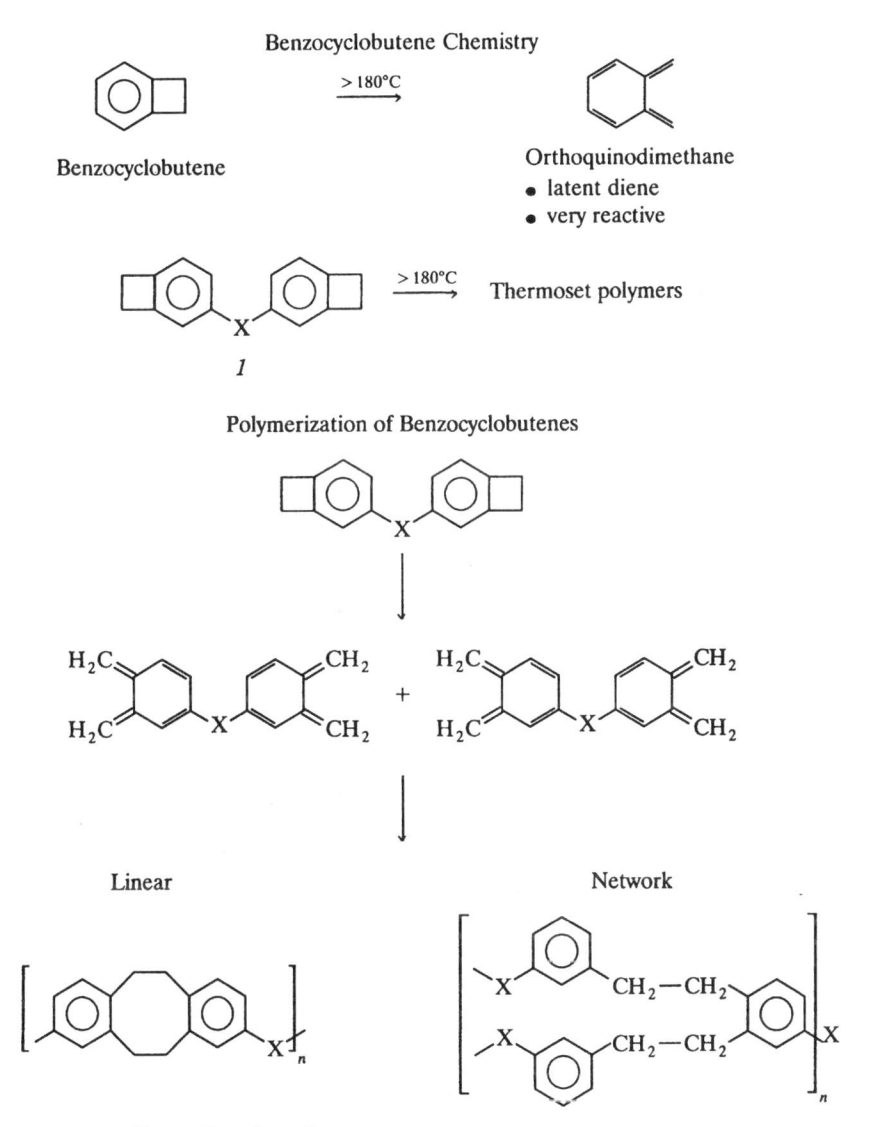

Linear   Network

Figure 5.14 Properties of parylenes.

is based on the thermal rearrangement process, BCB requires no catalyst and there are no by-products during the curing process. The properties of BCB can be modified by the substituted group X in structure 1 of Fig. 5.14. Dimethysiloxane groups, the substituted X groups, are usually used as copolymer in the BCB to enhance its adhesion and reduces its modulus. As such, BCB monomers are normally "B-stage" (partially thermally cross-linked) to enhance their viscosity stability.

TABLE 5.7 **Typical Properties of Benzocyclobutene**

Flexural Modulus, psi	498,000
Dielectric constant, 10^4–10^7 Hz	2.68
Dielectric strength	10,000 V/mil
Dissipation factor, 10^4–10^7 Hz	$<10^{-3}$
Coefficient of thermal expansion (25–3000°C), ppm	34
Safe use temperature (in nitrogen)	348°
0% weight loss temperature, °C	450
T_g, °C	350
Resistivity	$10^{19}/\Omega/cm$
Water absorption	0.2 (24-h water boil)
Flexural strength	10,300 psi
Elongation	<1%
Hardness	3H (pencil lead test)

The commercially available BCB is usually in a "B-stage" diluted with xylene for spin-coating application. BCB has excellent physical, chemical, and electrical properties, which make its use in microelectronic applications similar to polyimides, with its low dielectric constant (2.7), low moisture absorption (<1%) and good adhesion properties (see Table 5.7). BCB is a potential IC-passivating encapsulant and interlayer dielectrics for the current multichip-module applications [65]. More recently, higher thermal oxidative stability with added antioxidant additive and photosensitive BCBs are also available.

5.4.3.8 Sycar (a silicon-carbon hybrid) polymers. Hercules (at Wilmington, Delaware) has developed a new class of silicon-carbon hybrid materials. This class of material consists of a backbone of the siloxane (-Si-O-Si-O-) structure with hydrocarbon substitute groups that crosslink with the silicon backbone (see Fig. 5.15); as such, it provides excellent mechanical and solvent properties, which are the drawbacks of silicones, yet maintains its silicone-like excellent electrical properties. Applications in electronics such as glob-top, molding compounds, and high-performance PWB are being made by this class of material. However, it is not commonly used yet, but eventually will have potential for electronic applications [66].

5.4.3.9 Bis-maleimide triazine (BT) resins. A new resin used to prepare a high-temperature printed wiring board (PWB) [versus the conven-

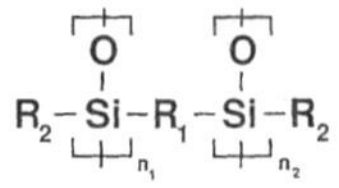

Figure 5.15 Typical structure of silcar.

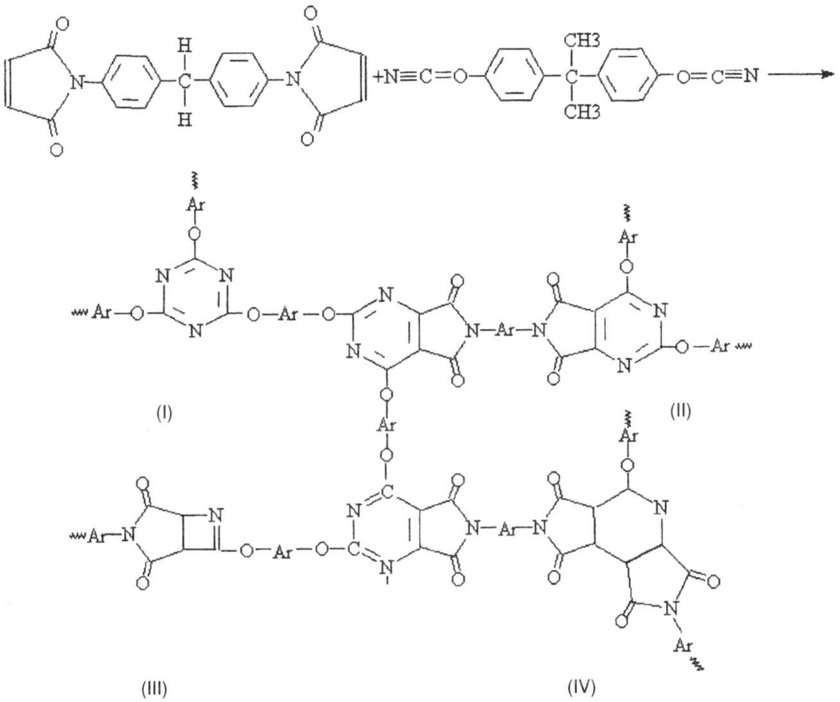

Figure 5.16 Structure of bismaleimide triazine (BT).

tional FR (fire retardant)-4 PWB] is called *bis-maleimide triazine* (BT). The triazine polymer is mainly produced by Mitsubishi Chemicals in Japan. (See Fig. 5.16 and Table 5.8.) The trimerization of the monomers forms a high-temperature, high T_g (>230°C) triazine (Cyanurate ring) resin that is generally mixed with epoxy resin to form a high-performance PWB substrate for Ball Grid Array (BGA) and Multichip Module-Laminate (MCM-L) substrate applications. The mixture of BT/epoxy also has good electrical and good thermomechanical properties. Epoxy is blended into the BT resin to provide tough BT/epoxies, which have a regular T_g of 180 to 190°C. The process is compatible with the PWB board manufacturing process and PWB wire-bonding processes. Furthermore, BT/epoxies have a long history of good resistance to ionic conductive growth (CAF) and popcorn testing results [67].

5.4.3.10 Polycyclicolefins. BF Goodrich has recently developed a new class of polycyclicolefins (PCO) based on the principle of polynorbornene chemistry. A transistion metal catalyst is used to provide a tightly control polymerization of the monomers to saturated polymers with excellent T_g (>350°C), low dielectric constant (2.45), low moisture

TABLE 5.8 Common Properties of BT Polymers

Item	Unit	Conditioning procedure	Mitsubishi CCL 4830	Mitsubishi CCL-H832
Insulation resistance	Ω	C-96/20/65 C-96/20/65 + D-2/100	5×10^{14}–10^{15} 5×10^{12}–10^{14}	5×10^{14}–10^{15} 5×10^{12}–10^{14}
Volume resistivity	Ω-cm	C-96/20/65 C-96/20/65 + C-96/40/90	10^{15}–10^{16} 10^{14}–10^{15}	10^{15}–10^{16} 10^{14}–10^{15}
Surface resistance*	Ω	C-96/20/65 C-96/20/65 + C-96/40/90	5×10^{13}–10^{14} 10^{12}–10^{14}	5×10^{13}–10^{14} 10^{12}–10^{14}
Dielectric constant (1 MHz)	—	C-96/20/65 C-96/20/65 + D-48/50	4.5–4.8 4.5–4.8	4.5–4.8 4.5–4.8
Dissipation factor (1 MHz)	—	C-96/20/65 C-96/20/65 + D-48/50	0.0050–0.0090 0.0050–0.0090	0.0050–0.0090 0.0050–0.0090
Hot solder resistance	sec	A	300°C more than 30	300°C more than 30
Heat resistance	—	A	230°C · 30 min No change	230°C + 30 No change
Glass transition temperature	°C	A	170–215	170–215
Thermal expansion coefficient**	cm/cm/°C	A X Y Z	$1.3 \sim 1.5 \times 10^{-5}$ $1.5 \sim 1.7 \times 10^{-5}$ $4.6 \sim 5.6 \times 10^{-5}$	$1.3 \sim 1.5 \times 10^{-5}$ $1.5 \sim 1.7 \times 10^{-5}$ $4.6 \sim 5.6 \times 10^{-5}$
Thermal conduction coefficient	cal/cm · s · °C	A	$4 \sim 5 \times 10^{-4}$	$4 \sim 5 \times 10^{-4}$
Peel strength (35 μm)	kgf/cm	A S_4 E-1/105	1.3–1.8 1.4–1.9 1.3–1.8	1.3–1.8 1.4–1.9 1.3–1.8
Water absorption	%	E-24/50 + D-24/23	0.02–0.10	0.02–0.10
Flexural strength (C.W.)	Kgf/mm^2	A	45–60	45–60
Flame resistance (UL)	sec	A, E-168/70	94V-O	94V-O
Chemical resistance 1,1,1.-trichloroethane MEK 20% NaOH 20% HCl	— — — —	Boiling 2 min 23°C dipping 5 hrs 	No change	No change
Color	—	A	Natural	Black

NOTE: Above values are standard values and not guaranteed values.
Test method: JIS C6481
(1) Above values are standard value of 1.6-mm thickness. *CCL-HL820: Thickness All grades are guaranteed the JIS standardized value. ** Operating Temperature
(2) The meaning of marks and figures in Conditioning Procedure:
A: Acceptance condition
C: Treatment in fixed temperature and humidity
D: Dipping in fixed-temperature water
E: Treatment in fixed-temperature air
1st number: Time (hrs)
2nd number: Temperature (°C)
3rd number: Humidity (%)
S_4: Float on solder at 260°C for 20 sec.

(<0.1%), and low thermal coefficient expansion (50 ppm). All the drawbacks of polyimide may be alleviated by this new material. Furthermore, the material has isotropic physical properties—the same properties in x, y, z directions—that are lacking in other high performance materials such as polyimides, BCB, and so forth [68]. The incorporation of siloxane segments into this PCO provides excellent adhesion property to copper, gold, silver, silicon, and other oxides. It can also be reactive ion etched (RIE) using a mixed gas plasma of oxygen and CHF_3 at 450-m torr at 300 W RF power for 2 min with a 12-mm-thick film. This material could be synthesized by fewer steps versus BCB or polyimides that result in a low-cost, high-performance material for MCM applications. BF Goodrich is marketing these materials under the tradename of Avatrel. Their structures and properties are shown on Fig. 5.17 and Table 5.9, respectively [68].

5.4.4 High-performance no-flow and fast-flow underfills for low-cost flip-chip applications

Underfill encapsulant is critical to the reliability of the flip-chip solder joint interconnects [47, 69]. No-flow underfill encapsulant [70–71] is one type of underfill encapsulants and is now becoming increasingly attractive to production engineers due to the simplification of the no-flow underfilling processing—to develop the no-flow underfill materials suitable for no-flow underfilling processing for low-cost flip-chip solder joint interconnects. Several catalysts for no-flow underfill formulations have been studied. These catalysts were reacted with epoxy resins (cycloaliphatic-type epoxy), cross-linkers or hardeners, and other additives, such as adhesion promoters, silica fillers, self-fluxing agents, and surfactants, to form the low-cost, high-performance underfills. The effects of concentration of hardeners and catalysts as well as other factors on the curing profile and physical properties of the cured formulations were studied. The kinetics and reaction heat of the

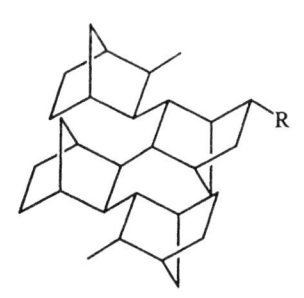

Figure 5.17 Conventional underfilling process.

Generic Polynorbornene Structure

TABLE 5.9 Avatrel Dielectric Polymer versus Current Materials

Dielectric	Moisture	Coefficient	Glass transition	Dielectric
Avatrel	<0.1	50	>350	2.4–2.6
BCB	0.23	52	350	2.7
Polyimide (PI)	2–3	20	430	3.5
Low Stress PI	0.5	3	330	2.9
Fluorinated PI	1.5	60	290	3.0
Silica	—	2–4	>800	3.8–4.2

curing reactions of these formulations were investigated by differential scanning calorimeter (DSC). Glass transition temperature (T_g) and thermal coefficient of linear expansion (TCE) of these cured resins were investigated by thermomechanical analyzer (TMA). Dynamic moduli of cured formulations are measured by dynamic-mechanical analyzer (DMA).

5.4.4.1 Introduction of no-flow underfills. Flip-chip area array is the emerging interconnect technology for the next generation of high-performance electronics. An important trend of developing a modern semiconductor device is miniaturization, high-performance, and low-cost electronic products. As such, controlled collapse chip connection (C^4) or flip-chip and direct chip attachment (DCA) technology are drawing increasing attention and developments. As C^4 and DCA techniques become more popular, the problem with TCE mismatch between chip and substrate becomes more and more serious, particularly with the larger IC chips and smaller solder joints.

Underfill encapsulant is one of the polymeric materials used to reinforce physical, mechanical, and electrical properties of the solder joints between the chip and the substrate during operation and temperature cycling. The encapsulant not only provides dramatic fatigue life enhancement with minimal impact on the manufacturing process flow, but also extends its use to a variety of organic and inorganic substrate materials resulting in ten- to hundredfold improvement in fatigue life compared to an unencapsulated package. Therefore, the underfill encapsulation has been the key to the development of flip-chip DCA technology [47]. A typical normal flip-chip underfill formulation is listed on Table 5.10.

The current flip-chip underfill process requires dispensing liquid encapsulants on one or two edges of the assembled flip-chip package. This allows capillary action to draw the underfill into the gap between the chip and substrate of the assembled package to complete the encapsulation process, as shown in Fig. 5.18a. This process has two disadvantages: (1) The process involves flip-chip solder joint fluxing, reflowing

TABLE 5.10 Ideal Properties of Flip-Chip Underfill Formulations

Solids	100%
Viscosity	<20 Kcps
CTE	<40 (ppm/°C)
T_g	>125°C
Modulus	4–12GPa
Fracture toughness	>1.3 Mpa-m$^{1/2}$
Cure temperature	<130°C
Filler size	<25 μm
Filler content	<70 wt%
Ionics (Cl$^-$)	<10 ppm
Extractable chlorinated solvents	<10 ppm
Alpha activity (needed for memory)	<.001 counts/hr/cm^2
Shelf life (at –40°C)	>6 months
Pot life	>16 hrs
Electrical resistivity	>1.0 × 10^{12} ohm/cm
Dielectric constant	<4
Good chemical resistance	Against process solvents
Low moisture absorption	Essential
Flow under larger chips	Essential
C^4 life improvement	5–10×
T&H reliability	2000 hrs

solder bumping, deflux cleaning, underfilling, and curing the encapsulants. These multistep processes result in lower production efficiency. (2) The underfilling process takes a long time (15–30 minutes for typical material) to complete the flow of the underfill. Furthermore, the ever increasing larger IC die-size requires even longer underfill flow time and long cure time.

The no-flow underfilling process was invented to dispense the underfill materials on the substrate or the semiconductor devices first, then perform the solder bump reflow and underfill encapsulant curing simultaneously, as shown in Fig. 5.18b. Therefore, the no-flow underfilling process not only eliminates the strict limits on the viscosity of underfill materials, process temperature, and package size, but also improves the production efficiency [70–71]. Pennisi et al. described the no-flow underfilling process [72]. Until now, however, the no-flow underfilling process has not been widely used in production. The reason mainly lies in the lack of successful no-flow underfill materials.

The successful no-flow underfill material should meet the following primary requirements:

1. Little curing reaction occurs at the temperature below solder bump reflow temperature (~210–230°C).

2. Rapid curing reaction takes place after maximum solder bump reflow temperature.

Fast-Flow Process

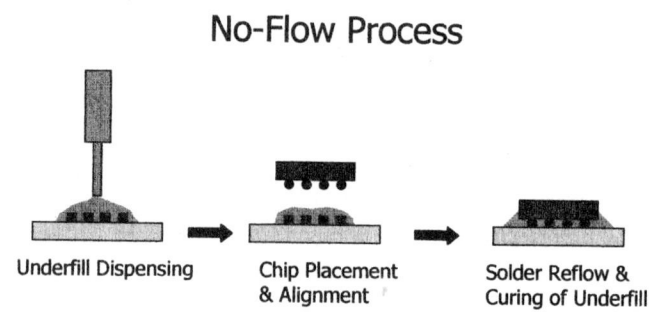

Figure 5.18 Novel no-flow underfilling process.

Chip Placement & Alignment → Solder Reflow → Underfilling → Curing of Underfill

No-Flow Process

Underfill Dispensing → Chip Placement & Alignment → Solder Reflow & Curing of Underfill

3. Good adhesion of underfill material to chip, substrate, and solder joints.

4. Lower shrinkage of the material during curing, lower TCE, and reasonable modulus to minimize the thermal stress result from the curing process and consequent cooling; and finally, having the self-fluxing ability to passivate the substrate oxide conductor lines prior to the solder reflow.

Based on these requirements, epoxy-based materials are widely used for no-flow underfill applications. Some advantages of epoxy resin are low cost, adjustable curing temperature and curing rate by selecting proper catalyst, and good adhesion to most substrates.

5.4.4.2 Underfill chemical formulations. Table 5.11 lists the chemical structures of epoxy resin, hardener, and catalysts used in the no-flow underfill experiments. Epoxy resin is 3,4-epoxy cyclohexyl methyl-3,4-epoxy cyclohexyl carboxylate provided by Union Carbide under the tradename ERL-4221D and was used as received. The molecular weight and epoxy equivalent weight (EEW) of the epoxy resin is 252.3 g/mol and 133 g, respectively. The hardener or cross-linker is hexahydro-4-methylphthalic anhydride (HMPA) from Aldrich Chemical Company,

Inc., and was used as received. HMPA molecular weight is 168.2 g/mol and its purity is more than 97 percent. As for curing catalysts, we used metal acetylacetonate salts, known to be effective in accelerating the curing reaction of bisphenol A/anhydride systems [73]. All these catalysts were used as received. Their concentrations are listed in Table 5.11.

The specified quantity of hardener was added into the epoxy resin and then the mixture was stirred for more than 2 hours at ~60 to 70°C. Hereafter, a specified quantity of catalysts was added into the mixture and stirred for an additional 2 hours at ~60 to 70°C until the catalyst was homogeneously dissolved (see Table 5.12). The formulations were then stored in a freezer at −40°C. When fillers or other additives are used in these formulations, the frozen formulations would be warmed to room temperature and the desired fillers or additives could be incorporated into the master formulations.

5.4.4.3 Underfill characteristics.
The following methods are used to characterize the no-flow underfills.

Curing profile and DSC glass transition temperature (T_g^{DSC}). To study the curing profile and T_g of our underfill formulations, DSC was used for this

TABLE 5.11 Ingredients of Formulation

Name of chemicals	Structure of chemicals	Usage quantity (parts by weight)
Cycloaliphatic epoxy resin	(structure)	100
Curing hardener	(structure)	30 ~ 100
Curing catalysts	See the list in Table 5.12	0.1 ~ 1

TABLE 5.12 List of Curing Catalysts Used in the Experiments

Name of catalysts	Chemical structure	Some specifications
Cobalt (II) acetylacetonate	$[CH_3COCH = C(O^-)CH_3]_2Co$	Red solid (Tm = 165 ~ 170°C)*
Cobalt (III) acetylacetonate	$[CH_3COCH = C(O^-)CH_3]_3Co$	Green solid (Tm = 211°C (dec.))*
Iron (III) acetylacetonate	$[CH_3COCH = C(O^-)CH_3]_3Fe$	Red solid (Tm = 182 ~ 185°C)*

NOTE: * means the data were provided by the chemical vendor.
Also sodium, potassium, and lanthanide acetylacetonates are also capable of acting as latent catalysts for the epoxy cure systems.

study. A sample of ~10 mg of the prepared formulation, which had been equilibrated to room temperature from −40°C, was placed into a hermetic DSC sample pan. The prepared sample was then heated in the DSC cell at 5°C/min to around 300°C to obtain the curing profile. To obtain the DSC T_g of the sample, the cured sample was left in the DSC cell and was cooled down to room temperature at 5°C/min. Then the sample was reheated to 280°C at 5°C/min to obtain the T_g^{DSC} of the cured sample.

Thermal coefficient of expansion (TCE) and thermal mechanical analyzer (TMA) glass transition temperature (T_g^{TMA}) of no-flow underfills. Measurement of TCE and TMA T_g of the cured formulations were usually performed on a thermomechanical analyzer. The specimen for TMA test was made by placing an uncured sample in an aluminum pan at a preheated convective oven at 80°C, then heating to 250°C at about 3°C/min, and postcure it in oven at 250°C for 15 minutes. It was then removed from the oven and cooled to room temperature. A diamond saw was used to cut the cured sample into strips with dimensions of about 16 × 6.3 × 2 mm. After placing a specimen in the TMA instrument, heat was applied from room temperature to about 250°C at a programmable rate of 5°C/min. The thermal coefficient of linear expansion was obtained from the thermal expansion quantity versus temperature. The inflection point of thermal expansion was defined as T_g^{TMA}.

Dynamic modulus and DMA glass transition temperature (T_g^{DMA}) of no-flow underfills. The preparation of specimen used for the DMA test is the same as that for TMA and the experiments were performed on a DMA. However, the dimension for the DMA sample was approximately 32 × 11 × 3 mm. The measurement was performed in single cantilever mode under 1-Hz sinusoidal strain loading. Storage modulus (E'), loss modulus (E''), and tanδ were calculated by the preinstalled software. The peak temperature of tanδ was defined as the T_g^{DMA}.

Comparison of curing profile of underfills and heating profile of solder bump reflow. The basic requirement of no-flow underfill materials is that the curing profile of the materials must fit into the heating profile of reflow of specified solder bump metallurgy; that is, little curing reaction occurs before solder bump reflow but rapid curing reaction takes place after solder bump reflow. Figure 5.19 describes the typical heating profile of a low-temperature reflowable eutectic tin/lead solder bump. It can be seen that the reflowing temperature (or melting point temperature of eutectic solder material) of the bump is higher than the melting point of eutectic solder (37 Pb/53 Sn), 186°C. Usually, the maximum reflowing temperature is set at the higher melting point of the solder bump materials by approximately 20 to 40°C to ensure the ease and completion of the formation of the interconnect structure between solder

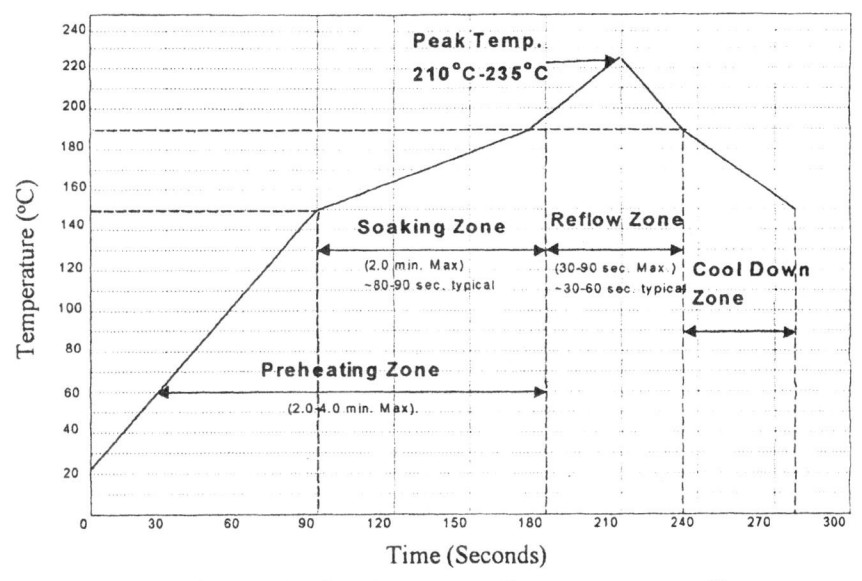

Figure 5.19 Typical conventional surface-mount reflow temperature profile.

bump and metallic pads on the substrate. Accordingly, the no-flow underfill materials should not gel, with a preference of maintaining its low viscosity so that the solder can self-align easily, even by misplacement of the flip-chip from the contact pad during solder bump reflow. At the same time, the underfill must rapidly complete the curing within a couple of minutes when the temperature goes a little bit higher than the maximum reflowing temperature.

Figure 5.20 shows the curing profile of several no-flow underfill materials, including two commercial samples and one generic sample for comparison. Most no-flow underfill formulations fit the heating profile of eutectic tin/lead solder bump. Since some of these formulations have a higher curing peak temperature, they can be used for high lead solder such as 95/5 Pb/Sn solder bump materials. Figure 5.21 shows the T_g^{DSC} of these no-flow underfill formulations. Again, except for one sample in which T_g is around 93°C, all other formulations have T_g greater than 150°C, which are good for underfill applications. Low T_g underfills, with lower module and low thermal expansion coefficient before and after T_g (low α_1 versus α_2), are potential good underfills for temperature reliability performance. Moreover, most of these no-flow underfill materials show much higher T_g (200°C and more), which can be explored for other applications such as multilayer dielectric substrate and conductive adhesives.

Effects of concentration of underfill hardener. Figure 5.22 shows the relationship between the curing peak temperature and the concentration

434   Chapter Five

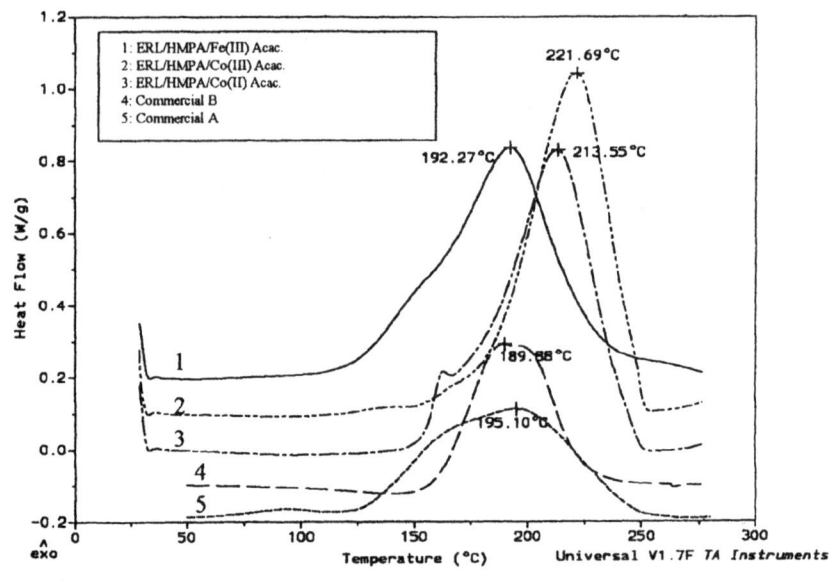

Figure 5.20   No-flow underfill DSC curing-temperature profile.

of hardener. It is seen that the concentration of hardener has little effect on the curing peak temperature. However, increasing the concentration of hardener can significantly decrease the TCE of the cured formulations in the lower hardener concentration region (<75%) and then level off in the higher hardener concentration region (>75%) as

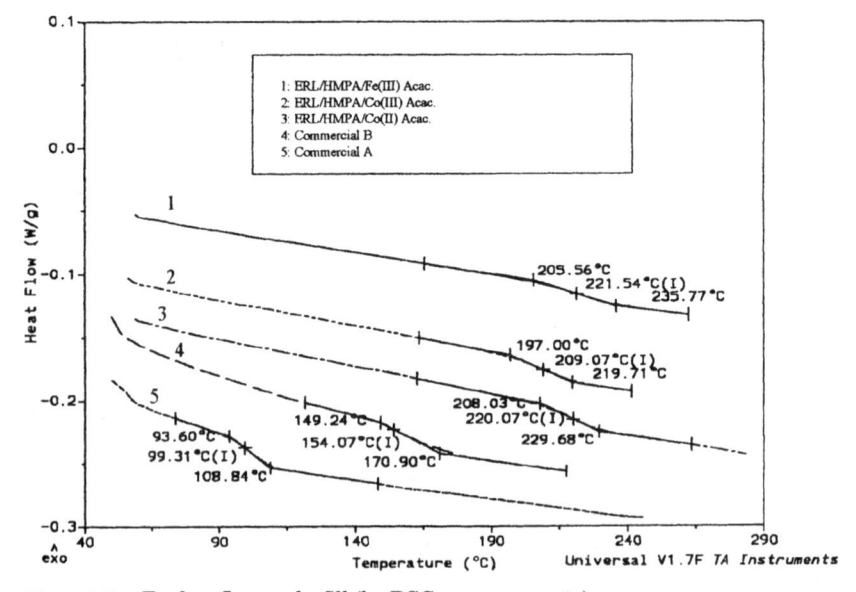

Figure 5.21 T_g of no-flow underfill (by DSC measurements).

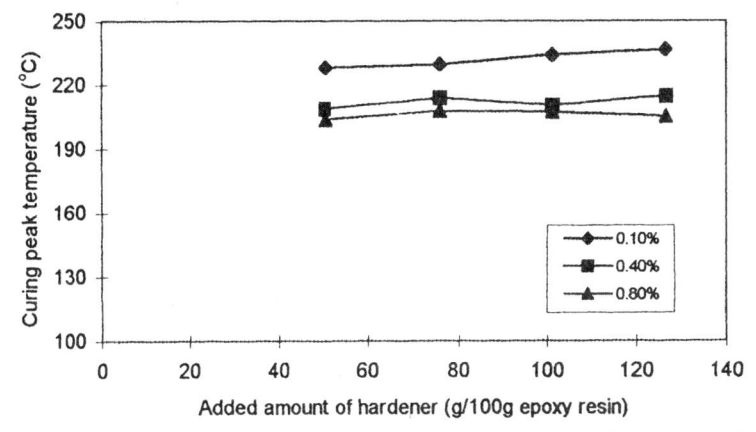

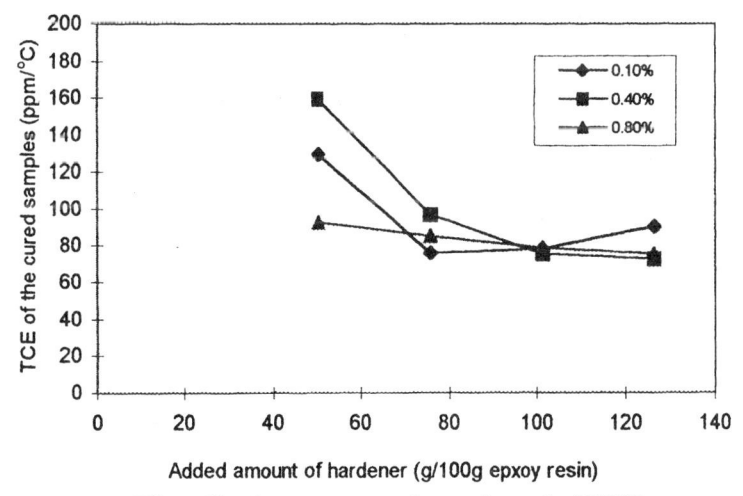

Figure 5.22 Effect of hardener concentration on the underfill curing peak temperature.

shown in Figure 5.23. The effect of the concentration of hardener on T_g of the cured samples is complicated. The measured T_g of the cured samples is slightly different when using different measuring methods. Figure 5.24 plotted the T_g^{DSC} versus the concentration of hardener. It shows that there is an optimal hardener concentration at which the T_g^{DSC} of the cured formulation achieves the highest (optimal) value. In the lower hardener concentration region (<75%), increasing the concentration of the hardener can greatly increase the T_g^{DSC} of the cured formulations; in the region higher than this concentration, however, increasing the hardener concentration only results in decrease of T_g^{DSC}. However, T_g^{TMA} of the cured samples increases with increasing the concentration of hardener within the whole tested concentration range,

Figure 5.23 Effect of hardener concentration on the underfill TCE.

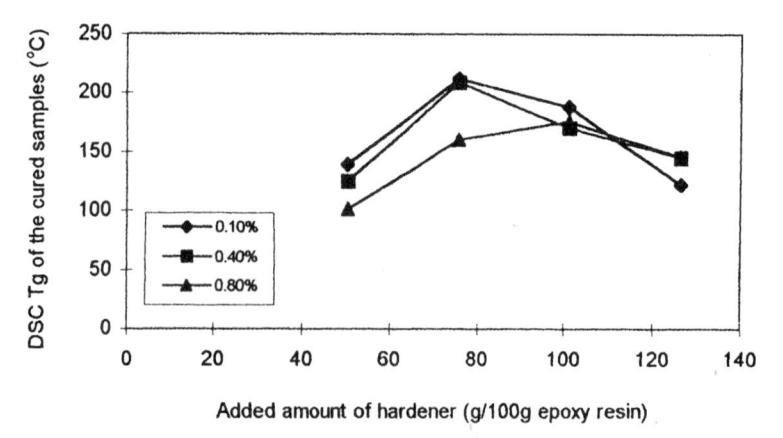

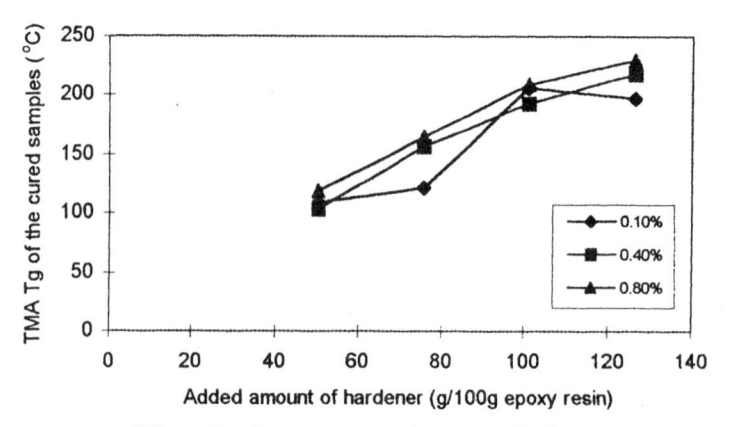

Figure 5.24 Effect of hardener concentration on the T_g (by DSC).

which is shown in Fig. 5.25. The various dynamic mechanical properties of the cured samples are tabulated in Table 5.13. Figure 5.26 shows the relationship between the T_g^{DMA} and the concentration of hardener. It is seen that there is a hardener concentration corresponding to the highest T_g^{DMA}. Below this hardener concentration, the T_g^{DMA} significantly increases with increasing the concentration of hardener, whereas above this hardener concentration (~75%), the T_g^{DMA} slightly decreases with further increasing of the hardener concentration. The last column in Table 5.13 is the calculated cross-linking density using the kinetic theory of rubber elasticity [74]. The cross-linking density $\rho_{E'}$ can be determined by the following equation:

$$\rho_{(E')} = E'/3\phi RT$$

where E' is the storage elastic modulus of the cured formulation at peak temperature of $\tan\delta + 40°C$, ϕ is a front factor (assumed as $\phi = 1$),

Figure 5.25 Effect of hardener concentration on the T_g (by TMA).

TABLE 5.13 Dynamic Viscoelastic Properties and Cross-linking Densities of the Cured Formulations

$C_{catalyst}^{c}$	$C_{hardener}^{d}$	Tanδ Peak Value	Tanδ Peak Temp (°C)	E'(MPa)^a 30°C	E'(MPa)^a $T_g - 60$ (°C)	E'(MPa)^a $T_g + 40$ (°C)	$\rho(E')^b$ (10^{-3}/mol/cm^3)
0.1	50.58	0.665	149.3	3146	2119	12.48	1.08
0.1	75.89	0.494	211.4	2434	1645	21.51	1.64
0.1	101.18	0.519	210.7	2428	1620	22.63	1.73
0.1	126.46	0.461	175.8	2565	1650	30.46	2.50
0.4	50.58	0.815	129.2	2145	1860	11.38	1.03
0.4	75.89	0.546	201.3	2514	1723	26.69	2.08
0.4	101.18	0.557	240.2	2594	1587	24.07	1.74
0.4	126.46	0.460	212.7	2677	1548	33.69	2.57
0.8	50.58	0.878	124.4	2801	2142	10.01	0.92
0.8	75.89	0.719	191.3	2598	1775	14.24	1.13
0.8	101.18	0.542	232.8	2315	1537	21.51	1.58
0.8	126.46	0.560	211.4	2733	1572	29.70	2.27

^{a} E' = storage elastic modulus.
^{b} $\rho(E')$ = cross-linking density.
^{c} Catalyst concentration (g/100 g epoxy/hardener mixture).
^{d} Hardener concentration (g/100 g epoxy resin).

R is the gas constant, and T is the absolute temperature (K). The assumption behind this calculation is that the cured epoxy resin is in real rubbery state at the temperature $T_g^{DMA} + 40°C$. Figure 5.27 shows the calculated $\rho_{(E')}$ versus hardener concentration. It is seen that the cross-linking density of the cured samples increases with increasing the hardener concentration within the investigated concentration range.

Based on the preceding experimental results, it is quite clear that the hardener concentration may play dual roles in affecting the final properties of the cured formulations: hardener and plasticizer. On one hand, increasing the hardener concentration from 50 to 126 percent

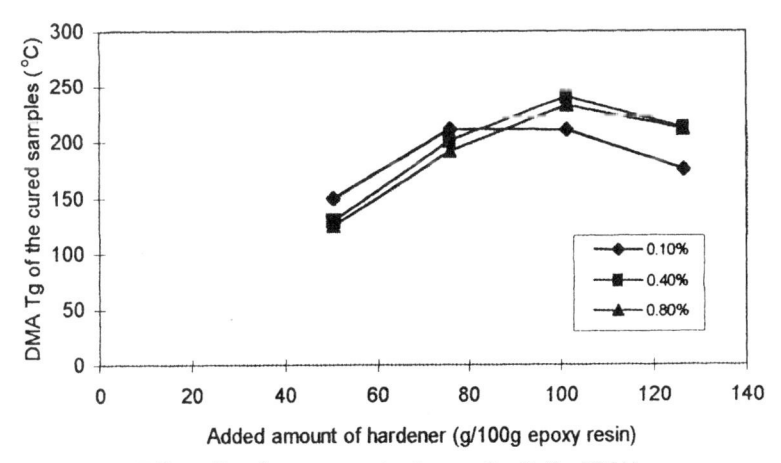

Figure 5.26 Effect of hardener concentration on the T_g (by DMA).

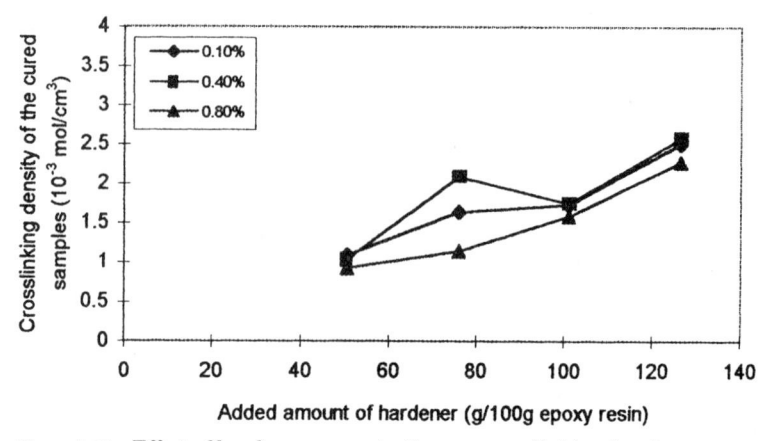

Figure 5.27 Effect of hardener concentration on cross-linking density.

drives the added amount of anhydride more closely to its stoichiometric quantity with epoxy resins. Thus, the cross-linking density increases with increasing the hardener concentration. Consequently, the TCE and T_g^{TMA} of the cured samples increase with increasing hardener concentration, since the TCE and T_g^{TMA} are mainly affected by the cross-linking density. Not all the anhydride molecules can effectively open the epoxy ring and become a part of the network; however, increasing the concentration of hardener can increase the concentration of small molecules in the cured network. These small molecules can act as plasticizer to decrease the T_g^{DSC} and T_g^{DMA}, since the free volume increases with increasing small molecule quantity. Thus, the net effect of the hardener concentration on T_g^{DSC} and T_g^{DMA} is determined by the competition between the effect from increasing cross-linking density and the effect from increasing small molecule quantity due to the increase of the hardener concentration.

Effects of concentration of underfill catalyst. Figure 5.28 shows that increasing the concentration of catalyst significantly decreases the curing peak temperature. This effect appears more obvious in the low catalyst concentration region. However, increasing the catalyst concentration does not show noticeable effects on the TCE, T_g^{DSC}, T_g^{TMA}, T_g^{DMA} and cross-linking density, which are plotted in Figs. 5.29–5.33, respectively.

The underfill catalyst, such as metal acetylacetonate, must first dissociate at an elevated temperature close to its melting point temperature and form a reactive enolate anion; then it reacts with the anhydride hardener to form active species which react with or open the epoxy ring. Metal acetylacetonate or anhydride by itself cannot effectively react with or open the cycloaliphatic epoxy ring, even if the temperature reaches 230°C. Considering that the added concentration of

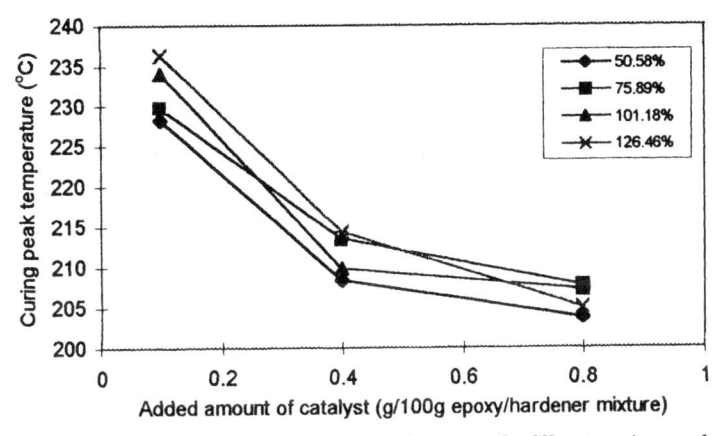

Figure 5.28 Effect of catalyst concentration on underfills at curing peak temperature.

hardener is far beyond the added amount of catalyst, increasing the catalyst concentration will only increase the level of active species, and consequently decrease the curing peak temperature. However, the concentration of active species does not affect the cross-linking density of the cured samples, but may have the plasticizer effect. This is due to its low-level concentration. Therefore, the catalyst concentration up to a 2 wt% does not show any noticeable effect on the final properties of the cured samples.

Effect of absorbed water on the latent catalyst on underfill curing profile. The effect of absorbed water of the latent metal acetylacetonate catalyst is

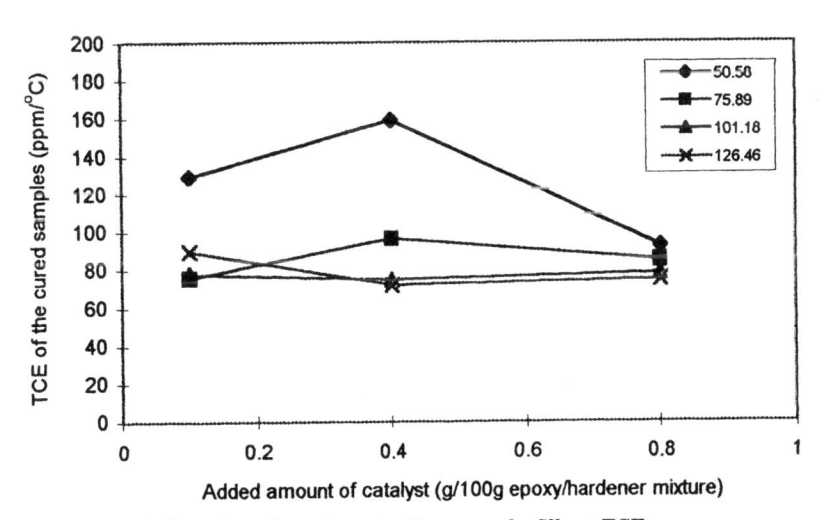

Figure 5.29 Effect of catalyst concentration on underfills at TCE.

440 Chapter Five

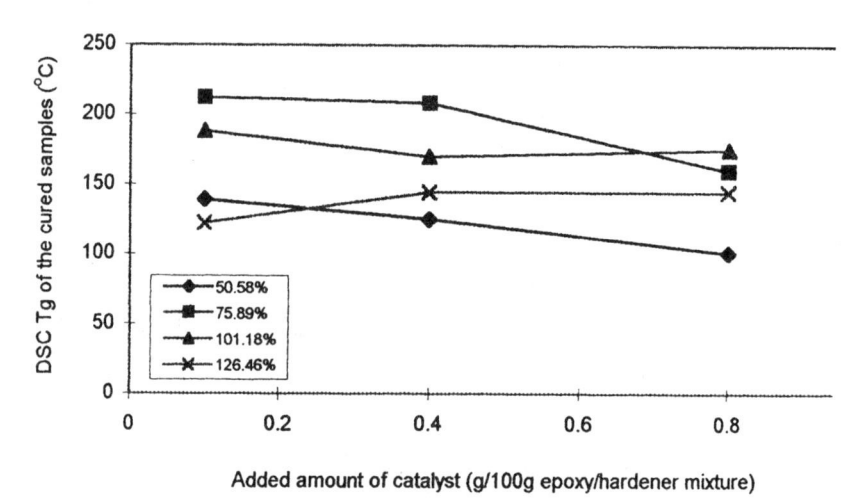

Figure 5.30 Effect of catalyst concentration on underfills at T_g.

interesting, as it is shown on Fig. 5.34. The curing peaks for aged cobaltous acetylacetonate catalyzed formulations look like an overlay of three maximum curing peaks and the curing peak temperature is around 210°C. Further investigation shows that this phenomenon is caused by absorbed water in cobaltous acetylacetonate. There are two experimental evidences supporting this assumption. The first evidence is from the DSC study. We examined the stoichiometric amount of dry, aged, and hydrated cobaltous acetylacetonates. DSC curing profile of

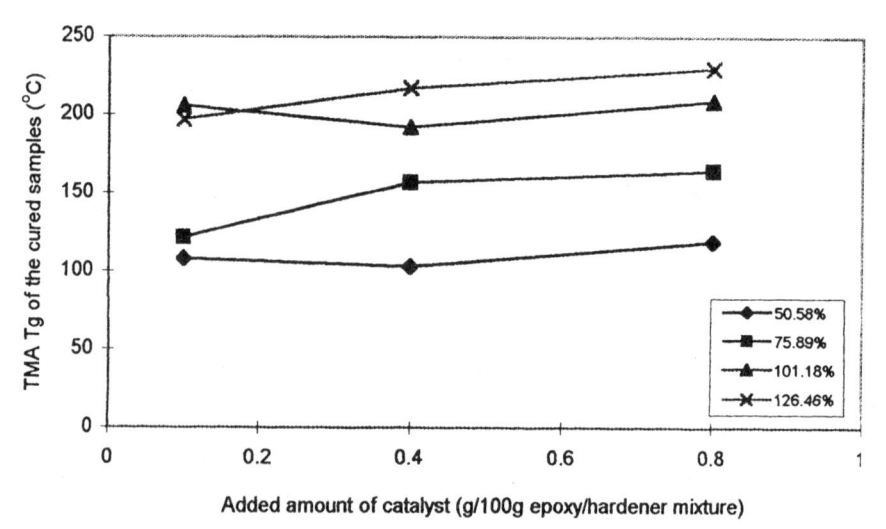

Figure 5.31 Effect of catalyst concentration on underfills at T_g (by TMA measurements).

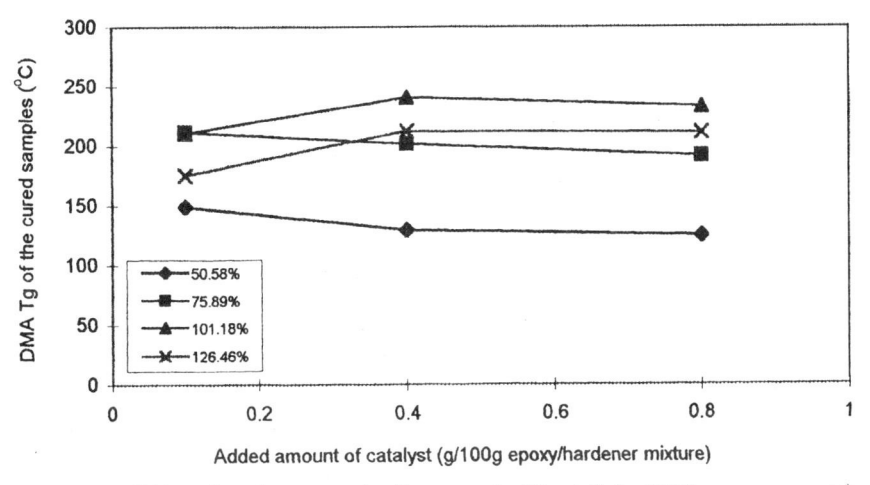

Figure 5.32 Effect of catalyst concentration on underfills at T_g (by DMA measurements).

the dry cobaltous acetylacetonate catalyzed formulation shows one clean exothermic peak, whereas the hydrated and aged cobaltous acetylacetonate catalyzed formulations show very similar curing profiles as more complex DSC, which are shown in Fig. 5.34. Another evidence of the water absorption effect is by FTIR study. In Fig. 5.35, the FTIR spectrum shows the difference between aged, dry, and hydrated cobaltous acetylacetonates. Although there is a weaker -OH vibration peak for dry cobaltous acetylacetonate, the stronger absorption intensity of the same OH vibration peak for aged and hydrated cobaltous acetylacetonate. This indicates that cobaltous acetylacetonates contain

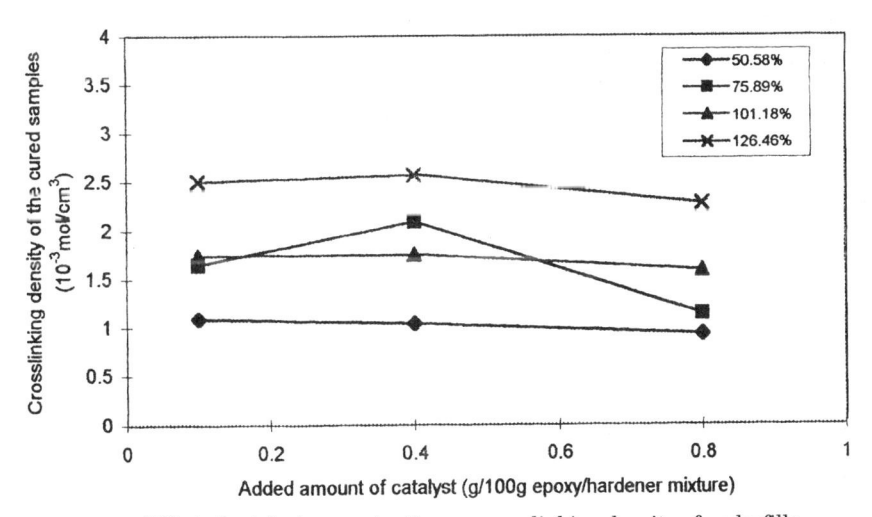

Figure 5.33 Effect of catalyst concentration on cross-linking density of underfills.

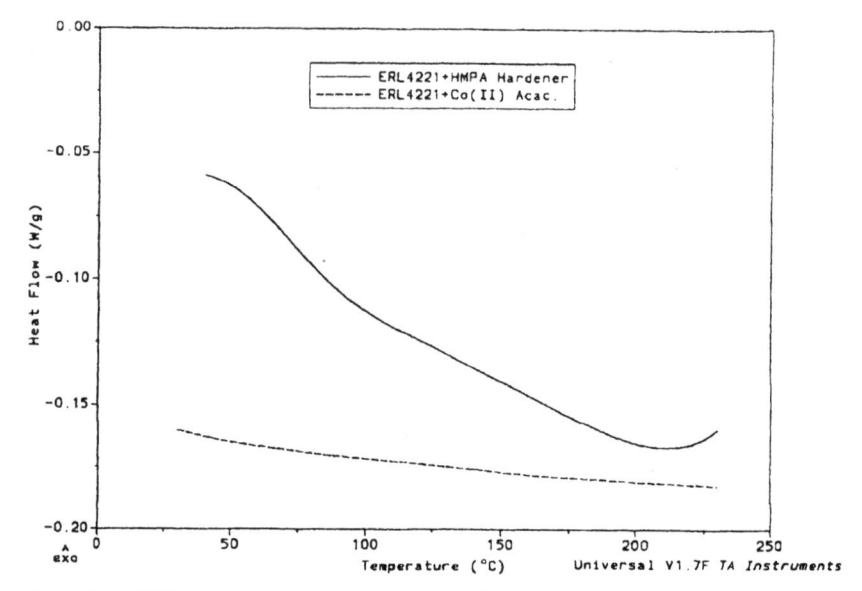

Figure 5.34 DSC spectrum of epoxy reaction with latent metal catalyst and hardener.

absorbed moisture. The affinity of the dry cobalt complex to expand its coordination by absorbing water is quite evident in this set of studies.

The effect of absorbed water on the curing profile may be explained as follows: The absorbed water molecule forms a weak complex ligand with the Co^{2+} ion as additional one or two ligands, which are released at around 110°C by thermogravimetric analyzer prior to the release of the acetylacetone when samples are heated, since the ligand-field coordination between the Co^{2+} ion and the water is weaker than the Co^{2+} ion and acetylacetone ligand. The three curing peaks may imply that two water molecules can at the same time complex with one Co^{2+}, and the release of the first water molecule can be followed by the release of the second water molecule, and finally followed by the release of the acetylacetone. The expansion of cobaltous ions from the coordinate is in $Co(acac)_2$ to six coordinate $Co(acac)_2 \cdot 2H_2O$ is the answer to this DSC study.

No-flow underfill curing mechanism. The curing mechanism of epoxy/anhydride/metal chelate formulations: One is epoxy (ERL-4221) plus anhydride, the other is ERL-4221 plus cobaltous acetylacetonate. Other than the transition metals, lanthanide and actinide metal acetylacetonates or their various counter anions such as acetate, nitrate, and chloride anions provide the same curing formulations.

5.4.5 Conclusions on no-flow underfills

Metal acetylacetonate is a potent latent curing catalyst for an epoxy/anhydride-based no-flow underfill system. The curing reaction peak

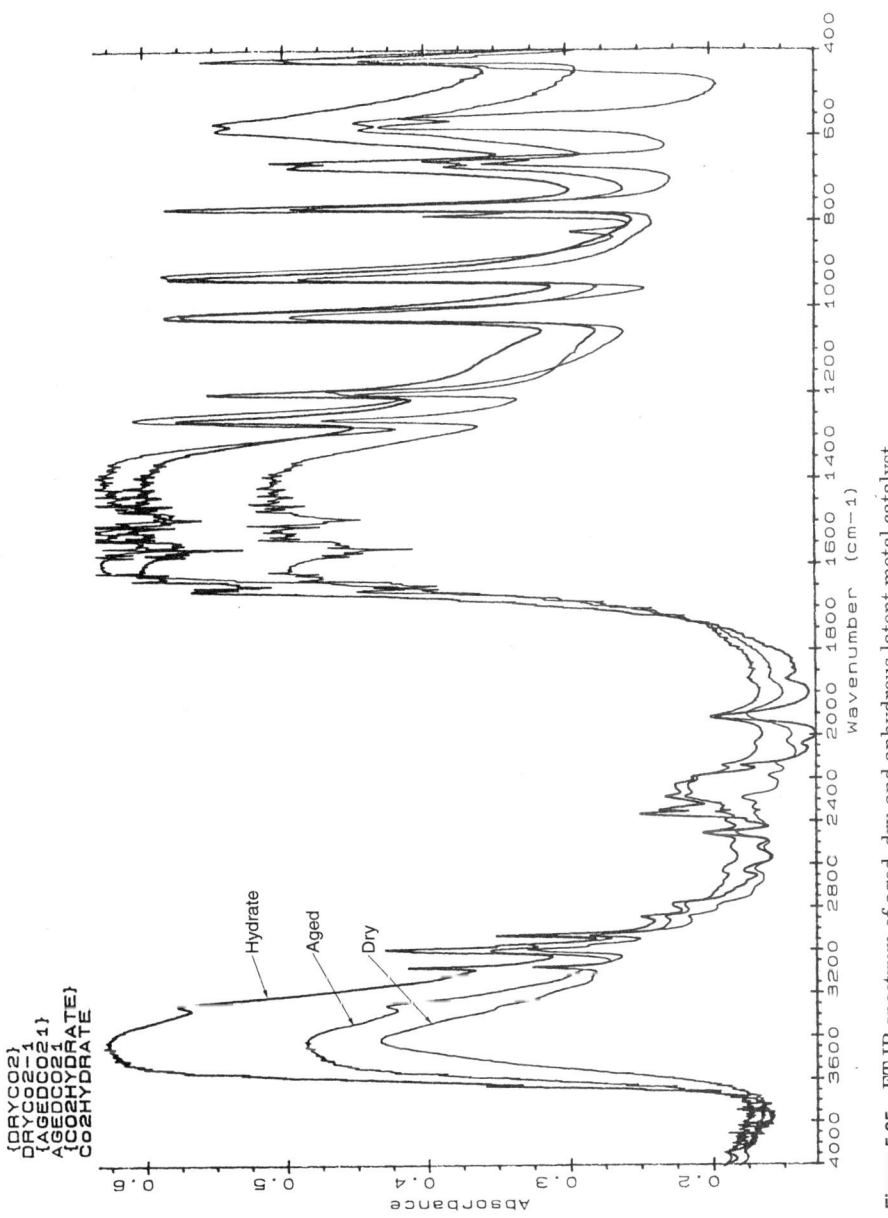

Figure 5.35 FT-IR spectrum of aged, dry, and anhydrous latent metal catalyst.

temperature of these types of catalysts can be manipulated within the range from 150 to ~300°C, which is very useful for designing no-flow underfills for various solder metallurgies in surface mount reflows. The concentration of catalyst mainly affects the curing reaction peak temperature, but it has little effect on TCE, T_g^{DSC}, T_g^{TMA}, T_g^{DMA} and cross-linking density of the cured formulations. The concentration of hardener (cross-linker) does not noticeably affect the curing profile of the formulations, but does affect the TCE, T_g^{DSC}, T_g^{TMA}, T_g^{DMA}, and cross-linking density of the cured formulations. Increasing the concentration of hardener results in a decrease of TCE and an increase of T_g^{TMA} and cross-linking density of the cured samples. In the low hardener concentration region, increasing the hardener concentration can increase the T_g^{DSC} and T_g^{DMA}. Based on the studies on the bisphenol A/anhydride/metal acetylacetonate system, Smith et al. [73] thought that the decomposition fragments of metal acetylacetonate are the most likely active species responsible for the initiation of polymerization in epoxy/anhydride resins. They further speculated that the initial polymerization mechanism is the one involving electron transfer between carboxylic anhydride and the liberated metallic cations to give a reactive initiating species, which can be described in Fig. 5.36 a and b. But this initiation mechanism cannot explain the experimental result that Co^{2+} and Co^{3+} acetylacetonates both show the same catalytic reactivity, and the characteristic color of Co^{2+} and Co^{3+} did not change before or after curing. Therefore, Wong and Shi [70] have recently proposed another initiation mechanism to explain all the experimental results that they had obtained, which can be schematically shown in Fig. 5.36 c and d. The acetylacetone anion (enolate) is first released or dissociated from the metal complex when it is heated to its melting point temperature. The enolate acetylacetone anion is a strong base and can effectively undergo nucleophilic reaction with the carboxylic carbons of the anhydride and form a carboxylic anion, which can further open the epoxy ring or another anhydride molecule to initiate an anionic polymerization that forms a cross-linked network. Such a reaction is very fast and leads to rapid cure of the no-flow underfill after the surface mount reflow process [70].

5.5 Material Process Techniques

Material process techniques include off-chip packaging and encapsulation techniques that consist mainly of (1) cavity-filling and (2) saturation and coating. These are described as follows:

5.5.1 Cavity-filling processes

The cavity-filling process consists of (1) molding, (2) potting, and (3) casting.

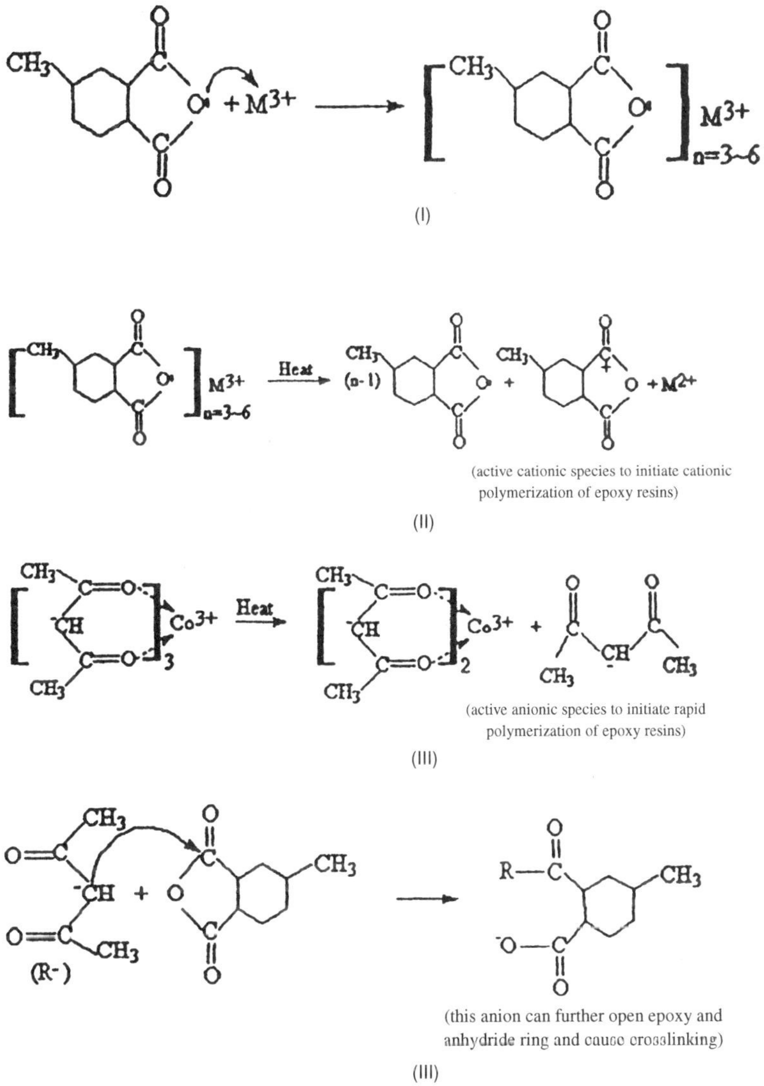

Figure 5.36 Novel no-flow flip-chip underfill chemical reaction mechanism.

5.5.1.1 Molding. Molding is the most cost-effective and high-performance plastic packaging of IC devices. It involves injecting a polymeric resin (one of the thermosetting or thermoplastic resin molding compounds) into a mold and then curing. The process involves the following steps: (1) The molding compound is preheated until it melts and the resin flows through runners, gates, and finally fills up the cavities. (2) The resin is then cured and released from the mold to the predetermined shapes. The exact control of the mold pressure; viscosity of the molten molding compound; and the delicate balance of runners, gates, and cavity designs is very critical in optimizing the increasing molded plastic IC. Finite element analysis of the plastic molding process is becoming an integral part in improving this process. Since the shear stress of the IC chip molded component could cause wire-bond sweep, device passivation cracks, top-layer metallization deformation, and multilayer oxide and nitride cracks, improved molding compounds and processes could eliminate the damage to the molded IC devices. These molding techniques are well documented in the literature. Pressure, injection, and conformal moldings are some of the current molding processes. Figure 5.37 illustrates the typical conformal molding process. With the new advances of low-stress molding compounds, techniques such as the new transfer molding, aperture plate molding, and reactive injection molding are in production use and provide economic ways to encapsulate and package IC devices. A thorough review article on epoxy molding compounds by Kinjo and coworkers [75], and a recently published plastic packaging text by Manzione [76] are excellent sources for the molding materials and processes.

5.5.1.2 Potting. Potting is the simplest part of the process. It involves placing the electronic component within a container, filling the container with a liquid resin, and then curing the material as an integral part of

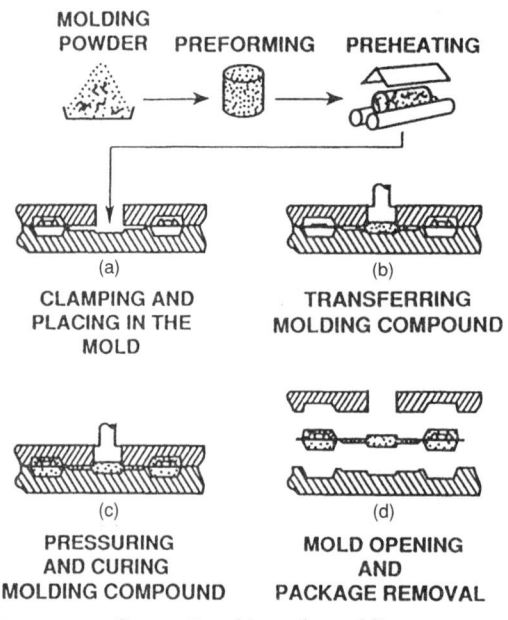

Figure 5.37 Conventional transfer molding process.

the component. Polymeric resins (such as epoxies, silicones, polyurethanes, etc.) are usually used as potting materials. Containers such as metal cans or rugged polymeric casing made from high-performance engineering thermoplastic polymers enhance the effectiveness of the encapsulant. However, the adhesion between the potted material and the casing is essential in achieving a long-lasting reliable package. In the fast growing automated manufacturing process, rugged, machine insertable components, such as the surface mounted chip carriers, plastic thin quad flat packages (TQFP), ball grid array (BGA), chip size packagers (CSP), dual-in-line (DIP), single-in-line (SIP) packages, molded and potted packages, and discrete components are highly desirable components for automation manufacturing processes.

5.5.1.3 Casting. Casting is similar to potting, except the outer case is removed after the polymer cavity-filling process is completed and cured. No heat or pressure is applied in the process. However, this labor-intensive casting process is not commonly used as compared to potting of modern electronic packaging.

5.5.2 Saturation and coating processes

Saturation coatings consists of (1) impregnation, (2) dip, and (3) conformal and surface coatings.

5.5.2.1 Impregnation coating. Impregnation coating is performed by the saturation of a low-viscosity resin to the component, which also includes a thin film coated on the component surface. This process is usually used with a cavity filling or conformal coating process.

5.5.2.2 Dip coating. Dip coating is performed by dipping the component into an encapsulating resin. The component is then withdrawn, dried, and cured. Coating thickness is usually a function of resin viscosity and withdrawal rate and coating speed. This process also depends on the resin reactivity, curing rate, curing temperature, and so forth. This dip-coating process is widely used in glass-laminated printed circuit board and optical fiber coatings.

5.5.2.3 Conformal and surface coatings. Conformal and surface coatings are the common techniques used in IC-device encapsulation. They include spin coating, spray coating, and flow coating of the encapsulant onto the component. Suitable rheological properties of the encapsulant such as dynamic viscosity (η^*), yield stress, G' (storage modulus), and G'' (loss modulus) are critical in obtaining a good flow coating package (see Fig. 5.4), especially in hybrid IC encapsulation, where the encapsulant tends to run-over from the substrate and wick onto the leads of the hybrid devices [77]. In addition, fluidized epoxy powder bed coatings of single-in-line (SIP) hybrid ICs and printed circuit boards (PWBs) are a very attractive conformal coating processes. Surface-mounted components on PWB are routinely encapsulated by this conformal coating process for high-performance applications.

5.6 Anisotropically Conductive Adhesives for Microelectronic Assembly

Conductive adhesives consist of three major types: (1) isotropic, (2) anisotropic, and (3) intrinsic conductive adhesives. Isotropic (nondirectional) conductive adhesives have been used in the electronics industry for a number of years to attach chips to package lead frames. These materials have also been used for pad interconnection by stenciling or screen printing them onto the pads to be connected. The conduction mechanism is based on the percolation theory that electrical conduction by means of the physical contact of the conductive particles (such as silver-flake-loaded epoxies, cyanurate esters, and polyimides) are some of the most popular isotropic conductive adhesives. They are used to interconnect electronic components such as plastic molded IC onto PWB that replace lead-containing solder joints. In the last decade, however, a new class of adhesives that are conductive in only one direction have been developed. These are referred to as *anisotropic conductive adhesive films* (ACAFs). These anisotropically conductive adhesives can provide electrical as well as mechanical interconnections between conductive pads on parts to be permanently assembled. The conductivity of these materials is restricted to the z-direction (perpendicular to the plane of the board), with electrical isolation provided in the x-y plane. Thus, the ACAF materials offer an alternative method for fine-pitch interconnection.

Currently, at least a couple dozen ACAF materials are commercially available. In this section, we describe a methodology for evaluating these materials for their mechanical and electrical properties and interconnection use in the fine (~50 µm) pitch range. In addition, we characterize the materials according to their physical properties and cure characteristics. This section details some findings, with a comparison of physical form to assembly/cure and final electrical properties. Scanning electron microscopy, thermal analysis of the ACAFs, and cure and assembly studies on mixed substrate test vehicles are discussed in detail. Information on electrical testing and long-term reliability testing is also given. Furthermore, the recent advances of ACAF in microelectronic applications is reviewed.

The most commercially significant ACAFs are based on the single particle bridging concept. These materials are under investigation by numerous companies and several preliminary studies have been reported [78–85]. Examples of typical interconnections using ACAFs are shown in Fig. 5.38. Figure 5.38a shows the interconnection of a flexible circuit to a liquid crystal display (LCD) using a narrow strip of ACAF, and Fig. 5.38b shows the attachment of a tape automated bonded (TAB) circuit connected to a second flexible circuit using a picture frame of ACAF. In general, the adhesive material is applied as a film to one interconnection surface, such as a display with a pad array. A part, such as the flexible circuit, is aligned to the display with standard placement equipment and then bonded by the simultaneous application of heat and pressure. Obviously, a wide variety of assembly options are possible using ACAF materials.

5.6.1 Materials and components

5.6.1.1 The anistropic conductive adhesive (ACA). In general, the ACAF materials are prepared by dispersing electrically conductive particles in an adhesive matrix at a concentration far below the percolation

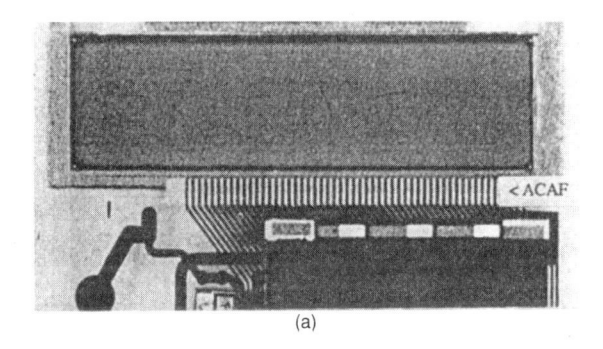

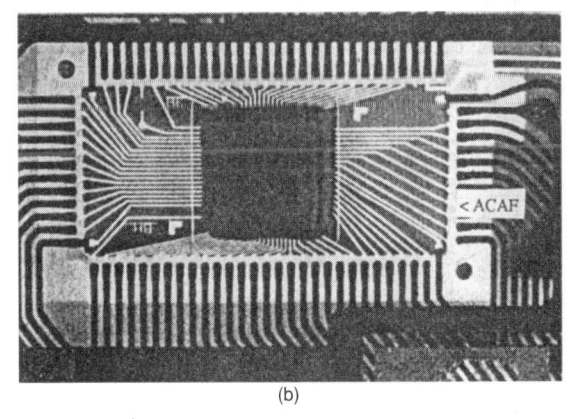

Figure 5.38 Flexible circuit on anisotropic conductive adhesive film.

threshold. The concentration of particles is controlled such that enough particles are present to ensure reliable electrical conductivity between the assembled parts in the z-direction, while too few particles are present to achieve percolation conduction in the x-y plane.

5.6.1.2 ACAF types. When designing materials to achieve fine-pitch interconnections, several important variables must be considered and are application-dependent. These variables include adhesive characteristics as well as particle type.

Two basic types of adhesives are available: thermoplastic and thermosetting. *Thermoplastic adhesives* are rigid materials at temperatures below the glass transition temperature (T_g) of the polymer. Above this temperature, polymer flow occurs. When using this type of material, assembly temperatures must exceed the T_g to achieve good adhesion. Thus the T_g must be sufficiently high to avoid polymer flow during use conditions, but the T_g must be low enough to prevent thermal damage to the electronic circuits during assembly. The principal advantage of thermoplastic adhesives is the relative ease with which the interconnection can be disassembled for repair operations.

Thermosetting adhesives, such as epoxies and silicones, form a three-dimensional cross-linked structure when cured under specific conditions. Curing techniques include: heat, UV light, and added catalysts. As a result of this irreversible cure reaction, the initial non-cross-linked material is transformed into a rigid solid. The curing reaction is not reversible. This fact may hinder disassembly and interconnection repair. The ability to maintain strength at high temperature and robust adhesive bonds are the principal advantages of these materials.

The principal criterion used for selecting the adhesive is that robust bonds are formed to all surfaces involved in the interconnection. Numerous material surfaces can be found in the interconnection region including: SiO_2, Si_3N_4, polyester, polyamide, FR-4, glass, gold, copper, and aluminum. Adhesion to these surfaces must be preserved after standard tests such as temperature-humidity-bias aging and temperature cycling. Some surfaces may require chemical treatments to achieve good adhesion. In addition, the adhesive must not contain ionic impurities that would degrade electrical performance of the interconnections.

The materials used as conductive particles must also be carefully selected. Silver offers moderate cost, high electrical conductivity; as a matter of fact, silver oxide is the only metal oxide that is still conductive, has high current-carrying ability, and low chemical reactivity—but problems with silver electromigration may occur. Nickel is a lower-cost alternative, but corrosion and oxidation of nickel surfaces have been found during accelerated aging tests. The material that offers the best properties is gold; however, costs may be prohibitive for large-volume applications. Plated particles may offer the best combination of properties at moderate cost. In particular, nickel, gold, or other conductive elements coated with a thin layer of elastomers are being used at present.

1. *Frequency response.* Because of the composite structure of ACAFs, the possibility of signal-frequency-dependent phenomena is a concern. The fundamental principle in achieving high-frequency performance in any interconnection system is to minimize inductance and capacitance, or have uniform controlled impedance. The shortest interconnection length gives the lowest inductance. The interconnection length of an ACAF material is determined by the diameter of conductive particles, which can range from 0.5 to 30 µm, significantly shorter than pins or wire-bonds. Therefore, the inductance of an adhesive interconnection should be negligible. The capacitance resulting from the adhesive's dielectric is mainly dependent upon the physical layout of the interconnection. Only the dielectric located between the signal and ground lines will show stray capacitance. Although it has not been experimentally demonstrated, z-axis conductive adhesive's high-frequency performance should be better than conventional pin or wire-bonded interconnections.

2. *Current-carrying capability.* Since the amount of conductive metal is small and varies in ACAFs, one must evaluate the current-carrying capability of each material and its application. AC current-carrying capability has been demonstrated using an active bipolar IC bonded with an ACAF to a flexible circuit. The results show ACAFs filled with plated metal particles can carry at least 20-mA root mean square (RMS) current at 2 MHz. For carrying high radio frequency current, it appears that the surface of the conductive particles should be smooth.

DC current-carrying capability has been demonstrated with ACAFs containing silver-coated nickel particles at the 200-mA level. Other test results on ACAFs loaded with silver-coated polymer spheres showed a limited current-carrying capability of less than 5 mA. In general, DC current-carrying capability can be increased and through resistance, decreased by increasing the conductive particle loading.

5.6.1.3 Parts considerations. Adhesives are often used to attach the copper conductors onto the polyamide (Kayton) or polyester (Mylar) core materials of flexible circuits. These substrate adhesives have to be dimensionally stable at the bonding temperature of the z-axis adhesive. If rigid particles are used in the ACAF, the substrate adhesive layer also has to soften enough to provide for some compliance under the conductor. This compliance is required to provide for particle size variation and, thus, assure bridging contact for all particles. Compliance can also be gained by increasing the conductor line width and the die attach adhesive thickness.

Design consideration for the substrate conductors includes the height (or thickness), width, and spacing between adjacent conductors. The width and spacing are determined by pad size and spacings on the parts. The narrowest conductor spacing affects the maximum conductive particle size for the adhesives. For a specific conductor pitch, slight increases in conductor widths result in decreasing conductor spacing. As a result, it becomes more difficult to maintain electrical x-y isolation with specific ACAF materials. The primary benefits of enlarging conductor widths are to accommodate any lateral misregistration between pad and conductor during the bonding operation and increase the number of particles in the interconnection. The pad heights and particle size, together with any compliance in substrate or particles, determine the needed ACAF thickness. By matching this thickness to the final gap between parts, voids in the assembly interconnection layer can be minimized. Voids between the adhesive and the parts lower the strength of the bond and provide sites for moisture accumulation. Both of these phenomena can degrade the performance and reliability of the interconnection.

Because oxides on conductors can interfere with electrical contact, the ideal conductor surface finish is a gold layer plated over a nickel flash on the copper contact. For the adhesives studied, bare copper without any surface treatment gave poor interconnection. In general, the key is to minimize surface oxides before the bonding operation.

5.6.2 Characterization of ACAFs

After an extensive review and some initial experimentation of various characterization techniques, we determined that the procedures listed in Table 5.14 would give us the best information for comparison of ACAFs. The table also contains the information provided by the techniques. Next, we describe the individual techniques in more detail and discuss the types of information that can be obtained with their use. To demonstrate the techniques, we chose four ACAFs and subjected them to our characterization and evaluation process.

5.6.2.1 Microscopy. Various types of microscopy can be used to examine the ACAF materials and the type is chosen based on the type of information being sought. Overall surface detail, particle shape, particle type, and a rough measure of the loading of particles in the ACAF requires low magnification and is best done with optical microscopy of

TABLE 5.14 Characterization Methods

Technique	Information
Optical microscopy	Morphology
Scanning electron microscopy (SEM)	Detailed morphology
Energy dispersive X ray (EDX)	Material identification
Differential scanning calorimetry (DSC)	Cure temperature and glass transition
Thermogravimetric analysis (TGA)	Decomposition and stability
Dielectrometry	Cure kinetics
Fourier transform infrared spectroscopy (FT-IR)	Cure time and polymer type
Mechanical push-off	Adhesion
Assembly test (at ambient and 85°C/85%RH)	Electrical properties and reliability

low magnification scanning electron micrograph (SEM). Figure 5.39a shows this type of microscopy. For higher magnification of a film, SEM is ideal and gives more detailed information on the conductive particles and other possible fillers. Figure 5.39b shows a typical photomicrograph of an ACAF particle. This data was obtained by SEM.

5.6.2.2 Energy dispersive X ray (EDX). EDX systems are often coupled with SEMs, because they can use the electron imaging beam of the SEM to excite X-ray emissions from elements in the sample. The identity of the elements and a semiquantitative measurement of the amounts of those elements can be determined from the energy and flux of the emitted X rays. This technique can be used to analyze the surface composition of the conductive particles or other fillers.

Table 5.15 shows the types of data that can be obtained from the commercial ACAFs we chose to examine. This information gives a general view of the types of particles, coatings, and particle sizes that occur in a range of the adhesive films.

5.6.2.3 Fourier transform infrared spectroscopy (FTIR). Absorption of infrared radiation is widely used in science to analyze polymeric materials. FTIR is a very sensitive tool for measuring the vibrational energy of the reactive functional groups of the polymer films such as heat-curable epoxies and silicones. Furthermore, the presence of various polymer types and their modification can be identified with FTIR. In addition, FTIR can be used to measure completeness of cure in some polymer systems by measuring the change in an absorption band. For example, the strong absorption of Si-H at 2100 cm^{-1} shows the presence of an uncured silicone in a sample. During a hydrosilation cure of the sample, the decrease of Si-H absorption is an indication of the degree of the silicone curing and could be easily measured by its peak height or

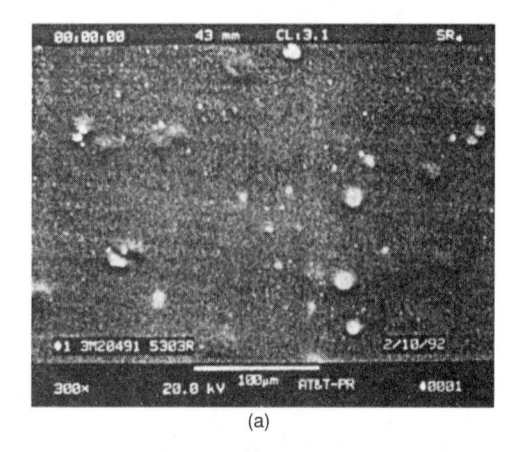

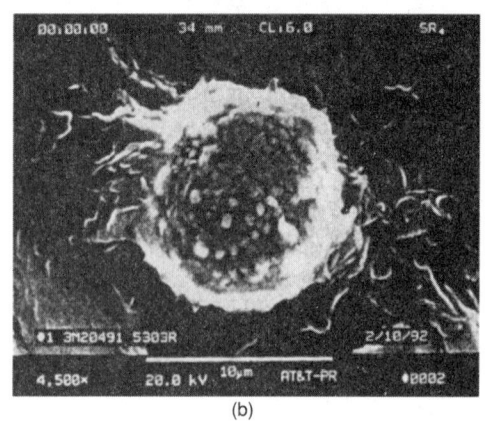

Figure 5.39 SEM of ACAF particle shape.

peak area. Stabilization or disappearance of the Si-H absorption is a good indication of the cure state of the materials.

Figure 5.40 shows the IR spectra of the four polymer films and demonstrates the wide variety of polymer variations found in the films. One can, however, pick out chemical moieties that are common to all these materials and could compare those moieties to standards to determine the polymer variations present.

TABLE 5.15 Particle Characterization

Material	Particle size	Particle type
A	5–7 µm	Au/Ni
B	8–12 µm	Ag/glass
C	12 µm	Ni/polymer
D	20 µm	Au/Ni

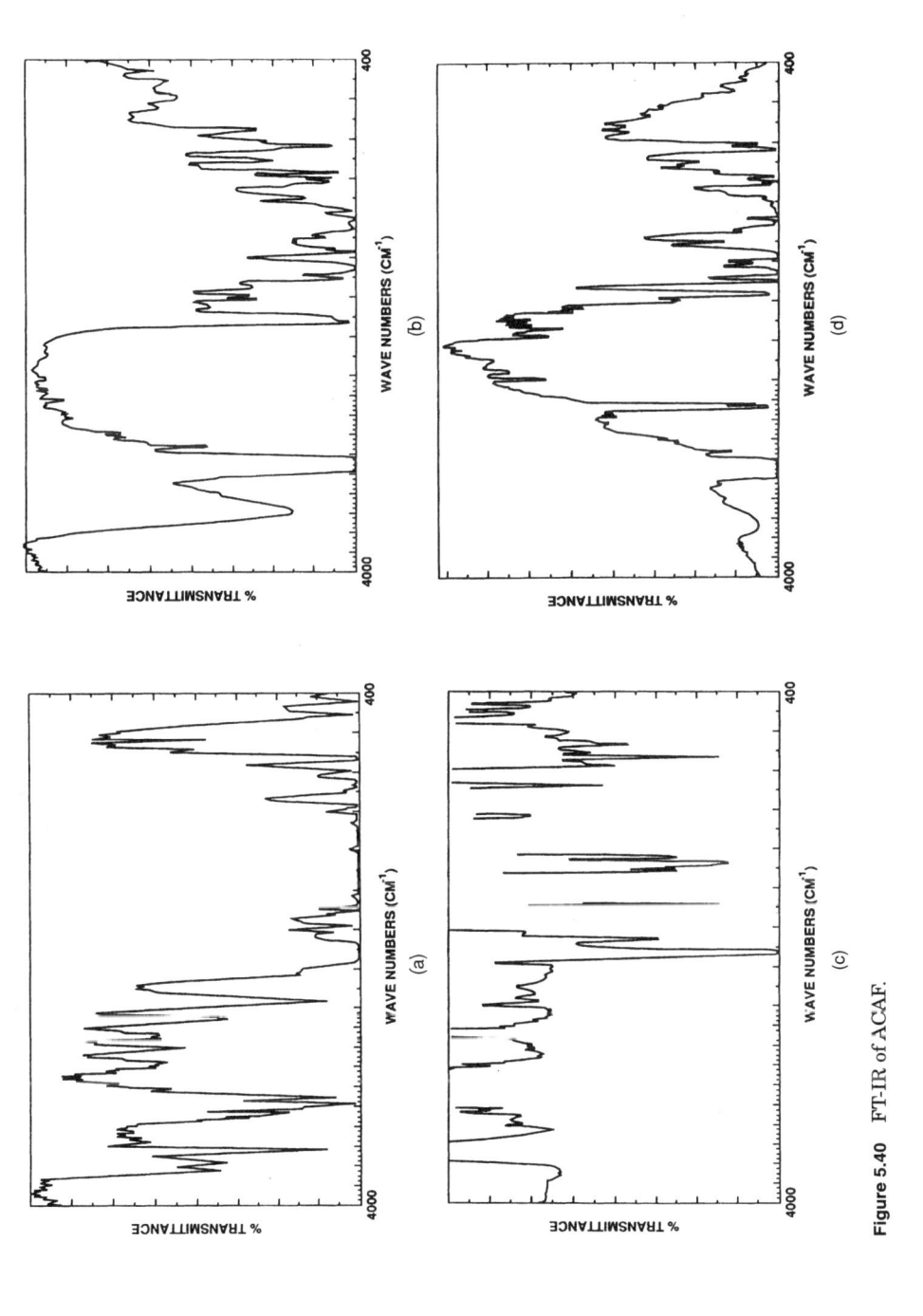

Figure 5.40 FT-IR of ACAF.

5.6.2.4 Differential scanning calorimetry (DSC). *Differential scanning calorimetry* (DSC) measures the heat change in a material during a thermal change. It measures the heat capacity of both endothermic and exothermic transitions of the sample, and provides quantitative information regarding the enthalptic changes in each material. A DSC scan plots energy supplied against average programmable temperature. The peak areas can be directly related to the enthalptic changes quantitatively. T_g (glass transition temperature) can be readily obtained and kinetic information can also be calculated. For the commercial ACAF materials, DSC scans from ambient to 250°C were taken at a prescribed rate. Figure 5.41 shows the DSC scan of typical uncured ACAF materials. A general exothermic peak, such as peak I in the figure, provides qualitative information regarding each of the ACAF materials' cure temperature. When the materials are fully cured, there should be no heat loss or gain in the region of the cure temperature during subsequent DSC scans. The four ACAFs being evaluated showed cure temperatures to be in the range of 140 to 160°C. Peak II shows the point at which polymer degradation begins. These data provide additional information regarding the thermal stability of the ACAF materials. For more detail regarding the thermal stability of the ACAF, thermogravimetric analysis (TGA) is regularly used to study these materials. (See Sec. 5.6.2.6.)

5.6.2.5 Microdielectrometry. *Dielectrometry* is the measurement of dielectric change in a sample at a particular frequency during a physi-

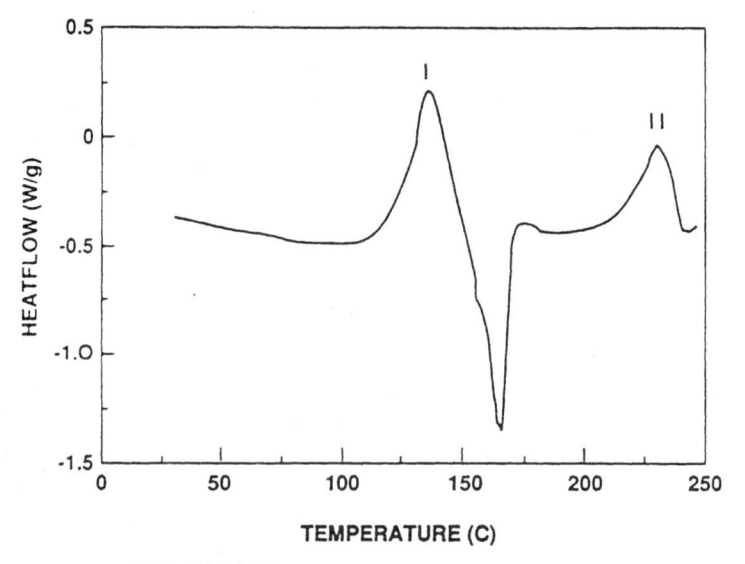

Figure 5.41 DSC of ACAF.

cal or chemical change in that sample. The recent development of highly sensitive microdielectrometry, capable of making measurements in the frequency range of 0.05 Hz to 100 kHz, allows one to study dielectric changes in polymeric films [81]. A Micromet Instruments Eumetric System II or TA Instrument Dielectric Analyzer (DEA) model 2960 microdielectrometer with a miniature IC sensor is capable of using a wide range of frequencies (from 0.05 Hz to 100 kHz) to monitor the loss factor (E'') in the ACAF films. To make the measurement, a layer of the uncured film was placed on a miniature IC sensor and put inside a programmable oven. The temperature of the oven was set to a temperature (i.e., 130, 140, or 150°C) and the loss factor (E'') at various frequencies (0.05, 1, 100, 1000, 10,000 Hz) was monitored periodically during the curing time. Results of the microdielectric loss factor (E'') measurements are shown in Fig. 5.42. During an isothermal cure of the ACAF film, using low-frequency sweep experiments, we could use this sensitive tool for measuring a material degree of cure and thus determine its exact cure time at a given temperature. As an example, Fig. 5.42a shows an ACAF's E'' measurement versus curing time at 140°C. The initial increase of E'' with various frequency scans (0.05, 0.1, 1, 100 Hz) are due to the thermal randomization of dipoles within the ACAF resin during the heating period from ambient to 140°C. When the material's temperature reaches 140°C, all E''s begin to decrease rapidly: this indicates rapid materials cure at this temperature. When E'' reaches its equilibrium state—that is, no further change in E'' with time—the material is completely cured. The low-frequency sweep experiments (0.05, 0.1, 1 Hz) provide a sensitive method of determining a polymer's cure time [86]. Figure 5.42b and c show results obtained for the ACAF films A and D, respectively. Curing times were approximately 35 min for the A materials and about 15 min for the D material at 140°C. The length of the cure time is inversely proportional to the cure temperature.

5.6.2.6 Thermal gravimetric analysis (TGA).
The stability of a polymer film can be evaluated by heating it through a range of temperatures and measuring its weight loss. The temperature where weight loss begins is the temperature at which change (loss of water, degradation, etc.) of the polymer starts. *Thermal gravimetric analysis* (TGA) is a thermal analytical technique that automates this procedure. A TGA output is displayed in Fig. 5.43. As can be seen, this ACAF begins to lose weight at 140°C and shows a large loss above 150°C. This is an example of an only partially cured resin to demonstrate the effect of low stability. When completely cured, the four materials analyzed all showed the start of degradation between 190 and 220°C and no material resisted major degradation above 240°C, in air.

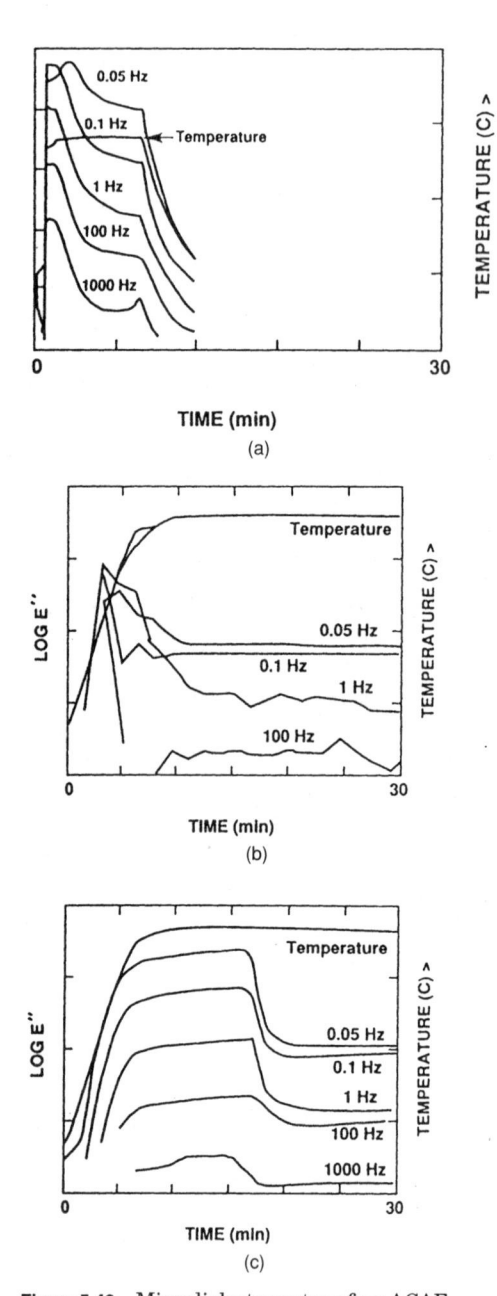

Figure 5.42 Microdielectrometry of an ACAF.

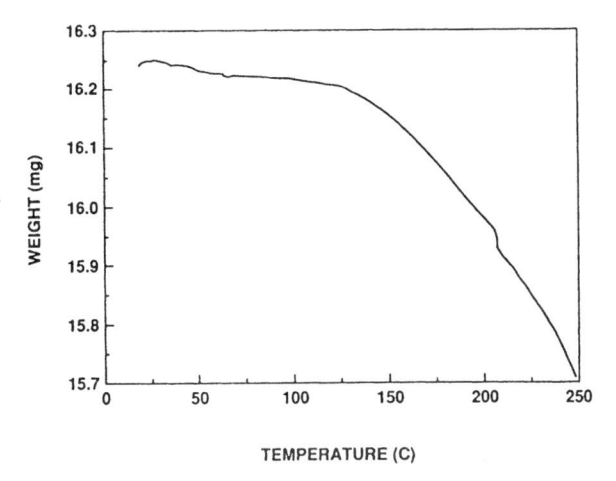

Figure 5.43 TGA of an ACAF.

5.6.2.7 Mechanical adhesion. Adhesion of parts connected with ACAFs was measured using a push-off-type tester. A set of parts is assembled with an ACAF and one part is securely held in the tester while a measured force is applied to the edge of the other part; when separation of the parts occurs, the force is recorded. The ACAFs in the set of four under examination were tested using a silicon chip with circuitry connected to an FR-4 printed wiring board. In all cases, the push-off force was greater than 8 lbs and in some cases greater than 16 lbs. Adhesion testers such as Sebastion, Royce, and Dage as well as Instron are widely used in this testing.

5.6.3 ACAF assembly

5.6.3.1 Registration. Placement accuracy is a critical element in using ACAFs. Unlike the reflow solder bump interconnect process, conductive adhesives do not have the self-alignment capability that corrects minor misregistration between pad and substrate conductors. Thus, the placement accuracy is determined by the pad size of the parts to be assembled.

Particle type will also contribute to planarity variations. For compliant particles, such as plated polymer spheres, the large spheres will be compressed more than small spheres. Therefore, coplanarity between interconnecting parts can still be maintained and all spheres will have equal probability of making contact between pad and substrate conductor. However, for particle types which are less compliant, such as silver-coated nickel, the differences in sphere sizes can be compensated for with thicker pad metallization or by a thicker adhesive layer under the substrate conductors.

5.6.3.2 Assembly parameters. Important process parameters for ACAF assembly are temperature, load, tacking time (the time needed for the adhesive to soften and flow), and bonding time (final cure time). One of the interconnecting parts is preheated to a temperature below the ACAF's bonding temperature, but high enough to partially soften the film so that it has the ability to flow and fill void areas. The bonding load should be high enough to allow the conductive spheres to make good physical contact between the assembly conductors, but not high enough to damage any of the parts. Finally, the tacking time should be sufficient to give adequate time for the film to flow before cure begins so that it seals the contact area during the final bonding process.

5.6.3.3 Assembly test. To evaluate the ability of our selected ACAFs to form electrical interconnections and to evaluate their robustness, the four ACAFs were used to connect a polyester film circuit to a standard FR-4 printed wiring board. A photograph of the two circuits is presented in Fig. 5.44. When the circuits are successfully interconnected, they form a group of six daisy chains in three pairs. Probe points for testing the through resistance are located on the FR-4 board and half of each daisy chain is located on each part. This design pro-

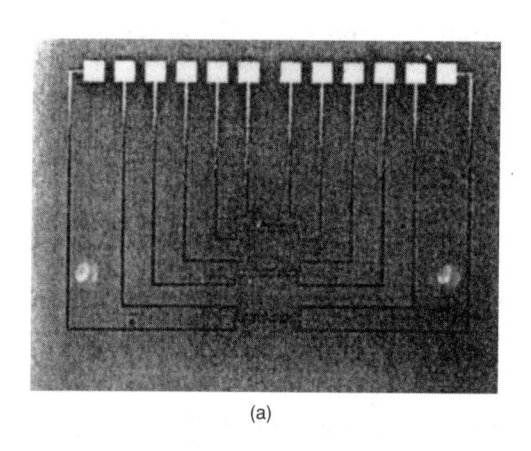

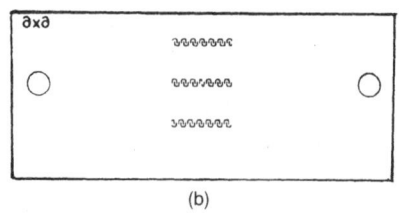

Figure 5.44 ACAF assembly circuit.

vides 14 contact points per daisy chain and the pairs of daisy chains are interdigitated to allow future measurements of isolation resistance. All circuitry is copper with a nickel and then gold plate. Contact pads are 6 mils square and are separated by 6-mil spaces. Alignment is accomplished mechanically, using two close-fitting dowel pins in a metal fixture. Once the stack up of parts with adhesive and alignment fixture is made, we cure the assembly in a laboratory hydraulic press with heating and cooling capabilities. For all the assemblies described here, the cure was run at 160°C for 10 min at 100 lbs of load. The test vehicles were then cooled to room temperature before the load was released. Those that showed interconnection of all daisy chains (resistances per chain were normally less than 0.5 W as measured with a digital multimeter) were stored at ambient conditions for a minimum of two days and is some cases for as long as seven days, before being remeasured. No change of greater than 5 percent was found. It is obvious from this experiment that short-term ambient storage is not useful for differentiating between ACAF assembly capabilities.

5.6.4 Accelerated life testing

The performance of ACAF interconnections in an accelerated temperature and humidity environment can yield valuable information for the evaluation of ACAFs and their use.

We conducted one such test where the samples previously assembled, using a flex-to-rigid interconnection, were evaluated in an elevated temperature and relative humidity (RH) environment. These samples were subjected to an 85°C/85%RH environment for 100 h. Interconnection functionality was measured at the end of the test and circuit changes were evaluated. The results indicate that an 85°C/85%RH environment is appropriate for differentiating ACAF materials. A compilation of the results is shown in Table 5.16. While all samples showed some increase in some daisy chains, the A ACAF material has many fewer failures than the D material. These extrapolations, from limited data sets, will not give completely accurate lifetime predictions; however, they are useful for qualitative evaluations in the evaluation and screening of commercial ACAFs and the assembly processes associated with them.

5.6.5 Conclusions on ACAF

In this section, we have outlined various evaluation techniques that are useful in the characterization of anisotropically conductive adhesive films (ACAFs) for use in fine pitch assembly. Robust interconnections can be achieved only if both materials and processes are optimized for a particular application. A methodology for the evalua-

TABLE 5.16 Accelerated Aging Tests of ACAF Assembly Vehicles at 85°C and 85%RH

Material	ID	Chain resistance Same	Increased	Open
A	1	5	1	—
	3	6	—	—
	4	6	—	—
	5	4	1	1
	10	6	—	—
Totals		27/30	2/30	1/30
B	1	—	4	2
	4	1	2	3
	5	4	1	1
	7	2	4	—
	9	6	—	
	10	2	4	—
	12	4	1	1
	13	4	2	—
	16	—	4	2
	17	—	—	6
Totals		23/60	22/60	15/60
C	1	—	4	2
	3	2	4	—
	5	1	5	—
	6	1	5	—
	7	1	3	2
	8	2	4	—
	9	2	4	—
	10	4	2	—
Totals		13/48	31/48	4/48
D	1	—	2	4
	2	—	1	5
	4	—	2	4
	6	—	—	6
	7	—	—	6
	8	—	2	4
	9	2	1	3
	10	2	2	2
	13	—	2	4
Totals		4/54	12/54	38/54

tion and trial assembly of ACAF materials has been developed which demonstrates the dependence of interconnection reliability on material properties and morphology.

In conclusion, there is a need for both suppliers of adhesives, as well as manufacturers of assembly equipment, to develop improved materials and methods for ACAF assembly. At present, these materials are used only in niche-market products such as liquid crystal displays, lead on chip- low-voltage, and low-profile miniaturization products. However, its low curing temperature that results in low residue stress of components and the environmentally sensitive demand for no-lead products will eventually drive these products in popular applications. Nevertheless, the conductivity fatigue that causes the conductivity drops in these ACAF materials, particularly after accelerating aging in 85°C/85% relative humidity, and the toughness problems of the adhesives still require further investigations.

5.7 High-Performance Plastic Packages with Reliability Without Hermeticity: A Case Study in Hermetic versus Nonhermetic Packaging

There is a natural division of definitions between the two most common packaging approaches for integrated circuits: hermetic packaging, which presents a sealed, impermeable package to the operating environment; and what we term nonhermetic, which is permeable. The *hermetic package* usually consists of a cavity package made of ceramic, metal, glass, and so forth, that are for the most part resistant to permeation by fluids and gases. On the other hand, *nonhermetic packages* are generally associated with organic coatings or seals such as epoxies, silicones, polyimides, and the like, which are not capable of preventing penetration of fluids and gases. There are also new technologies that are designated *hermetic-equivalent packages* that use novel approaches to obtaining hermeticity through depositions of metal films to seal the electronics package [87]. Both hermetic and nonhermetic packages are able to house single devices or multichip systems. Considering the issues, the packaging engineer's objective is to evaluate the criticality of the application, the types of devices and assembly methods that will be contained in the package, and the operational environment the package will be used in, to determine which type of packaging approach provides the best trade-off between cost, reliability, and performance. In general, the most critical applications requiring the highest, long-term reliability will utilize hermetic packaging, but today the choice is not always so clear-cut, as there are many factors that must be weighed—including cost, size, and weight. With the great strides that have been made in packaging with polymers in the last 10 years, more and more applications are open to its use. As different as the two packaging approaches are, so are the screening and qualification testing used to evaluate their performance. Hot, wet tests such as 85/85 or HAST do not have much effect on a properly designed and assembled hermetic package. The opposite is true with nonhermetic packages, as organic materials tend to break down, or depolymerize and hydrolyze in these environments. The remainder of this chapter provides an overview of both packaging methods, as well as some of their disadvantages, advantages, and technical details of their design and production-related processes.

5.7.1 Hermetic packages

Condensed moisture on a device's surface during operation has historically been shown to lead to the principal causes of failure in the field. The extremely small geometries involved in integrated circuits, differing galvanic potentials between metal structures, and the presence of high electric fields all make the device susceptible to interactions with moisture. One packaging method designed to prevent performance degradation due to moisture's deleterious effects is the hermetic package. Ceramic or metal is used to form an enclosure to isolate the electronic devices from the ambient operating environment. By eliminating condensible moisture from the cavity of the package during the sealing process, and preventing the ingress and egress of moisture at the package perimeter during its operating life, excellent long-term reliability can be achieved. The word *hermetic* is defined as completely sealed by fusion, solder, and the like, so as to keep air or gas from getting in or out; in other words, airtight. In hermetic packaging practice such seals are nonexistent. Small gas molecules will enter the package over time through diffusion and permeation. Eventually these gases will reach equilibrium within the cavity of the package. In light of this permeation, long life can still be obtained in the field because of the extremely slow nature of this activity. Accelerated tests and leak testing for screening and qualification are specified in the military standard, MIL-STD-883, and represent the foundation and industry-accepted approach to testing for reliability in hermetic packages.

5.7.2 Nonhermetic packages—a case study on flip-chip processes and reliability

The widespread availability of nonhermetic packaging followed that of hermetic packaging by many years due to problems associated with the initial polymer materials used for accelerated testing, which is usually the means by which nonhermetic packaging is assessed during screening and qualification in the manufacturing process, with hot-wet environmental tests the most common. Temperature cycling is the most common thermomechanical environmental test. Temperature cycling does not test the corrosion-resistant properties of the package and polymer system, but tests the ability of the assembly to endure the stresses imparted by the various materials that comprise the device, interconnect, and polymer packaging. An accurate correlation of screening and qualification testing to field life has yet to be accomplished, but individual manufacturers have historical, and most often proprietary, data that prove the technology achieves acceptable reliability in specific markets and end-use environments. As more and more is understood in relation to the fundamental principles surrounding the use of polymers in packaging, its domination is expected to increase, and more difficult markets, such as the military, will open to it.

An example to demonstrate the use of silicones to encapsulate the high-voltage flip-chip device in achieving reliability without hermeticity is AT&T's No. 5 Electronic Switching System (ESS). This ESS is a state-of-the-art electronic switch that uses solid-state gated-diode-crosspoint (GDX) for high-speed switching of telephone calls. These GDXs operate at 375 volts and require exceptional protection and reliability performance of these packages. This solid-state switch is used in an electronic switch system (ESS) that routes phone calls from customer premises to and from the telephone central switching offices. An ultrahigh reliability switch is required for this system. Conventionally, the GDX is packaged in a hermetic leadless ceramic package, which is then surface-mounted on a ceramic hybrid IC (HIC) [88–90]. In order to reduce the cost of these GDXs, a new type of flip-chip GDXs was developed, structurally bonded with 95/5 Pb/Sn solder joints. These flip-chip GDXs were surface-mounted on a conventional ceramic substrate. A thixotropic silicone gel was used as an underfill and overfill glob-coating to provide the ultrahigh reliability performance of these high voltage devices, and a silicone elastomer was used to pot the hybrid surface-mounted GDX multichip module structure to further enhance the robustness of its structure. The flip-chip solder (95/5 Pb/Sn) bumping process, the reflow, cleaning, encapsulation, and reliability testing of the GDX HIC are described as follows.

5.7.2.1 Flip-chip solder (95Pb/5Sn) bumping process.
To reduce the cost of the 5 ESS switch, the expensive hermetic ceramic chip carrier is being replaced by direct flip-chip solder bonding of the Si ICs onto the ceramic hybrid IC, followed by nonhermetic gel encapsulation. The flip-chip bonding process developed for this purpose has to be compatible with all wafer codes used in the HIC, including gated-diode-crosspoint (GDX), linear CMOS, and digital bipolar devices. The process has to satisfy the following criteria: (1) approaching 100 percent *bumping yield* (defined as the ratio of the number of good devices after bumping to the number before bumping); (2) passing the specified THB (temperature-humidity-bias) qualification test; (3) transparency to different device codes; and (4) robustness in manufacturing.

The solder composition selected for this process is 95%Pb/5%Sn, which has a higher melting point than eutectic solders. This is appropriate for chip-level bonding to ensure integrity of joints during subsequent board-level reflows. The solder joint itself is the sole mechanical and electrical interconnection between the Si IC and the ceramic HIC. These different substrate materials, with vastly different coefficients of thermal expansion, will induce stress in the solder joint during thermal cycling. To ensure sufficient thermal-fatigue life in the solder joints, a joint height of 75 µm has been specified. The smallest bump pitch among the wafer codes is on the order of 150 µm. For simplicity, the entire volume of solder is to be deposited on the Si chip. The ceramic HIC will have contacts finished with the appropriate interface metals for wetting and joining to the solder.

The entire process from a finished Si wafer to solder-mounted and encapsulated flip chips on the ceramic HIC is shown in Fig. 5.45. This section addresses Fig. 5.45a: the act of forming solder bumps of appropriate dimensions and properties on individual Si wafers. The best known wafer scale solder bumping process in production is IBM's C[4] (controlled collapse chip connection) process, which is based on evaporating base metal layers and solder through a metal mask aligned onto a wafer. Electroplating of solder is also in production, although on a smaller scale. Currently, there are many processes for this type of flip-chip interconnection technology [91–97].

5.7.2.2 Assembly of flip-chips onto ceramic HICs. Bumped Si ICs are flip-chip bonded onto ceramic HICs for multichip modules, which are then assembled into boards for the 5 ESS. The landing site for the Si chips consists of a layout of contact pads resembling the layout on the Si chip. The base metals on the ceramic are a multilayer stack of Ti/Pd/Cu/Ni/Au, with layer thicknesses typically higher than those used on

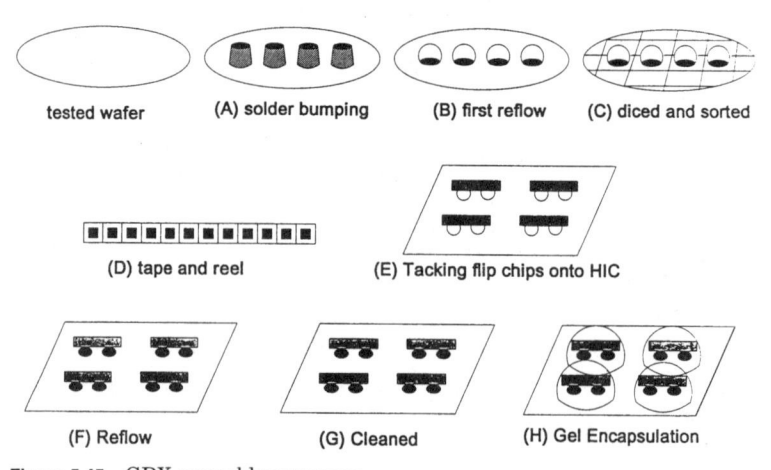

Figure 5.45 GDX assembly processes.

Si. The bumped Si chips are first temporarily mounted onto the ceramic module using a procedure known as *tacking,* a process in which the ceramic substrate is held on a bottom platen at a temperature about half of the reflow temperature, and a fluxed Si chip is held on a mechanical pick-up tool. Optics are arranged such that the layout of the solder bumps is superimposed on the layout of the contact pads on the ceramic. Alignment is done, and the Si chip is brought into contact with its ceramic mate, with a slight pressure being imposed in the process. With the correct settings of temperature, pressure, and tacking time, the solder will soften enough to initiate adhesion with the contact pads. The sticky flux also helps hold the ensemble together. The alignment has to be accurate only to within one pitch of the pads, because of the property of self-alignment, which is described later. The temporary assembly is then sent through a reflow oven similar to the one described earlier.

During this reflow, which is aided with the applied flux, the solder bump from the Si chip wets the base metal layers on the ceramic HIC and spreads to the extent defined by the base metal pad. The bump shape transforms from a sphere truncated on one side to one truncated to two sides. If the bump is initially misaligned with respect to the bond pad on the ceramic HIC, surface energy forces will bring the chip into alignment such that a nearly perfect truncated sphere can be formed. This property is known as *self-alignment* and is an important benefit of the solder reflow process. It relaxes the tolerance for mechanical alignment of chip to module tremendously.

With the proper reflow parameters, flipped chips will self-align onto their mating substrates and solder joints will approach their equilibrium shape as dictated by the relative dimensions of wettable diameters and joint height. Under these conditions, the solder joints will have an equilibrium density and uniform distribution of porosity. These microvoids have been shown to exist in healthy joints and do not appear to have any effect on the mechanical strength of the joint. On the other hand, a poor reflow process could result in chip misalignment, distorted joint shapes, which lead to localized areas of concentrated stress, and macrovoids within the solder. These effects could promote drastically different modes of joint failure and result in much reduced joint strength and fatigue life. The GDX HIC reflow process is described on Fig. 5.45.

5.7.2.3 Preencapsulation cleaning of GDX HIC process. Since flux (Alpha 100) was used in the GDX HIC reflow process, cleaning was required prior to encapsulation to ensure the long-term device reliability. A non-CFC, terpene-based EC-7R (manufactured by Petroferm, Inc.) was used to remove any residue flux contaminant. An automatic

TABLE 5.17 Contact Angles of
Measurements (Θ) of GDX HIC

Sample	Θ
As received (before cleaning)	46
After EC-7R clean	32
EC-7R plus UV-03	0

semiaqueous washing machine (Detrex model) was used in this process. EC-7R was sprayed on the DGX HIC first, then with DI water through a conveyer-belt washer. Results are listed on Table 5.17.

5.7.2.4 Glob-coating of flip-chip GDX HIC with silicone gel. To ensure the high-voltage nonhermetic flip-chip GDX HIC and long-term reliability, an ultrahigh purity, low-modulus silicone gel with vinyl and hydride cure siloxane gel was used to glob-coat the GDX HIC. The thixotropic silicone gel was dispensed (glob) on the GDX devices, vacuum applied at 28 in mercury pressure for 2 min to remove any entrapped bubble and ensure the complete coating of the flip-chip solder joints. The vacuum-deaired silicone gel was then cured at 135°C for 2 h. FTIR and microdielectrometry were used to monitor the gel curing to optimize its performance.

5.7.2.5 Reliability testing of the nonhermetic flip-chip GDX HIC. Given the stringent reliability requirements of the AT&T No. 5 ESS, the GDX HIC underwent a thorough preproduction qualification to ensure its flip-chip nonhermetic packaging reliability testing. Temperature-humidity-bias (85°C/85%RH, 600 V) for 1000 h was used to accelerate the electrical performance of this new process. Temperature cycling (−40 to 125°C) at 1000 cycles was used to test the solder joint mechanical reliability; the MIL-STD-883C method was also added to ensure the reliability of the process. Results of this testing are shown on Table 5.18.

TABLE 5.18 Reliability Testing of Nonhermetic
Flip-Chip GDX HIC

Test method	Results (no. of failure)
Temperature cycle (−40°C → 125°C) 1000 cycles	0/125
Temperature humidity bias (85°C/85%RH/600V d-c)	0/125
High temperature storage (125°C for 30 days)	0/125
MIL-STD-883C	0/125

5.7.3 Results and discussions

5.7.3.1 Flip-chip bumping and reflow of GDX HIC. We have developed a solder bumping process based on solder evaporation, but which uses a resist mask instead of a rigid metal mask. The tedious mechanical alignment of the metal mask scheme, with its associated problems such as mask warpage and tolerance, is now replaced by the well-established and accepted discipline of photolithography. This improvement brings the entire solder bumping operation into closer alignment with the practices of standard IC facility. For the thickness of the solder involved, we have decided to use liftoff, instead of etching, to define the solder bumps. In the conventional understanding of liftoff, the thickness of the liftoff resist mask has to be higher than the thickness of the intended deposit, which, in our case, is on the order of 75 μm. We have chosen dry films (such as Dupont's Riston) for this application. These dry films are laminated onto a clean Si wafer, exposed in a conventional aligner/exposure tool, and developed in either solvent or aqueous-based chemicals. The resultant structure is a polymer layer with vias at the interconnection points where bumps are to be formed. The dry film we have chosen behaves as a negative resist, with an important property—the vias have a reentrant profile such that deposits do not cover the sidewall of the via, allowing solvents to enter the via and interact directly with the resist during the liftoff process. Profile control depends on the optimization of exposure and development parameters.

A reliable solder joint can be viewed as consisting of two parts: the solder joint itself, which supplies the mechanical strength, the electrical connection, and the appropriate clearance between the two surfaces to be joined; and the base metals on the two substrates on either side of the joint, which serve as the basis for wetting and adhesion. The base metals must have sufficient adhesive strength to both the underlying substrate and to the solder. Since solders typically wet metal films through chemical interactions such as dissolution and intermetallic formation, and the thicknesses of solder deposits are much larger than the thicknesses of the metal thin films, it is possible that the entire base metal layer is totally consumed during multiple reflows, thus leading to joint delamination. The base-metal layers chosen for the Si IC consist of a multilayer stack designed for sustained adhesion through multiple reflows and prolonged temperature and humidity testing. It is based on the Cr/Cu system and includes a codeposited layer, where both metals coexist in a fine-grained structure. The Cu content is intended for interacting with the solder during reflow and Cr, which is inert to solder, serves as a background three-dimensional mesh, which interlocks mechanically with the penetrated solder and holds it in place. The fine structure of Cr means that such an adhesion mechanism could be very strong, provided the appropriate layer thickness and material ratio are chosen. The entire stack consists of the following: Cr/CrCu/Cu/Au. The top Au finish provides oxidation protection for the Cu.

All base metal layers are deposited in one pumpdown in a sputtering system. Sputtering is chosen as the deposition method for the base metals because of the uniformity, ease of control, and ease of scaling. Just prior to sputtering, the resist-coated wafers are treated with ion-milling, which is integrated with the loadlock of the sputtering system. The ion milling cleans the bottom of the resist via and removes the oxide from the Al bond pads on the Si wafers. The via is typically slightly larger than the Al bond pad so that Al is not exposed at all after the solder bump is formed. This helps eliminate corrosion problems during THB testing.

The wafer then proceeds into another vacuum chamber where solder is being evaporated. For this process, we have chosen inductively coupled thermal evaporation sources with alloy solder charges. The solder charges have compositions designed to produce a 95Pb/5Sn solid deposit on the wafers. The charges are preweighed and evaporated to completion. Because of the much higher vapor pressure of Pb compared to Sn, Pb evaporates first and Sn evaporates last. Thus, the solder deposit formed on the wafer is segregated, with the bulk of the solder being Pb at the bottom, while the top of the solder deposit is mostly pure Sn. This is only an interesting aspect of the evaporation of the Pb/Sn system, and does not have any implications on subsequent processing.

The solder-covered wafers are then immersed in a hot bath of solvents designed to strip the resist, along with the liftoff sheet of solder, without affecting the solder. Most commercial strippers can be used for this purpose, with slight modifications to improve the cleanliness of the liftoff. At this point, the wafer surface will be covered with solder bumps with a slight conical shape, which is a function of the reentrant profile of the resist via. Since a single layer of resist is used for both the sputtering of base metals and for the evaporation of solder, the footprint of the solder matches closely with the footprint of the base metals. This does not always have to be the case. One patterning step can be used for the liftoff of the base metals and another, with a different via diameter, can be used to lift off the solder. In this case, the second patterning step can be used to produce a via with a much larger diameter, and thus, a solder deposit which is larger in diameter than its wettable base metal area. The advantage of this is that, with an enlarged diameter, a smaller solder deposit height can provide a similar volume of solder (see Fig. 5.46). During reflow, the solder will withdraw from the unwettable areas and confine itself to within the wettable base,

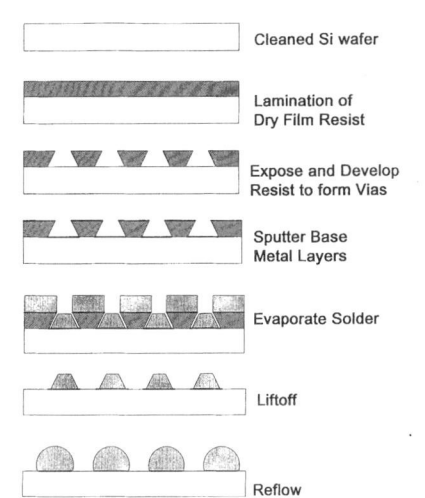

Polymers for Electronic Packaging: Materials, Processes, and Reliability 471

- Cleaned Si wafer
- Lamination of Dry Film Resist
- Expose and Develop Resist to form Vias
- Sputter Base Metal Layers
- Evaporate Solder
- Liftoff
- Reflow

Figure 5.46 Formation of flip-chip solder by lift-off process.

producing a bump similar in characteristics to the one produced by the first instance. The modified scheme is useful for relieving the bottleneck in evaporation, which is typically the step with the lowest throughput in the entire sequence.

The reflow step is used to consolidate the solder bumps from a stack of segregated layers into an alloy, establish wetting between the solder and the base metals, and to transform the bump shapes from truncated cones into truncated spheres. A flux is dispensed onto the wafer surface and the wafer sent through a belt oven, set with a peak temperature slightly higher than the liquidus temperature of 95Pb/5Sn. The flux (Alpha 100) we used is based on organic materials and contains no ionic-corrosive agents. Its purpose is to reduce the oxides covering the solder deposits and allow the bumps to settle to their lowest surface energy state. Thus consolidated, the bumped wafers are then cleaned, tested, diced, sorted, and packaged for subsequent assembly onto ceramic HICs. The sequence outlined earlier is shown schematically in Fig. 5.45.

Micrographs of solder bumps produced using the process described previously are shown in Figures 5.47 and 5.48, for pre-reflowed and post-reflowed bumps respectively. The as-deposited bump has the shape of a truncated cone and the reflowed bump, that of a truncated sphere. The bump heights within a wafer lot have a uniformity of about ±7 percent, both before and after reflow. This is an indication that the entire bump shape, not just its height, is extremely uniform. Otherwise, reflow will bring about a wider distribution of bump heights than that obtained after deposition. This uniformity is controlled primarily in the processing of the dry film and in the design of wafer-holding fix-

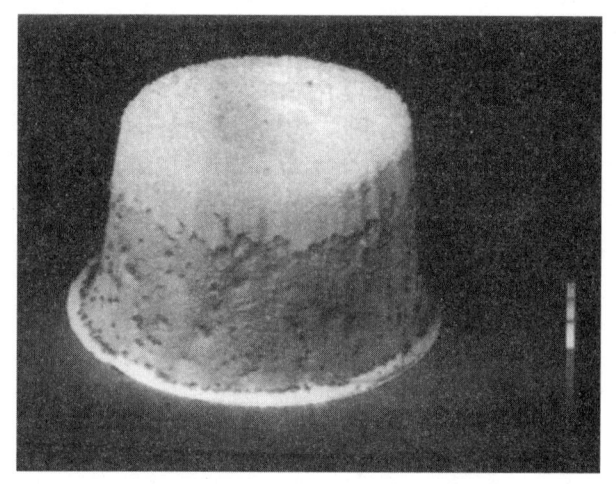

Figure 5.47 Solder bump prior to reflow.

tures within the evaporation chamber. The bumping yield, as defined earlier, typically approaches 100 percent and sometimes even surpasses it. The apparent anomaly has to do with good chips being initially pronounced defective because of errors in testing, usually because of poor contacts. Bumped chips offer better contacts for test probes; thus, nominally "defective" chips can now be pronounced "good," and bumping yield can surpass 100 percent in principle, and sometimes in practice.

We have initiated an independent isothermal low-cycle fatigue testing of typical joints produced with the standard process. The complete testing scheme and its results are reported elsewhere [98]. The high-

Figure 5.48 Solder bump after reflow.

lights of the test reveal that cylindrically shaped joints tend to distribute the cycling-induced damage throughout the solder volume and eventual fracture occurs through intergranular cracking, with a strong component of grain boundary sliding. The more natural barrel-shaped joints have areas of reduced cross section at both the top and the bottom of the joint, where the induced stress (and strain) is concentrated. Thus, the damage is not shared uniformly by the rest of the joint; it is localized to parallel fracture planes along these regions of reduced cross section. Nevertheless, the barrel-shaped joints have sufficient fatigue life for most intents including this effort. The point here is that a higher fatigue life in solder joints may be obtained through proper design and control of the joint shape.

Test vehicles with flip-chip solder joints of all relevant wafer codes have been fabricated with the standard bumping process. A parallel effort is undertaken to optimize the gel composition and the treatment process. Fully assembled and encapsulated test vehicles are subjected to qualification testings which include THB and temperature cycling.

5.7.3.2 Preencapsulation cleaning of GDX HIC.
Since Alpha 100, a mild activated flux, was used to reflow the flip-chip GDX onto the ceramic substrate, preencapsulation cleaning prior to silicone gel coating is needed for this process to ensure the long-term reliability of the GDX HIC. An EC-7R, a terpene-based orange peel extract, was used to remove the reflux residence and DI water rinse and dry prior to the gel coating [99]. This EC-7R semiaqueous cleaning solution provides excellent cleaning results. Contact angle and surface analysis shows low angle (~30°) and low contaminant after this cleaning process.

5.7.3.3 Glob-coating of flip-chip GDX HIC.
A thixotropic silicone gel with silica filler was used to glob-coat the flip-chip GDX HIC. This silicone gel was formulated to have an extremely low modulus (<1 psi) by intentionally under-cross-linking the polymer network, so that during temperature cycling and thermal shock testing, no failure was observed. Table 5.18 shows the results of these performances. The silica filler also acted as a thixotropic agent that controls the flow of the silicone gel. The glob-coating flowed evenly under those solder joints and stopped wicking or bleeding excess outside the globbed GDX devices. Furthermore, the glob-coating GDX HIC was placed inside an Ultem elastomeric (from GE Plastics Inc.) plastic casing and potted with a silicone for further reliability protection of film IC for a robust GDX HIC structure (see Fig. 5.49).

5.7.3.4 Reliability testing of the nonhermetic flip-chip GDX HIC.
To ensure the superior performance of the AT&T No. 5 ESS, extensive testing was performed to ensure its reliability. Temperature cycling (−40 to

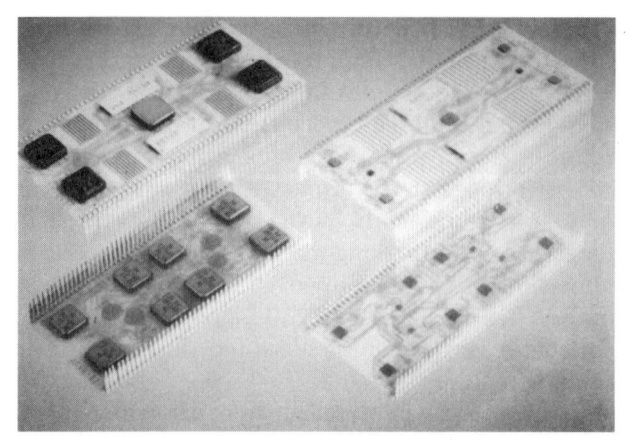

Figure 5.49 Hermetic and nonhermetic GDX.

125°C) at 1000 cycles was used to ensure the reliability of the flip-chip solder joints. Temperature humidity bias (at 85°C/85%RH and 600V) for 1000 h was used to ensure the anticorrosion reliability of the modules; MIL-STD-883C with standard salt spray, highly accelerated stress test (HAST) with 121°C, 2 atm pressure, at 125°C were used to ensure the module reliability. The gel-coated GDX HIC performed perfectly. Results are shown on Table 5.18.

5.7.4 Conclusions

The original GDX devices were packaged in hermetic ceramic leadless chip carriers, which were bulky, heavy, and expensive. However, we have demonstrated that a high reliability flip-chip bonded 95/5/Pb/Sn solder joints for the high-voltage GDX HIC electronic switch for use in AT&T's No. 5 ESS, and a robust nonhermetic packaging structure, which consists of a coating of an ultrasoft silicone gel and an elastomeric silicone. Silicone, with its superior electrical property, can nonhermetically protect the high-voltage (375-V versus normal 3–5-V device) GDX devices. When proper materials are chosen and processes are carefully monitored and controlled, a reliable flip-chip high-voltage device can be manufactured with tremendous cost savings.

5.8 482Reliability Testing

To ensure the long-term reliability of the packaged IC devices, reliability testing is an integrated part of the manufacturing process. Effects of moisture, mobile ions, and so forth, directly affect the reliability [100]; accelerated testings such as the conventional 85°C/85%RH and the highly accelerated stress test (HAST) are critical to predict this manufacturing process. These are discussed as follows.

5.8.1 Correlation of 85°C/85%RH to HAST

Reliability testing results are used to give a degree of confidence of a device's field reliability to the electronics merchant and the customer. The only way this degree of confidence can be given is if there is some correlation between the accelerated test time-to-failure (TTF) data that has been obtained from qualification testing and the expected field TTF rates. This can be accomplished by formulating reliability models that relate accelerated environmental tests to actual field life. The objective for any reliability model is to be able to estimate device performance accurately outside the range of experimental observations. This can be done in either of two ways: relate the accelerated test TTF results of a specific device to that of the same device at different accelerated test or field conditions; or through comprehensive, fundamental modeling of the physical activities at the molecular level leading to degradation and failure. The latter technique is commonly termed the *physics-of-failure approach*. The former of the two approaches incorporates statistically significant data obtained from well-planned and consistently executed environmental test evaluations to form a probabilistic model, allowing extrapolation to other use conditions. Though the physics-of-failure approach is the most desirable because of its fundamental nature, models based on these theories are still at a research level in industry, and are just beginning to be investigated in-depth. This section summarizes the evaluation of two existing predictive models for the time-to-failure of nonhermetic packages subjected to constant temperature and relative humidity.

Two of the leading candidate models, one proposed by Peck [101] and the other by Denson and Brusius [102], are essentially empirical in their formulation. To assess the relative consistency of the two models, their functional forms are considered and compared relative to a given set of data. Again, the drive behind these two models and others like them [103] is the need to extrapolate beyond the range of available observations. The main application appears to be for the estimation of acceleration factors from laboratory test conditions, in which temperature and humidity are much higher than operational, or field, conditions. If the acceleration factors are reasonably accurate, tremendous savings in time and testing costs can be realized. If the extrapolation holds into the range of operating conditions, then the model could have great impact for long-term reliability estimations and the cost of reliability testing.

The underlying assumptions in the Peck model are that component times-to-failure are lognormal random variables, and that their variances are equal and constant over a wide range of temperatures (C) and relative humidities (%RH). With these assumptions, an estimated sample median time-to-failure, tm(C,%RH) was obtained. Peck's model for this sample median is:

$$tm(C,\%RH) = A(\%RH)^{-n} \exp[Ea/k\ (°C + 273)^{-1})$$

where k is Boltzmann's constant, $(8.615 \times 10^{-5}\ eV,K)$; A is a scaling constant; and n and Ea are the model parameters. No doubt, much of the success of the model is due to the fact that a power law for %RH, and an Arrhenius law for temperature seems quite reasonable. Since 85°C/85%RH (85/85) is an industry standard test condition, Peck defined his observed ratio, Ro, between different test conditions to be:

$$Ro = tm(°C,\%RH)t_m(85,85)$$

Ro was compared to a calculated ratio, Rc, derived from tm(C,%RH) in a similar manner to Ro. Specifically, Rc is found to be:

$$Rc = \left(\frac{85}{\%RH}\right)^n \exp\left\{\frac{Ea}{k}\ [(°C + 273)^{-1} - 358\ K^{-1}]\right\}$$

If the graph of Ro versus Rc is linear with slope one and intercept zero, then the model is an excellent representation of the experimental data.

Another issue is fundamentally at the heart of the role of acceleration factors. Peck has chosen 85°C/85%RH as the reference condition in order to extrapolate to other accelerated test conditions. However, for normal operating conditions, the extrapolation must be from the accelerated reference condition to the decelerated conditions of the operating environment. For the latter, Peck's model does not fit as well, based on statistical criteria. A final issue that may be somewhat theoretical is the choice of a statistical criterion for parameter estimation. Since a linear model is assumed for Ro versus Rc, the criterion for a best fit could be based on mean square error, best slope, or best intercept. Each yields slightly different results. All things considered, Peck's model has utility for first order approximations. Nevertheless, caution should be exercised for reliability estimation when accuracy and high levels of confidence are required. It usually takes a large data base with large field return failure parts, careful failure mode analysis, and a well thought-out model to correlate the predicted reliability results of one electronic component.

5.8.2 Recent advances in materials packaging

The greatest advancements have been made in the materials used for encapsulation. These new-generation materials are intrinsically cleaner with respect to ionic contamination and possess far better physical and electrical properties. While plastic-molded packages have been around for some time and their materials are fairly well developed, more demanding and novel packaging approaches such as direct chip attach (DCA), chip-on-board (COB), and multichip modules (MCMs) in the flip-chip area array interconnects require not only new areas of understanding but also entirely different material requirements.

For overcoat applications such as COB, flip-chip (FC) underfills, tape automated bonding (TAB), and chip-scale packaging (CSP), polymer manufacturers are actively pursuing materials that have the specific properties required of these applications. Coating materials for COB, FC, CSP, and TAB have very different requirements from plastic molding compounds, not only because of the vastly different processing used for encapsulation, but also from the standpoint of physical properties and performance requirements. In general, material requirements for COB, FC, or TAB applications would entail a material that:

- Is easy to dispense with existing robot-syringe equipment
- Is compatible with the remaining assembly processes
- Possesses an acceptable combination of viscosity and thixotropic properties for the application
- Has sufficient adhesion to the constituents that are to be protected
- Has a consistent batch-to-batch ionic content of less than 2 parts per million each of Cl^-, K^+, and Na^+
- Requires a short duration cure with a maximum temperature of 150°C
- Has a T_g of 150°C or greater
- Possesses a residual stress after curing of less than 5 Mpa
- Does not cause stress-related failures of the assembly during the temperature cycling required for qualification
- Provides sufficient mechanical protection to the device and interconnect so additional packaging is not required
- Is able to stand up to hot-wet-electrically biased environmental tests
- Maintains acceptable electrical properties in hot and humid environments
- Exhibits less than 1 percent linear shrinkage during cure
- Is nonvolatile and nontoxic
- Has at least an hour pot life and at least a six-month shelf life
- Is comparatively low in cost

Given that many of these properties are application-specific and would therefore require materials that are also application-specific, there does not exist today a single-dispense material that fits the above criteria for commercial versions of COB, and certainly not for the more advanced MCMs.

With the numerous materials on the market, each application requires different considerations when it comes to decisions on how to best provide environmental protection. Vast ranges of physical properties are available to the packaging engineer. Silicones in general possess extremely high CTEs and extremely low moduli, whereas epoxies have fairly low CTEs and relatively high moduli. For example, a well-known silicone gel on the market has a CTE of well over 800 to 1000 ppm/C and a modulus of less than 0.001 GPa, and one of the better epoxies for COB and flip-chip underfill has a CTE of 22 ppm/C (with high loading of silica) and a modulus of 10.5 GPa. Also, silicone gels are very clean with respect to Cl^-, K^+, and Na^+. This specific silicone gel is reported to have 2 ppm of each. Epoxies, by the nature of their manufacture, are much higher in ionic content, and this epoxy is one of the best with 20 ppm of each. The two materials in this example are at opposite ends of the spectrum with respect to their physical properties, yet both can be used to coat a wire-bonded COB and flip-chip solder joints device and provide both thermomechanical stress resistance and corrosion resistance. The silicone gel needs additional packaging in most cases, because it does not provide much in the way of mechanical protection with its extremely low modulus; the epoxy will not survive nearly as well as the silicone gel in corrosion tests such as biased 85°C/85%RH, or biased HAST due to its ionic content and adhesion characteristics [104]. Continuing improvements in epoxy resins have resulted in epoxies that have stronger polymer matrices and are better able to withstand the difficult-to-pass hot, wet environments common in accelerated testing. These improved epoxies would appear to be more apt to provide a higher degree of reliability in testing and the field, but they are at a developmental stage today.

New materials have been formulated in an attempt to incorporate the best of both worlds. There are recent resins that combine the physical properties of two materials to produce a superior composite resin like the silicone modified epoxies and polyimide siloxanes. If you combine or form a copolymer or a polymer blend a silicone and an epoxy into one formulation, you obtain a mix of their respective physical properties that may in some cases provide better overall performance. The same is true for the polyimide siloxanes. Material-dispense characteristics have been improved to facilitate more consistent dispensing, and are formulated with a specific application in mind. TAB is beginning to see more use in commercial electronic assemblies, certainly in consumer-destined electronics coming out of Japan, and normally incorporate an encapsulation coating on the surface of the die and inner lead bond area. TAB encapsulation materials are required to have different properties than those for COB, in that smaller particles and higher flow are needed to fully encapsulate around the inner lead bonds, whereas for COB, overcoat materials need to have viscosity and thixotropic properties that help hold the desired shape of the coating over the device and interconnect through cure. Flip-chip area interconnects draw more attention as high-performance IC requires high I/O. Underfill encapsulant with fast-flow, fast-cure, no-flow, and reworkable properties are needed by the industry for the production of low-cost flip-chip devices.

Cleaning of the assembly prior to the coating application has been identified as a critical process step in obtaining long-term reliability. As CFCs are phased out, new cleaning surfactants and processes will be available to the packaging engineer. Cleaning must remove not only ionic species, but organic contaminants as well. What removes ionic contaminants, such as a wet-cleaning process, does not always remove organic impurities. Surfactant cleaners may be good at eliminating ionic species, but they can leave a residue themselves that may require another cleaning process to remove. Reactive oxygen and DC-hydrogen plasma combined with conventional cleaning processes would be an answer to all cleaning requirements.

With all the advances in polymer materials technology, a rapid increase in the area of reliability performance has been realized. With this has come the increased difficulty of designing reliability tests that provide the necessary acceleration factors for reasonable test times in the manufacturing environment. Electrically biased 85°C/85%RH (85/85) has been the standard test in the plastic-molded package industry for many years. With new molding compounds, 85/85 tests can last several thousand hours before failures begin to appear. These long test times are not conducive to the fast pace of today's electronics design and production floor. The highly accelerated stress test, better known as HAST, is quickly becoming an accepted alternative to 85/85. HAST utilizes a pressurized chamber to allow the temperature to be elevated while maintaining humidity control, to produce the additional activation energy for more accelerated testing. Electrical bias in 85/85 or HAST provides additional activation energy so that corrosion failures and mechanisms can be identified in a shorter time. Alternative environmental tests such as Battelle's Flowing Mixed Gas Test (FMG) introduce parts-per-million or parts-per-billion quantities of contaminants commonly found in the atmosphere, to simulate corrosive processes found in some electronics applications. FMG was designed to test connector technologies and has not yet been used with any regularity in the testing of nonhermetic packages due to the relatively long test times required. Correlation of 85/85, or HAST, to field life is still an issue of considerable discussion, as well as the acceleration factors between 85/85 and HAST. A great deal of ongoing work in the correlation and testing area is needed to be able to accurately estimate reliability in the field from models based on these accelerated tests.

Other work has been done where environmental testing is performed in a sequential series of steps. Temperature cycling has historically been used in testing plastic packages as a single-step test to evaluate the thermomechanical stresses arising from mismatches of CTEs. 85/85, HAST, or similar corrosion/polymer degradation tests are also usually run as separate evaluations. Recent studies have been performed using temperature cycling and salt atmosphere as preconditioning tests for 85/85 or HAST. The temperature cycling will induce cracks and delaminations of the polymer matrix and polymer-to-substrate interface. Next in the sequence, a salt atmosphere adds an ionic contaminant (NaCl) to the newly created cracks and voids in the package. Finally, placing the parts in either biased 85/85 or biased HAST is done to accelerate the corrosion mechanism that is now in place with moisture, NaCl, and electrical potentials. Even with a test sequence this severe, the biased HAST portion of the test sequence has run in excess of 1000 h without significant performance degradation of the wire-bonded and coated test chip. With the desire to run ultraclean 85/85 and HAST chambers for the purposes of repeatability and control, among other things, a sequence such as this does have its dangers and unknowns.

On the device side of the test spectrum, the fabrication of test chips that are designed to amplify the failure mechanisms that have been observed by the polymer packaging industry, has been greatly advanced in recent years. Sandia National Labs, for one, has taken the sophistication of test chips to new levels [105–107]. Sandia's family of Assembly Test Chips, designated ATCs, have incorporated assorted detectors and sensors that are of primary interest to those in the polymer encapsulation industry. The ATC family now has five versions in production. The ATC01 is the first-generation test chip designed for studying aluminum corrosion. The ATC01 incorporates eight different triple tracks, three corrosion ladders, aluminum sheet resistance, in addition to other corrosion-detection structures. The ATC04 is the most recent and advanced test chip in the family, with passivated and unpassivated triple tracks in both single- and double-level metal, and some triple tracks have underlying polysilicon heaters for heat dissipation studies, diode thermometers, and 48 addressable piezoresistive stress sensors. Both the ATC01 and ATC04 have features sized down to 1.25-µm line and space widths. These chips have been made available to industry in an effort to provide an industrywide standard test chip family for comparative studies and data correlation. In addition, Auburn University has developed a new test chip that could detect the z-direction (out-of-plane) stress as well as x-y (in-plane) direction.

The unquestioned superior reliability of hermetic packages will remain a factor in their continued use for military and critical civilian applications despite their higher cost, size, and weight. The predicted increases in device densities will influence failure susceptibilities in several ways. The increased device density will mean increased power densities, and hence, generally higher operating temperatures, larger voltage gradients; and less conductor material means there is less material loss required for opens or shorts. Also, the packing of more devices on the same chip, in concert with physically increasing the sizes of chips, greatly increases the probability of chip-level failure. The cumulative effect of these factors serves to increase the susceptibility of failure by an order of magnitude for each doubling in circuit density. Certainly, the group of technologies that form COB, MCM, flip-chip, and chip-size assemblies will continue to evolve, and with it so will polymer encapsulation. These considerations, in combination with the continuing advances in materials and processes for polymer packaging, will constantly redefine the limits of use for nonhermetic versus hermetic packaging.

5.9 Conclusions and Future Developments

The rapid development of IC technology has continuously created a critical need for electronic packaging, advanced polymeric materials as device interlayer dielectrics, passivation layers, encapsulants, and packagings. Recent advances in high-performance polymeric materials, such as improved silicone elastomers and/or ultrasoft silcone gels, low-stress epoxies, high-performance flip-chip underfills, and low thermal expansion coefficient photo-definable polyimides and benzocyclobutenes have provided polymers that are compatible with the deep submicron ULSI technology. The advances of high-density multichip modules packaging and the next-generation DCA such as flip-chip, chip-size packages will undoubtedly require nonhermetic packaging for the ever increasing large chip and modules. However, the demands for improved properties of materials in the areas of low dielectric constants, high-breakdown voltage strength, high sheet resistance, and less dielectric change with humidity, yet low cost, will continue to require the developing of high-performance polymeric materials. Their application in on-chip and off-chip interconnections, wafer-level packaging, and wafer-scale integration architecture structures with very fast interconnecting networks and high-performance packaging such as chip scale, ball grid array, and wafer-level packaging will become apparent. It is a challenge that the collaborative efforts between polymer chemists, material scientists, and device engineers will face in the near future. Finally, electronic packaging and material development in the 21st century will play a critical role in the next generation of semiconductor technology. The ultrahigh density packages with embedded integrated passives, opto-electronics for inter- and intrachip interconnects and high-heat flux, high thermal management, and low-cost DCA flip-chip assembly will be needed.

5.10 Acknowledgments

The author acknowledges his colleagues at Bell Labs and Georgia Tech, particularly his student, S. Shi, for contributions to our continuous efforts in both materials and processes development and R&D described in various parts of the chapter.

5.11 References

1. Tummala, R., and C. P. Wong. "Materials in Next Generation of Packaging," *IEEE Proceedings from the 3rd Int. Symp. on Adv. Packaging Materials Processes, Properties and Interfaces,* 1977, pp. 1–3.
2. Schaller, R. R. "Moore's Law: Past, Present and Future," *IEEE Spectrum,* 1997, p. 53.
3. Grove, A. S. *Physics of Semiconductors.* New York: John Wiley & Sons, 1967.
4. Wong, C. P., and R. McBride. "Pre-encapsulation Cleaning and Control for Microelectronic Packaging," *IEEE Trans. on CPMT,* 17(4), 1994, p. 542.
5. Wong, C. P., W. O. Gillum, D. Powell, R. Walters, and P. Sackech. "Reactions of High Lead Solder with EC-7R Semiaqueous Cleaning Agents," *IEEE Trans. on CPMT,* 19(2), 1996, p. 119.
6. Korner, N., E. Beck, A. Domman, N. Onde, and J. Ramm. "Surface and Coatings Technology," *Coatings Technology,* 1995, p. 731.
7. Noll, W. "Chemistry and Technology of Silicones," New York: Academic Press, 1968.
8. Rochow, E. G. *Silicon and Silicones.* Berlin: Springer-Verlag, 1987.
9. Bruins, P. F. *Silicon Technology.* New York: Wiley, 1970.
10. Voronkov, M. G., V. P. Mileshkevich, Y. Yuzhelevskii, and A. Yu. *The Siloxane Bond.* New York: Consultants Bureau, 1978.
11. Eaborn, C. *Organosilicon Compounds.* London: Butterworth Science Publishers, 1960.
12. Lynch, W. *Handbook of Silicone Rubber Fabrication.* New York: Van Nostrand Reinhold, 1978.
13. Yilgör, J., and J. E. McGrath. "Polysiloxane Containing Copolymers: A Survey of Recent Developments," in *Advances in Polymers Science,* D. Olivé, ed., vol. 86, Berlin: Springer-Verlag, 1988.
14. Filas, R. W., B. Johnson, and C. P. Wong. "Index-Matching Elastomers for Fiber Optics," *IEEE Trans. on CHMT,* 13(1), 1990, p. 133.
15. Wong, C. P., and D. M. Rose. "Alcohol Modified RTV Silicone for IC Device Packaging," *IEEE Trans. on Components, Hybrids, and Manufacturing Technology,* 6(4), 1983, p. 485, and references therein.
16. White, M. L. "Encapsulation of Integrated Circuits," *IEEE Proc. 19th Electronic Components Conference,* 1969, p. 1610.
17. Mancke, R. G. "A Moisture Protection Screening Test for Hybrid Circuit Encapsulants," *IEEE Trans. Components, Hybrids and Manufacturing Technology,* 4(4), 1981, p. 429.

18. Jaffe D., and N. Soos. "Encapsulation of Large Beam Leaded Devices," *IEEE Proc. Electronic Components Conference*, 1978, p. 213.
19. C. P. Wong, "High Performance RTV Silicones as IC Encapsulants," *The International Journal for Hybrids and Microelectronics*, 4(2), 1981, p. 315.
20. Wong, C. P., and D.E. Maurer. "Improved RTV Silicone for IC Encapsulants," National Bureau of Standards, Special Publication 400-72, *Semiconductor Moisture Measurement Technology*, 1982, p. 275.
21. Wong, C. P. "Improved Room-Temperature Vulcanized Silicone Elastomers as Integrated Circuit Encapsulants," in *Polymer Materials for Electronic Applications* (Feit, E., and C. Wilkins, Jr., eds.), Washington, DC: American Chemical Society, 1982.
22. Wong, C. P. "Thermogravimetric Analysis of Silicone Elastomers as Integrated Circuit Device Encapsulants," *American Chemical Society, Organic Coatings and Applied Polymer Science Proceedings*, vol. 42, 1983, p. 602.
23. Wong, C. P., and D. M. Rose. "Modified RTV Silicone as Device Packagings," *IEEE Proc. of 33rd Electronic Components Conference*, 1983, p. 505.
24. Otsuka, K., Y. Shirai, and K. Okutani. "A New Silicon Gel Sealing Mechanism for High Reliability Encapsulants," *IEEE Trans. Comp. Hybrids, Manuf. Tech., CHMT-7*, 1984, p. 249.
25. White, M. L., J. W. Serpeillo, K. M. Striny, and W. Rosenzwig. "The Use of Silicone RTV Rubber for Alpha Particle Protection of Silicon Integrated Circuits," *IEEE Proc. Int. Reliability Physics*, 1983, p. 43.
26. C. P. Wong. "Thermogrametric Analysis of Silicone Elastomers and IC Device Encapsulants," chapter 23 in *Polymers in Electronics, American Chemical Society Symposium Series* (T. Davidson, ed.), no. 242, p. 285, Washington, DC: American Chemical Society, 1984.
27. Wong, C. P. "Effect of RTV Silicone Cure in Device Packagings," chapter 43 in *Polymers for High Technology and Photonics* (Bowden, M. J., and S. R. Turner, eds.), Symposium Series, vol. 346, p. 511, Washington, DC: American Chemical Society, 1987.
28. Wong, C. P. United States Patent 4,278,784 (July 14, 1981).
29. Wong, C. P. United States Patent 4,318,939 (March 9, 1982).
30. Wong, C. P. United States Patent 4,330,637 (May 18, 1982).
31. Wong, C. P. United States Patent 4,396,796 (Aug. 2, 1983).
32. Wong, C. P. United States Patent 4,508,758 (April 2, 1985).
33. Wong, C. P. United States Patent 4,552,818 (Nov. 12, 1985).
34. Wong, C. P., J. M. Segelken, and J. W. Balde. "Understanding the Use of Silicone Gel for Non-hermetic Packaging of ICS," *IEEE Trans. on Components, Hybrids and Manufacturing Technology*, 4, 1989, p. 419.
35. Kookoostcdes, G. J. "Silicone Gel for Semiconductor Applications-Chemistry and Properties," chapter 20 in *Polymeric Materials for Electronics Packaging and Interconnection* (Lupins, J. H., and R. S. Moore, eds.), American Chemical Society Symposium Series, vol. 407, p. 230, Washington, DC: American Chemical Society, 1989.
36. Riley, J. E. "Ultrahigh Sensitive Uranium Analyses Using Fusion Track Counting: Further Analysis of Semiconductor Packaging Materials," *J. Radioanalytical Chemistry*, vol. 72, 1982, p. 89, and references therein.
37. Wong, C. P. United States Patent 4,564,562 (Jan. 14, 1986).
38. Wong, C. P. United States Patent 4,592,959 (June 10, 1986).
39. Wong, C. P. United States Patent 4,888,226 (Dec. 19, 1989).
40. Wong, C. P. "High Performance Screen Printable Silicone as Selective Hybrid IC Application," *IEEE Trans. on Components, Hybrids and Manufacturing Technology*, 13(4), 1990, p. 759.
41. Lee, H., and K. Neville. *Handbook of Epoxy Resins*. New York: McGraw-Hill, 1967.
42. May, C. A., and Y. Tanaka. *Epoxy Resins*. New York: Marcel Dekker, 1973.
43. *Encyclopedia of Polymer Science and Technology*, vol. 6, p. 209, New York: John Wiley & Sons, 1967.
44. Blyer, L. L., Jr., H. E. Blair, P. Hubbauer, S. Matsuoka, D. Pearson, G. W. Poelzing, and R. C. Pragelhof. "A New Approach to Capillary Viscometry of Thermoset Transfer Molding Compounds," *Polymer Engineering Sci.*, 26(20), 1986, p. 1399.
45. Kuwata, K., K. Ito, and J. Tabata. "Low Stress Reson Encapsulant for Semiconductor Devices," *IEEE Proc. of the 35th Electronic Components Conference*, 18, 1985.

46. Yamada, T. "Low Stress Design of Flip-chip Technology for Si on Si Multichip Modules," *Proc. Int. Symp. on Electronic Packaging,* Orlando, Fla., 1986.
47. Suryanarayana, D., R. Hsiao, T. P. Gall, and J. M. McCreary. "Flip-chip Solder Bump Fatigue Life Enhanced by Polymer Encapsulation," *IEEE Proc. of 40th Electronic Components and Technology Conference,* 1990, p. 338.
48. Wong, C. P., S. Shi, and G. Jefferson. "Novel No Flow Underfill for Low-cost Flip-chip Applications," *IEEE 47th Electronic Components and Technology Conference,* 1997, p. 850.
49. Saunder, J. H., and K. C. Frisch. *Polyurethanes: Chemistry and Technology, vols. I & II.* New York: Interscience, 1962.
50. Hepburn, C. *Polyurethane Elastomers.* New York: Applied Science Publishers, 1982.
51. Mittal, K. L., ed. *Polyimides: Synthesis, Characterization and Applications, vols. 1 & 2,* New York: Plenum Press, 1984.
52. Rubner, R. "Production of Highly Heat-Resistant Film Patterns from Photoreactive Polymeric Precursors," *Siemens Forche-u, Entwicki-Ber Bd.,* 5, p. 92, Springer-Verlag, 1976.
53. Pfeifer, J., and O. Rhode. "Direct Photoimaging of Fully Imidized Solvent-Soluble Polyimide," *Proceedings of Second International Conference on Polyimides,* Ellenville, New York, 1976, p. 130.
54. Banba, T., E. Takeuchi, A. Tokoh, and T. Takeda. "Positive Working Photosensitive Polymers for Semiconductor Surface Coating," *IEEE Proc. of the 41st IEEE Electronic Components and Technology Conference,* 1991, p. 564.
55. Numata, S., K. Fujisaki, D. Makino, and N. Kinjo. "Chemical Structures and Properties of Low Thermal Expansion Polyimides," *Proceedings of Second International Conference on Polyimides,* Ellenville, New York, 1985, p. 492.
56. Berger, A. U.S. Patent No. 4,319,547 (1979) and U.S. Patent No. 4,395,527 (1983).
57. Lee, C. J. *The First International Society for Advanced Materials and Process Engineers, vol. 1,* 1987, 576.
58. Lee, C. J. U.S. Patent No. 4,586,997 (1986); U.S. Patent No. 4,690,497 (1986).
59. Maudgal, S., and T. S. St. Clair. "Preparation and Characterization of Siloxane-Containing Thermoplastic Polyimides," *Proc. 2nd Int. Polyimides Conf.,* 1982, p. 48.
60. Davis, G. C., B. A. Heath, and G. Goldenblat. "Polyimiesiloxane: Properties and Characterizations for Thin Film Electronic Applications," *Proc. of 1st Intl. Conf. on Polyimides,* 1984, p. 847.
61. Gorham, W. F. "A New General Synthetic Method for the Preparation of Linear Poly-p-xylenes," *J. Polymer Science,* **4,** 1966, p. 3027.
62. Bachman, B. J. "Poly-p-xylenes as Dielectric Material," *The First International Society for Advanced Materials and Processes Conference,* 1, 1987, p. 431; Tong, H. M., L. Mol, K. R. Grebe, H. L. Yeh, K. K. Srivastava, and J. T. Caffin, "Parylene Encapsulation of Ceramic Packages for Liquid Nitrogen Application," *IEEE Proc. of 40th Electronic Components and Technology Conference,* 1990, p. 345.
63. Kirchoff, R. U.S. Patent 4,540,763 (1985).
64. Burdeaux, D., P. Townsend, J. Carr, and P. Garrou. "Benzocyclobutene Dielectrics for the Fabrication of High Density, Thin Film Multichip Modules," *Journal of Electronic Materials,* **19**(12), 1990, p. 1357.
65. Tessier, T., G. Ademon, and I. Turlik. "Polymer Dielectric Options for Thin Packaging Applications," *IEEE Proc. of 39th Electronic Components Conference,* 1989, p. 127.
66. Bard, J. K., R. L. Brady, and J. M. Schwark. "Processing and Properties of Silicon-Carbon Liquid Encapsulants," *IEEE 43rd Electronic Components and Technology Conference,* 1993, p. 742.
67. Coombs, C. F., Jr. *Printed Circuits Handbook,* Fourth edition. New York: McGraw Hill, 1996.
68. Shick, R. A., B. L. Goodall, L. H. McIntosh, S. Sayaraman, P. A. Kohl, S. A. Bidstrup-Allen, and N. R. Grove. "New Olefinic Interlevel Dielectric Materials for Multichip Modules," *IEEE Proceedings on Multichip Conference,* 1996, p. 182.
69. Zoba, D., and M.E. Edwards. "Review of Underfill Encapsulant Development and Performance of Flip-chip Development," *Proceedings of 1995 ISHM,* 1995.

70. Wong, C. P., S. Shi, and G. Jefferson. "High Performance No Flow Underfills for Low-cost Flipchip Applications," *3rd International Symposium on Adv. Packaging Materials,* 1997, p. 42.
71. Wong, C. P., M. B. Vincent, and S. Shi. "Fast Flow Underfill Encapsulant: Flow Rate and Coefficient of Thermal Expansion," *Advances in Electronic Packaging,* vol. 1, 1997, p. 301.
72. Pennisi, R. W., and M.V. Papageorge. U.S. Patent 5,128,746 (1992).
73. Smith, J. D. "Metal Acetylacetonates as Latent Accelerators for Anhydride-Cured Epoxy Resins," *Journal of Applied Polymer Science,* vol. 26, 1981, p. 979.
74. Tobolsky, A. V. *Properties of Structure of Polymers.* New York: John Wiley, 1960.
75. Kinjo, N., M. Ogata, K. Nishi, and A. Kaneda. "Epoxy Molding Compounds as Encapsulation Materials for Microelectronic Devices," in *Advances in Polymer Science* (K. Dusek, ed.), vol. 88, 1989, p. 1.
76. Manzione, L. T. *Plastic Packaging of Microelectronic Devices.* New York: Van Nostrand Reinhold, 1990.
77. Wong, C. P., D. Clegg, A. H. Kumar, K. Otsuka, and B. Ozmat. "Packaging and Encapsulation," chapter 14, pp. 873–930; and *Microelectronic Packaging Handbook,* R. R. Tummala, E. J. Rymaszewski, and A. G. Klopfenstein, eds., New York: Chapman Hall, 1977.
78. Basavanhally, N. R., D. D. Chang, and B. H. Cranston. "Direct Chip Interconnect with Adhesive-connector Films," in *Proc. Electron. Components Technol. Conf.,* May 18–20, 1992, pp. 487–491.
79. Chang, D. D., P. Crawford, J. A. Fulton, R. McBride, M. Schmidt, R. Sinitski, and C. P. Wong. "An Overview and Evaluation of Anisotropic Conductive Adhesive Film for Fine Pitch Electronic Assembly," *IEEE Trans. on Components, Packaging and Manu. Tech.,* **16**(8), 1993, p. 828.
80. Tsukagoshi, I., A. Nakajima, Y. Goto, and K. Muto. *Hitachi Tech. Rep.* 16, 1991, p. 23.
81. Chang, D., J. Fulton, H. Ling, M. Schmidt, R. Sinitski, and C. P. Wong. "Accelerated Life Test of Z-axis Conductive Adhesives," *IEEE Trans on Components, Packaging, and Manufacturing Technology,* **18**(8), 1993, p. 836.
82. Liu, J., K. Boustedt, and Z. H. Lai. "Development of Flip-chip Joining Technology on Flexible Circuitry Using Anisotropically Conductive Adhesives and Eutectic Solder," *Proceedings of the Surface Mount International,* San Jose, Calif., 1995, p. 102.
83. Ogunjimi, Y., S. Mannan, D. Whalley, and D. Williams. "The Assembly Process for Anisotropic Conduction Joints," *Proceedings of the International Seminar on Conductive Adhesive Joining in Electronics Packaging,* Eindhoven, Phillips, The Netherlands, ISBN No 91-630-3729-7, 1995, p. 127.
84. Rörgren R., and J. Liu. *IEEE Transactions of Components, Packaging and Manufacturing Technology,* editor-in-chief, J. I. Kroschwitz; editors, H. F. Mark, N. M. Bikales, C. G. Overberger, and G. Menges, New York: John Wiley and Sons, Inc., 1985.
85. Liu, J., L. Ljungkrona, and Z. H. Lai. "Development of Conductive Adhesives Joining for Surface-mounting Electronics Manufacturing," *IEEE Transactions of Components, Packaging and Manufacturing Technology* Part B, vol. 18, no 2, 1995, p. 313.
86. Wong, C. P. "Effect of RTV Silicone Cure in Device Packaging," American Chemical Society Symposium Series, vol. 346, 1987, p. 511.
87. Wong, C. P., D. Clegg, and K. Osatke. "Encapsulation and Sealing," chapter 14, *Handbook of Microelectronic Packaging,* R. Tummala, ed., Chapman-Hill, 1977.
88. Wong, C. P., and R. McBride. "Robust Titanate-modified Encapsulant for High Voltage Potting of MCM/HIC IC," *IEEE Trans. on CPMT,* vol. 16, 1993, p. 868.
89. Wong, C. P., J. M. Segelken, and J. Balde. "Understanding the Use of Silicone Gels for Non-hermetic Plastic Packaging," *IEEE Trans. on Components, Hybrid and Manufacturing Technology,* vol. 12, 1989, p. 421.
90. Suhir, E., and J. M. Segelken. "Mechanical Behavior of Flip-chip Encapsulants," *Journal of Electronic Packaging,* vol. 12, 1990, p. 327.
91. Koopman, N., T. Reiley, and P. Totta. "Chip to Package Interconnections," *Microelectronics Packaging Handbook,* R. Tummala and E. Rymaszewski, eds., New York: Van Nostrand Reinhold, 1989, p. 366.

92. Yung, E., and I. Turlik. "Electroplated Solder Joints for Flip-chip Applications," *IEEE Transactions on Components, Hybrids and Manufacturing Technology*, vol. 14, no. 3, 1991, pp. 549–559.
93. Zakel, E., J. Gwiasda, J. Kloeser, J. Eldring, G. Engelmann, and H. Reichl. "Fluxless Flip-chip Assembly on Rigid and Flexible Polymer Substrates Using the Au-Sn Metallugy," *Proceedings of the IEEE/CPMT International Electronics Manufacturing Technology Symposium*, 1994, pp. 177–184.
94. Kelly, M., and J. Lau. "Low Cost Solder Bumped Flip-chip MCM-L Demonstration," *Proceedings of the 1994 IEEE/CPMT International Electronics Manufacturing Technology Symposium*, 1994, pp. 147–153.
95. Ingraham, A., J. McCreary, and J. Varcoe. "Flip-chip Soldering to Bare Copper Circuits," *IEEE Transactions on Components, Hybrids, and Manufacturing Technology*, vol. 13, no. 4, 1990, p. 656.
96. Wong, C. P. "Recent Advances in Plastic Packaging of Flip-chip MCM," *Materials Chemistry and Physics*, vol. 42, 1995, p. 25.
97. Lau, J., M. Heydinger, J. Glazer, and D. Uno. "Design and Procurement of Eutectic Sn/Pb Solder-bumped Flip-chip Test Die and Organic Substrates," *Proceedings of the IEEE/CPMT International Electronics Manufacturing Symposium*, 1994, p. 132.
98. Dudderar, T. D., N. Nir, A. R. Storm, and C. P. Wong. *J. Experimental Mechanics*, **32**(1), 1992, p. 11.
99. Wong, C. P., W. O. Gillum, R. A. Walters, P. Sackach, and B. Boomer. "Reactions of High Lead Solder with EC-7R Semiaqueous Cleaning Agents," *IEEE 45th ECTC Proceedings*, 1995, p. 106.
100. Wong, C. P. "Thermal Mechanical Behavior of High Performance Encapsulants in Microelectronic Packaging," *IEEE Trans. on Components Packaging and Manufacturing Technology*, vol. 18, 1995, p. 270.
101. Hallberg, O., and D. S. Peck. "Recent Humidity Accelerations, A Base for Testing Standards, Quality and Reliability," *Engineering International*, 1991, pp. 169–180.
102. Denson, W., and P. Brusius. "VHSIC/VHSIC-like Reliability Modeling," *RADC-TR*, 1991, pp. 89–177.
103. Klinger, D. J. "Humidity Acceleration Factor for Plastic Packaged Electronic Devices," *Quality and Reliability Engineering International*, 1991, pp. 365–370.
104. Balde, J. W. "The Final Report of the IEEE Computer Society Computer Packaging Committee Special Task Force on the Effectiveness of Silicone Gels for Corrosion Prevention of Silicon Circuits," *IEEE Transactions CHMT*, June 1991.
105. Sweet, J. M., M. R. Tuck, D. W. Peterson, and D. W. Palmer. "Short and Long Loop Manufacturing Feedback Using a Multisensor Assembly Test Chip," *IEEE Transactions CHMT*, vol. 14, 1991, p. 529.
106. Sweet, J. M., M. R. Tuck, D. W. Peterson, and D. J. Renninger. "Assembly Test Chip Version 01 Description and User's Manual," Sandia National Laboratories Report, SAND90-0755, 1990.
107. Sweet, J. M., D. W. Peterson, M. R. Tuck, M. J. Kelly, and T. R. Guillinger. "Evaluation of Chip Passivation and Coatings Using Special Purpose Assembly Test Chips and Porous Silicon Moisture Dectectors," *IEEE Proceedings of the 41st ECTC*, 1991, pp. 731–737.

Index

(*Italics* indicate illustrations.)

ACA (*see* Anisotropic conductive adhesive)
ACAFs (*see* Anisotropic conductive adhesive films)
Accelerated life testing, 461
Accelerated stress tests, on packaged devices:
   85°C/85%RH (85/85) humidity test, 396, 463, 480
   HAST (highly accelerated stress test), 397–398, 463
   HAST correlated with 85/85, 475–476
ACF (*see* Anisotropic conductive film)
Active (driven) line, 77, 78
Aerogel-type materials, 393
Air-flow measurements, in thermal design/management:
   acoustic noise, 183
   flow resistance, 181
   flow visualization, 182–183
   velocity, 181
Aldrich Chemical Company, 430
Alloys (*see* Solder alloys)
Alpha-particle radiation, 400
Amplitude reduction, as type of signal distortion, 22
Anisotropic conductive adhesive (ACA), 6, 7, 449–450
Anisotropic conductive adhesive films (ACAFs), 448–449
   accelerated life testing of, 461, 462 (table)
   assembly of, 459–461
   characteristics of, 452–459
   conclusions on, 461–463
   current-carrying capability of, 451
   DSC of, *456*

Anisotropic conductive adhesive films (ACAFs) (*Cont.*):
   frequency response/performance in, 451
   FTIR of, *455*
   parts considerations for, 451–452
   TGA of, *459*
   types of, 450–451
Anisotropic conductive film (ACF), 6, 7, 8
Apple Computer, 5
Application-specific integrated circuits (*see* ASICs)
Area array flip-chip interconnect assembly, growing acceptance of, 393
Area-array solder-bumped flip-chip technology, 7, 11, 13
Arrhenius' law, 237, 476
ASICs (application-specific integrated circuits), 5–6
   pin count of, vs. number of gates per chip, *84*
Assembly test chips (ATCs), 480–481
Asymmetric model, of heat dissipation from naturally cooled board array, *143, 144*
AT&T, 465

Backward cross talk, dangers of, 82
Ball grid array (BGA), 122
   module, scaling in thermal analysis/design of, *158*
   types of, 7–11
Battelle's flowing mixed gas (FMG) test, 479
Bauschinger effect, 225
Bayer, Otto, 414
Bell Laboratories, 1
Bending moment–curvature rate relation, 249, *250–251*

488   Index

Bending/twisting interaction curves, 247–248
Benzocyclobutenes (BCBs), type of organic encapsulant, 422–424
Bessel functions, 318
BF Goodrich, 403, 404, 425
BGA (see Ball grid array)
Biot number (Bi), 135
Bipolar transistors, invention of, 1
Bis-maleimide triazine (BT) resins, 424–425
   common properties of BT polymers, 426 (table)
   as substrate, characteristics of, 291–294, *304–305*
Boiling coolants, fluorocarbons, properties of, 149 (table)
Boiling/immersion cooling, physical types of, *149*
Boundary condition, in thermal analysis/design modeling, 164–165
Boundary layer, 126–129
   on flat plate, *127*
   thickness, general rules for, 127–128
Building-block solutions, and superposition method, 167–168
Bumping yield, 465
Buried microstrip structure, *29,* 30
Busses:
   clock frequency for, 17–18, *20*
   cost-performance criterion for, *20*
   high-performance criterion for, 18, *20*

C4 (controlled collapse chip connection), *299,* 428, 466
Cache memory, 5, 6
Capacitance, 31–33
Cartesian coordinate system, 195, 212
   transformation of coordinates, 213
Cavity-filling processes:
   casting, 447
   molding, 445–446
   potting, 446–447
CBGA (see Ceramic ball grid array)
Ceramic ball grid array (CBGA), 7, *9,* 11, 13
   on PCB, example of, 312
Ceramic pin grid array (CPGA), 7, 11
CFCs (chlorofluorohydrocarbons), environmental concerns with use of, 401
CFD (see Computational fluid dynamics)
Chain rule of differentiation, 54

Characteristic curve, in analysis of air-moving device, 171, *172*
Characteristic impedance, 30–31, 54
Chip-embedded sensors, 179
Chip-level connection, as zero-level package, 2
Chip-on-board (COB), 477
Chip-scale package (CSP), 6, *8,* 12–13, 477
   advantages of, over DCA, 13
   on PCB, example of, 306–310
Ciba Geigy, 418
CISC (complex instruction-set computing), 5
Clock frequency:
   and concern over EMI/EMC control, 23
   and pin counts, 5
   problem of electromagnetic radiation/reception efficiency for, 19
   for various busses, increase in, 17–18, *20*
Closed-form solutions, for boundary-value problems, 194, 197
CMOS (complementary metal-oxide semiconductor) process, 2, *3, 12*
   gate current driver capability, 23,
   significance of noise for, 86
   system wiring, basic type of, 65
   Toshiba's technological road map of, 4 (table)
   transmission line voltage requirements for, 63–64
Coefficient of linear thermal expansion, 194, 197, 224
Computational fluid dynamics (CFD):
   simulations, 169
   in thermal analysis/design, 164
Conductive adhesives, types of:
   anisotropic (ACA), 448, 449
   intrinsic, 448
   isotropic, 448
Conformal coating, 447–448
Conjugate mode, of heat transfer, 134–136
Constitutive equations:
   of incremental theory, 229–232
   of total strain theory, 232–233
Continuous-heat approximation, 117
Control (lump) volume, 161, 167
   for writing balance equations, *166*
Coolant flow, in system box, 136–139
Coplanar structure, *29*

Index

Cost-performance criterion:
  for busses, 17, *20*
  of end product cost history, *20*
  impact on electrical design of packaging structures, 17–22
  for single-chip package, *21*
Coupled noise (cross talk), 77–82
CPGA (*see* Ceramic pin grid array)
Cray-2 supercomputer, cooled by FC77, 147
Creep strain, 233
Cross talk (*see* Coupled noise)
CSP (*see* Chip-scale package)
Curing profile, of no-fill underfills, 432–433
Current density spatial distribution, as function of frequency, *102*
Cyrix, 5

DCA (*see* Direct chip attach)
Decoupling capacitors, uses/effects of, on noise control, 104–108
Delay adder, 23–24, *25,* 68
  modeling of, due to discontinuities, 72–75
Dense flatpacks, *131*
Deviatoric stress tensor, 243
Dielectric constants, 30 (table)
Die shrink, as greatest enemy of DCA, 13
Die size, and chip complexity, 17, *19*
Differential scanning calorimetry (DSC), 456
  of solder alloys, 257–259
Digital Equipment Corporation, 5
Dip coating, 447
Direct chip attach (DCA), 6, *7,* 12–14
  technology, 428, 477
Discontinuities, 57–58
  capacitive, 68–70
  effect of, on modeling of delay adders, 72–74
  effect of, on signals, 74–75
  inductive, 70–72
Distance to neutral point (DNP), 300
  prediction of solder interconnect fatigue life using, examples of, 306–314
  role of, on solder interconnect reliability, 301–306
Distortion energy theory, 227
Distributed capacitive loading, 74
Distributed-element analysis, 45–47
  vs. lumped-element analysis, *47–51*

Distributed-element modeling, 24
Distributed-element simulation, 26
Distributed inductance loading, 74
Dittus-Boelter correlation, 147
DMA glass transition temperature, of no-flow underfills, 432
Doppler effect, 80
Dow Chemical, 403, 422
Dow Corning, 405
DRAM (*see* Dynamic random access memory)
Driving waveform, 56–57
DYCOstrate, 260
Dynamic mechanical analysis, of solder alloys, 253–257
Dynamic modulus, of no-flow underfills, 432
Dynamic pressure, 138
Dynamic random access memory (DRAM), 5, *8,* 12, 13
  devices, alpha-particle radiation damage to, 400
Dynamic viscosity, of encapsulants, 408–409

85°C/85%RH (85/85) accelerated stress test, 396, 463, 480
  correlated with HAST, 475–476
Elastomers, 404–405
Electromagnetic compatibility (EMC), 22–23
Electromagnetic interference (EMI), 22–23
Electronic packaging:
  basic characteristics of transmission lines in, 43–83
  basic electrical parameters/phenomena in, 17–43
  complexity factors of heat transfer in, 112
  definition/purpose of, 398–401
  electrical design goals of, 22–24
  functions of, *399*
  governing equations for boundary-value thermal stress/strain in, 195–197
  hardware devices for cooling in, 168–177
  heat transfer in, thermal design challenges of, 111–112
  hierarchy, 1, *2*
  major functions of, 1
  materials, mechanical behavior of, 222–271

490 Index

Electronic packaging (*Cont.*):
  materials, thermal properties of, 121 (table)
  mechanical vibration of, 314–326
  modeling, types of, 24, *27*
  polymeric materials for, 402–444
  polymers for, 393–482
  prediction of electrical performance of, 24
  recent advances in material packaging for, 476–481
  simulation, types of, 26
  structural performance goals of, 22–24
  technology trends in, 2–6
  thermal analysis/design in, 154–168
  underfill materials used in, 261–271
  updates in, 6–13
  (*See also specific technologies*)
Electronic Switching System (ESS), 465
Electrostatic induction coefficients, as elements of Maxwell Matrix, 33
EMC (*see* Electromagnetic compatibility)
EMI (*see* Electromagnetic interference)
Emissivity, 145–146
Encapsulation:
  and preencapsulation cleaning, 401–402
  purposes of, 398
  typical properties of selected encapsulants, 406–407 (table)
  use of polymers for, 404–427
Energy dispersive X ray (EDX), 453
Epoxies, type of organic encapsulant, 413–414
  structure-property relationships of, 416 (table)
Equivalent transport properties, in thermal analysis/design, 162–164

Fairchild Semiconductor, 1, 394
Far-end waveforms, *48–49*
FET channel, 118–119
Film redistribution layer (FRL), 12, 261
Finite-element analysis:
  examples of, 275–301
  nonlinear finite-element formulation, 271–275
Finite-element method, of approximation, 197
Flash memory, 12, 13
Flexural loss modulus, 255
Flexural modulus, 254
Flexural storage modulus, 254–255

Flip-chip area array, as emerging interconnect technology, 428
Flip-chip bumping, and reflow of GDX (gated-diode-crosspoint) HIC, 469–473
Flip-chip GDX (gated-diode-crosspoint) HIC:
  glob-coating of, 468, 473
  preencapsulation cleaning process for, 467–468, 473
  reliability testing of, 468, 473–474
Flip-chip solder bumps:
  on ceramic substrates, 466–467
  formation by lift-off process, *471*
  and reflow process, 465–473
  reliability testing of, 468, 473–474
Flip-chip underfill encapsulant, 393, 427–428
  chemical formulation of, 430, 431 (table)
  ideal properties of, 429 (table)
  no-flow, 427–444
Floating conductors, 31
Flow regimes:
  laminar, *128*
  transitional, *128*
  turbulent, *128*
Flow resistance, components of, *139*
Fluorinated polyimides, increased use of, in electronic packaging, 393
Fluorinert coolant FC77, 146–147
Flux tube model, 122–123
Forced-convection boiling, *149*
Forced convective heat transfer, 130–133
Forced vibration analysis, 314
Formulation, as representation of heat transfer process, 165–166
Fourier component frequencies, and problem of radiated noise, 18–19
Fourier's law, 163
FR (fire retardant)-4 PWB, 425
Free vibration analysis, 314–315
FRL (*see* Film redistribution layer)
FTIR (Fourier transform infrared spectroscopy), 453–455
  spectrum, 441, *443*
Full-wave simulation (*see* Non-TEM simulation)

Garofalo-Arrhenius creep equation, 239–240
Gated-diode-crosspoint (GDX), 465
  devices, conclusions on, 474

Gauss-Seidel iteration, 167
General Electric, 404
Gold bumped flip chip, 6, 7
Grid volume, in thermal analysis/design, 161
Ground voltage (bus) disturbances, as system-limiting noise source, 83–108

Hardener concentration, effects of, on no-flow underfill, 433–438
Harmonic heat source, thermal stress in package subjected to, 208–212
HAST (highly accelerated stress test), 397–398, 463, 474, 480
  correlated with 85/85 test, 475–476
HDI (*see* High-density interconnect)
Heat balance equation, 137
Heat capacitance, 114, *115*
Heat conduction:
  equation, 195
  from integrated circuits, 119, *120*
Heat exchangers, 151–154
Heating curve, 379, 381
Heat pipes, 175–176
Heat sinks, air-cooled:
  disk fins, *168*, 169
  impairment of benefits of, 171
  performance of, with increasing air velocity, *170*
  pin fins, *168*, 169
  plate fins, *168*, 169
  serrated fins, *168*, 169
Heat spreader, model of, *134, 135*
Heat spreading function, estimation of, 135
Heat-transfer coefficient(s), 127
  curves of, vs. air velocity, *131*
  for forced air convection air-cooling, interpretation of data for, 131–132
  Nusselt number as representation of, 128
  ranges of, 157 (table)
Heat transfer/diffusion:
  complexities of, in electronic equipment, 111–113
  conjugate mode of, 134–136
  correlation, constants in, 129 (table)
  forced, from flat surfaces, 126–130
  by heat exchangers, 151–154
  across interfaces, 121–125
  by liquid cooling, 146–154
  measurements, 177–183

Heat transfer/diffusion (*Cont.*):
  nomenclature concerning, 187–189
  principles of, 113–154
  problems, analysis of, 155–156
  by radiation, 145–146
  and thermal analysis, 154–168
  and thermal design considerations, 168–177
  from transistor junctions, 116–119
  in wiring substrates, 125–126
Hencky's theory, of plasticity, 245–246
Hermetic-equivalent packages, 463
Hermetic microwave packages:
  design/fabrication of, 327–329
  hermeticity performance of, 329–334
  qualification tests of, 337–338
  thermal stress performance of, 334–337
Hermetic packaging, 396, 463–464
Hewlett-Packard General Semiconductor Specifications (GSS), 326, 337
High-density interconnect (HDI), 13, 261
Highly accelerated stress test (*see* HAST)
High-performance criterion, for busses, 18, *20*
High-performance plastic packages, 463–474
Hitachi Chemical, 404, 419
HMPA (hexa-4-methylphthalic anhydride), 430–431
Hooke, Robert, 222
Hooke's law, 225

IBM, 5
IBSS (*see* Interpenetrating polymer build-up structure system)
ICs (*see* Integrated circuits)
Impedance curve, in analysis of air-moving device, 172
Impingement flow (jet), 127, *128, 129*
Impregnation coating, 447
Inductance, 34–41
  bond wire, 39
  effect of "skin effect" on, 41
  external, 37
  formal definition of, 35
  internal, 37
  "partial," 35–36, *38*
Infrared (IR) thermal imaging, 179–180
Inhomogeneous dielectrics, 55
Instantaneous heat source, thermal stress in package subjected to, 201–208

Integrated circuits (ICs):
  effects of hostile environments on, 400–401
  encapsulants/packaging, stricter requirements for, 394
  global revolution in production and use of, 1–2
  heat conduction from, 119, *120*
  and major functions of electronic packaging, 1
  miniaturization in, resulting from increased integration, 394–395
  passivation materials for packaging of, 403–404
  performance of, adverse elements/conditions affecting, 398–401
  reliability testing on, 474–481
  silicon, invention of, 1
  steady increase in chip size for, 394
  trends in, 2–6
Intel, 5, 19, 20
Interconnects, 1
  input/output (I/O), 393–394
  reliability of, 427–428
  solder, in area-array assemblies, 303, 306
  as "transmission lines," 43
Interlayer dielectrics, 403
Intermediates, 414–415
Interpenetrating polymer build-up structure system (IBSS), 12–13, 261
Interpenetration network (IPN), of polyimides, 419
Intrinsic delay, 28–30
Isochronous stress-strain curves, 235–237
Isotherm imaging, in thermal analysis/design, *160, 161*
Isotropic hardening, *228*

JAN-S environmental testing specifications, 326, 337–338
Jet impingement boiling, *149*, 151
Junction temperature, 112
  and thermal management of electronic equipment, 116–119, 154
Junction-to-case thermal resistance, 183

Kilby, Jack, 1
Kinematic hardening, *228*
Known good die (KGD), 12, 13

Lame's constant, 197
Land grid array (LGA), 339
  NuCSP as new type of, 339–352
Lattice diagram, 59, *60*, 61
Levy-Mises equation, of elastic deformation, 231
Linear polymers, unsuitability of, for encapsulants, 408
Liquid cooling:
  boiling, 147–151
  devices, 173–175
  single-phase, 146–147
  typical performance of, *148*
Lithography capability, in semiconductor chips, 17
Loading, 229
Local free stream temperature, 132–133
Loss modulus, of encapsulants, 408
Lossy signal trace, *24*
Lossy transmission lines, 75–77
Lumped-element analysis, 46
  vs. distributed-element analysis, *47–51*
Lumped-element modeling, 24
Lumped-element simulation, 26
Lumped equivalent circuits, *29*
Lump volume (*see* Control volume)

Magnetic field, calculation of, 34
Magnetic flux density, 34–35
M&T Chemicals, 404
Material process techniques:
  cavity-filling processes, 444–447
  saturation and coating processes, 447–448
Matrix inversion, 166
Maximum shear theory, 227
Maxwell Matrix, 32–33
Maxwell's equations, for electric and magnetic fields, 31
MBGA (*see* Metal ball grid array)
Mechanical vibration, of electronic packaging:
  horizontal, 315–322
  vertical (out-of-plane), of PCBs, 322–326
Memory devices:
  DRAM, 5, *8*, 12, 13, 410, 412
  packaging for, 400
  SRAM, 5–6, 12, 13
  testing components of, 412
Mesh volume, in thermal analysis/design, 161

Metal ball grid array (MBGA), 7, *10,* 11, 13
Microelectrometry, 456–457
Microprocessors, 5–6
  increased signal requirements of, 393
  package cost history, 20, *21*
  pin count of, vs. number of gates per chip, *84*
Microstrip, *45*
  geometry, 41, *42*
  structure, *29,* 30
Microwave frequency, performance of, 19–20
MIL-STD-883, military standard for electronic packaging, 326, 396, 400, 474
Minimum creep rate, 235
Mitsubishi Chemicals, 425
Mobile ion contaminants, as source of IC-device destruction, 399–400
Modeling, in thermal analysis/design:
  boundary condition in, 164–165
  equivalent transport properties in, 162–164
  model of geometry in, 161–162
Model of geometry, in thermal analysis/design, 161–162
Modules on board, forced convective heat transfer from, 130–133
Moisture permeability, as major source of IC-device corrosion, 399
Moore, Gordon, 1, 394
Moore's Law, 17, *18,* 394
Motherboard, as third-level package, *2*
Motorola, 5
Multichip module (MCM), 477–478
  applications of, 427
  as first-level package, *2*
  flip-chip assembly onto ceramic HICs for, 466–467
Multichip Module-Laminate (MCM-L), 425
Multiple board manufacturing (MBM), 275
Multiple transmission line (MTL) systems, 44

National Starch and Chemical, 404
Natural convection:
  from flat surfaces, 139–142
  heat transfer, in constrained space, 142–144, *145*
N-channel MOSFET, 87–90, *91,* 94–95, 106

Near-end waveforms, *50–51*
Neutral loading, 229
Newton-Raphson method, 275
NMOS, 88, 94
No-flow underfill encapsulants, 427–430
  characteristics of, 431–442
  conclusions on, 442–444
Noise, as descriptor of distortion types, 22
Nonhermetic (plastic molded) packaging, 396–397, 463–465
Nonlinear finite-element formulation, 271–275
Non-TEM (very high frequency) modeling, 24
Non-TEM simulation, 26, 28
Norton steady-state creep equation, 346–347
Novolacs, 414
Noyce, Robert, 1
NuBGA:
  design concepts of, 353–354
  design examples, 354–357
  electrical performance of, 357–364
  experimental measurement of, 368–373
  finite-element analysis of, 366, *367,* 373–377
  package family, 357
  package parasitic parameters of, 358–361
  solder-joint reliability of, on PCB, 382
  standard, cross section of, *354*
  thermal performance of, 364–382
NuBus controller, 5
NuCSP:
  design concept of, 339–340
  design examples of, 340–344
  electrical performance of, 344
  solder-joint reliability of, on PCB, 345–350, *352*
  thermal performance of, 344–345
Nusselt number (Nu), 143, 147
  correlated with Rayleigh number (Ra), 141
  correlated with Reynolds (Re) and Prandtl (Pr) numbers, 129, 167
  as nondimensional representation of heat transfer coefficient, 128–129

Occidental Chemicals, 404
Opto-electronic devices, UV-VIS sensitivity of, 400

Organic encapsulants:
  benzocyclobutenes (BCBs), 422–424
  bis-maleimide triazine (BT) resins, 424–425, 426 (table)
  categories of, 404–405
  epoxies, 413–414
  Parylenes, 420–422
  polycyclicolefins (PCOs), 425, 427
  polyimides, 418–420
  polyurethanes, 414–417
  silicone-polyimide (new modified polyimides), 420
  silicones, 405–412
  Sycar polymers, 424
Organic polymers, moisture permeability of, 399

Parallel flow, 127, *128*
Parasitic impedance, 86
Partial mutual inductance, 35–36, 39, 107
Partial self-inductance, 35–36
Parylenes (poly-para-xylylene), 404, 420–422
Passivation layer, 398, 403–404
Passivation materials, for IC device packaging, 403–404
PBGA (*see* Plastic ball grid array)
PCB (*see* Printed circuit board)
P-channel MOSFET, 104, 106
PCMCIA card, 396
Peck's model, predictive of time-to-failure for nonhermetic packages, 475–476
Peltier effect, 177
Pentium microprocessors, 5, 11
Perfluorinated coolants (FC), 146–147
Perimeter-arrayed PBGAs, 292
PERL (*see* Plasma-etched redistribution layer)
Permeability, of dielectric medium, 30, 31, 36
Permittivity, of dielectric medium, 30, 31
Physics-of-failure approach, and electronic packaging design, 155, 184, 475
Pin counts:
  of ASICs, vs. number of gates per chip, *84*
  and clock frequency, *5*
  of microprocessors, vs. number of gates per chip, *84*
Pitot-static tube, 181
Plasma-etched redistribution layer (PERL), 12, 260–261

Plastic ball grid array (PBGA), 7, *8, 9,* 11, 13
  material properties for, on PCB, 294 (table)
  on PCB, example of, 313–314
  solder-joint reliability, design for, 290–293
Plastic deformation, 225
Plasticity, normality principle of, 229, 232
Plastic leaded carrier chip (PLCC) low-pin-count IC devices, 6
Plastic pin grid array (PPGA), 7, 11
Plastic quad flat pack (PQFP), 6, 11–12
  diagram of, in position for SSO noise, *100, 101*
PLCC (*see* Plastic leaded carrier chip)
Poisson's ratio, 197, 224, 273, 288–289
Polycyclicolefins (PCOs), 425, 427
  as interlayer dielectric materials, 403
Polyimides, type of organic encapsulants, 418–420
Polymers, for electronic packaging:
  materials, 402–444
  processes, 444–474
  reliability of, 474–482
Polyorganosiloxanes (*see* Silicones)
Polyurethanes, type of organic encapsulant, 414–416
  synthesis and structures of, *417*
Pool boiling, 147, *149*
  curve, for heat sink configurations, *152*
Power disturbances, as system-limiting noise source, 83–108
Power supply collapse, 106–108
PPGA (*see* Plastic pin grid array)
PQFP (*see* Plastic quad flat pack)
Prager's loading function, 231–232
Prandtl-Nadai creep law, *235,* 237
Prandtl number (Pr), 129
  correlation with Nusselt (Nu) and Reynolds (Re) numbers, 129
Prandtl-Reuss equation, 231
Prepackage (preencapsulation) cleaning processes, of IC devices:
  conventional solvent/aqueous cleaning, 401
  reactive DC-hydrogen plasma cleaning, 402
  reactive oxygen cleaning, 401–402
Pressure drop, measurement of, 138–139
Primary creep, 233–234
Principal axes, in stress analysis, 214, 220

Principal planes, in stress analysis, 214, 220
Principal stresses, in stress analysis, 214, 220
Printed circuit board (PCB), 13
  assembly of surface mount components in, 275–282
  boundary-value problem of, 277
  deflection of, boundary-value example of, 197–201
  as second-level package, 2
Printed wiring board (PWB), use of bismaleimide triazine (BT) resin in, 424–425
Pulse reflection, 59, 60–62, 67, 69

Quasi-TEM structures, examples of, 29, 30
Quasi-TEM waves, 28
Quiet (inactive) lines, 77

Radiation heat transfer, 145–146
Rayleigh number (Ra), 141, 143
  correlation with Nusselt number (Nu), 141
Rayleigh-Ritz frequency approximation method, 324
Reactive DC-hydrogen plasma cleaning, 402
Reactive oxygen cleaning, types of:
  microwave discharge of oxygen, 402
  plasma oxygen, 402
  UV-ozone, 401
Read only memory (ROM), 5
Reciprocity, principle of, 33
Reflection coefficient, 58
Reliability testing, of packaged IC devices:
  correlation of 85/85 with HAST, 475–476
Rent's Rule, 83
Resistance, of conductor, and skin effect, 42–43
Resistance coefficient, 139
  typical values of, 140 (table)
Resole, 414
Reuss-Prager equation, 232
Reynolds number (Re), 129
  correlated with Nusselt (Nu) and Prandtl (Pr) numbers, 129
RISC (reduced instruction-set computing), 5

ROM (see Read only memory)
Room-temperature vulcanized (RTV) silicones, 408–410, 412

Sandia National Labs, 480
Saturation and coating processes:
  conformal and surface coatings, 447–448
  dip coating, 447
  impregnation coating, 447
Scaling, in thermal analysis/design, 157–161
Secondary creep, 235
Seebeck effect, 178
Self-alignment, as benefit of solder reflow process, 467
Semiconductor chips:
  and Moore's Law, 17, 18
  technical advances, and quality of polymeric materials, 393–398
  thermal properties of, 121 (table)
Semiconductor Industry Association (SIA), IC technology road map of, 395 (table)
Shape distortion, as type of signal distortion, 22
Shear modulus, 197
Shear strain, 219–220
Signal distortion, types of, 22, 24, 25
Signal integrity, 22–23
Signal propagation delays, reduction of, 393
Signal trace:
  and dicretization of ground plane, 102
  positions of, relative to chip ground connections, 103
Silicone-polyimide (new modified polyimides), 420
Silicones (polyorganosiloxanes), 405
  applications of, in electronic coatings, 412
  gel formulations, 412 (table)
  general properties of, 405, 408–411
  heat-curable hydrosilation, 411–412
Silicon Graphics, 5
Simultaneous switching output (SSO) noise (= SSN), 39, 87–88, 356, 358
  modeling, 90–104
  in NuBGA, 361–364
  present/future concerns with, 88–90
Single chip module, as first-level package, 2

Singulation process, 403
63Sn37Pb (eutectic) solder:
  DMA (dynamic mechanical analysis) of, 253–257
  DSC (differential scanning calorimetry) of, 257–259
  TGA (thermal gravimetric analysis) of, 259–260
  TMA (thermal mechanical analysis) of, 253, *254, 255*
Skin depth, 40
Skin effect, 39–43
SLC (*see* Surface laminar circuit)
Small outline IC (SOIC) low-pin-count devices, 6
SMT (*see* Surface-mount technology)
SOIC (*see* Small outline IC)
Solder alloys:
  bending/twisting interaction curves for various, *247–248*
  thermal conductivity of, 121 (table)
  twisting moment vs. twist rate for various, *252–253*
Solder-bumped flip chips (SBFCs):
  effect of underfill materials on mechanical/electrical performance of, 260–271
  on PCB (DCA), *6,* 310–311
  solder interconnect reliability of PBGA packages with, 294–301
Solder bump reflow, heating profile of, 432–433
Solder creep, under combined loads, 239–251
Solder joint reliability, for PBGA assembly, 290–293
SOP (*see* Swiss outline package)
Space discretization, in thermal analysis/design, *159*
Sparce flatpacks, 130, *131*
SPICE simulation, in SSO noise modeling, 90–98
Split-via-connections (SVCs), 354, 356
Split-wrap-around (SWA), 353–354, 356
Squirrel cage blower, 171, *172*
SRAM (*see* Static random access memory)
Static random access memory (SRAM), 5–6, 12, 13
Steady-state creep, 235
Stefan-Boltzmann constant, 145, 241, 346, 476

Step responses, of capacitive and inductive discontinuities/terminations, 65–72
Storage modulus, of encapsulants, 408
Strains, analysis of:
  infinitesimal strains in Cartesian coordinates, 217–221
  infinitesimal strains in cylindrical coordinates, 222
Stress, analysis of:
  examples of, 215–217
  principal stress/direction in three dimensions, 213–215
  three-dimensional stress state, 212
  transformation of coordinates in, 213
  two-dimensional stress state, 216–217
Stress-strain relationship:
  creep, 233–238
  elastic, 222–224
  equations describing, 223–224
  plastic, 225–235
Stripline structure, 28, *29*
Sumitomo Bakelite Plastics, 404, 419
Sun Microsystems, 5
Superconducting ceramics, microstrip structures substrate matches for, 288–290
Superposition method, in physical systems analysis, 116–118, 167
Surface coating, 447–448
Surface laminar circuit (SLC), 12, 261
Surface-mount technology (SMT):
  packaging in, 11–12
  PQFP, 12
  (*See also specific technologies*)
Swiss outline J-leaded (SOJ) low-pin-count IC devices, 6
Swiss outline package (SOP) low-pin-count IC devices, 6
Sycar (silicon-carbon hybrid) polymers, 424
Symmetric model, of heat dissipation from naturally cooled board array, *143, 144*

Tacking, 467
Tangent delta (tanδ), 257
  of underfill materials, 265–266
Tangent modulus buckling load, 238, *239*
Tape automated bonding (TAB), 477, 479
Tape ball grid array (TBGA), 7, *10,* 11, 13

Index 497

Tape carrier package (TCP), 7, 11
TBGA (*see* Tape ball grid array)
TCP (*see* Tape carrier package)
Teflon, 393, 403
Telegraphers' Equations, 52, 56
Temperature excursion, 150–151
Temperature-humidity-bias (THB) testing, 410
Temperature measurements, in thermal design/management:
  chip-embedded sensors, 179
  infrared (IR) thermal imaging, 179–180
  purpose of, 177–178
  quick diagnostic measures, 180
  thermistors, 179
  thermocouples, 178–179
Temperature overshoot, 150–151
Temperature scales, and Biot number, 135
TEM (transverse electromagnetic) waves, 25–28, 44–45
  propagation of, in homogeneous structure, 80
Terpenes (orange peel extracts), vs. CFCs, 401
Tertiary creep, 235
Texas Instruments, 1, 394
Thermal analysis/design/management:
  breadth/depth of, 155–156
  cooling modes in, 156
  formulation, 165–166
  hardware devices for cooling in, 168–177
  measurements, as integral part of, 177–183
  modeling, 161–165
  nomenclature of terms in, 187–189
  objectives of, 154–155
  resources in, types of, 156
  scaling, 157–161
  solution, 166–168
Thermal coefficient of expansion (TCE), of no-flow underfills, 432
Thermal diffusivity, 140–141
Thermal gravimetric analysis (TGA), 457–459
  of solder alloys, 259–260
Thermal impedance measurement, of NuBGA, 379–382
Thermal mechanical analysis (TMA), of solder alloys, 253, *255*

Thermal mechanical analyzer (TMA):
  glass transition temperature, of no-flow underfills, 432
  schematic diagram of, *254*
Thermal mechanical boundary-value problems, examples of:
  deflection of PCB, 197–201
  thermal stress in package subjected to harmonic heat source, 208–212
  thermal stress in package subjected to instantaneous heat source, 201–208
Thermal resistance, 113–116, 120–125
  in forced convective heat transfer, 131–133
  fundamental forms of, *113*
  for heat flow from flat surfaces, 126–130
  measurements, in thermal design/management, 183, 187
Thermal strain, 193
Thermal stress analysis, of plastic leaded package, 282–288
Thermal vias, 126, *144*
Thermal wake, 133
Thermistors, 179
Thermocouples, 178–179
  pairs, 179 (table)
Thermoelectric coolers, 176–177
Thermomechanical stress, reduction of, 393
Thermoplastic adhesives, 450
Thermoplastic polymers, 404
Thermosetting adhesives, 450
Thermosetting polymers, 404
Thin-film packaging, 45
Thin quad flat pack (TQFP), 6
Thin small outline package (TSOP), 6, 11–12
Thixotropic agent, 408–409
3M Company, 146
Time delay penalty, 65
Time-domain reflectometer (TDR), 353
Time-domain transmission (TDT), 353
Time-of-flight (TOF) delay, 23–24, *25*
Time-to-failure (TTF) accelerated test, 475–476
Toshiba, 4
TQFP (*see* Thin quad flat pack)
Transistors, proliferation of, 1
"Transmission line" analysis (*see* Distributed-element analysis)

Transmission lines:
  basic characteristics of, 43–83
  behavior of, rule of thumb for, 45–46
  and coupled noise (cross talk), 77–83
  with distributed capacitive loading, 73
  distributed vs. lumped-element simulation for, 45–51
  lossless equations for, 52–55
  lossy, 75–77
  response of, to ramp and step inputs, 56–63
  step responses for terminated, 65, 66 structures, 29, 43–44
  voltage requirements on, for CMOS systems, 63–64
Tresca yield condition (maximum shear theory), 227
TSOP (see Thin small outline package)
Tube-axial fan, 171, 172
Turbulence model, in thermal analysis/design, 164
Twisting moment, vs. twist rate, for various solder alloys, 252–253

Ultra-large-scale-integration (ULSI) device, 1, 394–395
Underfill catalyst, 438–439
Underfill materials:
  advantages of encapsulant, 261
  curing conditions of, 262–263
  disadvantages of encapsulant, 261–262
  electrical performance of, 269–270
  flow rate of, 267–269
  material properties of, 263–267
  mechanical performance of, 269
  for SBFCs, 393
  types (A, B, C, D) of, 262

Union Carbide, 404, 420, 430
Unloading, 229
UV-VIS (ultraviolet-visible) light radiation, 400

Very large scale integration (VLSI) technologies, 1, 193
Von Mises stress contours, 283–284, 286, 336
Von Mises yield condition (distortion energy theory), 227, 230, 232

Waveforms:
  driving voltage, 56–57
  far-end, 48–49
  near-end, 50–51
  propagation of voltage and current, 53–54, 62
Wiring substrates, 122
  heat spreading function of, 125–126
  models of, 125

XY-model, of wiring substrate, 125, 126

Yield conditions:
  Tresca, 227
  Von Mises, 227
Yield surface, 225
  initial, 225–227
  subsequent, 227–229
Young's modulus, 197, 224, 233, 288–289, 307, 316, 347

"Zero gel," 403
Z-model, of wiring substrate, 125, 126